Fractals for the Classroom

Heinz-Otto Peitgen Hartmut Jürgens Dietmar Saupe

Fractals for the Classroom

Part One
Introduction to Fractals and Chaos

With 289 Illustrations

Evan Maletsky Terry Perciante Lee Yunker
NCTM Advisory Board

National Council of
Teachers of Mathematics

Springer-Verlag

Authors

Heinz-Otto Peitgen
Institut für Dynamische Systeme
Universität Bremen
D-2800 Bremen 33
Federal Republic of Germany and
Department of Mathematics
Florida Atlantic University
Boca Raton, FL 33432
USA

Hartmut Jürgens
Institut für Dynamische Systeme
Universität Bremen
D-2800 Bremen 33
Federal Republic of Germany

Dietmar Saupe
Institut für Dynamische Systeme
Universität Bremen
D-2800 Bremen 33
Federal Republic of Germany

NCTM Advisory Board

Evan Maletsky
Department of Mathematics and
Computer Science
Montclair State College
Upper Montclair, NJ 07043
USA

Terry Perciante
Department of Mathematics
Wheaton College
Wheaton, IL 60187-5593
USA

Lee Yunker
Department of Mathematics
West Chicago Community High School
West Chicago, IL 60185
USA

Cover design by Claus Hösselbarth.

TI-81 Graphics Calculator is a product of Texas Instruments Inc.
Casio is a registered trademark of Casio Computer Co. Ltd.
Macintosh is a registered trademark of Apple Computer Inc.
Microsoft BASIC is a product of Microsoft Corp.

Library of Congress Cataloging-in-Publication Data
Peitgen, Heinz-Otto, 1945–
Fractals for the classroom / Heinz-Otto Peitgen, Hartmut Jürgens, Dietmar Saupe.
p. cm.
Includes bibliographical references and index.
ISBN 0-387-97041-X
1. Fractals. I. Jürgens, H. (Hartmut) 1955– II. Saupe, Dietmar, 1954– III. Title
QA614.86.P45 1991
514'.74–dc20 91-11998

Printed on acid-free paper.

© 1992 Springer-Verlag New York, Inc.
Published in cooperation with the National Council of Teachers of Mathematics (NCTM).
All rights reserved. This work may not be translated or copied in whole or in part without the written permission of the publisher (Springer-Verlag New York, Inc., 175 Fifth Avenue, New York, NY 10010, USA) except for brief excerpts in connection with reviews or scholarly analysis. Use in connection with any form of information storage and retrieval electronic adaptation, computer software, or by similar or dissimilar methodology now known or hereafter developed is forbidden. The use of general descriptive names, trade names, trademarks, etc., in this publication, even if the former are not especially identified, is not to be taken as a sign that such names, as understood by the Trade Marks and Merchandise Marks Act, may accordingly be used freely by anyone.

Camera-ready copy supplied by the authors.
Printed and bound by Hamilton Printing Co., Rensselaer, NY.
Printed in the United States of America.

9 8 7 6 5 4 3 2 1

ISBN 0-387-97041-X Springer-Verlag New York Berlin Heidelberg
ISBN 3-540-97041-X Springer-Verlag Berlin Heidelberg New York

To
Karin, Iris, and Gerlinde

Preface

Research in chaos — the most interesting current area of research that there is. I am convinced that chaos research will bring about a revolution in the natural sciences similar to that produced by quantum mechanics.

Gerd Binnig
Winner of the Nobel Prize for Physics

Despite its title, this book which comes in two parts has not just been written for the classroom, but for everyone who, even without much knowledge of technical mathematics, wants to know *the details* of chaos theory and fractal geometry. This is not a textbook in the usual sense of the word, nor is it written in a 'popular scientific' style. Rather, it has been our desire to give the reader a broad view of the underlying notions behind fractals, chaos and dynamics. In addition, we have wanted to show how fractals and chaos relate both to each other and to many other aspects of mathematics as well as to natural phenomena. A third motif in the book is the inherent visual and imaginative beauty in the structures and shapes of fractals and chaos.

For almost ten years now mathematics and the natural sciences have been riding a wave which, in its power, creativity and expanse, has become an interdisciplinary experience of the first order. For some time now this wave has also been touching distant shores far beyond the sciences. Never before have mathematical insights — usually seen as dry and dusty — found such rapid acceptance and generated so much excitement in the public mind. Fractals and chaos have literally captured the attention, enthusiasm and interest of a world-wide public. To the casual observer, the color of their essential structures and their beauty and geometric form captivate the visual senses as few other things they have ever experienced in mathematics. To the computer scientist, fractals and chaos offer a rich environment in which to explore, create and build a new visual world, like an artist fashioning a fresh work. To the student, they bring mathematics out of the realm of ancient history and into the twenty-first century. And to the teacher, they provide a unique, innovative opportunity to illustrate both the dynamics of mathematics and its many interconnecting links.

But what are the reasons for this fascination? First of all, this young area of research has created pictures of such power and singularity that a collection of them, for example, has proven to be one of the most successful world-wide series of exhibitions ever sponsored by the Goethe-Institute[1]. More important, however, is the fact that chaos theory and fractal geometry have corrected an outmoded conception of the world.

The magnificent successes in the fields of the natural sciences and technology had, for many, fed the illusion that the world on the whole functioned like a huge clockwork mechanism, whose laws were only waiting to be deciphered step by step. Once the laws were known, it was believed, the evolution or development of things could — at least in principle

[1] Alone at the venerable London Museum of Science, the exhibition *Frontiers of Chaos: Images of Complex Dynamical Systems* by H. Jürgens, H.-O. Peitgen, M. Prüfer, P. H. Richter and D. Saupe attracted more than 140,000 visitors. Since 1985 this exhibition has travelled to more than 100 cities in more than 30 countries on all continents.

— be ever more accurately predicted. Captivated by the breathtaking advances in the development of computer technology and its promises of a greater command of information, many have put increasing hope in these machines.

But today it is exactly those at the active core of modern science who are proclaiming that this hope is unjustified; the ability to see ever more accurately into future developments is unattainable. One conclusion that can be drawn from the new theories, which are admittedly still young, is that stricter determinism and apparently accidental development are *not* mutually exclusive, but rather that their coexistence is more the rule in nature. Chaos theory and fractal geometry address this issue. When we examine the development of a process over a period of time, we speak in terms used in chaos theory. When we are more interested in the structural forms which a chaotic process leaves in its wake, then we use the terminology of fractal geometry, which is really the geometry whose structures are what give order to chaos.

In this sense, fractal geometry is first and foremost a new 'language' used to describe the complex forms found in nature. But while the elements of the 'traditional language' — the familiar Euclidean geometry — are basic visible forms such as lines, circles and spheres, those of the new language do not lend themselves to direct observation. They are, namely, algorithms, which can be transformed into shapes and structures only with the help of computers. In addition, the supply of these algorithmic elements is inexhaustibly large; and they are capable of providing us with a powerful descriptive tool. Once this new language has been mastered, we can describe the form of a cloud as easily and precisely as an architect can describe a house using the language of traditional geometry.

The correlation of chaos and geometry is anything but coincidental. Rather, it is a witness to their deep kinship. This kinship can best be seen in the Mandelbrot set, a mathematical object discovered by Benoit Mandelbrot in 1980. It has been described by some scientists as the most complex — and possibly the most beautiful — object ever seen in mathematics. Its most fascinating characteristic, however, has only just recently been discovered: namely that it can be interpreted as an illustrated encyclopedia of an infinite number of algorithms. It is a fantastically efficiently organized storehouse of images, and as such it is *the* example par excellence of order in chaos.

Fractals and modern chaos theory are also linked by the fact that many of the contemporary pace setting discoveries in their fields were only possible using computers. From the perspective of our inherited understanding of mathematics, this is a challenge which is felt by some to be a powerful renewal and liberation and by others a degeneration. However this dispute over the 'right' mathematics is decided, it is already clear that the history of the sciences has been enriched by an indispensable chapter. Only superficially is the issue one of beautiful pictures or of perils of deterministic laws. In essence, chaos theory and fractal geometry radically question our understanding of equilibria — and therefore of harmony and order — in nature as well as in other contexts and offer a new holistic and integral model which can encompass an edge of the true complexity of nature for the first time. It is highly probable that the new methods and terminologies will allow us, for example, a much more adequate understanding of ecology and climatic developments, and thus they could contribute to our more effectively tackling our gigantic global problems.

The title of this book reflects our intention to bring chaos theory and fractal geometry closer to an audience which is actively or passively involved with lessons or classes. In April, 1988, within the framework of the centennial celebration of the American Mathematical Society,

Heinz-Otto Peitgen had the honor of giving an address on the subject of this book at the annual meeting of the National Council of Teachers of Mathematics (NCTM) in Chicago.[2] This event resulted in a series of similar lectures at subsequent annual, and many regional, meetings of the NCTM. Two events will be unforgettable: the keynote lecture on the occasion of the Presidential Awards Ceremonies in Washington, D.C. in October, 1988, and a principal lecture at the NCTM Annual Meeting in Orlando, Florida in April, 1989, where more than 3,500 teachers filled the hall beyond capacity. From these contacts with the world of educational mathematics, there soon developed a deep association. Together we became aware that chaos theory and fractal geometry had a previously unforeseen potential for giving witness to how alive and up-to-date mathematics really is and what exciting, future-oriented questions it deals with.

The members of the NCTM soon expressed the desire for a small pamphlet which would summarize the lectures. Springer-Verlag and NCTM agreed to a co-production and hoped for its appearance in the fall of 1990. There were, however, many conversations with teachers. Evan M. Maletsky, Terence H. Perciante and Lee E. Yunker became our advisors and friends. Through them we came to understand that an introductory book only for schools would not be sufficient. Workbooks, with exercises which meshed smoothly into the existing curriculum and ready-for-the-classroom work sheets, would have to be produced. There were numerous changes in plans, and the small pamphlet became these two full-fledged books. They have been organized into thirteen chapters — seven in the first part focusing more on fractals, and six in the second part dealing more with chaos phenomena. Explicitly, the chapters in the second part are

8. Recursive Structures: Growing of Fractals and Plants
9. Pascal's Triangle, Cellular Automata and Attractors
10. Deterministic Chaos: Sensitivity and Mixing
11. Order and Chaos: Period Doubling and its Chaotic Mirror
12. Julia Sets: Fractal Basin Boundaries
13. The Mandelbrot Set: Ordering the Julia Sets

We have worked hard in trying to reveal the elements of fractals, chaos and dynamics in a non-threatening fashion. Each chapter can stand on its own and can be read independently from the others. Each chapter is centered around a running 'story' in *Times* typeset printed toward the outer margins. More technical discussions, which are produced in *Helvetica* typeset toward the inner margins, have been included to occasionally enrich the discussion and provide a deeper analysis for those who may desire them and who are prepared to work themselves through some mathematical notations. At the end of each chapter we offer a short BASIC program, the *Program of the Chapter*, which is designed to highlight one of the most prominent experiments of the respective chapter.

Naturally there remain many holes which we would have liked to fill in, but there are fortunately exceptional books already in print which can close these gaps. We list the following only as examples: For portraits of the personalities in the field and the genesis of the subject matter, as well as the scientific background and interrelationships, there are *Chaos Making a*

[2] The initial idea for the subject and forum for this address came from the then Chairman of the Joint Policy Boards of Mathematics, Professor Kenneth M. Hoffmann of the Massachusetts Institute of Technology.

New Science,[3] by James Gleick and *Does God Play Dice*,[4] by Ian Stewart. For the reader who is more interested in a systematic mathematical exposition or who is ready to advance into the depths, there are the following titles: *An Introduction to Chaotic Dynamical Systems*[5] and *Chaos, Fractals, and Dynamics*,[6] both by Robert L. Devaney, *Fractals Everywhere*,[7] by Michael F. Barnsley, and *Fractal Geometry*,[8] by Kenneth Falconer. And last but not least, there is the book of books about fractal geometry written by Benoit B. Mandelbrot himself, *The Fractal Geometry of Nature*.[9]

This book is complemented by strategic classroom activities, which come in several volumes co-authored by Evan M. Maletsky, Terence H. Perciante, and Lee E. Yunker. These activity books directly involve the students in constructing, counting, computing, visualizing and measuring using carefully designed work sheets. These additional volumes focus on the large number of mathematical interrelationships which exist between fractals and the contemporary mathematics curricula found in our schools, colleges and universities.

So eine Arbeit wird eigentlich nie fertig; (man muß sie für fertig erklären, wenn man nach Zeit und Umständen das mögliche getan hat.)[10]

Johann Wolfgang Goethe, 1787

Acknowledgements

Fractal geometry and chaos are often associated with *experimental mathematics* and its contributions to mathematics. We are very grateful that Benoit B. Mandelbrot accepted to write a foreword for our book addressing the historcal perspectives as well as the dangers and merits of experimental mathematics on the background of his profound experiences. Around 1982/83 we became more familiar with fractals through his great book. The impetus to delve into the subject was a discussion which Heinz-Otto Peitgen had about Julia sets with Leo Kadanoff in Salt Lake City, Utah on December 6, 1982. Soon thereafter we enjoyed the privilege of meeting Benoit personally on occasion of the vernissage for our Goethe Exhibition *Frontiers of Chaos: Images of Complex Dynamical Systems* in Bremen, and since then we have been accompanied by the loyal support and critical advice of a good friend. We hope that this book can contribute to making his great work even more accessible to the public.

We owe our gratitude to many who have assisted us during the writing of this book. Our student Torsten Cordes has produced most of the graphics very skillfully and with unlimited patience. Two more of our students, Ehler Lange and Lutz Voigt, came in near the end when the number of figures had simply outgrown a single man's capacity. Douglas Sperry has read our text very carefully at several stages of its evolution and, in addition to helping to get

[3]Viking, 1987.

[4]Penguin Books, 1989.

[5]Second Edition, Addison Wesley, 1989.

[6]Addison Wesley, 1990.

[7]Academic Press, 1989.

[8]John Wiley and Sons, 1990.

[9]W. H. Freeman, 1982.

[10]"A work such as this is never really finished; one must simply declare it finished when one has, within the limits of time and circumstances, done what is possible." (Goethe was referring to his *Iphigenie*).

our English de-Germanized, has served in the broader capacity of copy editor. Friedrich von Haeseler and Guentcho Skordev have read several chapters and provided valuable suggestions. We also thank Eugen Allgower and Richard Voss for reading parts of the original manuscript. Gisela Gründl has helped us with selecting and organizing third-party art work and the index. Claus Hösselbarth did an excellent job in designing the cover. Evan M. Maletsky, Terence H. Perciante and Lee E. Yunker read parts of our early manuscripts and gave crucial advice concerning the design of the book.

We were greatly inspired during our efforts by the continued encouragement and enthusiasm we found in many audiences of mathematics and science teachers at national and regional meetings of the NCTM. James D. Gates and Harry Tunis of the NCTM put confidence in our work very early on and did not lose it when we stretched their patience to its limits by not meeting several time schedules. When we delivered our manuscript to Springer-Verlag in August 1991 we were happy that the book was finally finished. However, we felt that it could well be expanded even further. Thus, we appreciated the warming aphorism by Goethe.

The entire book has been produced using the TEX and LATEX typesetting systems where all figures (except for the half-tone and color images) were integrated in the computer files. Even though it took countless hours of sometimes painful experimentation setting up the necessary macros it must be acknowledged that this approach immensely helped to streamline the writing, editing and printing.

Finally, we have been very pleased with the excellent cooperation of Springer-Verlag in New York.

Heinz-Otto Peitgen, Hartmut Jürgens, Dietmar Saupe

Bremen, August 1991

Authors

Heinz-Otto Peitgen. *1945 in Bruch (Germany). Dr. rer. nat. 1973, Habilitation 1976, both from the University of Bonn. Since 1977 Professor of Mathematics at the University of Bremen and between 1985 and 1991 also Professor of Mathematics at the University of California at Santa Cruz. Since 1991 also Professor of Mathematics at the Florida Atlantic University in Boca Raton. Visiting Professor in Belgium, Italy, Mexico and USA. Editor of several research journals on chaos and fractals. Co-author of the award winning books *The Beauty of Fractals* (with P. H. Richter) and *The Science of Fractal Images* (with D. Saupe)

Hartmut Jürgens. *1955 in Bremen (Germany). Dr. rer. nat 1983 at the University of Bremen. Employment in the computer industry 1984–85, since 1985 Director of the Dynamical Systems Graphics Laboratory at the University of Bremen. Co-author and co-producer (with H.-O. Peitgen, D. Saupe, and C. Zahlten) of the award winning video *Fractals: An Animated Discussion*

Dietmar Saupe. *1954 in Bremen (Germany). Dr. rer. nat 1982 at the University of Bremen. Visiting Assistant Professor of Mathematics at the University of California at Santa Cruz, 1985–87 and since 1987 Assistant Professor at the University of Bremen. Co-author of the award winning book *The Science of Fractal Images* (with H.-O. Peitgen)

Contents

Foreword

Fractals and the Rebirth of Experimental Mathematics

Benoit B. Mandelbrot[1]

To contribute to a book from the Dynamical Systems Laboratory of the University of Bremen is always a great pleasure I am unable to resist. But this pleasure is necessarily associated with a challenge: my admiration for their efforts and achievements is so well-known that simply to state it again publicly could seem an example of back-scratching among good friends.

I have been asked to respond to this book as I have responded to two earlier books from the same Laboratory in Bremen. That is, I have been asked to use this foreword to indulge in some more history, philosophy and also (when appropriate) autobiography and critique of current events. The broad issue I shall address here concerns the present standing and nature of concrete geometry. More generally, it concerns the new 'experimental mathematics' that is arising from some mathematicians' response to the computer, and has already brought mathematics (to quote David Mumford) 'to a turning point of its history.' There is now a *Journal of Experimental Mathematics*, showing that the field has recently reawakened with a vengeance, or perhaps has been reborn. Some of the events that have accompanied this rebirth have been attracting wide attention, which they certainly deserve.

In a rapidly changing world, a dubious privilege of age is that it gives historical perspective. The previous major change in mathematics started before I was born, but I was present when it created its own institutions and yesterday's order became firmly established. This is why I feel that the inclusion of some autobiography will be of help.

[1]Physics Department, IBM T. J. Watson Research Center, Yorktown Heights NY 10598, and Mathematics Department, Yale University, New Haven, CT 06520.

The Gray and the Green

We must begin by noting that experimental mathematics *does not* imply an attempted invasion of pure mathematics by the applied. *Applied mathematics* has always been permeated with science, hence with experiment. This feature greatly contributed to its being thoroughly unpopular with those believing that applied mathematics is bad mathematics. But *experimental mathematics* means something different: it means injecting experiment back into core parts of mathematics that need not — at least at present — have any contact with science.

Its most striking impact may be that it underlines the reality of an essential distinction we shall encounter repeatedly, between mathematical *fact* and mathematical *proof*. I realize that many fine mathematicians insist on defining their field narrowly, as beginning with proof, and give short shrift to facts. This may be because they have grown accustomed to seeing new mathematical facts almost exclusively suggested by the proofs of old mathematical facts. But the historian knows that in the past, the development of mathematics has relied upon many other sources, both of observation and of experimentation.

Today's experimental mathematics does not even spurn the kind of observation that has been characteristic of the least 'sophisticated' among empirical sciences: natural history. But it primarily relies upon active experimentation. Mathematical proof can, if mathematicians so choose, preserve much of itself in the form to which they have become accustomed in recent decades (and which we shall discuss here and there). That is, I neither wish nor expect — and never have either wished or expected — to see proof *replaced* by mere pictures. All that is happening now is that new methods of searching for new facts provide mathematics with a powerful 'front end' of unexpected character, one that involves more than just the proverbial pencil and paper. Thus, pictures have already demonstrated their astonishing power to *help* in early stages of both mathematical proof and physical theory; as this help expands, it may well lead to a new equilibrium and to changes in the prevailing styles of completed mathematical proof and of completed physical theory.

In other words, we may well be witnessing the re-emergence of a new active 'doublet' of an experimental and/or theoretical study. Experimental and theoretical physicists seldom live in perfect harmony, but they know they *must* not only coexist, but actually listen to each other and otherwise interact. Few in either party want to annihilate the other. In mathematics, the situation is very different: there has been a long history of conflict, as beautifully expressed in the following lines by a poet:

Gray, dear friend, is every theory.
And green the golden tree of life.

These words appear in the play *Faust*, by Goethe (1749–1832), in a famous scene in which Mephistopheles puts on the robes of old Professor Faustus, and describes the various academic programs to an awed passing student. The devil dwells on medicine (which has the virtue of bringing many fair maidens into a Doctor's life), then concludes (lines 2038–9) by his great description of two cultures. In the original,

Grau, teurer Freund, ist alle Theorie,
Und grün des Lebens goldner Baum.

For two centuries, practitioners of the hard theoretical sciences had every reason to resignedly acknowledge this devil's wisdom. Even though most scholarly institutions have ceased to compel their Dons to uniforms and to celibacy, many Dons continue to take pride in the fact that outsiders view their subjects of study as irredeemably gray.

Recently, however, a new tool has come into being: the computer. When taken by itself, it is as 'gray' as can be. But it has brought two gifts to science. Its first gift is vastly enhanced calculations. They will not concern us here, though it is worth mentioning that much of the early justification for the computer in the 1940's did not come from business users, but from those seeking new understanding about differential equations. One was John von Neumann; in his youth, he had been a near-'normal' mathematician, but by the 1940's he was no longer viewed as one and was deeply involved in weather predictions. Another pioneer of the use of computers was Enrico Fermi, a physicist's physicist, who was inspired by a desire to put the computer to use in understanding some other kinds of nonlinear mathematics.[2] Computer calculations have already caused many changes in mathematics, but these changes can be called quantitative, a matter of degree rather than of kind. Take number theory; it was an experimental discipline up to the time of Gauss and had been proclaimed experimental by Edouard Lucas, so no one could argue against experiment in this discipline.

The second gift of the computer is graphics, which tells an altogether different story and has brought a profound qualitative change, hence a fair amount of upheaval. Being myself far from the mathematical mainstream, as will be seen later in this Foreword, I welcomed computer graphics almost before it existed, and have had the good fortune of being able to demonstrate that the folk wisdom that the above verse by Goethe had expressed as an

[2]E. Fermi, J. Pasta, S. Ulam, *Los Alamos document LA-1940*, 1955. Reprinted in the Collected Papers of Enrico Fermi, Vol 2, pp. 978–88. Also in S. Ulam, *Sets, Numbers and Universes*, M.I.T. Press, 1974, pp. 490–501.

incontrovertible devil's truth is of more modest value. Its apparent universality simply resulted from a period when technology lagged behind abstract thought, followed by a period when mathematicians lagged in their acceptance of new technology. Computer graphics has repeatedly allowed me the privilege and the delight of taking up theories in mathematics and in physics whose grayness had seemed unimpeachable (and had in some cases been certified by a century of commentary), and of proving that if they are suitably transformed, these very same theories are enriched in their own mathematical or physical terms. And they also generate patterns that readily pass as forgeries of Life, of Nature, and even of Art, in their unfathomable complication. That is, not only does one part of old theory cease to be gray, but it becomes colorful enough to rejoice even the artist.

Seen from close by, the role of computer graphics covers a wide range. All too often, it boils down to *mere visualization.* This idea applies when a scientist hands his data to a specialist who knows how to put it into pretty pictures — the goal being in many cases simply to impress a visiting committee. In a way, this is how I started myself in the 1960's, before the emergence of tools anyone would call computer graphics. My very practical goal was to impress upon reluctant colleagues that some two-line formulas of mine might fail to be deep mathematics, but were indeed capable of generating 'forgeries' of the stock market, the maps of galaxies, and the weather. In doing so, a more interesting fact emerged at the opposite extreme of mere visualization. The use of computer graphics is now in the process of altogether changing the role of the eye. The hard theoretical sciences had banished the eye for a long time, and many observers used to believe, and even hope, that it would remain banished forever. But computer graphics is bringing it back as an integral part of the very process of thinking, search and discovery. Let us treat these two roles in greater detail.

I must confess a deep dislike for the term *visualization.* Of course, I am pleased that the loneliness I had experienced in the 1960's and 1970's has been replaced by a maddening crowd. I am pleased when visualization impresses a visiting review committee, and I look forward to the riches we shall all reap from the industrial trends that created this term. But to me, it is redolent of the bad old times from which we have recently emerged. To me, *visualization* seems a term invented by algebraists. Some algebraists think, for example, that the term 'circle' denotes the equation $x^2 + y^2 = r^2$. For them, that lovely curve shaped like the edge of the full moon does not exist by itself, but only to visualize this isotropic quadratic equation. Poincaré is reported to have written of his teacher that 'Monsieur Hermite never evokes a concrete image; yet you soon perceive that the most abstract entities are for him like living creatures.' This surprises me as it seemed to surprise Poincaré, but I do not deny that it may be true. When people like Hermite acquire too much political power over mathematical life, nothing can be left to survive from the times before Descartes injected analysis into geometry. Over a century ago, it was taken for granted (and expressed eloquently by Felix Klein and Henri Poincaré) that geometers and algebraists are two distinct kinds of scientists. Unfortunately, the academic tests that recruit new scientists have come during our century to give less and less credit for skills in geometry and increasingly full weight to skills in algebra. In that respect, the United States stood as an extreme case, because it never went for the serious study of geometry characteristic of all the countries of Europe. This may explain in part why the refugees from Russian and Germany found 'native' U. S. mathematics to be strong, but mostly pure and algebraic to a fault, well before this came to be also the rule in Europe.

Naturally, those who provoked geometry's fall from grace described it as inevitable, as yet another proof that there is progress, that history moves relentlessly forward, never to turn back. But in this area, as in many others, the notion that events are ruled by an inevitable destiny has been sharply contradicted by recent events. That is, it now seems that relentless algebraisation was not inevitable. To a large extent, it manifested on the part of all theoretical scientists a spontaneous, practical and appropriate adaptation to the lag in technology that has been mentioned. It was hard not to acknowledge the exhaustion of the old tools of geometry, and the lack of new ones, and to act accordingly. But now the computer is pushing this adaptation and the resulting expedients into a historical limbo.

To interrupt for an autobiographical aside, I happen to have made myself a student of this anti-geometric trend since the 1940's. Being a dyed-in-the-wool geometer and very much beholden to my eye, I consider it a boon that I took the notorious French examinations when geometry was still in the saddle. Watching from the perspective of fractal geometry, to which the reader must know that I have devoted most of my working life, I have seen new trends develop, and have kept searching through the past for events that led to the banishment of the eye from hard theoretical science. Therefore, allow me to recount a few old but lively stories I have read, and to reminisce about recent stories, one in physics and the other in mathematics, in which I have been a prime participant.

Pluralists and Utopians in the Greek Golden Age

The common feature of these stories is that they concern the conflict that arose during the Greek Golden Age, at a time when mathematics and science were being formulated in nearly their present form, and when the notion of proof was being developed. The two sides of this conflict can be called pluralistic and utopian.

The pluralistic view is wonderfully expressed in these words: 'Certain things first became clear to me by a mechanical method, although they had to be demonstrated by geometry afterwards because their investigation by the said mechanical method did not furnish an actual demonstration. But it is of course easier, when we have previously acquired, by the method, some knowledge of the questions, to supply the proof than it is to find it without any previous knowledge. This is a reason why, in the case of the theorems that the volumes of a cone and a pyramid are one-third of the volumes of the cylinder and prism respectively having the same base and equal height, the proofs of which Eudoxus was the first to discover, no small share of the credit should be given to Democritus, who was the first to state the fact, though without proof.' The author of these words might easily have been a man near our time, but in fact he was Archimedes.[3] Please don't let your eye glaze over all these names of ancient heroes. Please, read on!

The reason why the views of Archimedes deserve to be called pluralist is because he acknowledges a proper balance between the role of proof and the role of experiment, including the role of the senses. He sees no harm in acknowledging experiment and the senses as tools in what must be described as a search for new *mathematical facts*. The existence of mathematical facts has long seemed to me undeniable, but experience proves that other authors deny any meaning to the very notion, and view it as internally self-contradictory. Thus, a loud buzz was heard in mathematical circles after a mathematical congress that met in Kyoto in 1990 gave the Fields medal to the physicist E. Witters. Letters to the Editor went flying, describing *mathematics without theorems* as something that should not be accepted as a part of 'real mathematics.'

Thundering against experience and the senses is a satisfying idea in our culture, but it is surely not a new one. Would it not be nice, therefore, to be able to identify the first person to have expressed this idea? Judging from the tone of the above quotation, it seems that Archimedes was responding to someone else's already well stated opinion. It must have been the Utopian view held by Plato (427–347 BC), a man of very great power, both in intellect and in influence. Yes, it happens that the curses I hear too often being cast today against the return of the eye into the hard sciences are not new, and do little but echo Plato. And the pluralists who welcome and praise the return of the eye, thinking of themselves as down-to-earth and modern, may not know much of Plato, yet are actively fighting his shadow.

[3] Archimedes (287–212 BC), Democritus (460–370 BC), Eudoxus (408–355 BC).

The most widely quoted evidence on Plato's views occurs in Plutarch's *Life* of the Roman general and politician, Marcellus, who led the siege of Syracuse where a soldier killed Archimedes. Quoting from the Dryden translation, 'Eudoxus and Archytas had been the first originators of this far-famed and highly prized art of mechanics, which they employed as an elegant illustration of geometrical truths, and as means of sustaining experimentally, to the satisfaction of the senses, conclusions too intricate for proof by words and diagrams ... But ... Plato [expressed] indignation at it, and [addressed] invectives against it as the mere corruption and annihilation of the one good in geometry, which was thus shamefully turning its back upon the unembodied objects of pure intelligence to recur to sensation and to ask help (not to be obtained without base supervision and deprivation) from matter'.

Since Plutarch's anecdote about Plato's bossy manners was written 400 years after the fact, it should be approached with caution. But it is true to Plato's own words, that geometers 'talk in most ridiculous and beggarly fashion ..., as though all their demonstrations have a practical aim ... But surely the whole study is carried on for the sake of knowledge.'

When Plato's double curse, against physics and against the eye, first came to my attention, I had been toiling for decades at the task of rebuilding the destroyed icons. I found myself wildly rooting for Eudoxus. It was a delight to know that Eudoxus was a pioneer, not only in mechanics and astronomy (as implied by Plutarch), but also in geometry (as stated by Archimedes). In fact, he is often viewed as the most creative of the ancient Greek mathematicians, while Euclid — who flourished about 300 BC, was an encyclopedist. As to Plato, he was not at all a creative geometer. To quote Augustus de Morgan, 'Plato's writings do not convince any mathematician that this author was strongly addicted to geometry.' In the name of purity, he wanted to restrict geometry to operations with the rules and the compass. Plato was an ideologue and author of more than one harmful Utopia. In fact, Plato's curses against physics and the senses were very much in line with his political ideal of an authoritarian state published in his *Republic*.

Under Plato's influence, Greek mathematics underwent a remarkable transformation — anti-empirical and anti-visual — which elicits the most extreme reactions. It is praised by some as the greatest and the most durable achievement of Greece, and harshly faulted by others. Thus, de Santillana[4] blames it for the failure of Greeks to develop physics in parallel with mathematics, with such disastrous effects on Greco-Roman technology that it may bear a share of responsibility in the fall of Rome.

[4]de Santillana, G., *The Origins of Scientific Thought*, University of Chicago Press, 1961.

Plato (like Hermite many centuries later) believed in the full reality of Ideas, which implies that mathematical objects and truths are discovered, not invented. (This belief has few concrete consequences, but I agree wholeheartedly.) But Plato also believed that the physical world possesses only 'relative reality.' This is what led him to the formulation of a Utopia, in which mathematical truths must be discovered and studied without reference to anything concrete and without the use of the 'senses,' which certainly includes the eye and may include 'intuition.' The Utopians who drove the pictures out of mathematics were themselves driven by a fervor so close to religion that it is appropriate to call them *iconoclasts*. *Icon* meant *image* (as is known to many computer-literate people today), and had the connotation of *an idol*. *Clasts* are those who break or destroy.

For most of the time between Plato and the present, most mathematicians paid little heed to Plato's words. Matter and the senses were providing far too many exciting facts to play with and prove.

The Two Inseparable Faces of the Coin of Mathematics

Returning from Plato's days to ours, we find the situation different, new and fluid, and opinions sharply divided. We hear ringing acclaim from many, including the young and their teachers, but certainly not from everyone. Herein lies an interesting story. When I was younger, and experimental mathematics had not yet reawakened, the stage was occupied by just one kind of mathematics. This was already the case when I studied in Paris in the mid-1940's, first briefly at the Ecole Normale Supérieure, and then for the usual two years at Ecole Polytechnique. When in high school, I had become utterly fascinated by a very difficult subject called *geometry*, which occupied a large place in the curriculum. To me at least, this was the study of objects that had two properties that could have been contradictory, yet went together: like glove in hand, like the two faces of a single coin, or (a better image) like body and soul, each necessary to the other. One could reason on them in abstract style — perhaps dry, but noble beyond anything else: the style that had been pioneered by Euclid. But this was not all. For me, mathematics was concerned with completely real objects; they could actually be seen and manipulated, as drawings and also as plaster casts from the shelves in the office of the Mathematics Chairman. As a teenager, I loved it when I heard that certain numbers, originally introduced as formal square roots of negative reals (their origin led them to be called 'imaginary'), had soon enough turned out to be identical to points in the plane. This seemed to prove that the original way of introducing them was incomplete, and no one was promoting it — to my knowledge. To carry out algebra without this interpretation would truly deserve to be called a 'complex' procedure. I also loved the Euclidean representations for the non-Euclidean geometries, which revealed that another coin that in the 1830's had seemed to have a single face was, in fact, properly endowed with both. It became my wild but deep hope that occasions when a shape remained 'abstract' were simply proof of a temporary lack of visual imagination on the part of the geometers.

Jumping well ahead of the story, one can imagine my glee when fractal geometry found that many of the so-called 'monsters of mathematics' were as 'real' as can possibly be.

Needless to say, this view influenced my way of handling the heavy mathematics homework we were given. After I had become acquainted with the basic outline of a new problem, I did not rush to worry about the questions being asked, but instead hastened to draw some kind of picture. When the problem was stated geometrically, this task was straightforward. When the problem was stated algebraically or analytically, this was the hardest step. Once a picture was available, it received my undivided attention; I played with it and introduced all kinds of changes. In particular, I modified it, trying to make it (somehow) richer, more attractive and more symmetric. At some point, 'geometric intuition,' something we shall discuss momentarily invariably rewarded me with a sudden shower of observations. Only then did I look up the questions we had been asked, and as a rule found that all the answers were 'intuitively' obvious. Invariably, again, the formal proofs of these guesses were the quickest and easiest steps in the process. I do not recall having been stumped once, but of course this fact helps describe the mathematics taught in France in those years. The same situation prevailed throughout Europe, but apparently never took hold in the USA.

Geometric intuition is a much maligned ability that deserves a moment of attention. I have heard all too many people assert that it does not exist, never imagining that they may be describing their own disabilities. Other people delight in warning against intuition's pitfalls and its inadequacies. They do not realize that intuition is not something fixed, but rather the fruit of past experience; it is easily destroyed, yet can be trained.

Returning to my student years, a combination of intrinsic interest and of formalism was to me necessary for a full enjoyment of mathematics. But, even in high school, I heard the rumor that there were problems in my paradise. Then the end of World War II brought back to Paris an uncle who was a professor of mathematics at the famed Collège de France. He made me one of the best advised among 20-year-old mathematics students in Paris. He started by informing me (gently but firmly) that, as a topic of active research, the geometry I loved was dead. Worse even than reduction to some small but lively stream, it had been dead for nearly a century, except in mathematics for children. He too had been a whiz at geometry when in high school. But he felt that in order to make a genuine contribution to mathematics, one had to outgrow geometry. I heard all this in the 1940's, but Brooks (1989) echoes my uncle's opinion when he describes my 'mathematical sensibility' (then as today) as being 'rather infantile and somewhat dull.'

More precisely, I was told that *geometry*, as a word, was very much alive, but had lost the last trace of its old hands-on applications. For example, algebraic geometry was being saved from a bunch (mainly Italians) who could not define or prove anything properly, and was one of the bright new kids on the block, reborn as a purely algebraic enterprise destined to a future better than its present.

As a first alternative, my uncle suggested that I move into his own field of complex analysis, which he described as being farthest from the growing mood towards abstraction. For example, he told me of the Fatou-Julia theory of iteration, and suggested that a bright new mathematical idea might enable me to do something truly worthwhile — and worth rewarding. He was one of the few to be aware of the Fatou-Julia theory. He viewed it as admirable, and was deeply annoyed that it had not advanced much in the thirty-odd years between 1917 and 1945 or so. He gave me the original reprints they had given him. Unfortunately, reading these authors' great works definitely showed me that this was not the geometry I loved. Besides, Gaston Julia himself was very much around (he was in his fifties and after moving to Polytechnique, I had him as a teacher — of differential geometry). Yet hardly anyone but my uncle seemed to know about J-sets (this was before the term, 'Julia sets,' for which I bear part of the responsibility). In fact, hardly anyone but my uncle had a half good word for Julia.

The second and more obvious alternative to studying geometry was to fall in line behind a group of mathematicians who called themselves 'Bourbaki.' That 'fall in line' is the correct expression is confirmed by an attractive autobiographical essay by E. Hewitt.[5] 'From Stone and his fellow mathematicians at Harvard, I learned vital lessons about our wonderful subject:

> Rule #l. Respect the profession.
> Rule #2. In case of doubt, see Rule #1.'

Who was Stone? Aside from being a great creative mathematician, Marshall Stone had been raised to exert the natural authority of the son of a future Chief Justice of the USA. And he described 'the

[5]Hewitt, E., *Math. Intelligencer* 12, 4 (1990) pp. 32–39.

profession' in the following ringing tones.[6]

'While several important changes have taken place since 1900 in our conception of mathematics and in our points of view concerning it, the one which truly involves a revolution in ideas is the discovery that mathematics is entirely independent of the physical world...

'When we stop to compare the mathematics of today with mathematics as it was at the close of the nineteenth century we may well be amazed to see how rapidly our mathematical knowledge has grown in quantity and in complexity, but we should also not fail to observe how closely this development has been involved with an emphasis upon abstraction and an increasing concern with the perception and analysis of broad mathematical patterns. Indeed, upon close examination we see that this new orientation, made possible only by the divorce of mathematics from its applications, has been the true source of its tremendous vitality and growth during the present century.'

The Bourbaki used the words 'structure' and 'foundation' on every imaginable occasion. For them, the terms were very 'positive,' associated as they were with the noble tasks of building or rebuilding. But the Bourbaki were inconsistent in stopping the search for foundations before it came to concern logic. More important to my mind (and I never saw a reason to change), they were working very far from the hardier ones who really lay foundations after having dug holes through messy and uncertain ground. They were moving furniture around, like decorators and not like builders. Even worse, they often seemed simple-mindedly committed to a task of compulsive house-cleaning, housekeeping and hectoring. My feelings towards them were strong and simple (which marks me as familiar with many mathematicians' propensity to strong feelings). Their 'formalisme à la française' was not a useless task, to be sure, but it was ridiculous to allow it to rule mathematics, to rule the choice of who was to become a mathematician, and to extend its influence wherever it could. Therefore, I loathed and feared Bourbaki.

The death of geometry and the emergence of Bourbaki were the reasons why I gave up the envied position of number 1 in the entering class of the exclusive Ecole Normale. (In mathematics and physics combined, my class was thereby reduced to 14 students, for the whole of France.) Later, I left France. As described,[7] both decisions have turned out to be very wise, because Bourbaki was growing and was about to take over, not only Ecole Normale, but much of French academia.

---

[6] Stone, M., *American Math. Monthly* 68 (1961) pp. 715-734.

[7] Mandelbrot, B. B., *Math. People,* D. J. Albers and G. L. Alexanderson eds., Birkhauser, 1985, pp. 205–225.

Eventually, Bourbaki died off, but only after it had trained many younger mathematicians who had known nothing else in their lifetimes. They find it difficult today to comprehend the intensity of the emotions Bourbaki evoked among friend and foe. For this reason, I wrote down a few facts and thoughts concerning Bourbaki.[8]

In any event, I would not have been a happy Normalien; and in later years I would not have been happy in France as a professor of mathematics, when the colleagues who belonged to the ruling club viewed me dimly, and certainly not as a gentleman. But a move out of academic departments of mathematics into IBM has allowed me to retain permanently the 'infantile sensibility' I enjoyed as an adolescent. As the computer became easier to use, and as primitive graphics became available to those willing to pay the very steep admission price in effort and aggravation, I made them, not a tool to be called only if needed, but a constant and integral part of my process of thinking.

This brings us to the question that is very old but has been asked especially sharply in this context: what are, in a discovery, the respective contributions of the tool and of its user? The puzzle is that different kinds of tools continue to be treated differently. Galileo wrote a book to complain bitterly about those who belittled his discovery of sunspots, claiming this was only due to his having lived during the telescope revolution. Fatou (a cripple) and Julia (a wounded war hero) are — quite rightly — praised for their theory of iteration, and no one would dream of belittling their work as being due to their having lived during World War I and in the age of Montel.[9] Today, there are some who belittle work based on the computer as solely due to the worker's having lived in the computer age.

If it were so, we would be faced with a mystery. Why should experimental mathematics have attracted so few practitioners for so long a time after von Neumann and Fermi (mentioned earlier in this foreword) had shown in which way mathematics can benefit from the computer. Their example was ignored. When I was new at IBM, which I joined in 1958, opportunities to use computers were knowingly and systematically spurned by every noted mathematician. Even the example of S. Ulam is interesting. He contributed to an (already-mentioned) famous early paper on experimental mathematics,[2] and might have been expected to become a herald of the new trend. Yet the preface he wrote in 1963 to a reprint of that paper asserts the following: 'Mathematics is not really an observational science and not even an experimental one.

[8]Mandelbrot, B. B., *Math. Intelligencer* 11, 3 (1989) pp. 10–12.

[9]In 1912, Paul Montel introduced Fatou's and Julia's key tool, the normal families of functions, and soon afterwards — like nearly all young people in French academia — he was called into the Army.

Nevertheless, the computation which [Paul Stein and I] performed were useful in establishing some rather curious facts about simple mathematical objects.'

The opinion that the tool was all that mattered is certainly not applicable in my case, because graphics became essential to my work well before the computer era started; I recall quite vividly the time I spent looking at a record of coin tossing to be found in a famous probability textbook. It is reproduced as Plate 241 in my *Fractal Geometry of Nature*. It led me to all kinds of useful models. William Feller, the textbook's author, once told me whether the random numbers were taken from a table or obtained by a physical coin, but I have forgotten his answer. But surely they were neither computer generated nor computer plotted, and no other textbook of probability felt such a figure was needed.

Computer assisted graphics first became critical to my work in the late 1960's when a series of papers I wrote with J. R. Wallis used a pen plotter to draw, next to one another, a series of actual weather records and of records generated by surreal models of weather variability. This turned out to be very important, but was very far from mathematics. The first seriously mathematical application occurred elsewhere, in a context of interest to the harmonic analysis specialists I. P. Kahane and J. Peyrière. Heuristic calculation helped in an essential way, but pictures led me to a series of mathematical conjectures about certain random singular measures (later called *multifractals*). I could only prove special cases, but Kahane and Peyrière proved the full conjectures, and went on to very interesting justifications.

A second serious application, published very belatedly in 1983, provided the first fast algorithm for the construction of the limit sets of certain Kleinian groups. This brings up a significant episode. At that time, I knew of no expert in that field, but I had long known Wilhelm Magnus of NYU. He had written a book about Kleinian groups, so I paid him a visit in 1978 or 1979 to inquire whether my algorithm was known to him and others. He said that it was, which greatly encouraged me. Then Magnus gave me a file of computer generated limit sets sent to him by various people. Not one among the authors of those illustrations had used them in the search for new mathematical facts! This came as a profound surprise to me, and also a strong source of encouragement.

Nonetheless, the above investigations were nothing but appetizers. From my own viewpoint (and also from a wider viewpoint, held by many people), mathematics took a sharp turn in 1979–1980, when the Fatou-Julia theory that I had spurned in the 1940's again became a wide awake component of mainstream mathematics. This event came about because the topic was thoroughly changed by a new tool. As already mentioned, the last time it had been thoroughly changed was in the 1910's, when Paul Montel's normal families set Fatou and Julia to work. But — to my uncle's repeated and bitter disappointment — the new tool was not 'purely mathematical.' It did not come from within, but from outside mathematics. The methods of fractal geometry had already allowed me to use the computer to bear upon many problems of physics, and it occurred to me that it could also be used in mainstream mathematics. More precisely, it has already been said that the Fatou-Julia theory of iteration had fallen out of the mainstream. But it came back into it with a bang after some investigations that I carried out in 1979–1980 on a set that is pedantically called *the locus of bifurcation of the map* $z \longrightarrow z^2 + c$. In my early papers, this set was called 'μ-map,' because physicists used to denote the constant c by $-\mu$, hence used to write the map in question as $z \longrightarrow z^2 - \mu$. In the same papers, the locus for the map $z \longrightarrow \lambda z(1-z)$ was called 'λ-map.'

My observations concerning this locus were presented in May 1980 at a special seminar at Harvard, then in November 1980 at a seminar run by David Ruelle in Bures near Paris, at the Institut des Hautes Etudes Scientifiques. The Bures seminar was widely attended, and it appears to have had a profound effect on Adrien Douady, who was there. He, and then his former student John H. Hubbard, dropped their previous work (he was still a leader of Bourbaki!), and they have since 1980 devoted themselves fully to the locus I had described to him at this seminar, and subsequently at many private meetings. Soon after that, Douady and Hubbard proposed denoting this locus by the term *MandelbrotSet* and the letter M.

The M-set

The M-set continues to draw attention. Many people (justly, in my opinion) view the work it has inspired as very special in the coming of the new experimental mathematics. This is also perhaps why its precise origins have come to attract an exceptional amount of attention. Less praiseworthy are the undocumented anecdotes that are being thrown about, either to prove that experimental mathematics is a terrible idea, or to prove that it is a great idea, but one that either happened just by historical necessity (e.g., solely because of the computer) or happened by other hands. Being told that assertions that are not denied are viewed as correct, I have reluctantly resolved to sketch these attempts at controversy. I conclude from their having backfired that there is no competitor for credit to the discovery of the first and most striking list of properties of M.

It should first be mentioned that several scholars have confided to me that the thought of studying M had entered their mind; but they never followed up, and make no claim. By 1988 (interestingly enough, not when the study of M was new), one scholar (who — charity reigns — will not be named) acted differently: believing perhaps that he *ought to* have performed this work, he took the unfortunate step of expressing his claim in print, but without providing any evidence one could see or judge. A second published claim is, on the contrary, documented: a paper by Brooks and Metelski[10] includes a rough drawing of M. An extensively circulated letter from Brooks to B. Branner brought this drawing to wide attention in 1988, and also pointed out that it dates from 1979.

Thus, the M-set was sighted nearly simultaneously in two places, in both instances through a deep fog. However, as I have tried to explain elsewhere,[11] the date of first sighting is devoid of interest. Friends and foes of experimental mathematics agree on one thing: by itself, a picture can have no interest whatsoever. (This is particularly so here, given that the drawing in Brooks' and Metelski's paper is grossly mislabeled.) For the experimental mathematician, what matters is not the first sighting, but the mathematical ideas — if any — that the pictures suggest.

Brooks and Metelski reported no mathematical idea. Precisely to the contrary, my first glimpse of M as a vague form raised an irresistible challenge to apply to iteration the same tricks that had worked so well for me over the preceding fifteen years. (This is why — as has already been said — Brooks later described my 'mathematical sensibility' as 'rather infantile and somewhat dull.') What followed in 1980 need not be repeated, having been described in sufficient detail in a feature I contributed to the famed book by Peitgen and Richter[12] (This feature was written in early 1985, for the Catalog of their *Frontiers of Chaos* exhibit.)

If Brooks wished to be heard, his wish may well have been over-fulfilled in 1989. His case was taken up by S. Krantz, who went on to achieve renown as a purveyor of mathematical anecdotes concerning events in which he was not a participant, or even a witness.[13] He has also won less favorable renown for the accuracy (or lack of it) of his tales.[14] In 1989, our anecdotist took up the Brooks-Metelski paper, to make assertions contrary to those of Brooks.[15] He claimed, first, that the drawing in question dated to 1978, and second, that it was well-known to the community of mathematicians. If the latter had been true, it would have saved

[10]Brooks, R., Metelski, J. P., *The dynamics of 2-generator subgroups of PSL(2,C)*, in: Riemann Surfaces and Related Topics, I. Kra and B. Maskit eds., Princeton U. Press, 1981.

[11]Mandelbrot, B. B., *Math. Intelligencer* 11, 4 (1989) pp. 17–19.

[12]Peitgen, H. O., Richter, P. H., *The Beauty of Fractals*, Springer-Verlag, 1986.

[13]Krantz S., *Math. Intelligencer* 12, 3 (1990) pp. 58–63.

[14]Krantz S., *Math. Intelligencer* 13, 4 (1990) p. 5.

[15]Krantz S., *Math. Intelligencer* 11, 4 (1989) pp. 12–16.

Brooks the need to do anything in 1988, and would have given Douady and Hubbard no reason to name the M-set after me.

In 1991, the very same anecdotist is being quoted again: he argues that the connectedness of the M-set proves that, 'even in [its] heartland,' experimental mathematics can fail. Numerous other accounts are flying around, based on the belief that experimental mathematics is both straightforward and ineffective. But the practitioners know better: in skilled and cautious hands, one experiment leads to another experiment, then combines with proven mathematical facts, and ultimately leads to new mathematical conjectures. In the particular instance of the M-set, the experimental method had led me substantially further than is credited in the diverse accounts one hears. The story ought to be more widely known since it warranted pages 155–157 in the already mentioned feature I wrote for *The Beauty of Fractals*. It may have been seen there by our chief anecdotist, since he quoted from my piece in his review[15] of Peitgen and Richter's book (if 'review' is the right word). Recent advances in experimental mathematics and its growing social acceptance make my old feature worth reading again as a realistic description of a successful use in mathematics of the experimental method I have been practicing for so long, and to which so many are rallying today.

The connectedness of M was only one among many empirical observations that I made concerning M, and that have led to splendid, fully proven theorems. In addition, what I called the hieroglyphic character of M was refined and proven by Tan Lei, and the fact that the boundary of M is of Hausdorff dimension 2 was proven very recently by M. Shishikura. I would have been quite incapable of providing even one of those proofs. But let me say it again: contrary to persistent anecdotes, I cherish equally the heuristics (graphical or other) and the proof. I do not systematically denigrate work I do not understand or cannot myself accomplish.

We hear that traditionalists (true to their role) fear that in embracing experiment again, mathematics may lose something special. It is indeed probable that it will lose something just as it will gain something else. One good thing is that it has already lost the monolithic structure that characterized it in the 1950's and the 1960's.

It is time to bring this story back to its key point without paying undue attention to these recent disputes. Their overly personal character used to annoy me greatly, but from a distance I see them as fresh episodes, however trivial, in a long battle for souls that has raged since Plato and Archimedes.

Chapter 1

The Backbone of Fractals: Feedback and the Iterator

I would therefore urge that people be introduced to the logistic equation early in their mathematical education. This equation can be studied phenomenologically by iterating it on a calculator, or even by hand. Its study does not involve as much conceptual sophistication as does elementary calculus. Such study would greatly enrich the student's intuition about nonlinear systems. Not only in research but also in the everyday world of politics and economics, we would all be better off if more people realized that simple nonlinear systems do not necessarily possess simple dynamical properties.

Robert M. May[1]

Unfortunately, May's emphatic message has remained largely unheard, at least as far as mathematics education is concerned. What are the phenomena to which he refers? And why is he convinced that they are of such outstanding importance? We will carefully imbed his message into a larger frame before we explore his results in greater detail.

When we think about fractals as images, forms or structures we usually perceive them as static objects. This is a legitimate initial standpoint in many cases, as for example if we deal with natural structures like the ones in figures 1.1 to 1.3.

But this point of view tells us little about the evolution or generation of a given structure. Often, as for example in botany, we like to discuss more than just the complexity of a ripe plant. In fact,

[1] R. M. May, *Simple mathematical models with very complicated dynamics*, Nature 261 (1976) 459–467.

Winter Sunrise

Figure 1.1 : Winter Sunrise, the Sierra Nevada, from Lone Pine, California, 1944. Photograph by Ansel Adams. Copyright (c) 1991 by the Trustees of the Ansel Adams Publishing Rights Trust. All rights reserved.

California Oak Tree

Figure 1.2 : California oak tree, Arastradero Preserve, Palo Alto. Photograph by Michael McGuire.

any geometric model of a plant which does not also incorporate its dynamic growth plan for the plant will not lead very far.

Fern

Figure 1.3 : This fern is from K. Rasbach, *Die Farnpflanzen Zentraleuropas,* Verlag Gustav Fischer, Stuttgart, 1968. Reproduced with kind permission by the publisher.

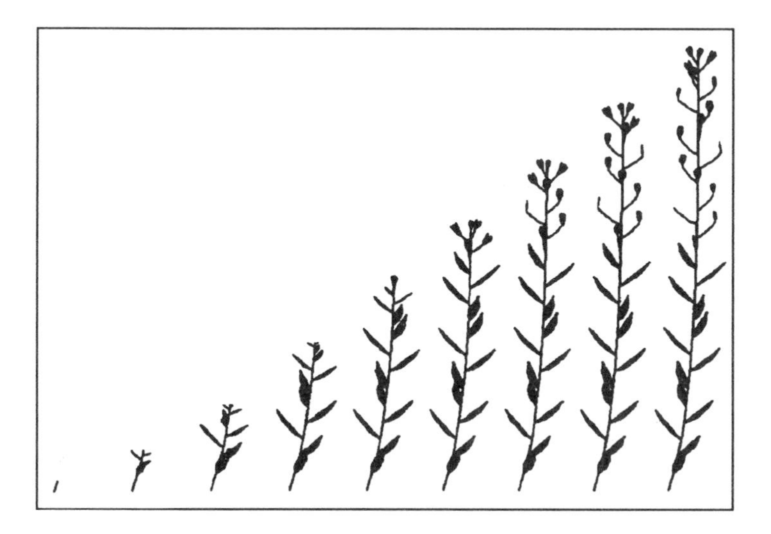

Plant Growth

Figure 1.4 : Plant growth simulated by a modified L-system (figure courtesy of P. Prusinkiewicz).

The same is true for mountains, whose geometry is a result of past tectonic activity as well as erosion processes which still and will forever shape what we see as a mountain. We can also say the same for the deposit of zinc in an electrolytic experiment.

Diffusion Limited Aggregation

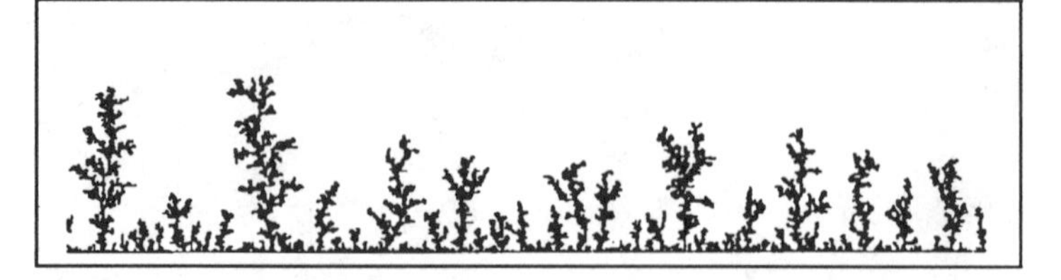

Figure 1.5 : DLA model of zinc deposition in an electrolytic process. The particles aggregate at the bottom line forming tree-like clusters (figure courtesy R. F. Voss).

Fractals and Dynamic Processes

In other words, to talk about fractals while ignoring the dynamic processes which created them would be inadequate. But in accepting this point of view we seem to enter very difficult waters. What are these processes and what is the common mathematical thread in them? Aren't we proposing that the complexity of forms which we see in nature is a result of equally complicated processes? This is true in many cases, but at the same time the long standing paradigm

> *Complexity of structure is a result of complicated interwoven processes*

is far from being true in general. Rather, it seems — and this is one of the major surprising impacts of fractal geometry and chaos theory — that in the presence of a complex pattern there is a good chance that a very simple process is responsible for it. In other words, the simplicity of a process should not mislead us into concluding that it will be easy to understand its consequences.

1.1 The Principle of Feedback

The most important example of a simple process with very complicated behavior is the process determined by quadratic expressions, like $x^2 + c$, where c is considered to be a fixed constant, or $p + rp(1 - p)$, where r is a constant. Before we enter an initial discussion of this phenomenon — a more systematic exploration is offered in chapter 10 — let us identify and discuss one of the central icons of our presentation.

Feedback processes are fundamental in all exact sciences. In fact, they were first introduced by Sir Isaac Newton and Gottfried W. Leibniz some 300 years ago in the form of dynamic laws; and it is now standard procedure to model natural phenomena using such laws. Such laws determine, for example, the location and velocity of a particle at one time instant from its values at the preceding instant. The motion of the particle is then understood as the unfolding of that law. It is not essential whether the process is discrete — i.e. takes place in steps — or continuous. Physicists like to think in terms of infinitesimal time steps: *natura non facit saltus.*[2] Biologists, on the other hand, often prefer to look at the changes from year to year or from generation to generation.

Iterator, Feedback and Dynamic Law

We will use the terms iterator, feedback, and dynamic law synonymously. Figure 1.6 explains the idea. The same operation is carried out repeatedly, the output of one iteration being the input for the next one.

The Feedback Machine

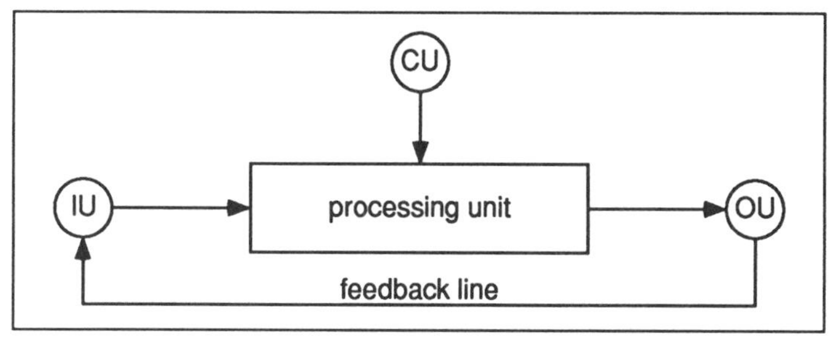

Figure 1.6 : The feedback machine with IU = input unit, OU = output unit, CU = control unit.

The Iterator, Principle of Feedback

The feedback machine has three storage units (IU = input unit, OU = output unit, CU = control unit, PU = processing unit), and one processor, all connected by four transmission lines, see figure 1.6. The whole unit is run by a clock, which monitors the action in each component and counts cycles. The control unit acts like a gear shift

[2]Nature does not make radical jumps.

in an engine. That is, we can shift the iterator into a particular state and then run the unit. There are preparatory cycles and running cycles, each of which can be broken down into elementary steps:

Preparatory cycle:

step 1: load information into IU
step 2: load information into CU
step 3: transmit the content of CU into PU

Running cycle:

step 1: transmit content of IU and load into PU
step 2: process the input from IU
step 3: transmit the result and load into OU
step 4: transmit the content from OU and load into IU

To initiate the operation of the machine we run one preparatory cycle. Then we start the running cycles and execute a certain number of them, the count of which may depend on observations which we make by monitoring the actual output. Execution of one running cycle is sometimes called one iteration.

What is a Feedback Machine?

When we refer to iterations we should imagine a proper feedback machine. The dynamic behavior of such a machine can be controlled by setting certain outside parameters, similar to control levers in an engine. We will discuss the basic principles guided by the simple example of video feedback, which in fact permits real experiments. This particular feedback machine can be built using particular pieces of equipment. It is a real machine in the original sense of the word. This case is rather the exception in this book. Here the term 'feedback machine' usually refers to an abstract machine, a 'Gedankenexperiment'. Such an abstract machine may be put into operation by executing an appropriate computer program, or by using a pocket calculator or merely paper and pencil to carry out the given feedback mechanism.

Video Feedback

Video feedback is a feedback experiment in the traditional sense of the word. Its basic configuration is probably as old as television. Nevertheless, the particular video feedback experiment which we will now present is so dramatic that its potential can excite even professionals from the television scene.[3] Figure 1.7 shows the basic setup. A video camera looks at a video monitor, and whatever it sees in its viewing zone is put onto the monitor. Apparently, there are quite a few controls which have an impact on what will be seen by an outside observer, e.g., the various control dials on the

[3]It was proposed by Ralph Abraham form the University of California at Santa Cruz in the 1970's. See R. Abraham, *Simulation of cascades by video feedback,* in: "Structural Stability, the Theory of Catastrophes, and Applications in the Sciences", P. Hilton (ed.), Lecture Notes in Mathematics vol. 525, 1976, 10–14, Springer-Verlag, Berlin.

Video Feedback Setup

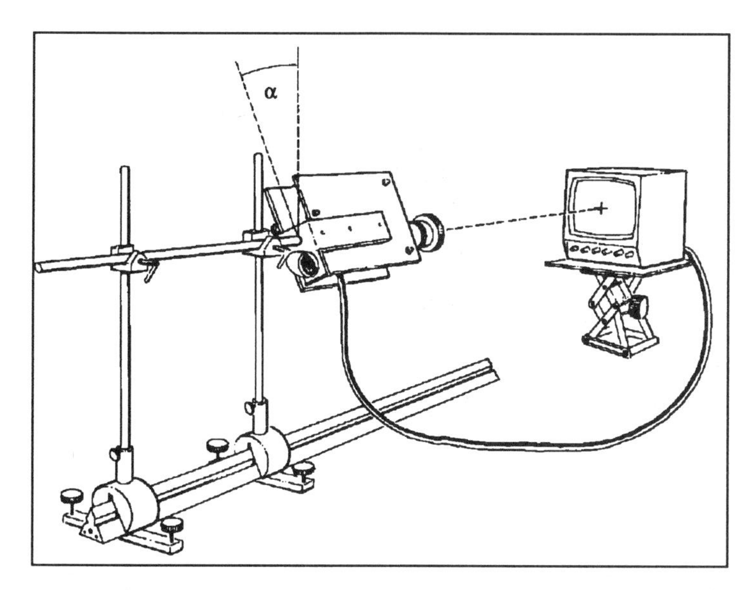

Figure 1.7 : The basic setup of video feedback.

monitor (contrast, brightness, etc.) and video camera (focus, iris aperture, etc.), but also the position of the camera with respect to the monitor. Below we collect some important tips which will help you to make a successful video feedback experiment yourself.

Hints for the Video Feedback Experiment

The experiment should be set up in an almost dark room. The distance between camera and monitor should be such that the mapping ratio is approximately 1 : 1. Turn up the contrast dial on the monitor all the way and turn down the brightness dial considerably. The experiment works better if the monitor is put upside down. Moreover, the tripod should be equipped with a head that allows the camera to be turned about its long axis, while it faces the monitor. Rotate the camera some 45°out of its vertical position. Connect the camera with the monitor. Now the basic setup is arranged. The camera should have a manual iris which is now gradually opened while the lens is focused on the monitor screen. Depending on the contrast and brightness setting you may want to light a match in front of the monitor screen in order to ignite the process.

It is quite obvious how we can imbed the experiment into our logo in figure 1.6 (input unit = camera, processing unit = camera and monitor electronics, output unit = monitor screen, control unit = focus, brightness, etc.). The feedback clock runs quite fast, i.e. about 30 cycles per second, or whatever number of frames per second your TV system generates.[4]

Dramatic Impact of Controls

Each of the controls has an impact on the process, some a very dramatic one. In this regard we can think of our setup as an *analog computer* with control dials. For some kinds of controls and variables it is relatively easy to understand their mechanisms. For others it is hard, and for still others it is hard as hell. In fact, many of the phenomena which can be observed are still very poorly understood. The physicist James P. Crutchfield has probably contributed most towards a deeper and systematic understanding of the process.[5]

The easiest variable which has a dramatic impact on the process of image generation is the position of the camera with regard to the monitor. When the distance from the camera to the monitor is long, the monitor is just a small part of the viewing field. Consequently, the monitor will be reproduced onto a small detail of itself, and this happens again, and again, and again, ad infinitum. In other words, we see a monitor inside a monitor inside a monitor, etc. (compare figure 1.8). The effect of the process can be described as compression, or dynamically, as a motion to the center of the monitor. Whatever image is initially on the monitor will be squeezed and put back onto the monitor, and that image will be squeezed again and so on. We would say that the mapping ratio is $1 : m$, where $m < 1$, i.e. something of unit length 1 on the monitor would reduce to something of length m in a single feedback cycle.

The *monitor-inside-a-monitor* effect is known by most people as video feedback. It is almost always easy to reproduce with any kind of equipment. But there is much more 'life' in this simple system than has been recognized because it is a little harder to reproduce with some equipment.

Next, let us discuss what will happen at the other extreme end of the positioning scale. When the distance between the camera and the monitor is so short that the viewing field of the camera is just a part of the monitor screen. That part is put back onto the entire screen of the monitor, and again, and again, ad infinitum (compare figure 1.9). We would say that the mapping ratio is $1 : m$, where $m > 1$, i.e. something of unit length 1 on the monitor would expand to something of length m in a single feedback cycle.

Now the action in the process is best described as expansion, or dynamically, as a motion to the border of the monitor. Whatever image is initially on the monitor, a small part of it will be expanded to the full screen, and of that a small part will again be expanded and so on. Since the TV refreshes its image about 30 times per

[4]NTSC is typically 30 frames per second at 480 lines per image.

[5]J. P. Crutchfield, *Space-time dynamics in video feedback*, Physica 10D (1984) 229–245.

Monitor Inside Monitor Inside ...

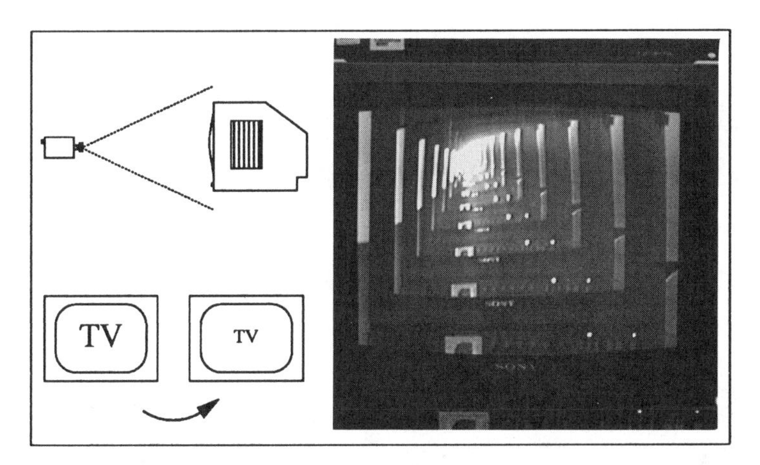

Figure 1.8 : Effect of long distance between camera and monitor. Basic setup and mapping principle (left), real feedback — monitor inside monitor (right).

Zoom into a Zoom into ...

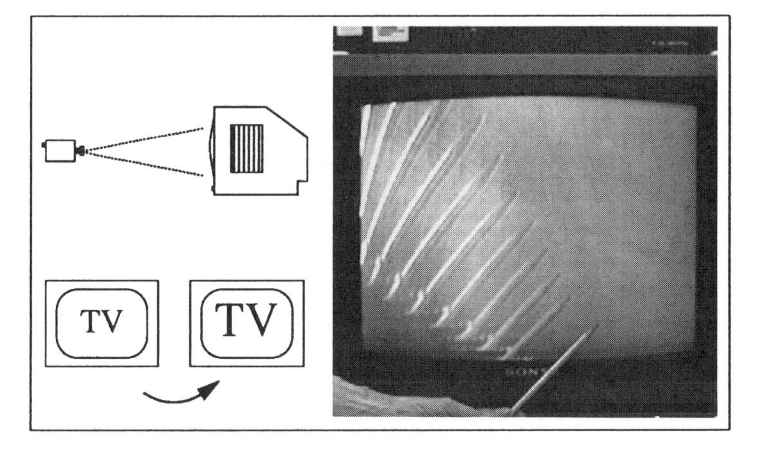

Figure 1.9 : Effect of short distance between camera and monitor. Basic setup and mapping principle (left), real feedback — repeated magnification of the image of a pencil (right).

second, it is impossible to see the individual steps in this process. The result of the close camera position can be a rather wild and almost turbulent motion on the screen.

Unchaining the Feedback

The more interesting effects occur when the position of the camera with regard to the monitor is carefully chosen to be such that the mapping ratio is nearly 1 : 1. The effect is increased dramatically if the camera is turned about its axis, i.e. an image on the monitor is seen by the camera as if rotationally changed by

Figure 1.10 : Some examples of real video feedback. There is a more or less pronouced periodicity in these pictures which depends on the angle of the video camera. From the upper left to the lower right we can see periods 3, 5, 5, 5, 8, 8, 11, 11, >11.

some angle. Thus it appears on the monitor (mapping ratio 1 : 1) in essentially the same size but rotated. From this point on, any simple description of the mechanisms for the wild and beautiful visual effects that can be observed breaks down. From what has been said so far, we would expect that in the rotated position we would eventually observe just a sequence of rotated images. But this prediction is far too simple. All kinds of peculiar effects occur due to many different characterisics innate to television image production. For example, the process of scanning the image on the monitor and in the camera is one of sequentially putting together a series of lines to compose the image. There is also the *memory effects* of the phosphorus on the monitor tube. In addition, there are electronic time chains and their delays in both the monitor and camera, as well as other factors.

In any event, this extremely simple feedback system demonstrates very dramatically how complicated structures can be the result of very simple feedback. In a way, this is the theme of the book. Our next set of experiments tries to bring more of a systematic light into this world of exciting phenomena. The basic principle is the same as with video feedback: an initial image is processed and then the resulting image is reprocessed by the same machine over and over again.

1.2 The Multiple Reduction Copy Machine

We now turn to a set of experiments which will provide us with a very intuitive access to the language of fractal geometry. In a sense, it is a continuation of the video feedback experiment.

First, let us consider a copy machine which is equipped with an image reduction feature. If we take an image, put it on the machine and push a button, we obtain a copy of the image. It is, however, reduced uniformly by say 50%, i.e. by a factor of 1/2. In the language of mathematics we say that the copy is *similar* to the original. The process to generate a copy is called a *similarity transformation* or *similitude*. The process just described embedded into the idea of figure 1.6 constitutes a feedback system[6] which would be very easily predictable in its long run effect: after some ten or so cycles any initial image would be reduced to just a point. In other words, running the machine would be a waste of paper (see figure 1.11).

Single Reduction Copy Machine

Figure 1.11 : Iteration by a copy machine with reduction applied to a portrait of Carl Friedrich Gauss (1777–1855).

We will now modify this principal setup. Remember, the basic action of our machine is the reduction of images. Such reductions, of course, are achieved by a lens system. As a simple modification of a stock copier, let us imagine that our custom copier has 2, or 3, or 7, or 14532231, or whatever number of reduction lenses. Each of them looks at the image on the copier, reduces it, and puts the result somewhere on the copy paper. One such design consists of the choice of the number of lenses, the reduction factors and the placements of the reduced images. It constitutes a particular feedback system which we can run to see what happens. We call such a machine a *Multiple Reduction Copy Machine*, abbreviated by the letters *MRCM*.

[6] Try to identify input-, processing- and output-unit.

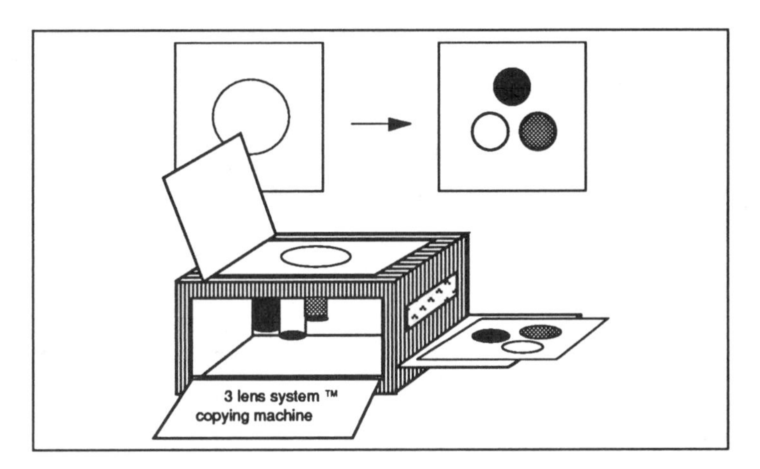

Figure 1.12 : The Multiple Reduction Copy Machine (MRCM): the processing unit is equipped with a three-lens system.

Multiple Reduction Copy Machine

Figure 1.12 shows a first example of an MRCM which incorporates just three reduction lenses, where each of them reduces to 50%, i.e. by a factor of 1/2.

What will we see emerging in the sequence of iterations as we run the feedback system? Will we see an arrangement of a smaller and smaller composite of images developing towards a point? Figure 1.13 gives the surprising answer, the consequences of which could potentially revolutionize almost everything we have thought about images in a technical sense. Let us start with a rectangle as an initial test image. We put it onto the multiple copier, obtaining three reduced copies which we color according to the respective lens system from which each copy is produced.

A First Hieroglyph: the Sierpinski Gasket

Then, indeed, we see $3 \times 3 = 9$ smaller copies, and then $3 \times 9 = 27$ even smaller copies, then 81, 243, 729, etc. copies which rapidly decrease in size, but the resulting compound images do not reduce to a point at all. Rather, they transform into a perfect *Sierpinski gasket*, which we will use as a major example exhibiting important aspects of fractals in general. Using the imagery of a language paradigm, we have just introduced a first hieroglyph in our new fractal dialect. From what we have said so far, it is clear that this basic principle will generate an infinite variety of images. All we have to do is convert the copier into one consisting of 4, or 5, or any other number of lens systems or with different reduction factors. We will be going into this matter in more detail in chapters 5 and 6, but there are two major surprises which are not immediately apparent and deserve some preliminary discussion here.

Rectangle in MRCM

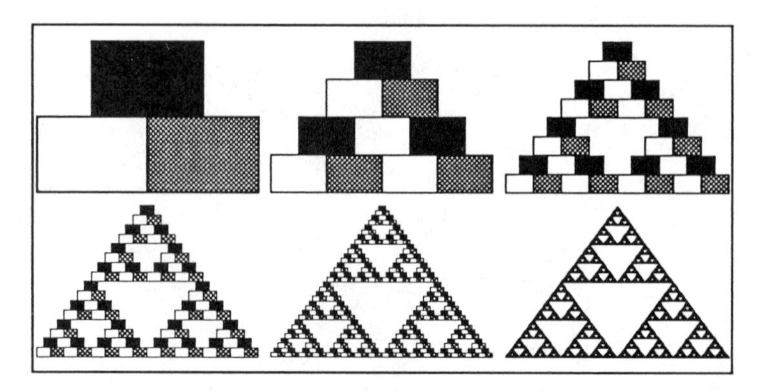

Figure 1.13 : Starting with a rectangle the iteration leads to the Sierpinski gasket. Shown are the first five steps and the result after some more iterations (lower right).

Looking at figure 1.13 again, we may be led to believe that the secret to the tendency toward the formation of the Sierpinski gasket is our choice of an appropriately dimensioned rectangle as the initial image in starting the feedback process. To show that this is not the case, let us assume that instead of a rectangle as the initial image, we choose a triangle or any arbitrary image, which may be represented well enough by the letters NCTM. The question is: what will then evolve in the process? Figure 1.14 gives the answer. The same final structure is approximated as we run the machine. Each step produces a composite of images which rapidly decrease in size. It doesn't matter in the least whether these images are rectangles, triangles, or the letters NCTM; the same final composite image is approached in each case — namely, the Sierpinski gasket. In other words, the machine produces one — and only one — final image in the process, and that final image is totally independent from the image with which we start! This magnificent behavior seems to be a miracle. But in mathematical terms it just means that we have a process which produces a sequence of results tending toward *one* final object which is independent from how we start the process. This property is called *stability*.

The second surprise is that the copy machine paradigm is not just a way to recover 'mathematical monsters' like the Sierpinski gasket or its relatives (soon we will see many of them). Let us ask what the images are which we can obtain this way. What can they look like? The answer is simply incredible. For many more natural pictures there is a copy machine of the above kind which generates the desired picture. How does one design the machine for a given picture? Well that is a difficult problem, as one can guess. But nevertheless, in chapter 5 we will introduce some of

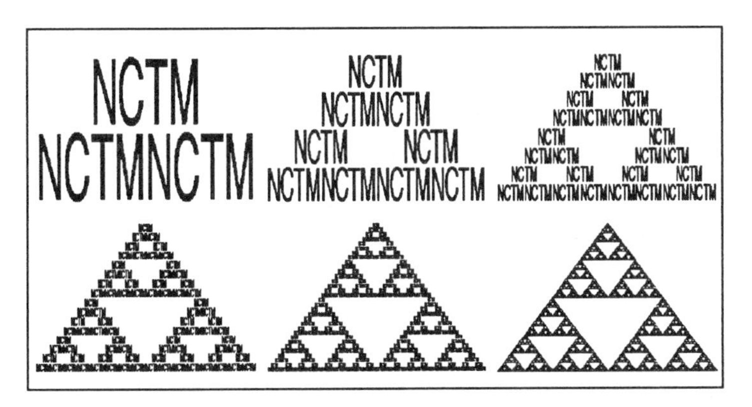

MRCM Applied to 'NCTM' and Other Shapes

Figure 1.14 : We can start with an arbitrary image — this iterator will always lead to the Sierpinski gasket.

the design principles which readily lead to the frontiers of current mathematical research.

The important point here is to see some of the variety of possible images obtained by very simple feedback processes, the elements of which are easily manipulated and completely under our control, quite unlike the video feedback experiment.

From Similar to Affine

In our first example, each lens system behaves like a similarity transformation, i.e. a rectangle is reproduced as a rectangle, a triangle with certain angles is reproduced as triangle with the same angles, and so on. The only thing which is changed is the scale of the image. If we pick any two points in the original image and compare their distances with that in the copy it will be scaled down by a constant factor. One principal direction for an extension will be to allow lens systems which reduce by different factors in different spatial directions. For example, the lens system may reduce by a factor of 1/2 in the horizontal direction and by a factor of 1/3 in the vertical direction. The effect of such a system is to destroy similarity: a square is reduced to a rectangle; a triangle with certain angles is reduced to a triangle with different angles. In mathematical terms we speak of *affine* transformations. Similitudes and affine transformations are, however, in one class of mathematical objects: *linear* transformations, i.e. transformations which when applied to a straight line reproduce a straight line. Only if we allow such extensions, will the metaphor of the copy machine develop its full power (see chapter 5).

From Linear to Nonlinear

Real lens systems are usually not perfect similitudes. They distort an image more or less: as a radical example, a straight line seen through a fisheye lens is reproduced as a curved line. In mathematical terms we speak of nonlinear effects. Let us simulate such an effect in a simplistic model. Let us consider the numbers

Nonlinear Transformation

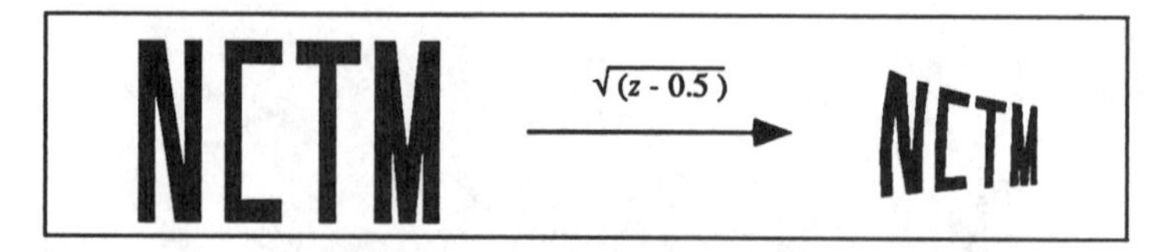

Figure 1.15 : The complex square root applied to the letters NCTM in the plane. Note that angles are preserved.

which are larger than 1. If we multiply such numbers by a factor of 1/3, for example, we have a perfect similitude. If we take the square root, however, we have a typical nonlinear effect: the segment between 1 and 10 is reduced to the segment between 1 and $\sqrt{10} \approx 3.16$, while the segment between 1 and 100, which is 11 times as long, is reduced to the segment between 1 and 10, which is only about 4 times as long as the segment between 1 and $\sqrt{10}$. The reduction factor changes, i.e. it depends on the location where the transformation is applied. Copy machines which are equipped with nonlinear lens systems are the content of chapter 12, and will lead to the famous Julia sets as well as the Mandelbrot set. Incidentally, the lens systems discussed there are related to similitudes in one important sense: they preserve angles. Figure 1.15 illustrates a typical transformation.

1.3 Basic Types of Feedback Processes

We will now turn to feedback machines which process numbers but before we get involved in the discussion of specific examples let us take an overview.

One-Step Machines

One-step machines are characterized by an iteration formula $x_{n+1} = f(x_n)$, where $f(x)$ can be any function of x. It requires one number as input and returns a new number — the result of the formula — as output (e.g. $f(x_n) = x_n^2 + 1$). The formula can be controlled by a fixed parameter (e.g. $x_n^2 + c$, i.e. with control parameter c), but in any case the output depends only on the input. The numbers are indexed in order to keep track of the time (cycle) in which they were obtained.

One-Step Feedback Machine

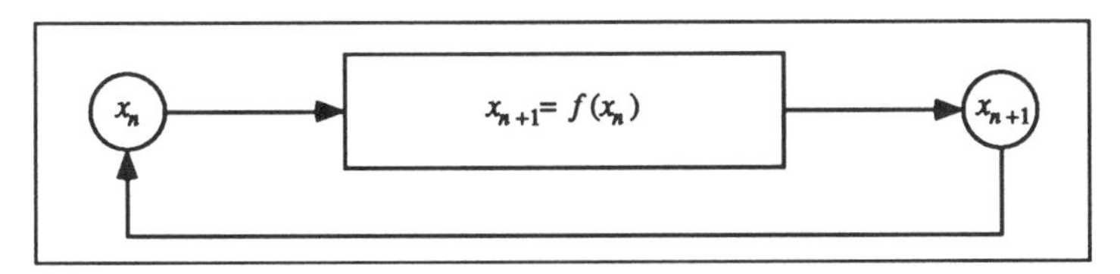

Figure 1.16 : Principle of the one-step feedback machine.

One-step machines are very useful mathematical tools and have been developed in particular for the numerical solution of complex problems. They have a tradition in mathematics which goes back at least a few thousand years.

Ancient Square Root Computation

The following example of a one-step feedback machine is an algorithm which was already known to the Sumerian mathematicians some 4000 years ago. It is a beautiful example of the strength and continuity of mathematics. Mankind has seen many advances and terrible setbacks since those times, while the power and beauty of mathematical thought has remained.

Given $a > 0$. Compute a sequence $x_1, x_2, x_3, \ldots$ such that the limit is $\sqrt{a}$, i.e. x_n approaches $\sqrt{a}$ closer and closer as we proceed to larger and larger n. Here is how x_n is defined. We begin with an arbitrary guess $x_0 > 0$ and continue with

$$x_{n+1} = \frac{1}{2}\left(x_n + \frac{a}{x_n}\right), \quad n = 0, 1, 2, \ldots \qquad (1.1)$$

Let us look at an example, $\sqrt{2}$. We guess $x_0 = 2$. Then

$$x_1 = \frac{1}{2}\left(x_0 + \frac{2}{x_0}\right) = \frac{1}{2}\left(2 + \frac{2}{2}\right) = 1.5$$

and

$$x_2 = \frac{1}{2}\left(x_1 + \frac{2}{x_1}\right) = \frac{1}{2}\left(1.5 + \frac{2}{1.5}\right) = \frac{17}{12} = 1.41666\ldots$$

and so on.

Let us give a brief argument why this method works in order to understand how well it works. To this end we introduce the *relative error* e_n of x_n, where e_n is defined by the equation

$$x_n = (1 + e_n)\sqrt{a}\,. \tag{1.2}$$

Replacing x_n by the equivalent $(1 + e_n)\sqrt{a}$ in eqn. (1.1) we arrive at

$$x_{n+1} = \sqrt{a}\left(1 + \frac{e_n^2}{2 + 2e_n}\right)\,.$$

Thus, using the definition in eqn. (1.2) again, we obtain an expression for the error e_{n+1}

$$e_{n+1} = \frac{e_n^2}{2 + 2e_n}\,. \tag{1.3}$$

Now $x_0 > 0$ and therefore $e_0 > -1$ and thus $e_n > 0$ for $n = 1, 2, 3, \ldots$ But then $x_n > \sqrt{a}$ for all $n > 0$. Finally, we can obtain estimates out of eqn. (1.3). If we drop the '2' in the denominator we obtain

$$e_{n+1} < \frac{e_n}{2}$$

and if we drop '$2e_n$' we obtain

$$e_{n+1} < \frac{e_n^2}{2}.$$

The first inequality and the definition of e_n by eqn. (1.2) shows that

$$x_1 > x_2 > x_3 > \ldots > \sqrt{a}$$

and that the limit is $\sqrt{a}$. The second inequality shows that if $e_n < 10^{-n}$ then $e_{n+1} < 10^{-2n}/2$, i.e. in each step of the sequence the number of correct digits is nearly doubled. This algorithm for the computation of the square root is an example of a more general method for the solution of nonlinear equations, which was discovered about 4000 years later and is nowadays called *Newton's method.*

Two-Step Feedback Methods

One-step feedback processes represent only a particular class of a whole family of feedback methods. Another class is known as *two-step methods*. Here the output is typically computed by a formula like

$$x_{n+1} = g(x_n, x_{n-1}).$$

Take for example the law which generates the *Fibonacci numbers*

$$g(x_n, x_{n-1}) = x_n + x_{n-1}.$$

Fibonacci Numbers and the Rabbit Problem

Leonardo Pisano, also known as Fibonacci[7] was one of the outstanding figures in medieval Western mathematics. He traveled widely in the Mediterranean world before settling down in his native Pisa. In 1202 he published his book, *Liber Abaci*, which changed Europe. It acquainted Europeans with the Indian Arabic ciphers 0, 1, 2, ... His book also contained the following problem, which has inspired people ever since. There is one pair of rabbits which is born at time 0. After one month that pair is mature and a month later gives birth to a new pair of rabbits and continues to do so (i.e., every month a new pair is born to the original pair). Moreover, each new pair of rabbits matures after one month and begins producing pairs of offspring every month after that ad infinitum. One assumes that the rabbits live forever. What is the number of pairs after n months?

Let us be careful and follow the evolution of rabbits step by step. In our rabbit population, let us distinguish between adult and young pairs of rabbits. A just-born pair is young, of course, and turns adult after one time step. Moreover, an adult pair gives birth to a young pair after one time step. Now let J_n and A_n be the number of young and adult pairs after n months. Initially, at time $n = 0$, there is only one young pair ($J_0 = 1$, $A_0 = 0$). After one month the young pair has turned into an adult one ($J_1 = 0$, $A_1 = 1$). After two months the adult pair gives birth to one young pair ($J_2 = 1$, $A_2 = 1$). Then again after the next month. Moreover, the young pair turns into an adult one ($J_3 = 1$, $A_3 = 2$). The general rule, of course, is that the number of newborn pairs J_{n+1} equals the previous adult population A_n. The adult population grows by the number of immature pairs, J_n, from the previous month. Thus, the following two formulas completely describe the population dynamics

$$\begin{aligned} J_{n+1} &= A_n, \\ A_{n+1} &= A_n + J_n. \end{aligned} \tag{1.4}$$

As initial values we take $J_0 = 1$ and $A_0 = 0$. From the first of the above equations it follows that $J_n = A_{n-1}$. Inserting this into the other equation we obtain

$$A_{n+1} = A_n + A_{n-1}$$

with $A_0 = 0$ and $A_1 = 1$. This is a single equation for the total rabbit population. Using this equation, the number of pairs in successive generations is easily computed:

$$0, 1, 1, 2, 3, 5, 8, 13, 21, 34, 55, 89, 144, 233, \ldots$$

Each number in this sequence is just the sum of its two predecessors. This sequence is called the *Fibonacci sequence*.

We have established another feedback system, but this one is a little different from the previous systems. In all the earlier feedback loops, the state at time n was determined only by the preceding

[7]Filius (=son) of Bonacci

state at time $n-1$. Such systems are called *one-step loops*. For the Fibonacci sequence the state at time $n+1$ requires information from states n and $n-1$. Such systems are called *two-step loops*. The simple and innocent looking Fibonacci sequence has a variety of interesting properties. Thousands of papers have been published about them, and there is even a *Fibonacci-Association* with its own periodical, *Fibonacci-Quarterly*, which reports on the never ending stream of new results. One property has been known for a long time and has led to amazing recent research in biology, as well as having had astonishing applications in architecture and the arts for many centuries.

Apparently the Fibonacci sequence can grow beyond all limits. Our rabbits exhibit a kind of a population explosion. We can ask, however, how the population progresses from generation to generation. For that purpose we look again at the Fibonacci numbers and compute the ratios of succeeding generations (rounded to six decimals).

n	A_n	A_{n+1}/A_n	in decimals
0	1	1/1	1.0
1	1	2/1	2.0
2	2	3/2	1.5
3	3	5/3	1.666666
4	5	8/5	1.6
5	8	13/8	1.625
6	13	21/13	1.615385
7	21	34/21	1.619048
8	34	55/34	1.617647
9	55	89/55	1.618182
10	89	144/89	1.617978
11	144	233/144	1.618056
12	233	377/233	1.617026

Apparently we are approaching steadily, if not exactly rapidly, some particular number. Have you seen that mysterious number

$$1.6180339887498948482 0\ldots$$

before? Let us open the curtain.

$$1.61803398\ldots = \frac{1+\sqrt{5}}{2},$$

which is the famous *golden mean*, or *proportio divina*,[8] as they called it in the middle ages. This number has inspired mathematicians, astronomers and philosophers like no other number in the history of mathematics.

---

[8]divine proportion (Latin).

At first it seems that processes of two-step methods are not covered by the concept of a feedback machine as we have discussed it so far. Indeed, the output x_{n+1} depends not only on the last step x_n, but also on the step preceding the last, namely x_{n-1}. Consequently, it may appear natural to extend the design of our feedback machines so that the concept incorporates a certain memory which conserves some information from the last cycles.

Feedback Machines with Memory

Machines with memory are typical for our computer age. While a machine without memory reacts on their inputs always in the same way, a machine with memory may react differently upon taking its own state or content of the memory into account. Take for example a soft drink machine. You will not be successful in getting a soda by just pushing a button. First you have to insert the right amount of money to make sure that the machine is in the appropriate state to accept your input.

Let us now extend the concept of a feedback machine by equipping the processing unit with an internal memory unit. Then the iteration of a two-step method $x_{n+1} = g(x_n, x_{n-1})$ can be implemented as follows. First note that to start the feedback machine two initial values x_0 and x_1 are required.

Preparation: Initialize the memory unit with x_0 and the input unit with x_1.

Iteration: Evaluate $x_{n+1} = g(x_n, x_{n-1})$ where x_n is in the input unit and x_{n-1} is in the memory unit. Then update the memory unit with x_n

Somehow it seems that feedback machines with memory should be more flexible in modeling different phenomena. But this is not at all the case. Rather, a machine with memory can be seen to be equivalent to a one-step machine which, however, works on *vectors* as input and output information. Input and output are given as pairs, or triples, or quadruples, and so on, of numbers. In other words, a pair of input variables (x_n, x_{n-1}) generates a pair of output variables (x_{n+1}, x_n).

One-Step Machines With Two Variables

Formally, we introduce a new variable, $y_n = x_{n-1}$, and extend the formula $x_{n+1} = g(x_n, x_{n-1})$ to the equivalent pair:

$$x_{n+1} = g(x_n, y_n)$$
$$y_{n+1} = x_n$$

This simple trick can easily be generalized. For example, let us assume that the formula, which determines the feedback depends on k preceding iterations, then one can rewrite this single formula as a one-step process which is given by a set of k formulas by introducing k independent variables. Usually, the independent variables are combined into a vector of variables. The

Two-Step Loop

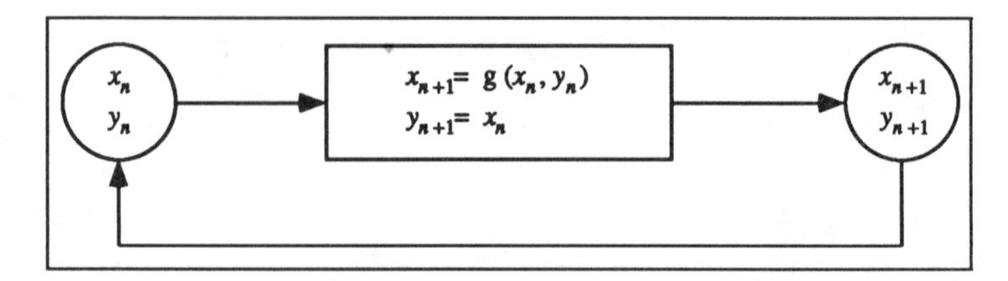

Figure 1.17 : Two-step loops are a special case of one-step feedback machines with two variables.

pair (x_n, y_n), for example, can be written symbolically as a single new variable Z_n. Moreover, we can then rewrite the set of formulas $x_{n+1} = g(x_n, y_n)$, $y_{n+1} = x_n$, by a single formula: $Z_{n+1} = G(Z_n)$. In other words, we do not have to go to the trouble of developing a special machine for two-step methods. They are perfectly covered by one-step machines.

The Rabbit Problem as One-Step Machine

Let us give one example, the Fibonacci numbers, defined by the two-step method

$$A_{n+1} = g(A_n, A_{n-1}) = A_n + A_{n-1}$$

with $A_0 = 0$ and $A_1 = 1$. The equivalent equations for a one-step method operating on pairs (x_n, y_n) is

$$x_{n+1} = x_n + y_n$$
$$y_{n+1} = x_n$$

with initial settings $x_0 = 0$ and $y_0 = 1$. This is exactly the same as in the derivation on page 35 setting $x_n = A_n$ and $y_n = J_n$.

One-Step Machines Based on Combined Formulas

Using the compact notation $G(X_n)$ for a whole set of formulas in the processing unit considerably simplifies the description of seemingly complicated feedback processes. Here is another example, which will become important in chapters 2 and 10:

$$x_{n+1} = \begin{cases} ax_n & \text{if } x \leq 0.5 \\ a(1 - x_n) & \text{if } x > 0.5 \end{cases}$$

Here a denotes a parameter, e.g., $a = 2$ or $a = 3$. Rather than introducing a feedback machine with two formulas and an additional switch, we will rewrite the above system of two equations as a one-step process of the form $x_{n+1} = f(x_n)$, where f is the transformation, whose graph — known as the *tent transformation* — is given in figure 1.18.

The Tent Transformation

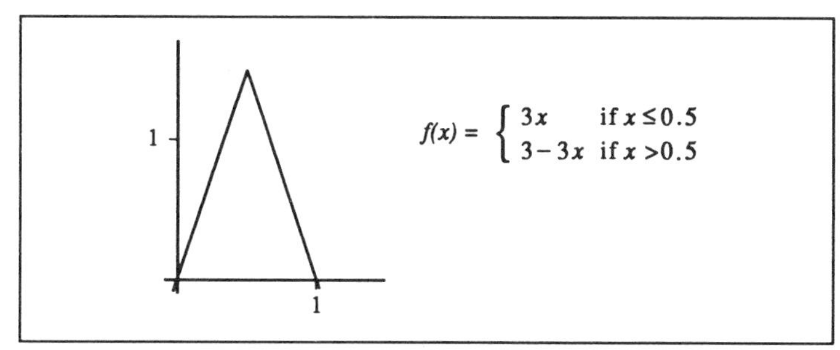

Figure 1.18 : The tent transformation is given by $f(x) = ax$ if $x \le 0.5$ and $-ax + a$ if $x > 0.5$. Here the parameter $a = 3$ has been chosen.

The $(3A + 1)$-Problem

This is an algorithm which produces a sequences of integers in a most simple way, but yet its unfolding is still not completely understood. Here is the original formulation due to Lothar Collatz:

Step 1: Choose an arbitrary positive integer A.
Step 2: If $A = 1$ then STOP.
Step 3: If A is even, then replace A by $A/2$ and go to step 2.
Step 4: If A is odd, then replace A by $3A + 1$ and go to step 2.

Let us try a few choices for A:

- 3, 10, 5, 16, 8, 4, 2, 1, STOP
- 34, 17, 52, 26, 13, 40, 20, 10, 5, 16, 8, 4, 2, 1, STOP
- 75, 226, 113, 340, 170, 85, 256, 128, 64, 32, 16, 8, 4, 2, 1, STOP

The obvious conjecture is: the algorithm comes to a stop no matter what the initial A is. Well, it seems that the larger the initial A is the more steps we have to run until we arrive at 1. Let us try $A = 27$ to verify that guess.

- 27, 82, 41, 124, 62, 31, 94, 47, 142, 71, 214, 107, 322, 161, 484, 242, 121, 364, 182, 91, 274, 137, 412, 206, 103, 310, 155, 466, 233, 700, 350, 175, 526, 263, 790, 395, 1186, 593, 1780, 890, 445, 1336, 668, 334, 167, 502, 251, 754, 377, 1132, 566, 283, 850, 425, 1276, 638, 319, 958, 479, 1438, 719, 2158, 1079, 3238, 1619, 4858, 2429, 7288, 3644, 1822, 911, 2734, 1367, 4102, 2051, 6154, 3077, 9232, 4616, 2308, 1154, 577, 1732, 866, 433, 1300, 650, 325, 976, 488, 244, 122, 61, 184, 92, 46, 23, 70, 35, 106, 53, 160, 80, 40, 20, 10, 5, 16, 8, 4, 2, 1, STOP

Apparently our guess was not correct. Moreover, seeing this example we can really begin to wonder whether all sequences will eventually stop. As far as we know this problem is still unsolved. However, the conjecture has been verified with the aid of computers up to at least $A = 10^9$. Such a test is not as straightforward as we might think, because in the course of the calculations the sequence may exceed the largest possible number which the computer is able to accurately represent. Thus, some variable precision routines must be programed in order to enlarge the range of numbers representable by a computer.

The algorithm can easily be extended to negative integers. Here are a few examples:

- -1, -2, -1, -2, ...CYCLE of length 2
- -3, -8, -4, -2, -1, ...runs into CYCLE of length 2
- -5, -14, -7, -20, -10, -5, -14, ...CYCLE of length 5
- -6, -3, -8, -4, -2, -1, ...runs into CYCLE of length 2
- -9, -26, -13, -38, -19, -56, -28, ...runs into CYCLE of length 5
- -11, -32, -16, -8, -4, -2, -1, ...runs into CYCLE of length 2

Are there other cycles? Yes indeed:

- -17, -50, -25, -74, -37, -110, -55, -164, -82, -41, -122, -61, -182, -91, - 272, -136, -68, -34, -17, ... CYCLE of length 18

If we modify our algorithm by removing the STOP in Step 1 we also obtain a cycle for $A = 1$:

- 1, 4, 2, 1, ... CYCLE of length 3

and if we also allow $A = 0$:

- 0, 0, ... CYCLE of length 1.

Moreover, we may now write the algorithm as a feedback system:

$$x_{n+1} = \begin{cases} x_n/2 & \text{if } x_n \text{ is an even integer} \\ 3x_n + 1 & \text{if } x_n \text{ is an odd integer} \end{cases} .$$

Thus, the general question is: what are the possible cycles of the feedback system and does any initial choice for x_0 generate a sequence which eventually runs into one of these cycles? This seems to be a moderate question which the enormous body of mathematics should have already answered — or at least be prepared to answer with no great difficulty. Unfortunately, this is not the case, which only shows that there is still a lot to do in mathematics and, moreover, simple looking problems may be awfully hard to solve. A truly important lesson for life.

MRCM as a One-Step Machine

A more subtle and surprising case is given by our MRCM machines from the last section. They also can be interpreted as one-step machines, which are mathematically described by a single formula of the kind $X_{n+1} = F(X_n)$. Incidentally, in this case F is called the *Hutchinson operator*. We will discuss the details in chapter 5.

Wheel-of-Fortune Machines

While all previous machines are strictly deterministic our last class of machines combines determinism with randomness. Similar to the previous examples, there is a reservoir of different formulas in the processing unit. In addition, however, there is a wheel of fortune, which is used to select one of the formulas at random. The input is a single number (or a pair of numbers), and the output is a new number (or a pair of numbers), which is the result of a formula with values determined by the input. The formula is chosen randomly from a pool at each step of the feedback process. In other words, the output does not just depend on the input, much like in the case of machines with memory. Unfortunately, however, there is no standard trick to rewrite the process as a (deterministic) one-step machine. If the number of formulas is N, then the wheel of fortune has N segments, one for each formula. The size of the segment can be different for each of them in order to accommodate for different probabilities in the random selection mechanism. Random machines like this will furnish extremely efficient decoding schemes for images, which are encoded by the metaphor of a copy machine. This is the content of chapter 6.

Wheel-of-Fortune Machine

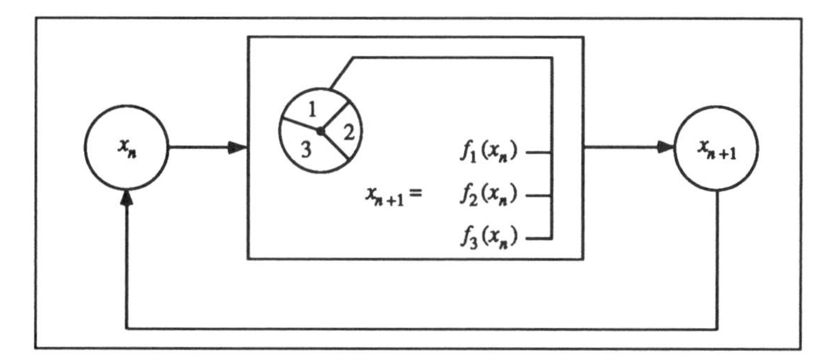

Figure 1.19 : Feedback machine with fortune wheel.

The Chaos Game

We want to touch upon a further exciting interpretation of what we learned in the multiple reduction copy machine, and this is another incredible relation between chaos and fractals.

The following 'game' has been termed *chaos game* by Michael F. Barnsley. At first glance, however, there seems to be no connection whatsoever with chaos and fractals. Let us describe the rules of the game. Well actually, there is not just one game; there is an infinite number of them. But they all follow the same scheme. We have a die and some simple rules to choose from. Here is one of the games:

Preparations: Take a sheet of paper and a pencil and mark three points on the sheet, label them 1, 2, and 3, and call them *bases*. Have a die which allows you to pick the numbers 1, 2, and 3 randomly. It is obvious how to manufacture such a die. Take an ordinary die and just identify the faces 6 with 1, 5 with 2, and 4 with 3.

Rules: Start the game by picking an arbitrary point on the sheet of paper and mark it by a small dot. Call it the *game point*. Now role the die. If number 2, for example, comes up, consider the line between the game point and base 2 and mark a dot exactly in the middle, i.e. halfway between the game point and base 2. This dot will be the new game point, and we have completed the first cycle of the game. Now repeat (i.e. role the die again) to randomly get the number 1, 2, or 3; and depending on the result, mark a dot halfway between the last game point and the randomly chosen base.

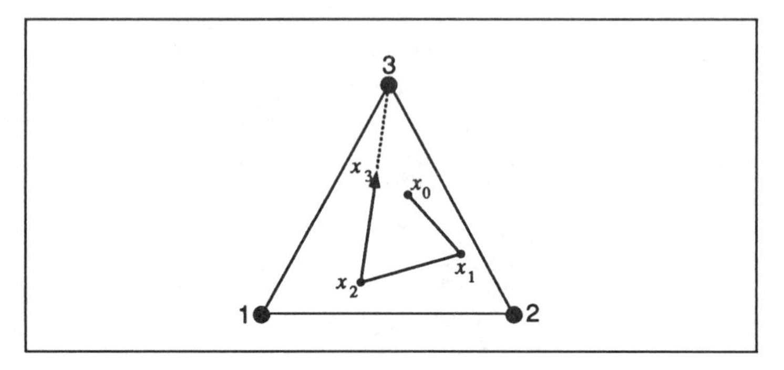

Figure 1.20 : The three base points (vertices of a triangle) and a few iterations of the game point.

Figure 1.20 shows the first results of the game, in which we have labeled the game points in the order of their generation by $x_0, x_1, x_2, \ldots$ We have just explained a very simple scheme to generate a randomly produced sequence of points in a plane; and as such the game appears to be rather boring. But this first impression will immediately change when we see what is going to evolve in this simple feedback system.

What do you guess the outcome of the game will be after a great many cycles, i.e., what is the picture obtained by the dots $x_0, x_1, \ldots, x_{1000}$? Note that once the game point is inside the triangle, which is defined by the three base points, the process will remain inside forever. Moreover, it is obvious that sooner or later the game point will land inside this triangle even if we start the game outside. Therefore, intuition seems to tell us that because of the random generation we should expect a random distribution of dots somehow arranged between base 1, 2, and 3. Yes, indeed, the distribution will be random, but not so the picture or image which is generated by the dots (see figure 1.21). It isn't random at all. We see the Sierpinski gasket emerge very clearly; and this an extremely ordered structure — exactly the opposite of a random structure.

At this point this phenomenon seems to be either a small miracle

Result of the Chaos Game

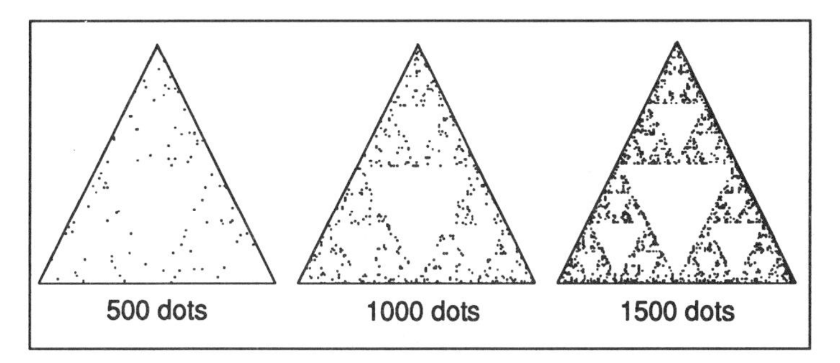

Figure 1.21 : 500, 1000 and 1500 dots generated by the chaos game.

or a funny coincidence, but it is not. Any picture which can be obtained by the multiple reduction copy machine can be obtained by an appropriately adjusted chaos game. In fact, the picture generation can generally be accelerated this way. Moreover, the chaos game is the key to extending the image coding idea which we discussed for the multiple reduction copy machine to grey scale or even color images. This will be the content of chapter 6, which will provide an elementary lesson in probability theory — though one filled with beautiful surprises.

1.4 The Parable of the Parabola — Or: Don't Trust Your Computer

Let us now turn to quadratic iterators. First, we implement the expression $x^2 + c$ in our iterator frameword. Here x and c are just numbers, however with different meanings. To iterate this expression for a fixed (control) value c means this: start with any number x, evaluate the expression, note the result and use this value as new x, evaluate the expression, and so on. Let's look at an example:

Preparation: Choose a number for c, say $c = -2$. Then choose a number x, for example $x = 0.5$.

Iteration: Evaluate the expression for x, obtaining $0.25 - 2 = -1.75$. Now repeat, i.e. evaluate the expression using the result of the first calculation as the new x, i.e. evaluate for $x = -1.75$, which yields 1.0625, and so on.

The Quadratic Iterator

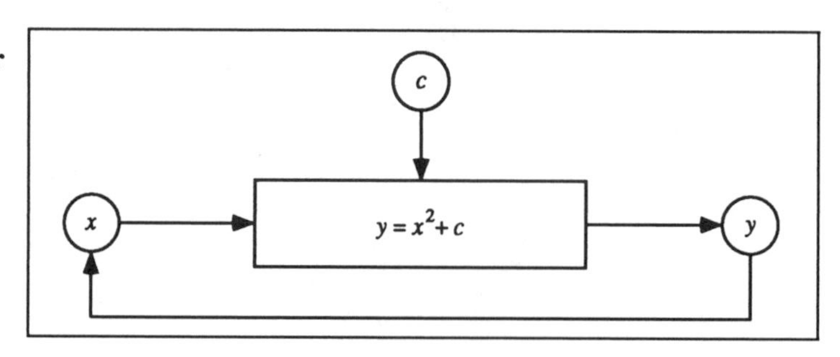

Figure 1.22 : The quadratic iterator interpreted as a feedback machine. The processing unit is designed to evaluate $x^2 + c$, given x and c.

The table summarizes the results for the first four iterations:

x	$x^2 + c$
0.5	−1.75
−1.75	1.0625
1.0625	−0.87109375
−0.87109375	−1.2411956787109375

Already after four cycles we are running into a problem. Because of the squaring operation the number of decimal places which are needed to represent the successive output numbers essentially double in each cycle. This makes it impossible to achieve exact results for more than a few iterations because computers and calculators work with only a finite number of decimal places.[9]

Do Minor Differences Matter?

This is, of course, a common problem in calculator or computer arithmetic, but we don't usually worry about it. In fact, the omnipotence of computers leads us to believe that these minor differences don't really matter. For example, if we compute $2 * (1/3)$ we usually don't worry about the fact that the number 1/3 is not exactly representable by our calculator. We accept the answer 0.6666666667, which, of course, is different from the exact representation of 2/3. Even in a messy calculation, we are usually inclined to take the same attitude, and some put infinite confidence in the calculator or computer in the hope that these minute differences do not accumulate to a substantial error.

Scientists know (or should we say knew) very well that this assumption can be extremely dangerous. They came up with methods, which go back to ideas of Carl Friedrich Gauss (1777–1855, see figure 1.11), to estimate the error propagation in their calculations. With the advent of modern computing this practice has somehow lost ground. It seems that there are at least two reasons for this development.

The Problem of Error Propagation

Modern computing allows scientists to perform computations which are of enormous complexity and are extensive to a degree that was totally unthinkable even half a century ago. In massive computations, it is often true that a detailed and honest error propagation analysis is beyond current possibilities, and this has led to a very dangerous trend. Many scientists exhibit a growing tendency to develop an almost insane amount of confidence in the power and correctness of computers.

If we go on like this, however, we will be in great danger of neglecting some of the great heroes of science and their unbelievable struggle for accuracy in measurement and computation. Let us remember the amazing story of Johannes Kepler's model of the solar system. Kepler (1571–1630) devised an elaborate mystical theory in which the six known planets Mercury, Venus, Earth, Mars, Jupiter, and Saturn[10] were related to the five Platonic solids, see figure 1.24.

Small Deviations with Consequences

In attempting to establish his mystical theory of celestial harmony, he had to use the astronomical data available at that time. He realized that the construction of any theory would require more precise data. That data, he knew, was in the possession of the Danish astronomer Tycho Brahe (1546–1601) who had spent 20 years making extremely accurate recordings of the planetary posi-

[9]For example, the CASIO fx 7000G has 10, and the HP 28S has 12 decimal digits of accuracy.

[10]These planets were known in ancient times before the invention of the telescope. The seventh planet Uranus was not discovered until 1781 by the amateur astronomer Friedrich Wilhelm Herschel, and Neptune was only discovered in 1846 by Johann Gottfried Galle at the Observatory in Berlin. The ninth and most distant planet Pluto was discovered in 1930 by Clyde William Tombaugh at Lowell Observatory in Flagstaff, Arizona.

Brahe and Kepler

Figure 1.23 : Tycho Brahe, 1546–1601 (left) and Johannes Kepler, 1571–1630 (right).

Kepler's Model of the Solar System

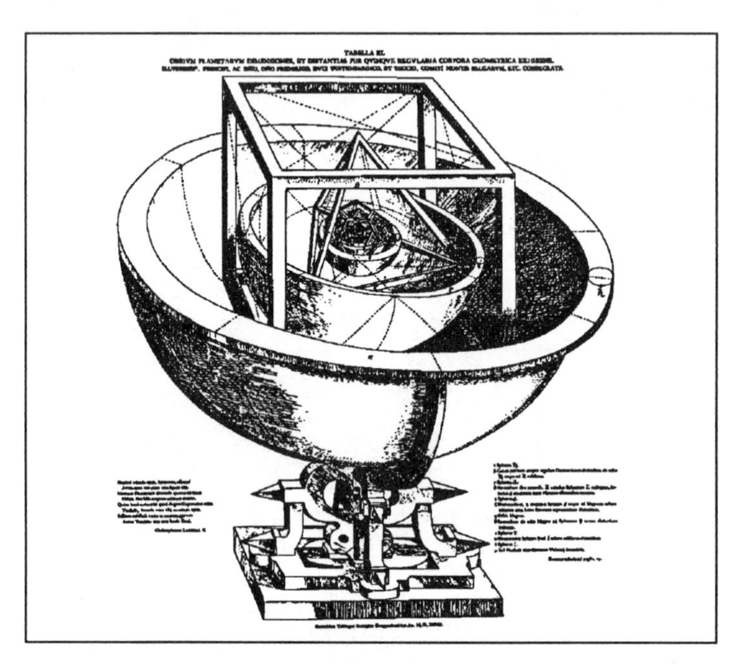

Figure 1.24 : Each planet determines a sphere around the sun containing its orbit. Between two successive spheres Kepler inscribed a regular polyhedron such that its vertices would lie on the exterior sphere and its faces would touch the interior sphere. These are the octahedron between Mercury and Venus, the icosahedron between Venus and Earth, the dodecahedron between Earth and Mars, the tetrahedron between Mars and Jupiter, and the cube between Jupiter and Saturn.

tions. Kepler became Brahe's mathematical assistant in February of 1600 and was assigned a specific problem: to calculate an or-

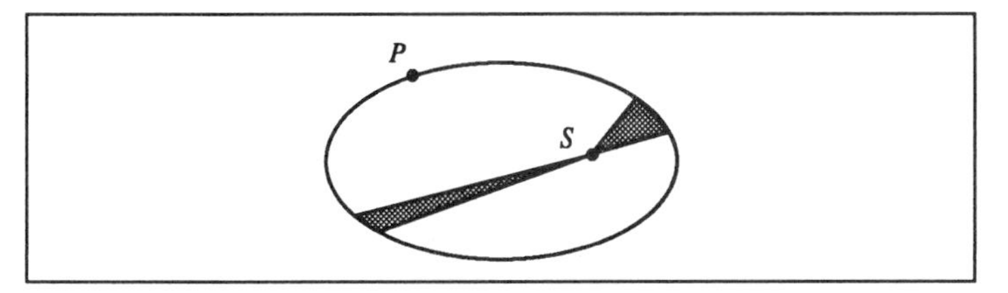

Figure 1.25 : Kepler's first and second law: (1) Law of elliptical paths. The orbit of each planet is an ellipse with the sun at one focus. (2) Law of areas. During each time interval, the line segment joining the sun and planet sweeps out an equal area anywhere on its elliptical orbit.

Kepler's First and Second Law

bit that would describe the position of Mars. He was given this particular task precisely because that orbit seemed to be the most difficult to predict. Kepler boasted that he would have the solution in eight days. Both the Copernican and the Ptolemaic theories held that the orbit should be circular, perhaps with slight modification. Thus, Kepler sought the appropriate circular orbits for Earth and Mars. In fact, the orbit for Earth, from which all observations were made, had to be determined before one could satisfactorily use the data for the positions of the planets. After years, Kepler found a solution that seemed to fit Brahe's observations. Brahe had died in the meanwhile. However, checking his orbits — by predicting the position of Mars and comparing it with more of Brahe's data — Kepler found that one of his predictions was off by at least 8 minutes of arc, which is about a quarter of the angle diameter of the moon. It would have been most natural to attribute this discrepancy to an error in Brahe's observations, especially because he had spent years in making his calculations. But having worked with Tycho Brahe, he was deeply convinced that Brahe's tables were accurate and therefore continued his attempts to find a solution. This led him in six more years of difficult calculations filling more than 900 pages, to his revolutionary new model, according to which the orbits of the planets are elliptical rather than circular. In 1609 he published his famous *Astronomica Nova*, in which he announced two of his three remarkable laws. The third law[11] was published later and helped Sir Isaac Newton formulate his law of gravity.

[11] The law of times: the square of the time of revolution of a planet about the sun is proportional to the cube of its average distance from the sun.

Elis Strömgren Computations for the Restricted Three-Body Problem

To demonstrate the enormous leaps which we have made through computers, we present the following instructive example. Figure 1.26 shows the result of computations, which were carried out by 56 scientists under Elis Strömgren at the Observatory of Copenhagen (Denmark) during a period of 15(!) years. The computations show particular solutions to the so-called restricted three-body problem (orbits of a moon under the influence of two planets) and were published in 1925.

Strömgren's Solution to the Three-Body Problem

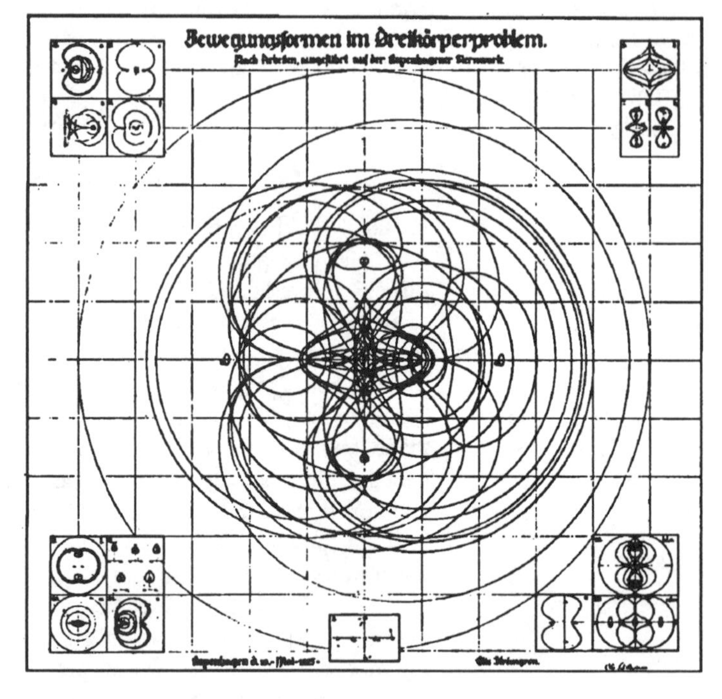

Figure 1.26 : Orbits of the restricted three-body problem.

Computations of this order of magnitude and complication would keep an ordinary PC busy for just a few days, if that long. This relation documents very well what some people call a scientific and technological revolution. The revolution, namely, which is fueled by the means and power of modern scientific computation.

The Problem of Black Box Software

More and more massive computations are being performed now using black box software packages developed by sometimes very well known and distinguished centers. These packages, therefore, seem to be very trustworthy, and indeed they are. But this doesn't exclude the fact that the finest software sometimes produces total garbage, and it is an art in itself to understand and predict when and why this happens. Moreover, users often don't have a chance to carry out an error analysis simply because they have no access to the black box algorithms. More and more decisions in the development of science and technology, but also in economy and politics, are based on large scale computations and simulations. Unfortunately, we cannot always take for granted that an honest error propagation analysis has been carried out to evaluate the results. Computer manufacturers find themselves in a race to build faster and faster machines and seem to pay comparatively little attention to the important issue of *scientific calculation quality control.*

Weather Paradigm from J. Gleick

To amplify the importance of such considerations we would like to quote from James Gleick's *Chaos, Making a New Science.*[12]

"The modern weather models work with a grid of points on the order of sixty miles apart, and even so, some starting data has to be guessed, since ground stations and satellites cannot see everywhere. But suppose the earth could be covered with sensors spaced one foot apart, rising at one-foot intervals all the way to the top of the atmosphere. Suppose every sensor gives perfectly accurate readings of temperature, pressure, humidity, and any other quantity a meteorologist would want. Precisely at noon an infinitely powerful computer takes all the data and calculates what will happen at each point at 12:01, then 12:02, then 12:03... The computer will still be unable to predict whether Princeton, New Jersey, will have sun or rain on a day one month away. At noon the spaces between the sensors will hide fluctuations that the computer will not know about, tiny deviations from the average. By 12:01, those fluctuations will already have created small errors one foot away. Soon the errors will have multiplied to the ten-foot scale, and so on up to the size of the globe."

This phenomenon has become known as the *butterfly effect,* after the title of a paper by Edward N. Lorenz *'Can the flap of a butterfly's wing stir up a tornado in Texas?'* Advanced calculation quality control in weather forecasting means to estimate whether the mechanisms which are at the heart of weather formation are currently in a stable or unstable state. Sooner or later the TV weather man will appear and say: 'Good evening; this is Egon Weatherbring. Because of the butterfly effect, there is no forecast this evening. The atmosphere is in an unstable state, making it impossible to take sufficiently accurate measurements for our computer models. However, we expect it to stabilize in a few days, when we will give you a prediction for the weekend.'

[12]James Gleick, *Chaos, Making a New Science,* Viking, New York, 1987.

Logistic Feedback Iterator

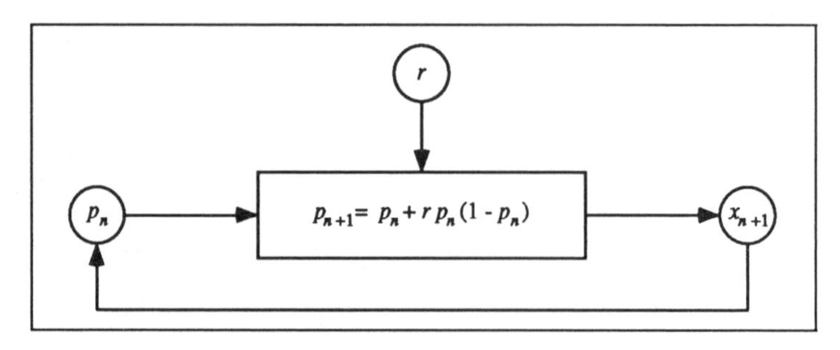

Figure 1.27 : Feedback machine for the logistic equation. The processing unit is designed to evaluate $p+rp(1-p)$, given p and r.

Back to the Quadratic Iteration

Let us now return to the iteration of quadratic expressions and look at the expression

$$p + rp(1 - p)$$

First, this expression can be built into an iterator as easily as we did with $x^2 + c$.

The quadratic expression $p + rp(1 - p)$ has a very interesting interpretation and history in biology. It serves as the core of a population dynamics model which in spirit goes back to the Belgian mathematician Pierre François Verhulst[13] and his work around 1845 and which led May to his famous article in *Nature*, see page 17.

A Population Dynamics Model

What is a population dynamics model? It is simply a law which, given some biological species, allows us to predict the population development of that species in time. Time is measured in increments $n = 0, 1, 2, \ldots$ (minutes, hours, days, years, whatever is appropriate). The size of the population is measured at time n by the actual number in the species P_n. Figure 1.28 shows a typical development.

Naturally, the size of a population may depend on many parameters, such as environmental conditions (e.g. food supply, space, climate), interaction with other species (e.g. the predator/prey relationship), but also age structure, fertility etc. The complexity of influences which determine a given population in its growth behavior is illustrated in the following medieval parable.

Of Mice and Old Maids

This year there are a lot of mice in the fields. The farmer is very concerned because he can harvest very little grain. That results in a period of very poor dowries, which leads to there being many more old maids. They all tend to love cats, which increases the cat population dramatically. That in turn is bad for the mice population. It rapidly decreases. This makes for happy farmers

[13]Two elaborate studies appeared in the *Mémoires de l'Académie Royale de Belgique*, 1844 and 1847.

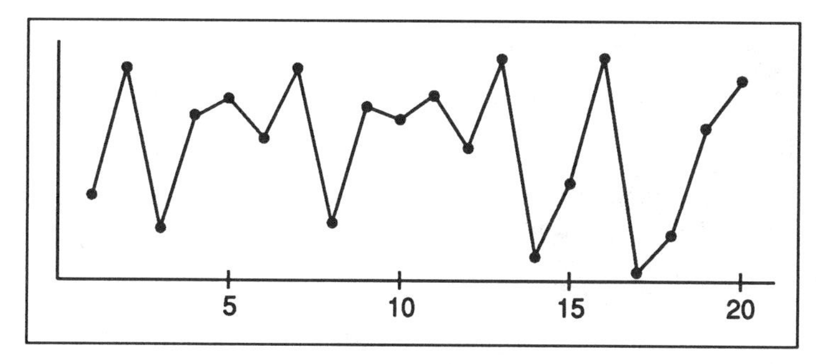

Time Series of a Population

Figure 1.28 : Time series of a population — a typical development. Successive measurements are connected by line segments.

and very rich dowries, very few old maids, very few cats, and therefore, back come the mice. And so it goes, on and on.

Though we shouldn't take this too seriously as a model for mice and maid populations, it indicates the potential complexity of population dynamics. It also shows that populations may display cyclic behavior: up → down → up → down →...

The Petri Dish Scenario

A natural modeling approach tries to freeze as many of the population parameters as possible. For example, assume that we have a species of cells which live in a constant environment, e.g. a petri dish with a constant food supply and temperature. Under such conditions we expect that there is some maximal possible population size N which is supported by the environment. If the actual population P at time n, which is P_n, is smaller than N, we expect that the population will grow. If, however, P_n is larger than N, the population must decrease.

Now we want to introduce an actual model. Just as *velocity* is one of the relevant characteristics for the motion of a body, so is *growth rate* the relevant characteristic for population dynamics. Now growth rate is measured by the quantity

$$\frac{P_{n+1} - P_n}{P_n} \tag{1.5}$$

In other words, the growth rate r at time n measures the increase of the population in one time step relative to the size of the population at time n.

Population Growth and Interest

If the population model assumes that the growth rate r is constant, then

$$\frac{P_{n+1} - P_n}{P_n} = r \tag{1.6}$$

for some number r independent of n. Solving for P_{n+1} we obtain the population growth law[14]

$$P_{n+1} = P_n + rP_n = (1+r)P_n \; .$$

In such a model the population grows by a factor of $1+r$ each time step. Indeed, the formula is equivalent to

$$P_n = (1+r)^n P_0 \tag{1.7}$$

where P_0 is the initial population with which we start our observations at time 0. In other words, knowing r and measuring P_0 would suffice to predict the population size P_n for any point in time without even running the feedback process. In fact, eqn. (1.7) is familiar from computing the accumulation of principal and compound interest, when the rate of interest is r.

The Verhulst Population Model

The most simple population model would assume a constant growth rate, but in that situation we find unlimited growth which is not realistic. In our model we will assume that the population is restricted by a constant environment, but this premise requires a modification of the growth law. Now the growth rate depends on the actual size of the population relative to its maximal size. Verhulst postulated that the growth rate at time n should be proportional to the difference between the population count and the maximal population size, which is a convenient measure for the fraction of the environment that is not yet used up by the population at time n. This assumption leads to the Verhulst population model

$$p_{n+1} = p_n + rp_n(1-p_n), \tag{1.8}$$

where p_n measures the relative population count $p_n = P_n/N$ and N is the maximal population size which can be supported by the environment. This is just a compact notation for our feedback process. We use integer indices to identify iterates at different time steps (p_n for input, p_{n+1} for output).

Derivation of the Verhulst Model

This population model assumes that the growth rate depends on the current size of the population. First we normalize the population count by introducing $p = P/N$. Thus p ranges between 0 and 1, i.e. we can interpret $p = 0.06$, for example, as the population size being 6% of its maximal size N. Again we index p by n, i.e. write p_n to refer to the size at time steps $n = 0, 1, 2, 3, \ldots$ Now growth

[14]Note that the concept of growth rate does not depend on N, i.e., if we use a normalized count $p_n = P/N$, then N cancels out in $r = (p_{n+1} - p_n)/p_n$, the equivalent of eqn. (1.6).

rate is measured by the quantity already given corresponding to the expression (1.5),

$$\frac{p_{n+1} - p_n}{p_n}.$$

Verhulst postulated that the growth rate at time n should be proportional to $1 - p_n$ (the fraction of the environment that is not yet used up by the population at time n). Assuming that the population is restricted by a constant environment the growth should change according to the following table.

population	growth rate
small	positive, large
about 1	small
less than 1	positive
greater than 1	negative

In other words,[15]

$$\frac{p_{n+1} - p_n}{p_n} \propto 1 - p_n,$$

or, after introducing a suitable constant r

$$\frac{p_{n+1} - p_n}{p_n} = r(1 - p_n).$$

Solving this last equation yields the population model eqn. (1.8)

$$p_{n+1} = p_n + rp_n(1 - p_n).$$

The Logistic Model

Following Verhulst this model given by eqn. (1.8) is called the *logistic* model[16] in the literature. There are several interesting remarks. First, note that it is in agreement with the table of growth rates in the technical section above. Second, it seems as if we again have a law which allows us to compute (i.e. predict) the size of the population for any point in time just as in the case of a constant growth rate. But there is a fundamental difference. For most choices of r, there is no explicit solution such as eqn. (1.7) for eqn. (1.6). That is, p_n cannot be written as a formula of r and p_0, as was previously possible. In other words, if one wants to compute p_n from p_0 one really has to run the iterator in figure 1.27 n times. We will begin our experiments with the setting $r = 3$.[17]

[15] The $\propto$ sign means 'proportional to'. The quantity on the left side is a multiple of the expression on the right side.

[16] from *logis* (french) = house, lodging, quarter

[17] It turns out that $r = 3$ is one of those very special choices for which there is an explicit formula of p_n in terms of r and p_0, see chapter 10.

The table below lists the first three iterates for $p_0 = 0.01$, i.e. the initial population is 1% of the maximal population size N.

p	$p + rp(1-p)$
0.01	0.0397
0.0397	0.15407173
0.15407173	0.545072626044...

For the same reasons as we noted when we iterated $x^2 + c$, we observe that continued iteration requires higher and higher computational accuracy if we insist on exact results. But that appears to be unnecessary in our population dynamics model. Isn't it enough that we get some idea for how the population develops? Shouldn't we be satisfied with an answer which is reliable up to three or four digits? After all, the third decimal place controls only some tenth of a percent in our model. Thus, it seems, there is no reason not to trust that a computer or calculator will do the job. But this is definitely not true as a general rule — computed predictions in our model can be totally wrong.

The Lack of Predictability

This is at the heart of what scientists nowadays call the presence of chaos in deterministic feedback processes. One of the first ones who became aware of the significance of these effects was the MIT meteorologist Lorenz in the late fifties.[18] He discovered this effect — the lack of predictability in deterministic systems — in mathematical systems which were designed and frequently used to make long range weather predictions.

The Lorenz Experiment

As so often is the case with new discoveries, Lorenz stumbled onto the effect quite by accident. In his own words,[19] the essential part of the events were as follows.

"Well, this all started back around 1956 when some [...] methods of [weather] forecasting had been proposed as being the best methods available, and I didn't think they were. I decided to cook up a small system of equations which would simulate the atmosphere, solve them by computers which were then becoming available, and to then treat the output of this as if it were real atmospheric observational data and see whether the proposed method applied to it would work. The big task here was to get a system of equations which would produce the type of output that we could test the things on because it soon became apparent that if the solution of these equations were periodic, the proposed method would be trivial; it would work perfectly. So we had to get a system of

[18] Lorenz, E. N., *Deterministic non-periodic flow*, J. Atmos. Sci. 20 (1963) 130–141.

[19] In: H.-O. Peitgen, H. Jürgens, D. Saupe, C. Zahlten, *Fractals — An Animated Discussion*, Video film, Freeman 1990. Also appeared in German as *Fraktale in Filmen und Gesprächen*, Spektrum der Wissenschaften Videothek, Heidelberg, 1990.

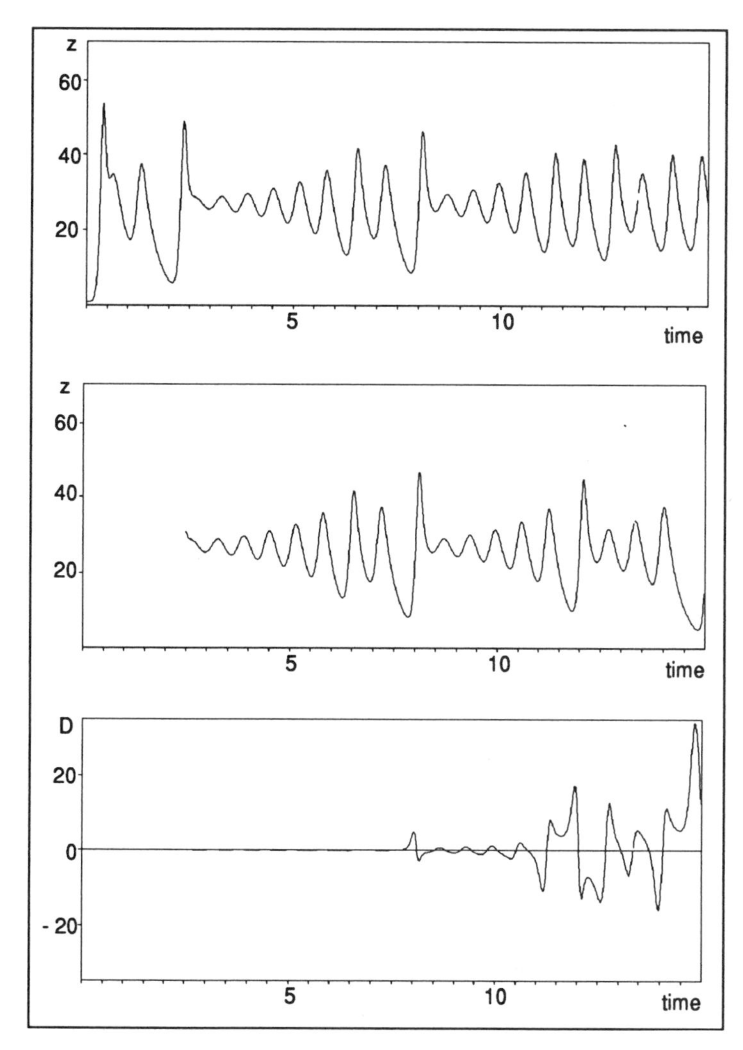

Figure 1.29 : Numerical integration of the Lorenz equation (top). It is recomputed starting at $t = 2.5$ with an initial value taken from the first integration, however, with a small error introduced (middle). The error increases in the course of the integration. The difference between the two computed results (signals) becomes as large as the signal itself (bottom).

The Original Lorenz Experiment

equations which would have solutions which would not be periodic, which would not repeat themselves, but would go on irregularly and indefinitely. I finally found a system of twelve equations that would do this and found that the proposed method didn't work too well when applied to it, but in the course of doing this I wanted to examine some of the results in more detail. I had a small computer in my office then, so I typed in some of the intermediate conditions which the computer had printed out as new initial conditions to start another computation and went out for a while. When I came back I found that the solution was not the same as the one I had before; the computer was behaving differently. I suspected computer trouble at first, but I soon found that the reason was that the numbers that I had typed in were not the same as the original ones, these [former ones] had been rounded off numbers and the small difference between something retained to six decimal places and rounded off to three had amplified in the course of two months of simulated weather until the difference was as big as the signal itself, and to me this implied that if the real atmosphere behaved as in this method, then we simply couldn't make forecasts two months ahead, these small errors in observation would amplify until they became large."

Sensitive Dependence on Initial Conditions

In other words, even if the weather models in use were absolutely correct — that is, as models for the physical development of the weather — one cannot predict with them for a long time period. This effect is nowadays called *sensitive dependence on initial conditions*. It is one of the central ingredients of what is called deterministic chaos.[20] Our next experiment imitates Lorenz' historical one in the simplest possible way. He had used much more delicate feedback systems consisting of twelve ordinary differential equations; we simply use the logistic equation.[21] We iterate the quadratic expression $p + rp(1 - p)$ for the constant $r = 3$ and the initial value $p_0 = 0.01$, see table 1.30. In the left column we run the iteration without interruption, while in the right column we run the iteration until the 10th iterate, stop, truncate the result 0.7229143012 after the third decimal place, which yields 0.722 and continue the iteration as if that were the last output. The experiment is carried out on a CASIO fx–7000G pocket calculator.

Trustworthy as Rolling Dice

Now, of course, the 10th iterates of the two processes agree only in 3 decimal places and it is no surprise that there is a disagreement also in the 15th iterates. But it is a surprise — and again that indicates chaos in the system, or in the words of Lorenz, it 'demonstrates lack of predictability' — that higher iterates appear totally uncorrelated. The lay-out of the experiment suggests that the column on the left is more trustworthy. But that is absolutely misleading, as we will see in the forthcoming experiments. Eventually the iterations of our feedback process become as trustworthy as if we had obtained them with a random number generator, or

[20]The term 'chaos' was coined in T. Y. Li's and J. A. Yorke's 1975 paper *Period 3 Implies Chaos,* American Mathematical Monthly 82 (1975) 985–992.

[21]In fact, later on, Lorenz himself discovered that his system is strongly related to the logistic equation.

The Lorenz Experiment Revisited

evaluations	without interrupt	with interrupt and restart
1	0.0397	0.0397
2	0.15407173	0.15407173
3	0.5450726260	0.5450726260
4	1.288978001	1.288978001
5	0.1715191421	0.1715191421
10	0.7229143012	0.7229143012
10	0.7229143012	restart with 0.722
15	1.270261775	1.257214733
20	0.5965292447	1.309731023
25	1.315587846	1.089173907
30	0.3742092321	1.333105032
100	0.7355620299	1.327362739

Table 1.30 : The Lorenz experiment for the population model. Two series of iterations with the same starting point are carried out. During the process one of the outputs of the second series is truncated to three decimal places and taken as input for the following iteration. Soon afterwards the two series of numbers lose all correlation. Underlined are those first digits which are the same on both sides.

rolling dice, or flipping coins. In fact the great Polish mathematician Stan Ulam discovered that remarkable property when he was interested in constructing numerical random number generators for the first electronic computer ENIAC in the late forties in connection with large scale computations for the Manhattan Project.

1.5 Chaos Wipes Out Every Computer

Being very skeptical, we might conclude that maybe the error — truncation after 3 decimal places — which we introduced in Lorenz' experiment was too large. Someone might conjecture that the strange behavior of the iteration would disappear if we repeated the experiment with much smaller errors in the starting values. We would not have wasted our time in calculating if that were the case. The fact is that no matter how small a deviation in the starting values we choose, the errors will accumulate so rapidly that after relatively few steps the computer prediction is worthless. To fully grasp the importance of the phenomenon, we propose a further experiment. This time we do not change the starting values for the iteration, but we use calculators produced by two different manufacturers. In other words, we conjecture that sooner or later their predictions will massively deviate from each other.

What happens if we actually carry out the iteration with two different fixed accuracy devices? What is the result after 10 iterations, or 20, or even 50? This seems to be a dull question. Doesn't one just have to evaluate 10, 20 or 50 times? Yes, of course, but the point is that the answer depends very much on the nature of the computation.

The Computer Race into Chaos

To demonstrate what we mean when we say that things depend on the computation, let us compare the results obtained by two different calculators, say a CASIO and an HP. Starting with 1 let's look at 2, 3, 4, 5, 10, 15, 20, ..., 50 repeated feedback evaluations (=iterations), see table 1.31 and figure 1.32.

While the first and second generation of our populations are predicted exactly the same by both calculators, they totally disagree at the 50th generation: the CASIO predicts that the population is about 0.3% of the maximum population while the HP tells us that the population should be about 22% of the maximum population! How is that possible?

We carefully check our programs, and indeed they both are correct and use exactly the same formula $p + rp(1 - p)$. The only difference is, of course, that the CASIO is restricted to 10 decimals, while the HP has 12. In other words, neither one is able to exactly represent the iterations 3 and higher. Indeed, the second iterate needs 8 decimals and therefore the third iterate would need 16, etc. Thus, there are unavoidable cut-off errors, which don't seem to matter much. At least that is suggested if we look at iterations 4 and 5. The results of the CASIO and HP agree in 10 decimal places. However, for the 10th iterate we observe that the CASIO and HP are in disagreement about the 10th decimal place: the CASIO proposes 2 and the HP insists on 7, see table 1.31. This suggests that we should look at the iterates between 5 and 10 in detail (table 1.34).

CASIO fx-7000G Versus HP 28S for $p + rp(1 - p)$

evaluations	CASIO	HP
1	0.0397	0.0397
2	0.15407173	0.15407173
3	0.5450726260	0.545072626044
4	1.288978001	1.28897800119
5	0.1715191421	0.171519142100
10	0.7229143012	0.722914301711
15	1.270261775	1.27026178116
20	0.5965292447	0.596528770927
25	1.315587846	1.31558435183
30	0.3742092321	0.374647695060
35	0.9233215064	0.908845072341
40	0.0021143643	0.143971503996
45	1.219763115	1.23060086551
50	0.0036616295	0.225758993390

Table 1.31 : Two different calculators at the same job do not produce the same results.

Differences in the Race

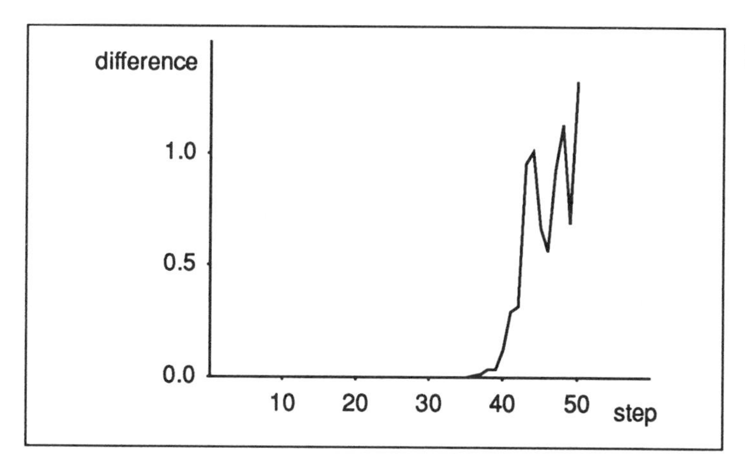

Figure 1.32 : Plot of the difference between the computed iteration values of HP and Casio.

Indeed, while for the 5th iterate both calculators agree at the 10th decimal, they mildly disagree at the 10th decimal for the 6th iterate. The difference being 2×10^{-11}, which is so minute that one certainly finds no reason to bother with it. Looking further at the 10th iterate, however, we see how this tiny disagreement has grown to 5×10^{-10}, which is still so small that one is inclined to neglect it. But for our records let's note that the disagreement has grown by an order of magnitude (a factor of 10).

When we go back to table 1.31 and now look at 15, 20, 25, ...

CASIO *fx*–7000G Program for Iteration of $p + rp(1-p)$

"P"? → P	input P
"R"? → R	input R
"N"? → N	input N (# of iterations wanted)
Lbl1	begin feedback loop
P+R*P*(1-P) → P	process and feedback
Dsz N	count iteration by reducing N by 1
Goto 1	end feedback loop
P△	displays P when N is reduced to 0

Table 1.33 : The program for one of the calculators that was used to generated the data for the tables.

The Critical Iterations

evaluations	CASIO	HP
5	0.1715191421	0.171519142100
6	0.5978201201	0.597820120080
7	1.319113792	1.31911379240
8	0.05627157765	0.056271577700
9	0.2155868393	0.215586839429
10	0.7229143012	0.722914301711

Table 1.34 : The critical iterations where the two calculators begin to show signs of differing bevavior.

iterations we seem to observe how the tiny little infection which we noticed in the 10th decimal for the 6th iterate has migrated through all decimal places, i.e. after 40 iterations the initial tiny disagreement has been amplified by a factor of 10^{10}!

But why do we say 'seem to observe'? Well, comparing the CASIO and the HP we are inclined to trust the HP more because it works with higher accuracy (two extra decimal places). In other words we tend to accept the HP answer for the 40th iterate and conclude that the CASIO is totally off. But this is a little premature.

If the CASIO is wrong — and of course at least one of the two must be totally wrong — we cannot assume that the error is due to a serious flaw of its design. Rather, the failure is due to a principal mathematical problem. And of course, for that reason the HP is subject to the same disease, but with a slight delay because of its higher accuracy. In other words, all we can say for sure is that one of the two calculators is totally wrong in its predictions despite the fact that the deterministic process is very simple. But it is also very likely that both calculators are off. This dramatic effect is the unavoidable consequence of finite accuracy arithmetic and would produce the same results and dramatic effects on multimillion dollar supercomputers.

The minute differences in the two calculators, i.e. their different accuracies, accumulate so rapidly that the predictive power of the calculators (computers) evaporates. But, believe it or not, this is still not the end of the story. Things are even wilder than we have seen so far.

We now run our example of the quadratic dynamic law, $p+rp(1-p)$, for $r = 3$ and the initial condition $p_0 = 0.01$ (as before) on one calculator (CASIO) in two comparative runs. So what is the difference? If we keep all data the same and use an identical calculator, the only thing we can possibly change is the programming code in the algorithm from table 1.33. And there the only thing we can possibly change is the way we evaluate the quadratic expression. And even this almost ridiculously small change matters as demonstrated in table 1.35.

$p + rp(1-p)$ Versus $(1+r)p - rp^2$

evaluations	$p + rp(1-p)$	$(1+r)p - rp^2$
1	0.0397	0.0397
2	0.15407173	0.15407173
3	0.5450726260	0.5450726260
4	1.288978001	1.288978001
5	0.1715191421	0.1715191421
10	0.7229143012	0.7229143012
11	1.323841944	1.323841944
12	0.03769529734	0.03769529724
13	0.146518383	0.1465183826
14	0.5216706225	0.5216706212
15	1.270261775	1.270261774
20	0.5965292447	0.5965293261
25	1.315587846	1.315588447
30	0.3742092321	0.3741338572
35	0.9233215064	0.9257966719
40	0.0021143643	0.0144387553
45	1.219763115	0.0497855318

Table 1.35 : Two different implementations of the same quadratic law on the same calculator are not equivalent.

Two Different Implementations of Quadratic Law

So far we have evaluated $p + rp(1-p)$, which is mathematically the same as $(1+r)p - rp^2$. After implementation (exchange 'P + R*P*(1-P)' by '(1+R)*P - R*P*P' in algorithm from table 1.33) we are anxious to see whether this almost ridiculously small change matters. We compare the results: there is total agreement until the 11th iterate. Then, in the 12th iterate a minute disagreement — check the last three places — 734 versus 724.

At first one doesn't trust one's eyes. Look at the 12th iterate. It is true. There it creeps in; the virus of unpredictability strikes again. Hereafter we are not surprised at all to see our prediction become completely unreliable.

Sooner or Later Predictability Breaks Down

If the first experiments didn't convince you that chaos is unbeatable the last experiment should have taught you the lesson. With finite accuracy computing there is no cure for the damaging effects of chaos. Predictability sooner or later breaks down.

Now you may argue that such phenomena are very rare, or easy to detect or to foresee. Wrong! Since chaos (= breakdown of predictability) has become fashionable in the sciences, there has been literally a flood of papers demonstrating that chaos is more like the rule in nature, while order (= predictability) is more like the exception. But doesn't this contradict the phenomenal success of space missions, as for example the Voyager II mission which left our planetary system after 12 years of travel when it passed Neptune, only a few kilometers off the predicted path? No it does not. There are strong hints that even the motion of celestial bodies is subject to the same phenomena — sooner or later ... Besides, since chaos has entered upon the scientific stage — and despite its amazing historical roots in the work of Henri Poincaréât the turn of last century, this is essentially an achievement made possible by the new powers provided to science by computers — there has been remarkable progress in the deeper understanding of phenomena such as turbulence, fibrillation of the heart, laser instabilities, population dynamics, climate irregularities, brain function anomalies, etc.

Chaos Will Not Resist Deeper Understanding

Moreover, and this is truly fascinating and gives rise for a lot of hope that chaos will not resist deeper understanding forever, it has recently become clear that chaos likes to follow certain very stable patterns. This again was discovered, strangely enough, by means of computers, which otherwise seem so vulnerable to chaos. This is the main subject of chapter 11 where we will discuss the ground breaking work of Mitchel Feigenbaum, Siegfried Großmann and Stefan Thomae, and Edward Lorenz, as well as Robert May, all of whom found order in chaos as well as routes from order into chaos.

The quadratic law $p + rp(1-p)$ which we have explored so far is just one of a universe of feedback systems which display very complicated behavior. The expression $x^2 + c$ is another example, only in a trivial sense, however. If we carried out experiments analogous to that in table 1.31 for $c = -2$, we would observe exactly the same behavior. The reason is simply that the two quadratic processes can be identified by means of a coordinate transformation, i.e. they really are the same.

Equivalence of $x^2 + c$ and $p + rp(1 + p)$

Using indices to identify iterates at different times (index n for input, index $n + 1$ for output), we can write the two quadratic laws as

$$p_{n+1} = p_n + rp_n(1 - p_n), \quad n = 0, 1, 2, 3, ... \tag{1.9}$$

and

$$x_{n+1} = x_n^2 + c, \quad n = 0, 1, 2, 3, ... \tag{1.10}$$

We now verify that with the setting of

$$c = \frac{1 - r^2}{4} \quad \text{and} \quad x_n = \frac{1 + r}{2} - rp_n, \tag{1.11}$$

the formulas (1.9) and (1.10) are identical.

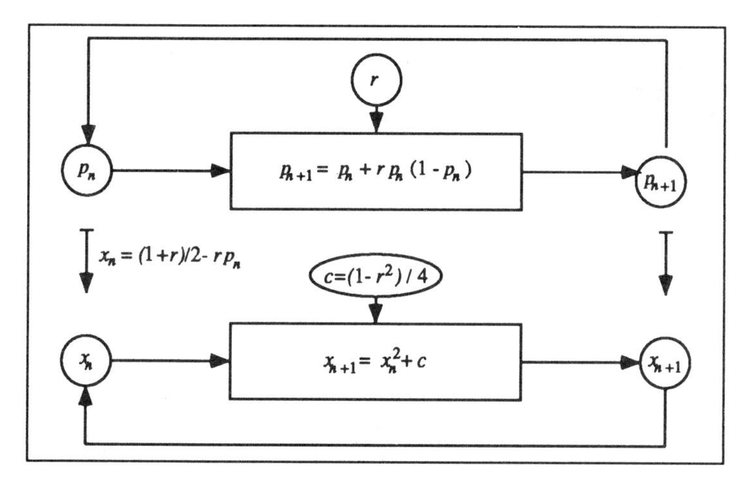

Figure 1.36 : Two quadratic iterators running in phase are tightly coupled by the transformations indicated.

We have to examine whether p_{n+1} from eqn. (1.9) can be transformed into x_{n+1} as in eqn. (1.10) when we make use of eqn. (1.11). If we apply eqn. (1.11) to p_{n+1} we get

$$x_{n+1} = \frac{1 + r}{2} - rp_{n+1}$$

and using eqn. (1.9) for p_{n+1}

$$x_{n+1} = \frac{1 + r}{2} - rp_n - r^2 p_n(1 - p_n).$$

On the other hand eqn. (1.10) with x_n and c transformed by eqn. (1.11) yields

$$x_{n+1} = \left(\frac{1 + r}{2} - rp_n\right)^2 + \frac{1 - r^2}{4}.$$

Upon resolving the right hand sides of both equations we see that they are in fact the same, namely

$$r^2p_n^2 - r(1+r)p_n + \frac{1+r}{2} \ .$$

Note that $r = 3$ corresponds to $c = -2$. This explains, indeed, that we may observe exactly the same behavior in both processes. Let us verify the equivalence of the two processes with some examples. If $r = 3$ and $p_0 = 0.01$, then $c = -2$ and $x_0 = 1.97$, according to eqn. (1.11). Computing x_n for $n = 10$ on a CASIO yields $x_{10} = -0.1687429036$. After transforming x_{10} according to eqn. (1.11) we obtain $p_{10} = 0.7229143012$, which is exactly the value we can read from table 1.31 for the 10th iterate.

If, however, we repeat the same for 50 rather than 10 iterations we obtain $x_{50} = 0.2310122906$ and $p_{50} = 0.2550655142$ (according to eqn. (1.11)), which is entirely different from the 50th iteration in table 1.31. This does not disprove the validity of the equivalence of the two processes, but rather reaffirms our earlier finding that even two different ways of numerical evaluation eventually lead to disagreeing results, i.e. chaos has hit again.

Why Study Different Quadratic Iterators?

Why should we look at $x^2 + c$ when the dynamics for iterators to this formula are the same (up to some coordinate transformation) as for $p+rp(1-p)$? There are many different problems to be solved with quadratic iterations, and indeed, in principle it does not matter which quadratic is taken because all are equivalent. However, the mathematical formulation of these problems and their solutions will be more illuminating (and perhaps less complex) depending on the particular quadratic we pick. Therefore, in each case we may choose the quadratic transformation which suits best the problem on hand.

Let us return for a moment to the question of whether there is an easy answer to why we see chaotic behavior? It seems to be obvious that whenever there is an inaccuracy in the feedback process, this error is amplified, i.e. the error propagation builds up dramatically, due to the quadratic character of the expressions. In other words, one might guess that the squaring operation is the cause of the problem. Yes, that is indeed the case; but in a much more subtle way than we might think. For the complete story, please refer to chapter 10. But let us convince ourselves that squaring alone does not explain anything! Let us look at two more simple experiments to illustrate the difficulties.

A Wild Iterator Becomes Very Tame

In our last experiment with the quadratic iterator $x_{n+1} = x_n^2 + c$ we fixed $c = -2$ and started with $x_0 = 1.97$. How about $x_0 = 1$, for example. Iteration now yields: $1, -1, -1, -1, \ldots$ Or $x_0 = 2$. Iteration yields $2, 2, 2, \ldots$ In other words, we have found initial values for x_0 with which the same wild iterator behaves perfectly tamely. We could demonstrate, however, that this is the exception, i.e. for almost all x_0 from $[-2, +2]$ one observes chaotic behavior. For example if we start with $x_0 = 1.999999999$, i.e. with a tiny deviation from $x_0 = 2$, then we will have the familiar messy behavior back again, provided we just allow sufficiently numerous iterates. This already shows that the error analysis problem is not a straight forward one, and this becomes even more apparent in our next experiment.

Let us now shift gears in our iterator by setting the control parameter to $c = -1$, rather than the previous value $c = -2$. If squaring alone were the secret to understand the lack of predictability, we should make very similar observations. Let us run an iteration where we start with $x_0 = 0.5$, see table 1.37.

Seventeen Iterations of $x^2 - 1$

evaluations	x	$x^2 - 1$
1	0.5	-0.75
2	-0.75	-0.4375
3	-0.4375	-0.80859375
4	-0.80859375	-0.3461761475
5	-0.3461761475	-0.8801620749
6	-0.8801620749	-0.2253147219
7	-0.2253147219	-0.9492332761
8	-0.9492332761	-0.0989561875
9	-0.0989561875	-0.9902076730
10	-0.9902076730	-0.0194887644
11	-0.0194887644	-0.9996201881
12	-0.9996201881	-0.0007594796
13	-0.0007594796	-0.9999994232
14	-0.9999994232	-0.0000011536
15	-0.0000011536	-1.0000000000
16	-1.0000000000	-0.0000000000
17	-0.0000000000	-1.0000000000

Table 1.37 : First seventeen iterates for the starting value $x_0 = 0.5$.

Here, we observe that after a number of iterations the process settles down to a repetition of two values: 0 and -1. In fact, repeating the iteration with other initial values, for example $x_0 = 1$, or $x_0 = 0.75$, or $x_0 = 0.25$ yields the same final answer. The feedback process is now in a perfectly stable mode.

Stable Cycle in Logistic Iterator

The same stability should occur in the iteration of the logistic equation if we choose the parameter r and the initial population p_0 appropriately. Solving eqn. (1.11) for r and p with the choice $c = -1$

yields

$$r = \sqrt{1-4c} = \sqrt{5}\,,$$
$$p = \frac{1+r}{2r} - \frac{x}{r} = \frac{1-2x+\sqrt{1-4c}}{2\sqrt{1-4c}} = \frac{1-2x+\sqrt{5}}{2\sqrt{5}}\,.$$

Thus, for this parameter setting there is a stable cycle of two points corresponding to $x = 0$ and $x = -1$, namely

$$p = \frac{1+\sqrt{5}}{2\sqrt{5}} = 0.723606797...$$

and

$$p = \frac{3+\sqrt{5}}{2\sqrt{5}} = 1.17082039...$$

We have seen this kind of behavior already in the discussion of the MRCM, where we always obtained a final image which was independent of the initial image. This property is called stability, and is very desirable in many cases. In these cases a process is predictable, and small errors along the way disappear or decay, i.e. they can be neglected. In other words, these are processes where a computer with finite precision arithmetic is a perfect tool and cannot fail.

So far we have been able to detect the stable or unstable state of an iteration by carefully monitoring the numerical values of competing runs of the feedback process. For the particular class of quadratic processes, there is another way to detect the different kinds of behavior, which is much more visual and immediate.

Graphical Iteration of Feedback Processes

We will restrict ourselves to the iteration

$$z_{n+1} = az_n(1-z_n).$$

Note that if we consider the graph corresponding to the function $y = ax(1-x)$, we just obtain a parabola, which passes through the points $(0,0)$ and $(1,0)$ independent of the choice of the parameter a. The vertex of the parabola, which is always located at $x = 0.5$, has height $a/4$. This quadratic iteration again is equivalent to the logistic equation, or to $x_{n+1} = x_n^2 + c$. We use it here because it produces iterates which always stay in the range from 0 to 1, provided the initial value x_0 is also in this range. There is an efficient way to construct the sequence $x_0, x_1, x_2, ...$ by a ruler based on the graph of the parabola leading to a nice graphical visualization of the iteration, called *graphical iteration.*

To describe the iteration we plot the graph of $y = ax(1-x)$ and draw the bisector (diagonal), see figure 1.38. We start by marking x_0 on the x-axis. Now we draw a vertical line segment from x_0

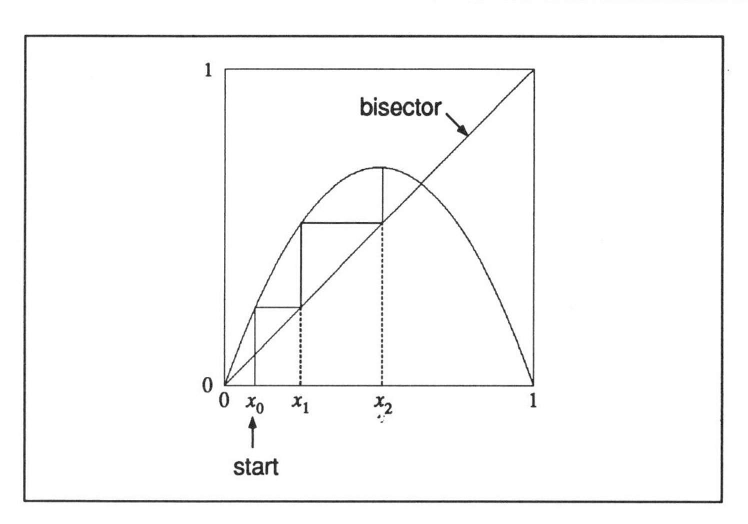

Principle of Graphical Iteration

Figure 1.38 : The first steps in the graphical iteration.

until we hit the graph. From that point we draw a horizontal line segment until we hit the bisector. From there we continue to draw a vertical line segment until we hit the graph, and so on.

Why does this procedure work? Simply because points on the bisector have the same distance from both axes. With the aid of this method one can literally see whether the elementary iterations are in the stable or unstable state. Figure 1.39 shows the graphical iteration method for three different values of a in the stable range of the process. For $a = 1.45$, we observe that the iteration creates a staircase which runs into the point of intersection between the graph and the bisector. For $a = 2.75$ the iteration generates a spiral which converges to the point of intersection between the graph and the bisector. For $a = 3.2$ we see how the iteration determines cyclic behavior.

Mixing

Figure 1.40 shows the iteration for $a = 4$ and one starting value x_0, however different numbers of steps of the iteration. From left to right we show the iteration after 10, 50 and 100 steps. Apparently the process does not come to rest. Rather, it occupies the entire available space. This phenomenon, called mixing, is an indicator for the unstable state of the system. However, a rigorous analysis has to use much more subtle means to distinguish genuine instability from just a cycle of very high order. For example, what is the difference between the cob webs in figure 1.39 ($a = 3.2$) and figure 1.40 ($a = 4$)?

Stable Behavior

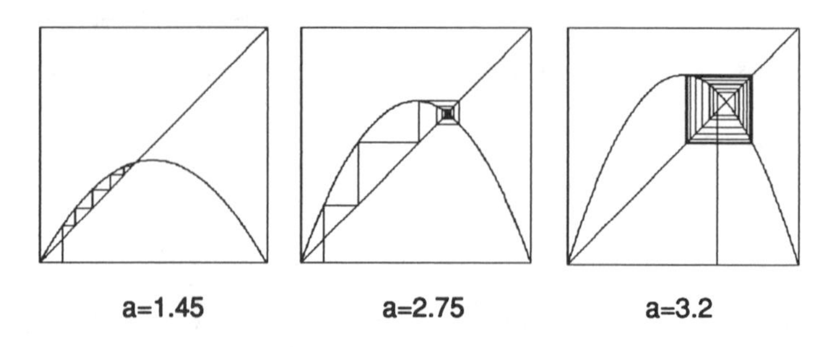

Figure 1.39 : Graphical iteration for three parameter values leading to stable behavior.

Unstable Behavior

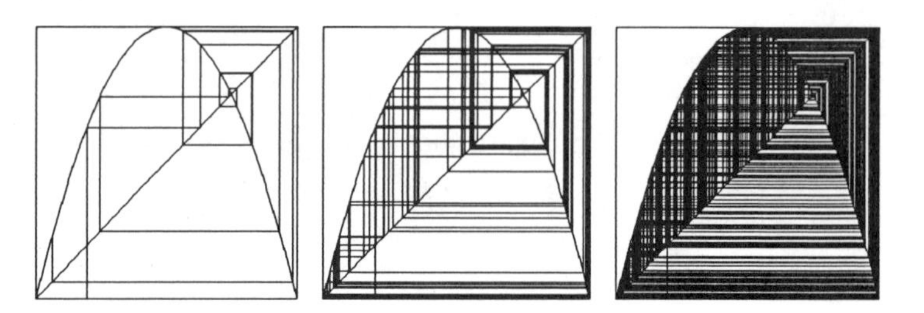

Figure 1.40 : Unstable behavior for $a = 4$. The same initial value is taken with differing numbers of iterations.

The Equivalence Graphical Iteration and the Population Model

We have already shown that the iteration process for the logistic equation is equivalent to the iteration of $x^2 + c$ (see page 63). Here we show the equivalence to the iteration based on $az(1-z)$ as used in the graphical method. Recall that

$$p_{n+1} = p_n + rp_n(1-p_n) \ . \tag{1.12}$$

We show that this is the same as

$$z_{n+1} = az_n(1-z_n) \tag{1.13}$$

when using the identification

$$z_n = \frac{r}{r+1}p_n \quad \text{and} \quad a = r+1 \ . \tag{1.14}$$

We compute z_{n+1} using eqn. (1.14) and the logistic iteration and then check if the result agrees with the iteration using eqn. (1.13). We have

$$\begin{aligned} z_{n+1} &= \tfrac{r}{r+1}p_{n+1} \\ &= \tfrac{r}{r+1}(p_n + rp_n(1-p_n)) \\ &= rp_n + \tfrac{r^2}{r+1}p_n^2 \end{aligned}$$

and on the other hand

$$\begin{aligned} z_{n+1} &= az_n(1-z_n) \\ &= (r+1)\tfrac{r}{r+1}p_n\left(1-\tfrac{r}{r+1}p_n\right) \\ &= rp_n + \tfrac{r^2}{r+1}p_n^2 \,. \end{aligned}$$

Thus we have that iterating $p_{n+1} = p_n + rp_n(1-p_n)$ is really the same as iterating $z_{n+1} = az_n(1-z_n)$. In fact, the iteration of any quadratic polynomials is equivalent to the iteration of the logistic equation (with properly chosen parameter). The proof of this assertion is similar to the above derivation.

Analysis of Chaos is Hard

The analysis of the quadratic feedback process is so difficult because the stable and unstable states are interwoven in an extremely complicated pattern. The feedback process can behave tamely or wildly depending solely on the setting of the control parameter.

This is much like the case with the systems used to predict weather. There are states where prediction is very reliable (like high pressure systems over the Utah deserts); and then again there are situations where any prediction breaks down, and where sophisticated multimillion dollar equipment and the brightest minds are as successful in their prediction as any Tom, Dick or Harry would be when predicting that the weather tomorrow will be the same as today. In other words one and the same system can potentially behave both ways and there are transitions from one into the other. This is the core of the mathematics or the science of chaos. The fact that this theme is also intimately connected with fractals is the content of chapters 10 and 11. The best way to express this relation is to say that fractal geometry is the geometry of chaos.

1.6 Program of the Chapter: Graphical Iteration

For every chapter of this book we have designed a computer program, the *program of the chapter* which highlights an important construction of the corresponding chapter. These programs are very short making it easy to enter the code into a computer and —more importantly — to understand what they are doing for you. We wrote the programs in BASIC. Yes we did! We can already hear people complaining because they would never consider touching this outdated, inefficient, and unstructured language which is known to prevent writing good programs. So why did we choose it?

First of all, BASIC is a language which should easily be available to all programmers. It comes built-in or free with many computers; in any case there is no problem getting it at very low cost. Furthermore, we believe that BASIC — or what we used of it — is so simple that even someone who is not familiar with programming should be able to understand what the programs are doing. Thus, it takes only a small effort to get ready to play with the programs and perhaps try out modifications. Finally, for all you lovers of more aesthetic computer languages, it should be no problem to translate these programs to your favorite dialect and type it on the fly into the machine.

Graphical Iteration

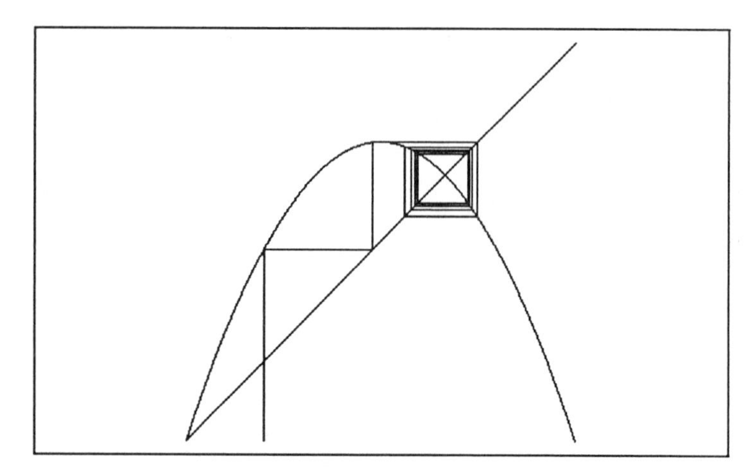

Figure 1.41 : Screen output of the program 'Graphical Iteration'.

All programs were developed under Microsoft BASIC on an Apple Macintosh computer. We also tried the programs on an IBM-compatible personal computer (read the special hints for PCs). In other words, they should also run on your computer without much trouble. All programs do mostly graphical output. They are set up to use a square region of the computer screen for this purpose. This region starts in the upper left hand portion at the point `(left,left)` and is `w` pixels wide. These parameters are set to `left = 30` and `w = 300` but can be changed easily to better fit the size of your computer screen. For the graphics we only use two very common BASIC commands: `LINE` and `PSET`. The command

```
LINE (x1,y1) - (x2,y2)
```

causes the drawing of a line from point `(x1,y1)` to point `(x2,y2)`.[22] The other command

```
PSET (x1,y1)
```

draws a point at `(x1,y1)`. Let us now discuss the first program.

Graphical iteration is an instructive way to visualize the dynamics of an iterator. Let us take the quadratic iterator

$$x_{n+1} = f_a(x_n), \quad f_a(x) = ax(1-x), \quad 0 \leq a \leq 4$$

as an example.[23] This program allows you to study the graphical iteration of f_a for different parameter values a and initial values x_0. Before starting the computation the user enters (at the `INPUT` statement) the numbers a and x_0.

The program first draws the graph of the function f_a (more precisely, its mth iteration f_a^m where $m \geq 1$) and the bisector of the coordinate system. You can change the setting of `m` directly in the program. Observe, that in the text of the program shown it is set to `m = 1`.

The graphical iteration starts at the initial value `x0`. Then the following steps are repeated:

- the value `xn` of the function f_a (or the mth iterate f_a^m) is computed,
- a vertical line is drawn up (or down) to the graph (i.e. to the value `xn`),
- from there a horizontal line is drawn to the bisector.

After `imax` repetitions (initially set to 10) of these steps, the program terminates. If you would like to display more or fewer iterations just set `imax` to a different value.

When experimenting with the program you should try several initial values between 0 and 1 for each choice of the parameter a. For the implemented quadratic iterator, the parameter should be chosen between 0 and 4. As a suggestion we recommend the parameter settings: 1.75, 2.0, 2.75, 3.1, 3.5, 3.6, 3.83, and 4.0.

---

[22] If the first point is not specified, the drawing of a line starts at the current point, which is the endpoint of the last `LINE` or `PSET` command.

[23] It is easy to replace this iterator by a different one. This requires changing only two lines of the code.

BASIC Program **Graphical Iteration**
Title Experiments for the quadratic iterator

```
INPUT "Parameter a, start x0",a,x0
left = 30
w = 300
m = 1
imax = 10

REM DRAW BISECTOR AND FUNCTION
LINE (left+w,left) - (left,left+w)
FOR i = 1 TO w
    xn = i/w
    FOR k = 1 TO m
        xn= a*xn*(1-xn)
    NEXT k
    LINE - (i+left,left+w*(1-xn))
NEXT i

REM START AT x0
xn = x0
PSET (left+w*xn,left+w)
FOR i = 1 TO imax
    REM EVALUATE FUNCTION
    FOR k = 1 TO m
        xn= a*xn*(1-xn)
    NEXT k
    REM DRAW VERTICAL AND HORIZONTAL LINE
    LINE - (left+w*x0,left+w*(1-xn))
    LINE - (left+w*xn,left+w*(1-xn))
    x0 = xn
NEXT i
END
```

Hints for PC Users

If you are using an IBM-compatible PC, you are probably used to the inconvenience that before you can do any graphics, you first need an appropriate graphics adapter, and then you have to enter the correct graphics mode. The code to accomplish this depends on the hardware of your computer. Note, that we are using only black and white graphics. This makes things at least a bit easier.

In BASIC the `SCREEN` command (not the function) allows you to change to several different graphics modes. For example, `SCREEN 1` will give you graphics capabilities at a resolution of 320 by 200 pixels. `SCREEN 2` gives you in x-direction an even higher resolution

(640 pixels), but this leads to images with an undesirable aspect ratio. `SCREEN 9` should give you a resolution of 640 by 350 pixels. For details please look up your user's manual. The programs are set up to draw square images which are approximately 300 pixels wide. This can be changed by assigning a different value to the variable `w` which can be found in most of our programs. If you have a graphics adapter which can display only 200 lines you can set `w = 200`. But having a higher resolution is certainly more satisfying.

You will have noticed that we have not labeled all the BASIC statements with line numbers. If you are using a BASIC dialect which still requires line numbering you are probably used to inventing these numbers while typing. The programs for the following chapters will use only a few labels. Make sure that your additional statement numbering does not interfere with these labels. For example, if the first label that we use is 100, then you can number the preceding statements with numbers from 1 to 99. If our next label is 200, you can use 101 to 199 for the numbering of the statements which lie in between, and so on.

Chapter 2

Classical Fractals and Self-Similarity

The art of asking the right questions in mathematics is more important than the art of solving them.

Georg Cantor

Mandelbrot is often characterized as the father of fractal geometry. Some people, however, remark that many of the fractals and their descriptions go back to classical mathematics and mathematicians of the past like Georg Cantor (1872), Giuseppe Peano (1890), David Hilbert (1891), Helge von Koch (1904), Waclaw Sierpinski (1916), Gaston Julia (1918), or Felix Hausdorff (1919), to name just a few. Yes, indeed, it is true that the creations of these mathematicians played a key role in Mandelbrot's concept of a new geometry. But at the same time it is true that they did not think of their creations as conceptual steps towards a new perception or a new geometry of nature. Rather, what we know so well as the Cantor set, the Koch curve, the Peano curve, the Hilbert curve and the Sierpinski gasket, were regarded as exceptional objects, as counter examples, as 'mathematical monsters'. Maybe this is a bit overemphasized. Indeed, many of the early fractals arose in the attempt to fully explore the mathematical content and limits of fundamental notions (e.g. 'continuous' or 'curve'). The Cantor set, the Sierpinski carpet and the Menger sponge stand out in particular because of their deep roots and essential role in the development of early topology.

Cauliflower Self-Similarity

Figure 2.1 : The self-similarity of an ordinary cauliflower is demonstrated by dissection and two successive enlargements (bottom). The small pieces look similar to the whole cauliflower head.

Abnormal Monsters or Typical Nature?

But even in mathematical circles their profound meaning had been somewhat forgotten, and they were seen as shapes, intended to demonstrate the deviation from the familiar rather than to typify the normal. Then Mandelbrot demonstrated that these early mathematical fractals in fact have many features in common with shapes found in nature. Thus the title *The Fractal Geometry of Nature*[1] of his book in 1982. In other words, we could say that Mandelbrot turned the manifested mathematical interpretation and value of these fantastic inventions upside down. But in fact, he did much more. The best way to describe his contribution is to say that, indeed, some characters, such as the Cantor set, were already there. But he went on to develop the language into which the characters could be embedded. In other words, he noticed that the seemingly exceptional is more like the rule and then developed a systematic language with words and sentences and grammar. According to Mandelbrot himself, he did not follow a certain grand

[1] Freeman, 1982.

plan when carrying out this program; but rather summarized, in a way, his complex — one is tempted to say nomadic — scientific experiences in mathematics, linguistics, economics, physics, medical sciences and communication networks, to mention just some areas in which he was active.

Self-Similarity

Before we open our gallery of classical fractals and discuss in some detail several of these early masterpieces, let us introduce the concept of self-similarity. It will be an underlying theme in all fractals, more pronounced in some of them and in variations in others. In a way the word self-similarity needs no explanation, and at this point we merely give an example of a natural structure with that property, a cauliflower. It is not a classical mathematical fractal, but here the meaning of self-similarity is readily revealed without any math. The cauliflower head contains branches or parts, which when removed and compared with the whole are very much the same, only smaller. These clusters again can be decomposed into smaller clusters, which again look very similar to the whole as well as to the first generation branches. This self-similarity carries through for about three to four stages. After that the structures are too small for a further dissection. In a mathematical idealization the self-similarity property of a fractal may be continued through infinitely many stages. This leads to new concepts such as fractal dimension which are also useful for natural structures that do not have this 'infinite detail'.

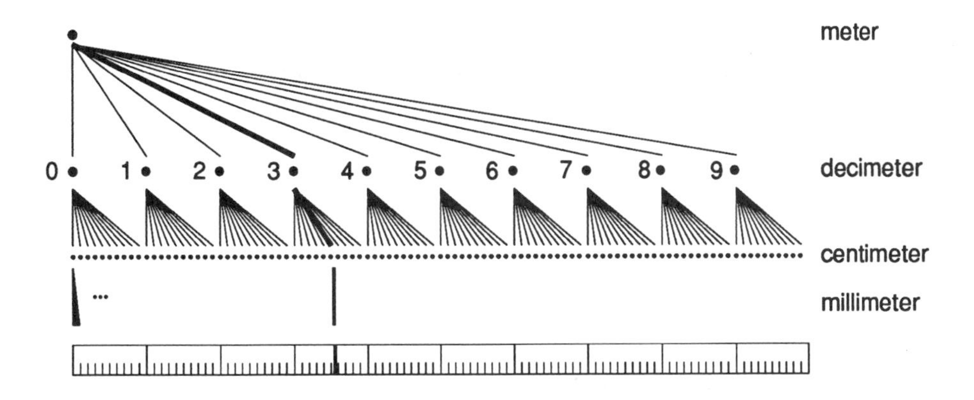

Figure 2.2 : The branches of the decimal tree leading to 357 are highlighted.

Self-Similarity in the Decimal System

Although the notion of self-similarity is only some 20 years old there are many historical constructions which make substantial use of its core idea. Probably the oldest and most important construction in that regard is our familiar decimal number system.[2] It is impossible to estimate where mathematics and the natural sciences would be without this ingenious invention. We are so used to the decimal number system that we are inclined to take it for granted. However, it evolved after a long scientific and cultural struggle and it is very closely related to the material from which fractals are made. It is also the prerequisite of the metric (measuring) system (for length, area, volume, weight, etc.). Let us look at a meter[3] stick, which carries markers for decimeters (ten make a meter), centimeters (ten make a decimeter; hundred make a meter), and millimeters (ten make a centimeter; a thousand make a meter). In a sense a decimeter together with its markers looks like a meter with its markers, however, scaled down by a factor of 10. This is not an accident. It is in strict correspondence with the decimal system. When we say 357 mm, for example, we mean 3 decimeters, 5 centimeters, and 7 millimeters. In other words, the position of the figures determines their place value, exactly as in the decimal number system. One meter has a thousand millimeters to it and when we have to locate position 357 only a fool would start counting from left to right from 1 to 357. Rather, we would go to the 3 decimeter tick mark, from there to the 5 centimeter tick mark, and from there to the 7 millimeter tick mark. Most of us take this elegant procedure for granted. But somebody who has to convert miles, yards and inches can really appreciate the beauty of this system. Actually finding a position on the meter stick corresponds to a walk on the branches of a tree, the decimal number tree, see figure 2.2. The structure of the tree expresses the self-similarity of the decimal system very strongly. Similar trees reflect the self-similarity of many fractal constructions considered in this chapter.

[2]Leonardo of Pisa, also known as Leonardo Fibonacci, helped introduce into mathematics the Hindu-Arabic numerals 0, 1, 2, 3, 4, 5, 6, 7, 8, and 9. His best known work, the *Liber abaci* (1202; 'Book of the Abacus') spends the first seven chapters explaining the place value, by which the position of a figure determines whether it is a unit, ten, hundred, and so forth, and demonstrating the use of the numerals in arithmetical operations.

[3]The metric system is now used internationally by scientists and in most nations. It was brought into being by the French National Assembly between 1791 and 1795. The spread of the system was slow but continuous, and, by the early 1970's only a few countries, notably the United States, had not adapted the metric system for general use. Since 1960 the definition of a meter has been: 1 meter = 1,650,763.73 wavelenghts of the orange-red line in the spectrum of the krypton-86 atom under specified conditions. In the 1790's it was defined as 1/10,000,000 of the circumference of the quadrant of the Earth's circumference running from the North Pole through Paris to the equator.

2.1 The Cantor Set

Cantor (1845–1918) was a German mathematician at the University of Halle where he carried out his fundamental work in the foundations of mathematics, which we now call *set theory*.

Georg Cantor

Figure 2.3 : Georg Cantor, 1845–1918.

The Cantor set was first published[4] in 1883 and emerged as an example of certain exceptional sets.[5] It is probably fair to say that in the zoo of mathematical monsters — or early fractals — the Cantor set is by far the most important, though it is less visually appealing and more distant to an immediate natural interpretation than some of the others. It is now understood that the Cantor set plays a role in many branches of mathematics; and in fact, in a very deep sense, in chaotic dynamical systems (we will touch upon this property at least a bit) and is somehow hidden as the essential skeleton or model behind many other fractals (for example Julia sets, as we will see in chapter 12).

The basic Cantor set is an infinite set of points in the unit interval [0,1]. That is, it can be interpreted as a set of certain numbers, as for example 0, 1, 1/3, 2/3, 1/9, 2/9, 7/9, 8/9, 1/27, 2/27, ... Plotting these and all other points (assuming we could know what they were) would not make much of a picture at all. Therefore, we use a common little trick. Rather than plotting

[4]G. Cantor, *Über unendliche, lineare Punktmannigfaltigkeiten V*, Mathematische Annalen 21 (1883) 545–591.
[5]The Cantor set is an example of a perfect, nowhere dense subset.

The Cantor Set

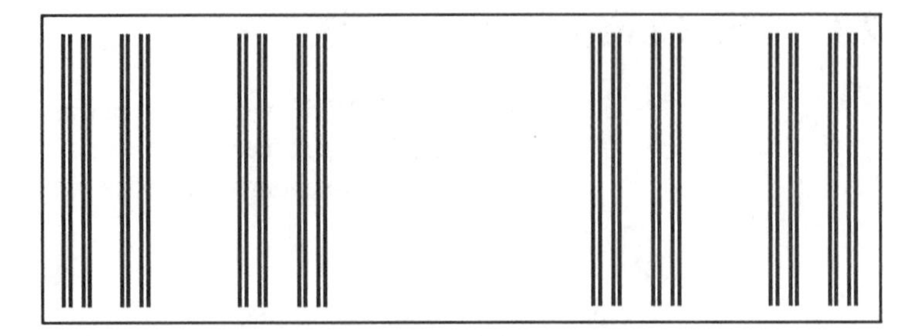

Figure 2.4 : The Cantor set represented by vertical lines whose base points are exactly at all the different points belonging to the set.

just points we plot vertical lines all of the same length whose base points are exactly at all the different points belonging to the Cantor set. By so doing, we are able to see the distribution of these points a bit better. Figure 2.4 gives a first impression. Rather than being able to actually see the Cantor set, it is probably much more important to remember its classical construction.

Construction of the Cantor Set

Start with the interval [0,1]. Now take away the (open) interval (1/3,2/3), i.e., remove the middle third from [0,1], but not the numbers 1/3 and 2/3. This leaves two intervals [0,1/3] and [2/3,1] of length 1/3 each and completes a basic construction step. Now we repeat, we look at the remaining intervals [0,1/3] and [2/3,1] and remove their middle thirds, which yields four intervals of length 1/9. Continue on in this way. In other words, there is a feedback process in which a sequence of (closed) intervals is generated — one after the first step, two after the second step, four after the third step, eight after the fourth step, etc. (i.e. 2^n intervals of length $1/3^n$ after the nth step). Figure 2.5 visualizes the construction.

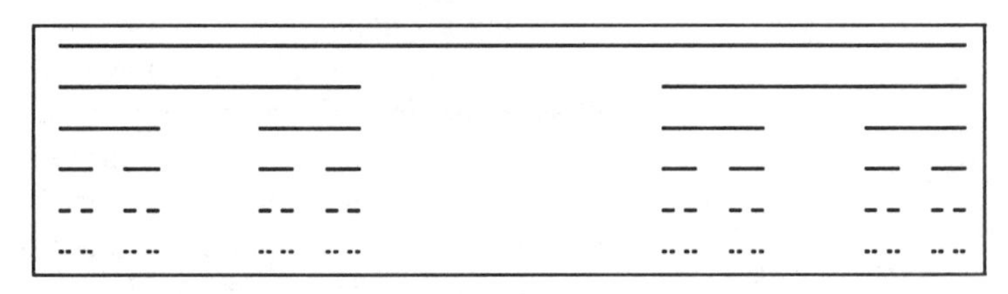

Figure 2.5 : Some initial steps of the construction.

Interval End Points Are in the Cantor Set ...

What is the Cantor set? It is the set of points which remain if we carry out the *removal* steps infinitely often. How do we explain *infinitely often*? Let us try. A point, say x, is in the Cantor set if we can guarantee that no matter how often we carry out the removal process, the point x will not be taken out. Obviously 0, 1, 1/3, 2/3, 1/9, 2/9, 7/9, 8/9, 1/27, 2/27, ... are examples of such points because they are the end points of the intervals which are created in the steps; and therefore, they must remain. All these points have one thing in common. Namely, they are related to powers of 3 — or rather, to powers of 1/3. This is an important observation which we will exploit later to understand the Cantor set. One is tempted to believe that any point in the Cantor set is of this kind, i.e. an end point of one of the small intervals generated in the process. This conclusion is categorically wrong. We will not give the complete argument but at least discuss the fact to some extent.

... But That's Not All

If the Cantor set were just the end points of the intervals of the generating process, we could easily enumerate them as shown in figure 2.6.

End Points of Intervals

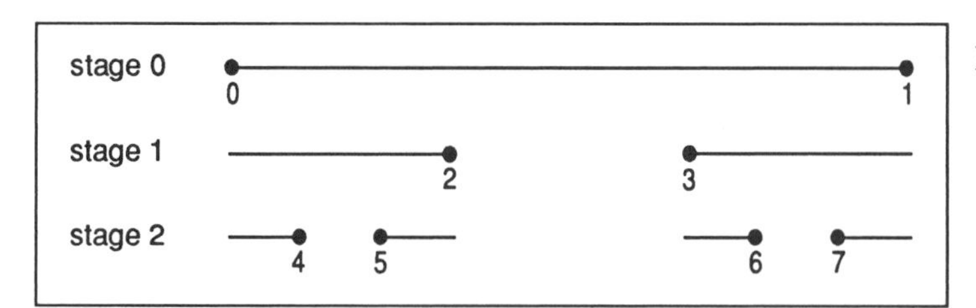

Figure 2.6 : Counting end points of intervals from the Cantor set construction. In stage $k, k > 0$ of the construction process 2^k new end points are added and enumerated as shown.

That means, the Cantor set would be a countable set, but it is known to be uncountable.[6] That is, there is no way to enumerate the points in the Cantor set. Thus, there must be many more points which are not end points. Can we give examples which are not end points? To name such examples, we will use a simple, but far reaching characterization of the Cantor set, namely by triadic numbers.

A Modification Using Decimals

But let us first see what can be done with the more familiar decimal numbers. Recall our discussion of the meter stick. Let us remove parts of the stick in several stages (see figure 2.7). Start with the meter and cut out decimeter number 5 in stage 1. This leaves 9 decimeters from each of which we take away centimeter number 5 in stage 2. Next, in stage 3, we consider the remaining 81 centimeters and remove millimeter number 5 from each one. Then we continue the process going to tenth of millimeters in stage 4 and so on. This clearly is very similar to the basic Cantor set construction. In fact, the set of points that survive all stages in the construction, i.e. which are never taken away, are a fractal which is also called a Cantor set.

[6]See further below, page 88.

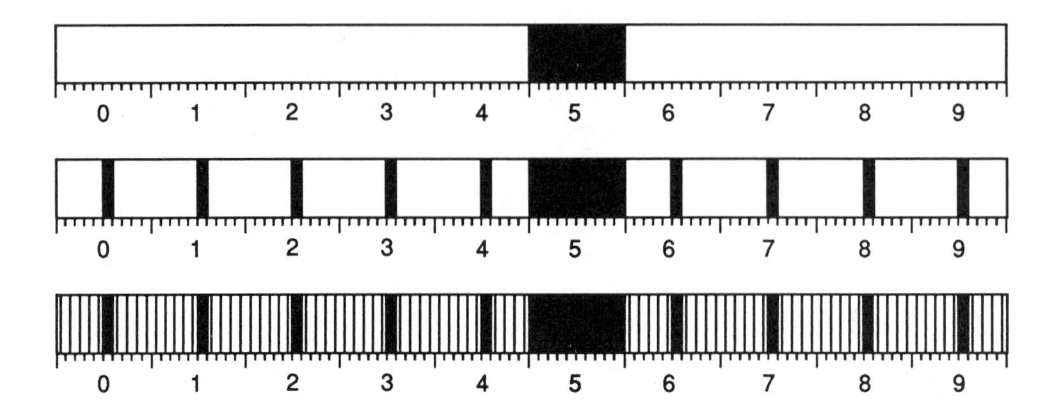

Figure 2.7 : In this meter stick the fifth decimeter (stage 1), the fifth centimeters (stage 2) and fifth millimeters (stage 3) are removed. This yields the first three stages of a modified Cantor set construction.

It is instructive to relate this modified Cantor set construction to the decimal number tree from figure 2.2. Removing a section of the meter stick corresponds to pruning a branch of the tree. In stage 1 the main branch with label 5 is cut off. In the following stages all branches with label 5 are pruned. In other words, only those decimals are kept which do not include the digit 5. Clearly, our choice to remove all fifth decimeters, centimeters, and so on is rather arbitrary. We could just as well have prefered to take out all numbers with a 6 in their decimal expansion, or even numbers with digits 3, 4, 5, and 6. For each choice we get another Cantor set. However, we will never obtain the classical Cantor set using this approach; this requires *triadic* numbers.

Characterization of the Cantor Set

Triadic numbers are numbers which are represented with respect to base 3. This means one only uses the digits 0, 1, and 2. We give a few examples in the following table.

decimal	in powers of 3	triadic
4	$1 \cdot 3^1 + 1 \cdot 3^0$	11
17	$1 \cdot 3^2 + 2 \cdot 3^1 + 2 \cdot 3^0$	122
0.333...	$1 \cdot 3^{-1}$	0.1
0.5	$1 \cdot 3^{-1} + 1 \cdot 3^{-2} + 1 \cdot 3^{-3} + \cdots$	0.111...

Table 2.8 : Conversion of four decimal numbers into the triadic representation.

Triadic Numbers

Let us recall the essence of our familiar number system, the decimal system, and its representation. When we write 0.32573 we mean

$$3 \cdot 10^{-1} + 2 \cdot 10^{-2} + 5 \cdot 10^{-3} + 7 \cdot 10^{-4} + 3 \cdot 10^{-5}.$$

In other words, any number x in $[0, 1]$ can be written as

$$x = a_1 \cdot 10^{-1} + a_2 \cdot 10^{-2} + a_3 \cdot 10^{-3} + \dots, \tag{2.1}$$

where the $a_1, a_2, a_3, \dots$ are numbers from $\{0, 1, 2, \dots, 9\}$, the decimal digits. This is called the *decimal expansion* of x, and may be infinite (e.g. $x = 1/3$) or finite (e.g. $x = 1/4$). When we say the expansion or representation is finite we actually mean that it ends with infinitely (redundant) consecutive zeros.

Binary Tree

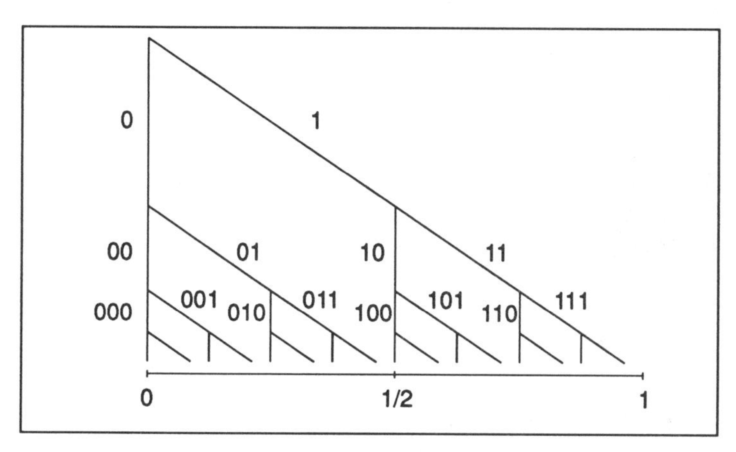

Figure 2.9 : Visualization of binary expansions by a two-branch tree. In contrast to real trees, we always draw address trees, so that the root is at the top. Any number in the interval [0,1] at the bottom can be reached from the root of the tree by following branches. Writing down the labels of these branches (0 for the left and 1 for the right branch) in a sequence will yield a binary expansion of the chosen real number. The tree has obvious self-similarity: any two branches at any node are a reduced copy of the whole tree.

You will recall that digital computers depend on *binary expansions* of numbers. In computers 10 as base is replaced by 2. For example 0.11001 is

$$1 \cdot 2^{-1} + 1 \cdot 2^{-2} + 0 \cdot 2^{-3} + 0 \cdot 2^{-4} + 1 \cdot 2^{-5}.$$

There is a little bit of ambiguity in these representations. For example, we can write $2/10$ in two ways 0.19999... or 0.20000..., or in base 2 we can have the example $1/4$ represented as $0.0011\overline{1}$ or $0.0100\overline{0}$, where the overlining means that the respective digit (or digits) will be repeated ad infinitum.

Triadic Tree

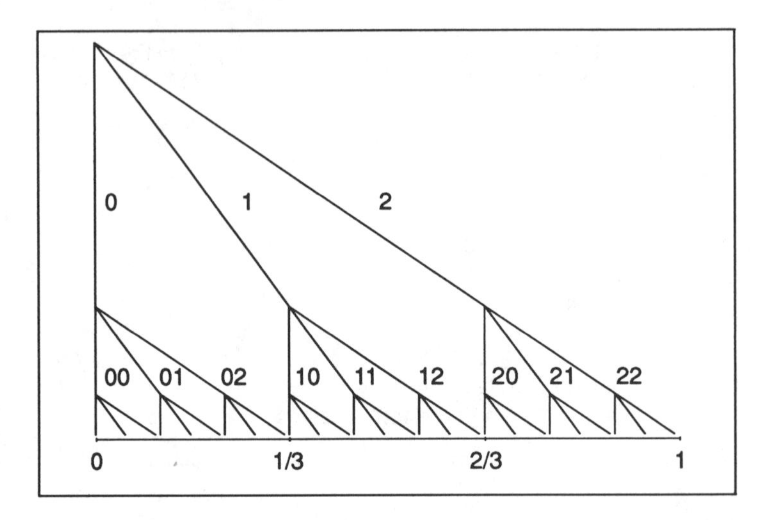

Figure 2.10 : A three-branch tree visualizes the triadic expansion of numbers from the unit interval. The first main branch covers all numbers between 0 and 1/3. Following down the branches all the way to the interval and keeping note of the labels 0, 1, and 2 for choosing the left, middle, or right branches will produce a triadic expansion of the number in the interval which is approached in this process.

Now we can completely describe the Cantor set by representing numbers from $[0, 1]$ in their triadic expansion, i.e. we switch to expansions of x with respect to base 3, as in eqn. (2.2).

$$x = a_1 \cdot 3^{-1} + a_2 \cdot 3^{-2} + a_3 \cdot 3^{-3} + a_4 \cdot 3^{-4} + \ldots \tag{2.2}$$

Thus, here the $a_1, a_2, a_3, \ldots$ are numbers from $\{0, 1, 2\}$.

Let us write some of the points of the Cantor set as triadic numbers: 1/3 is 0.1 in the triadic system, 2/3 is 0.2, 1/9 is 0.01, and 2/9 is 0.02. In general we can characterize any point of the Cantor set in the following way.

Fact. *The Cantor set C is the set of points in* $[0, 1]$ *for which there is a triadic expansion that does not contain the digit* '1'.

This number theoretic characterization eliminates the problem of the existence of a limit for the geometric construction of the Cantor set.

The above examples 2/3 and 2/9 are points in the Cantor set according to this statement, since their triadic expansions 0.2 and 0.02 do not contain any digits '1'. However, the other two examples 1/3 and 1/9 seem to contradict the rule, their expansions

0.1 and 0.01 clearly show a digit '1'. Yes, that is correct; but remember that we have ambiguity in our representations, and 1/3 can also be written as $0.0222\bar{2}$. Therefore, it belongs. But then, you may ask, what about $1/3 + 1/9$? This is a number from the middle third interval which is discarded in the first construction step of the Cantor set. It has a triadic expansion 0.11, and can't we also write this in different form and thus get into trouble? Yes, indeed, $1/3 + 1/9 = 0.1022\bar{2}$; but as you see, there appears a digit '1' no matter how we choose to represent that number in the triadic system. Thus, it is out and there is no problem with our description.

Distinguishing End Points from Others

Moreover, we can now distinguish points in C which are end points of some small interval occurring in the process of the feedback construction from those points which are definitely not. Indeed, end points in this sense just correspond to numbers which have triadic expansion ending with infinitely many consecutive 2's or 0's. All other possibilities, as for example

$$0.02002200022200002222000002222\ldots$$

or a number in which we pick digits 0 and 2 at random will belong to the Cantor set but are not end points, and those are, in fact, more typical for the Cantor set. In other words, if one picks a number from C at random, then with probability 1, it will not be an end point. By this characterization of the Cantor set we can understand that, in fact, any point in C can be approximated arbitrarily closely by other points from C, and yet C itself is a dust of points. In other words, there is nothing like an interval in C (which is obvious if we recall the geometric construction, namely the removal of intervals).

Addresses and the Cantor Set

Let us return for a moment to the intuitive geometric construction of the Cantor set by removing middle thirds in each step from the unit interval $[0, 1]$. After the first step, we have two parts, one is left and one is right. After the second step, each of these in turn splits into two parts, a left one and a right one, and so on. Now we design an efficient labeling procedure for each part created in the steps. The two parts after the first step are labeled L and R for left and right. The four parts after the second step are labeled LL, LR, RL, RR, i.e. the L part from step one is divided into an L and an R part, which makes LL and LR, and likewise with the R part. Figure 2.11 summarizes the first three steps.

As a result, we are able to read from a label with 8 letters like $LLRLRRRL$ exactly which of the 2^8 parts of length $1/3^8$ we want to pick. It is important when reading this address, however, to remember the convention that we interpret from the left to right, i.e. letters have a place value according to their position in a word much like the numerals in the decimal system.

Cantor Set Addresses

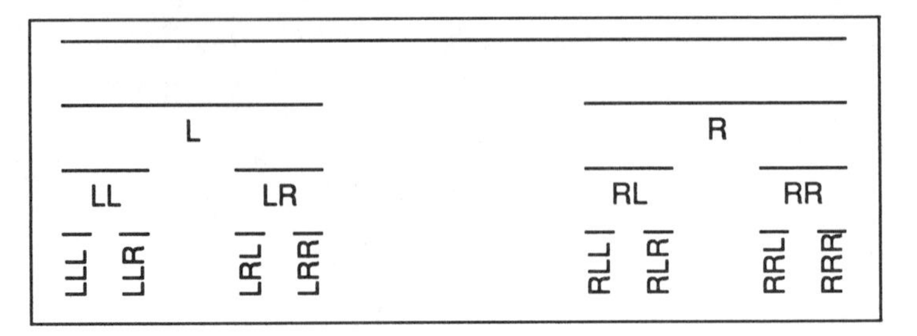

Figure 2.11 : Addresses for the Cantor set.

Addresses of Intervals Versus Addresses of Points

Finite string addresses such as $LLRLRRRL$ identify a small interval from the construction of the Cantor set. The longer the address, the higher the stage of the construction, and the smaller the interval becomes. To identify points in the Cantor set, such addresses obviously are not sufficient, as in each such interval, no matter how small it is, there are still infinitely many points from the Cantor set. Therefore we need infinitely long address strings to precisely describe the location of a Cantor set point. Let us give two examples. The first one is the point 1/3. It is in the left interval of the first stage, which has address L. Within that it is in the right interval of the second stage. This is $[2/9, 1/3]$ with address LR. Within that the point is again in the right subinterval with address LRR, and so on. Thus, to identify the position of the point exactly we write down the sequence of intervals from consecutive stages to which the point belongs: LR, LRR, $LRRR$, $LRRRR$, and so on. In other words, we can write the address of the point as the infinite string $LRRRR...$, or using a bar to indicate periodic repetition $L\overline{R}$. The point 2/3 is in the right interval of the first stage. Within that and all further stages it is always in the left subinterval. Thus, the address of 2/3 is $RLLL...$ or $R\overline{L}$.

Binary Tree for the Cantor Set

Another interesting way to look at the situation which is established by this systematic labeling is demonstrated in figure 2.12, where we see an infinite binary tree the branches of which repeatedly split into two branches from top to bottom. What is the connection between the tree and the Cantor set? Well, the tree consists of nodes and branches. Each level of the tree corresponds to a certain step in the construction of the Cantor set; and in this way, it is actually a genealogical tree. In other words, we can compare our situation with a cell division process, and the tree tells us exactly from where an individual cell of some future generation is derived. This is quite nice already, but there is much more to this simple idea. For example, rather than choosing the alphabet $\{L, R\}$ we can take another two letter alphabet and carry out a systematic replacement. Why not pick 0 and 2 as an alphabet, i.e. replace any

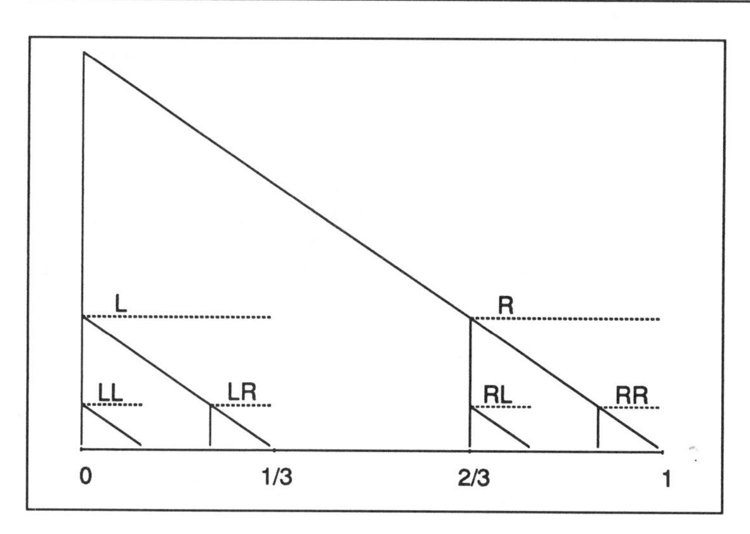

Figure 2.12 : Addresses for points of the Cantor set form a binary tree.

L by a 0 and any R by a 2. Then we obtain strings of digits like 022020002 in place of $LRRLRLLLR$. You have surely guessed what we are up to. Indeed, that string of digits can be interpreted as a triadic number by just putting a decimal point in front of it, i.e., 0.022020002. Thereby, we have demonstrated the connection between the triadic representation of the Cantor set and the addressing system. In fact, what we have just discussed provides an argument for the validity of the triadic characterization. In other words, if we want to know where in the Cantor set a certain number is located — up to a certain degree of precision, we just have to look at its triadic expansion and then interpret each digit 0 as L and each digit 2 as R. Then, looking up the resulting address, we can find the part of the binary tree in which the number must lie.

L and R Are Not 0 and 1

The relation of L, R addresses with triadic numbers might seem to suggest going even one step further, namely, to identify L and R with 0 and 1, i.e. with binary numbers. This is, however, somewhat dangerous. Let us consider again the example point 1/3. It is represented by the address string $L\overline{R}$. This would correspond to the binary number $0.0\overline{1}$, which as a binary number, is identical to 0.1. But translated back 0.1 corresponds to $R\overline{L}$, which in the Cantor set is the point 2/3! Clearly, this identification produces a contradiction. In other words, triadic numbers, but not binary numbers, are natural to the Cantor set. Or to put in another way, two-letter-based infinite strings are natural for the Cantor set, but these cannot be identified with binary numbers, despite the fact that they are strings made up of two letters/digits.

The Cardinality of the CantorSet

We can now see that the cardinality of the Cantor set must be the same as the cardinality of the unit interval $[0,1]$. We start with the interval $[0,1]$ and show how each point in it corresponds to one point in the Cantor set.

- Each point in the interval has a *binary* expansion.
- Each binary expansion corresponds to a path in the binary tree for binary decimals.
- Each such path has a corresponding path in the *ternary* tree for the Cantor set.
- Each path in the ternary tree of the Cantor set identifies a unique point in the Cantor set by an address in triadic expansion.

Therefore, for each number in the interval, there is a corresponding point in the Cantor set. For different numbers there are different points. Thus, the cardinality of the Cantor set must be at least as large as the cardinality of the interval. On the other hand, it cannot exceed this cardinality, because the Cantor set is a subset of the interval. Therefore, both cardinalities must be the same.

Self-Similarity

The Cantor set is truly complex, but is it also self-similar? Yes, indeed, if one takes the part of C which lies in the interval [0,1/3], for example, we can regard that part as a scaled down version of the entire set. How do we see that? Let us take the definition of the Cantor set collecting all points in [0,1] which admit a triadic representation not containing digit 1. Now for every point, say

$$\xi = \alpha_1 \times 3^{-1} + \alpha_2 \times 3^{-2} + \alpha_3 \times 3^{-3} + \alpha_4 \times 3^{-4} + \ldots,$$

(with $\alpha_i \in \{0,2\}$) in the Cantor set we find a corresponding one in [0,1/3] by dividing ξ by 3, i.e.

$$\frac{\xi}{3} = 0 \times 3^{-1} + \alpha_1 \times 3^{-2} + \alpha_2 \times 3^{-3} + \alpha_3 \times 3^{-4} + \ldots$$

Indeed, if $x = 0.200220\ldots$ and we multiply by $1/3 = 0.1$, that means that we just shift the decimal point one place to the left (i.e., we obtain 0.0200220...), which is in C again. Thus, the part of the Cantor set present in [0,1/3] is an exact copy of the entire Cantor set scaled down by the factor 1/3 (see figure 2.13). For the part of C which lies in the interval [2/3,1], essentially we can do the same calculation (we only have to include the addition of 2/3 = 0.2). In the same way, any subinterval in the geometric Cantor set construction contains the entire Cantor set scaled down by an appropriate factor of $1/3^k$. In other words, the Cantor set can be seen as a collection of arbitrarily small pieces, each of which is an exact scaled down version of the entire Cantor set. This is what we mean when we say the Cantor set is self-similar. Thus, taking self-similarity as an intuitive property means that self-similarity here is absolutely perfect and is true for an infinite range. Note that in our discussion of self-similarity we have carefully avoided the geometrical model of the Cantor set. Instead, we have exploited the number theoretic representation.

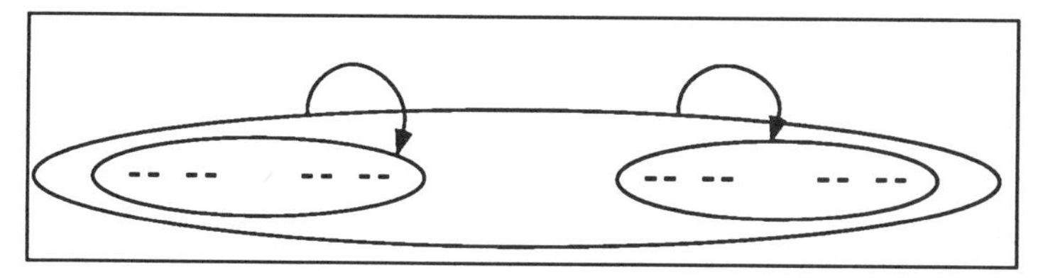

Figure 2.13 : The Cantor set is a collection of two exact copies of the entire Cantor set scaled down by the factor 1/3.

Note that the scaling property of the Cantor set corresponds to the following invariance property. Take a point from C and multiply it by 1/3. The result will be in C again. The same is true if we first multiply by 1/3 and then add 2/3. This is apparent from the triadic characterization, and it will be the key observation for chapter 5.

Before we continue our introduction to classical fractals with some other examples, let us touch one more property of the Cantor set which reveals an important dynamic interpretation and an amazing link with chaos.

Cantor Set as Prisoner Set

Let us look at a mathematical feedback system defined in the following way. If x is an input number, then the output number is determined by the following conditional formula (2.3).

$$x \rightarrow \begin{cases} 3x & \text{if } x \leq 0.5 \\ -3x+3 & \text{if } x > 0.5 \end{cases} \tag{2.3}$$

In other words, the output is evaluated to be $3x$ if $x < 0.5$ and is $-3x+3$ if $x \geq 0.5$.

Starting with an initial point x_0, the feedback process defines a sequence

$$x_0, x_1, x_2, x_3, \ldots$$

The interesting question then is: what is the long-term behavior of such sequences? For many initial points x_0 the answer is very easy to derive. Take for example a number $x_0 < 0$. Then $x_1 = 3x_0$, and $x_1 < 0$. By induction it follows that all numbers x_k from this sequence are negative, and, thus

$$x_k = 3^k x_0 \ .$$

This sequence then grows negatively without any bound, it tends to negative infinity, $-\infty$. Let us call a sequence with such a long-term behavior an *escaping sequence* and the initial point x_0 an *escaping point.*

Let us now take $x_0 > 1$. Then $x_1 = -3x_0 - 3 < 0$, and again the sequence escapes to $-\infty$. But not all points are escaping points. For example, for $x_0 = 0$ we have that all succeeding numbers in the sequence are also equal to zero. We conclude, that any initial point x_0, which at some stage goes to zero will remain there forever, and thus is not an escaping point. Such points we call *prisoners*. So far we have found that all prisoner points must be in the unit interval $[0, 1]$. This leads to the interesting question: which points in the unit interval will remain and which will escape? Let us look at some examples.

x_0	x_1	x_2	x_3	x_4	...	P/E
0	0	0	0	0		prisoner
1/3	1	0	0	0		prisoner
1/9	1/3	1	0	0		prisoner
1/2	3/2	−3/2	−9/2	−27/2		escapee
1/5	3/5	6/5	−3/5	−9/5		escapee

Clearly the entire (open) interval $(1/3, 2/3)$ escapes because when $1/3 < x_0 < 2/3$ we have $x_1 > 1$ and $x_2 < 0$. But then every point which eventually lands in that interval will also escape under iteration. Figure 2.14 illustrates these points and reveals the Cantor set construction for the points which will remain.

Fact. *The prisoner set P for the feedback system given by eqn. (2.3) is the Cantor set, while all points in* $[0, 1]$ *which are outside the Cantor set belong to the escape set E.*

This is a remarkable result and shows that the study of the dynamics of feedback systems can provide an interpretation for the Cantor set. This close relation between chaos and fractals will be continued in chapter 12.

Escaping Intervals

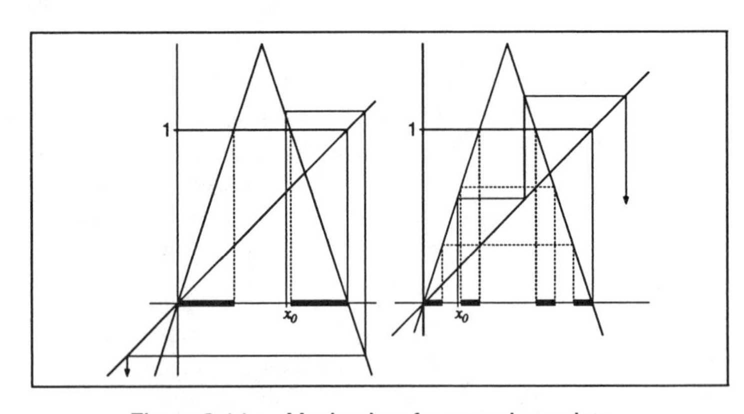

Figure 2.14 : Mechanism for escaping points.

2.2 The Sierpinski Gasket and Carpet

Our next classical fractal is about 40 years younger than the Cantor set. It was introduced by the great Polish mathematician Waclaw Sierpinski[7] (1882–1969) in 1916.

Waclaw Sierpinski

Figure 2.15 : Waclaw Sierpinski, 1882–1969.

Sierpinski was a professor at Lvov[8] and Warsaw. He was one of the most influential mathematicians of his time in Poland and had a worldwide reputation. In fact, one of the moon's craters is named after him.

The basic geometric construction of the Sierpinski gasket goes as follows. We begin with a triangle in the plane and then apply a repetitive scheme of operations to it (when we say triangle here, we mean a blackened, 'filled-in' triangle). Pick the midpoints of its three sides. Together with the old vertices of the original triangle, these midpoints define four congruent triangles of which we drop the center one. This completes the basic construction step. In other words, after the first step we have three congruent triangles

[7]W. Sierpinski, C. R. Acad. Paris 160 (1915) 302, and W. Sierpinski, *Sur une courbe cantorienne qui content une image biunivoquet et continue detoute courbe donneé* C. R. Acad. Paris 162 (1916) 629–632.

[8]Lvov, Ukrainian Lviv, Polish Lwów, German Lemberg, city and administrative center in the Ukrainian SSR. Founded in 1256 Lvov has always been the chief center of Galicia. Lvov was Polish between 1340 and 1772 until the first partition, when it was given to Austria. In 1919 it was restored to Poland and became a world famous university town hosting one of the most influential mathematics schools during the 1920's and 1930's. In 1939 it was annexed by the Soviets as a result of the Hitler-Stalin Pact, and the previously flourishing polish mathematics school collapsed. Later several of its great scientists were victims of Nazi Germany.

Sierpinski Gasket

Figure 2.16 : The basic construction steps of the Sierpinski gasket.

Sierpinski Pattern

Figure 2.17 : Escher's studies of Sierpinski gasket type patterns on the twelfth century pulpit of the Ravello cathedral, designed by Nicola di Bartolomeo of Foggia. Watercolor, ink, 278 by 201 mm. ©1923 M. C. Escher / Cordon Art – Baarn – Holland.

whose sides have exactly half the size of the original triangle and which touch at three points which are common vertices of two contiguous triangles. Now we follow the same procedure with the three remaining triangles and repeat the basic step as often as desired. That is, we start with one triangle and then produce 3, 9,

LRTT

Figure 2.18 : *LRTT* denote a subtriangle in the Sierpinski gasket which can be found following the left, right, top, top subtriangles.

27, 81, 243, ... triangles, each of which is an exact scaled down version of the triangles in the preceding step. Figure 2.16 shows a few steps of the process.

The Sierpinski gasket[9] is the set of points in the plane which remain if one carries out this process infinitely often. It is easy to list some points which definitely belong to the Sierpinski gasket, namely the sides of each of the triangles in the process.

The characteristic of self-similarity is apparent, though we are not yet prepared to discuss it in detail. It is built into the construction process, i.e. each part of the 3 parts in the kth step is a scaled down version — by a factor of 2 — of the entire structure in the previous step. Self-similarity, however, is a property of the limit of the geometrical construction process, and that will be available to us only in chapter 5. In chapter 9 we will explain how the Sierpinski gasket admits a number theoretic characterization from which the self-similarity follows as easily as for the Cantor set.

Addresses for the Sierpinski Gasket

Similar to our above discussion of the Cantor set we can introduce an addressing system for the subtriangles (or points) of the Sierpinski gasket. Here we must use three symbols to establish a system of addresses. If we take, for example, L (left), R (right) and T (top) we obtain sequences like $LRTT$ or $TRLLLTLR$ and read them from left to right to identify subtriangles in the respective construction stage of the Sierpinski gasket. For example, $LRTT$ refers to a triangle in the 4th generation which is obtained in the following way. Pick the left triangle in the first generation, then the right one therein, then the top one therein, and finally again the top one therein, see figure 2.18. We will discuss the importance of addresses for the Sierpinski gasket in chapter 6. They are the key to unchaining the chaos game introduced in chapter 1. We

[9]The Sierpinski gasket is sometimes also called the Sierpinski triangle.

Spider-Like Tree

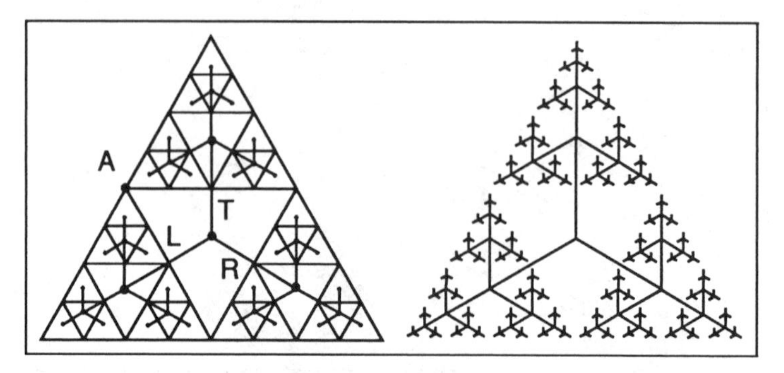

Figure 2.19 : This tree represents not only the structure of the Sierpinski gasket but also its geometry.

Sierpinski Carpet

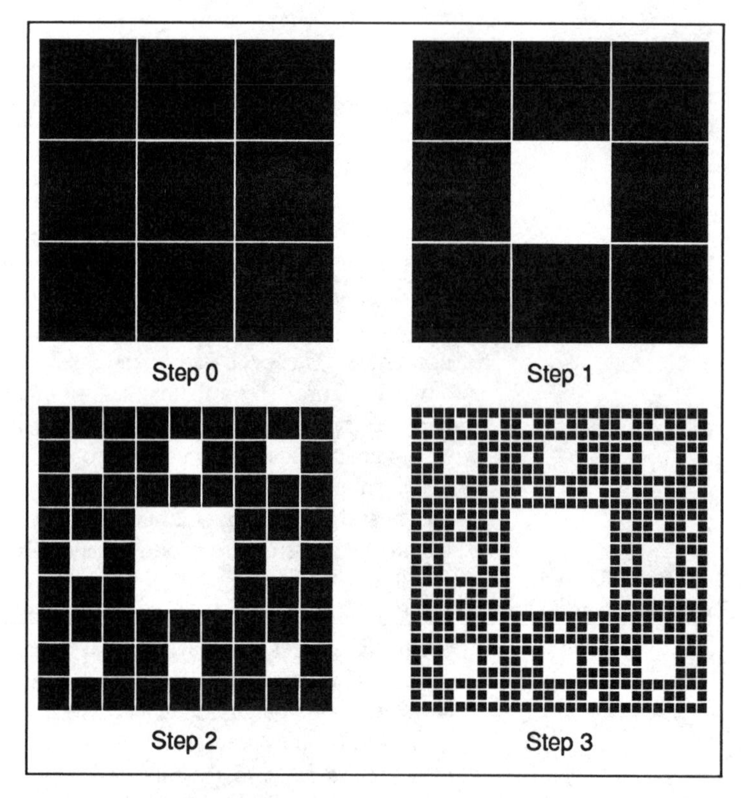

Figure 2.20 : The basic construction steps of the Sierpinski carpet.

should not confuse, however, our symbolic addresses with triadic numbers.

There are several ways to associate trees with symbolic addresses. A particular construction is based on the triangles which are taken away in the construction process. The nodes of the tree are the centers of these triangles. The branches of the tree grow generation by generation, as shown in figure 2.19. Observe that some of the branches touch when we go to the limit. For example, the branches corresponding to $LTTT...$ and $TLLL...$ touch in point A.

The Sierpinski Carpet

Sierpinski has added another object to the gallery of classical fractals, the Sierpinski carpet, which at first glance just looks like a variation of the known theme, see figure 2.20. We begin with a square in the plane. Subdivide into nine little congruent squares of which we drop the center one, and so on.The resulting object which remains if one carries out this process infinitely often can be seen as a generalization of the Cantor set. Indeed, if we look at the intersection of a line which is parallel to the base of the original square and which goes through the center we observe precisely the construction of the Cantor set.

We will see in section 2.7 that the complexities of the carpet and the gasket may at first look essentially the same, but there is in fact a whole world of a difference between them.

2.3 The Pascal Triangle

Blaise Pascal (1623–1662) was a great French mathematician and scientist. Only twenty years old, he built some ten mechanical machines for the addition of integers, a precursor of modern computers. What is known as the arithmetic triangle or *Pascal's triangle*, was not, however, his discovery. The first printed form of the arithmetic triangle in Europe dates back to 1527. A Chinese version of Pascal's triangle had already been published in 1303 (see figure 2.24). Pascal, however, used the arithmetic triangle to solve some problems related to chances in gambling, which he had discussed with Pierre de Fermat in 1654. This research later became the foundations of probability theory.

Blaise Pascal

Figure 2.21 : Blaise Pascal, 1623–1662.

The Arithmetic Triangle

The arithmetic triangle is a triangular array of numbers composed of the coefficients of the expansion of the polynomial $(x+1)^n$. Here n denotes the row starting from $n=0$. Row n has $n+1$ entries. For example, for $n=3$ the polynomial is

$$(x+1)^3 = x^3 + 3x^2 + 3x + 1 \ .$$

Thus, row number 3 reads $1, 3, 3, 1$ (see figure 2.22).

There are several ways to compute the coefficients. The first one inductively computes one row based on the entries of the previous row. Assume that the coefficients $a_0, ..., a_n$ in row n are given:

$$(x+1)^n = a_n x^n + \cdots + a_1 x + a_0 \ ,$$

and the coefficients $b_0, ..., b_{n+1}$ of the following row are required:

$$(x+1)^{n+1} = b_{n+1} x^{n+1} + \cdots + b_1 x + b_0 \ .$$

Pascal's Triangle

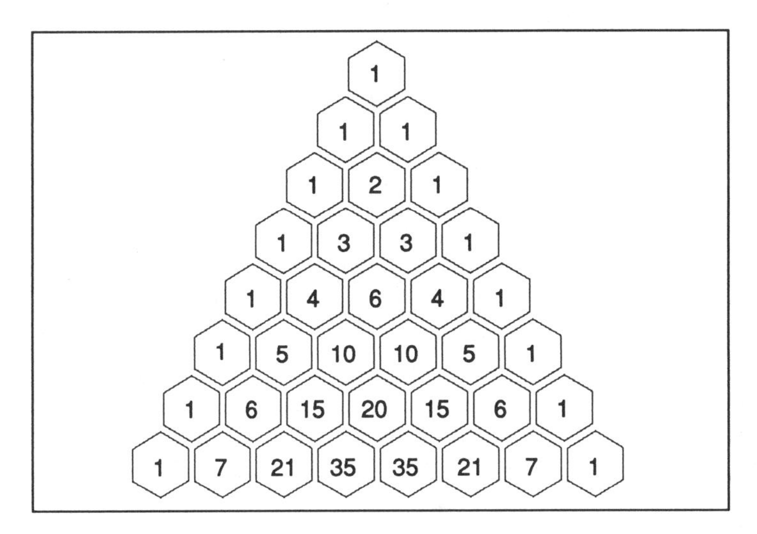

Figure 2.22 : The first eight rows of Pascal's triangle in a hexagonal web.

Pascal's Triangle, Eight Rows

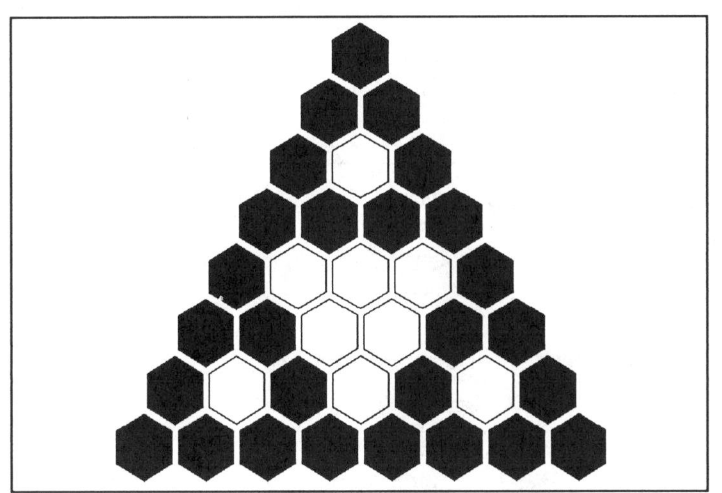

Figure 2.23 : Color coding of even (white) and odd (black) entries in the Pascal triangle with eight rows.

These are directly related to the known coefficients $a_0, ..., a_n$:

$$
\begin{aligned}
(x+1)^{n+1} &= (x+1)^n(x+1) \\
&= (a_n x^n + \cdots + a_1 x + a_0)(x+1) \\
&= a_n x^{n+1} + a_{n-1}x^n + \cdots + a_1 x^2 + a_0 x \\
&\quad + a_n x^n + \cdots + a_1 x + a_0 \\
&= a_n x^{n+1} + (a_n + a_{n-1})x^n + \cdots + (a_1 + a_0)x + a_0 \; .
\end{aligned}
$$

Chinese Arithmetic Triangle

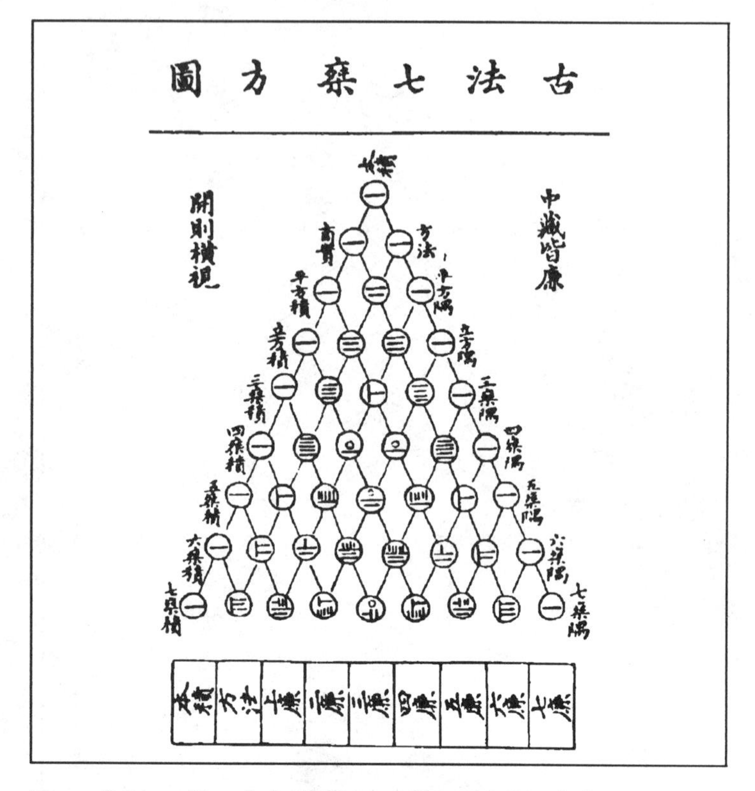

Figure 2.24 : Already in 1303 an arithmetic triangle has appeared in China at the front of Chu Shih-Chieh's *Ssu Yuan Yii Chien* which tabulates the binominal coefficients up to the eighth power.

Comparing coefficients we obtain the result

$$
\begin{aligned}
b_0 &= a_0, \\
b_k &= a_k + a_{k-1} \quad \text{for } k = 1, ..., n, \\
b_{n+1} &= a_n \,.
\end{aligned}
$$

The recipe to compute the coefficients of a row is thus very simple. The first and last numbers are copied from the line above. These will always be equal to 1. The other coefficients are just the sum of the two coefficients in the row above. In this scheme it is most convenient to write Pascal's triangle in the form with the top vertex centered on a line above it as shown in figure 2.22.

For computing small numbers of rows from Pascal's triangle the inductive method as outlined above is quite satisfactory. However, if the coefficients of some row with a large number n are desired, then a direct approach is preferable. It is based on the binomial

Pascal's Triangle in Japan

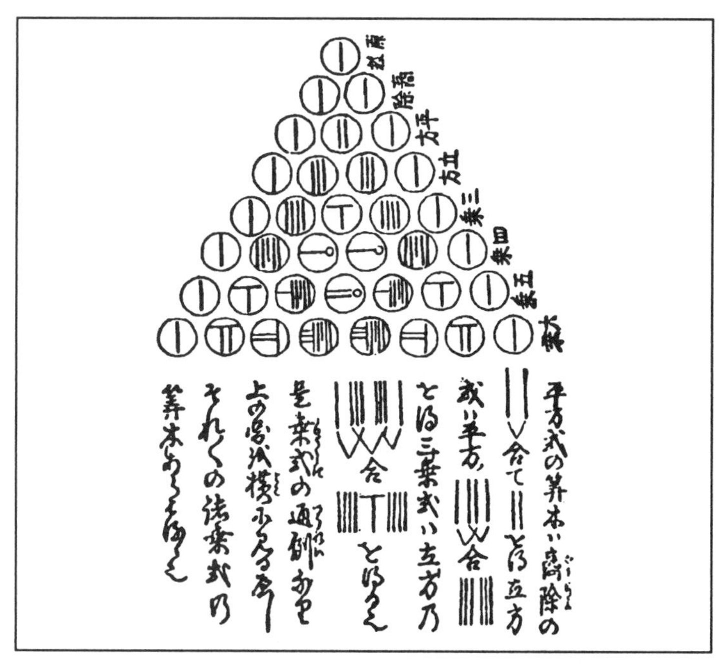

Figure 2.25 : Appeared 1781 in Murai Chūzen's *Sampō Dōshimon.*

theorem, which states

$$(x+y)^n = \sum_{k=0}^{n} \frac{n!}{k!(n-k)!} x^{n-k} y^k$$

where the notation $n!$ is 'factorial n' and defined as

$$n! = 1 \cdot 2 \cdots (n-1) \cdot n$$

for positive integers n and $0! = 1$. Setting $y = 1$ we immediately obtain the kth coefficient b_k (k runs from 0 to n) of row number n of Pascal's triangle via

$$b_k = \frac{n!}{k!(n-k)!} = \frac{n(n-1)\cdots(n-k+1)}{1 \cdot 2 \cdots k} .$$

For example, the coefficient for $k = 3$ in row $n = 7$ is

$$b_3 = \frac{7 \cdot 6 \cdot 5}{2 \cdot 3} = 35 ,$$

see the fourth entry in the last row in figure 2.22.

Another identity is easy to derive: the sum of all coefficients in row number n of Pascal's triangle is equal 2^n, which is seen by setting $x = y = 1$ in the binomial formula.

Figure 2.26 : Color coding of even and odd entries in the Pascal triangle with 16, 32, and 64 rows.

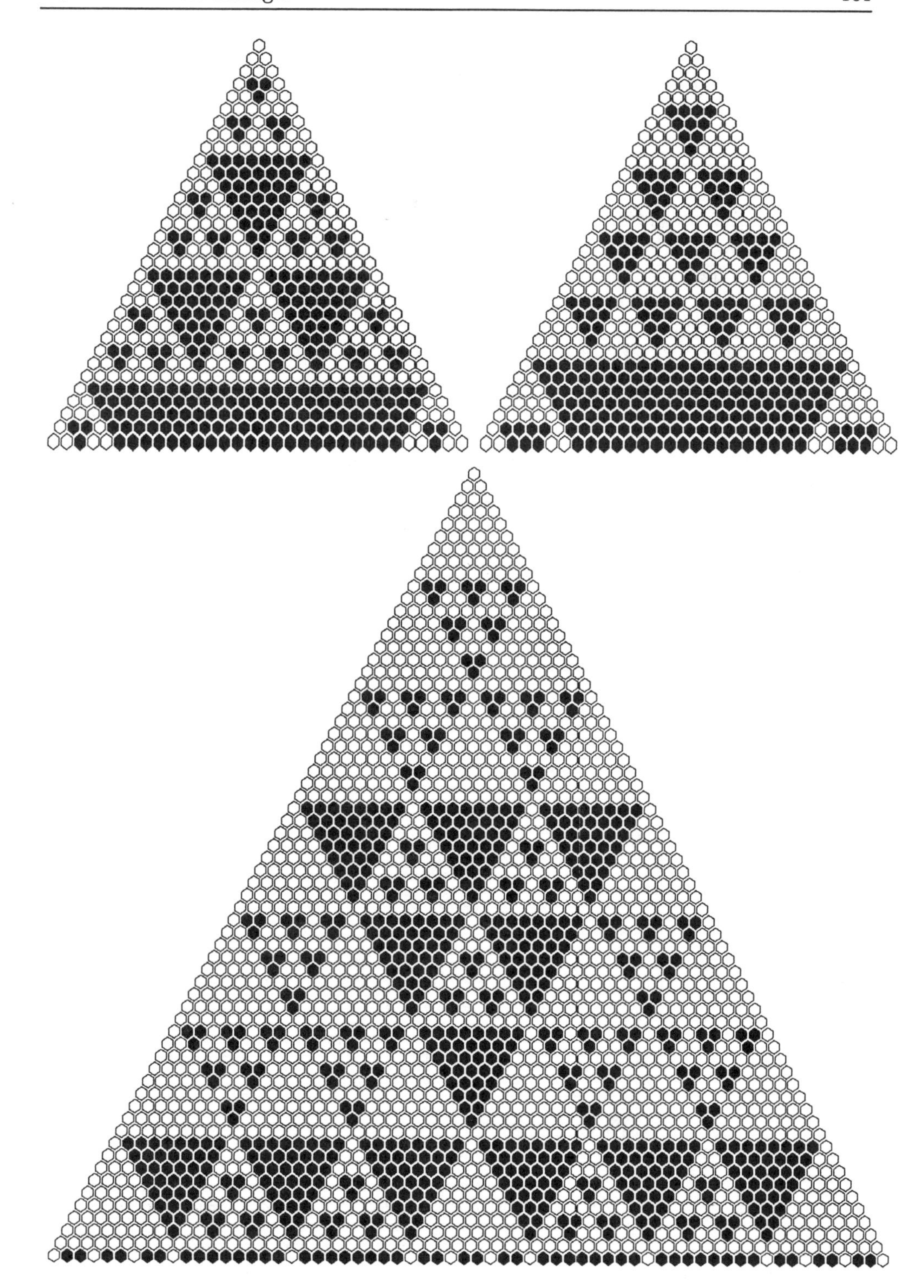

Figure 2.27 : Color coding the Pascal triangle. Black cells denote divisibility by 3 (top left), by 5 (top right) and by 9 (bottom).

To keep order we have put the first eight rows of a Pascal triangle into a hexagonal web (see figure 2.22). Now let us color code properties of the numbers in the triangle. For example, let us mark each hexagonal cell which is occupied by an odd number with black ink, i.e. the even ones are left white. Figure 2.23 shows the result.

It is worthwhile to repeat the experiment with more and more rows, see figure 2.26. The last image of that series already looks very similar to a Sierpinski gasket. Is it one? We have to be very careful about this question and will give a first answer in chapter 3. These number theoretic patterns are just one of an infinite variety of related ones. You will recall that even/odd just means divisible by 2 or not. Now, how do the patterns look when we color code divisibility by 3, 5, 7, 9, etc. by black cells and non-divisibility by white cells? Figure 2.27 gives a first impression.

Each of these patterns has beautiful regularities and self-similarities which describe elementary number theoretic properties of the Pascal triangle. Many of these properties have been observed and studied for several centuries. The book by B. Bondarenko,[10] lists some 406 papers by professional and amateur mathematicians over the last three hundred years.[11]

[10]B. Bondarenko, *Generalized Triangles and Pyramids of Pascal, Their Fractals, Graphs and Applications*, Tashkent, Fan, 1990, in Russian.

[11]In chapter 9 we will demonstrate how the fractal patterns and self-similarity features can be deciphered by the tools which are the theme of chapter 5.

2.4 The Koch Curve

Helge von Koch was a Swedish mathematician who, in 1904, introduced what is now called the *Koch curve*.[12] Fitting together three suitably rotated copies of the Koch curve produces a figure, which for obvious reasons is called the *snowflake curve*, see figures 2.29 and 2.30.

Koch's Original Construction

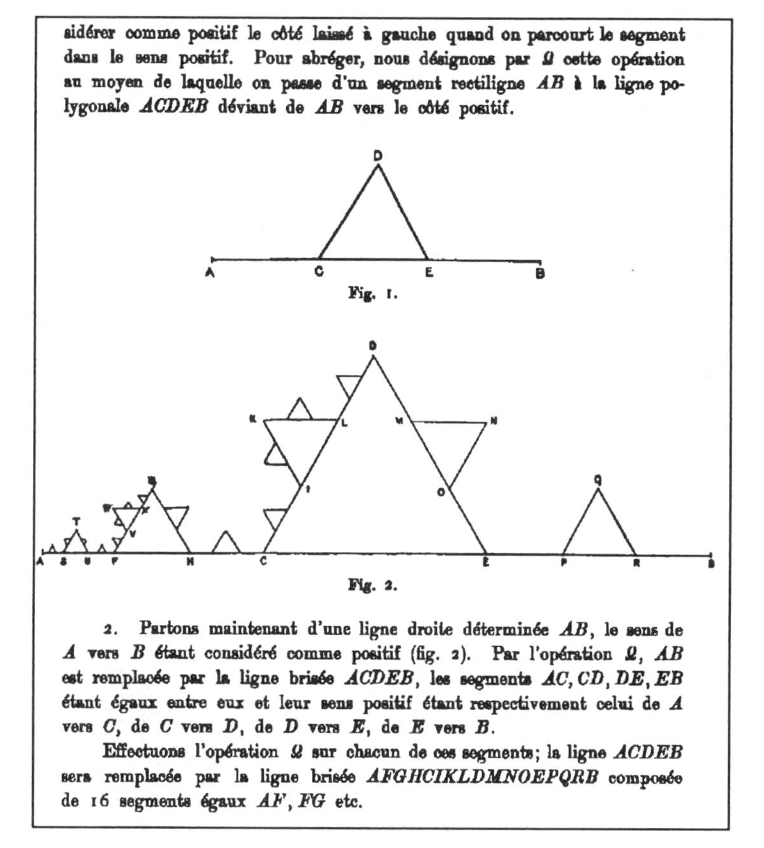
sidérer comme positif le côté laissé à gauche quand on parcourt le segment dans le sens positif. Pour abréger, nous désignons par Ω cette opération au moyen de laquelle on passe d'un segment rectiligne AB à la ligne polygonale $ACDEB$ déviant de AB vers le côté positif.

Fig. 1.

Fig. 2.

2. Partons maintenant d'une ligne droite déterminée AB, le sens de A vers B étant considéré comme positif (fig. 2). Par l'opération Ω, AB est remplacée par la ligne brisée $ACDEB$, les segments AC, CD, DE, EB étant égaux entre eux et leur sens positif étant respectivement celui de A vers C, de C vers D, de D vers E, de E vers B.

Effectuons l'opération Ω sur chacun de ces segments; la ligne $ACDEB$ sera remplacée par la ligne brisée $AFGHCIKLDMNOEPQRB$ composée de 16 segments égaux AF, FG etc.

Figure 2.28 : Excerpt from von Koch's original 1906 article.

Little is known about von Koch, and his mathematical contributions were certainly not in the same category as those of the stars like Cantor, Peano, Hilbert, Sierpinski or Hausdorff. But in this chapter on classical fractals, his construction must have its place simply because it leads to many interesting generalizations and must have inspired Mandelbrot immensely. The Koch curve is

[12] H. von Koch, *Sur une courbe continue sans tangente, obtenue par une construction géometrique élémentaire*, Arkiv för Matematik 1 (1904) 681–704. Another article is H. von Koch, *Une méthode géométrique élémentaire pour l'étude de certaines questions de la théorie des courbes planes*, Acta Mathematica 30 (1906) 145-174.

as difficult to understand as the Cantor set or the Sierpinski gasket. However, the problems with it are of a different nature. First of all — as the name already expresses — it is a curve, but this is not immediately clear from the construction. Secondly, this curve contains no straight lines or segments which are smooth in the sense that we could see them as a carefully bent line. Rather, this curve has much of the complexity which we would see in a natural coastline, folds within folds within folds, and so on.

The Koch Snowflake

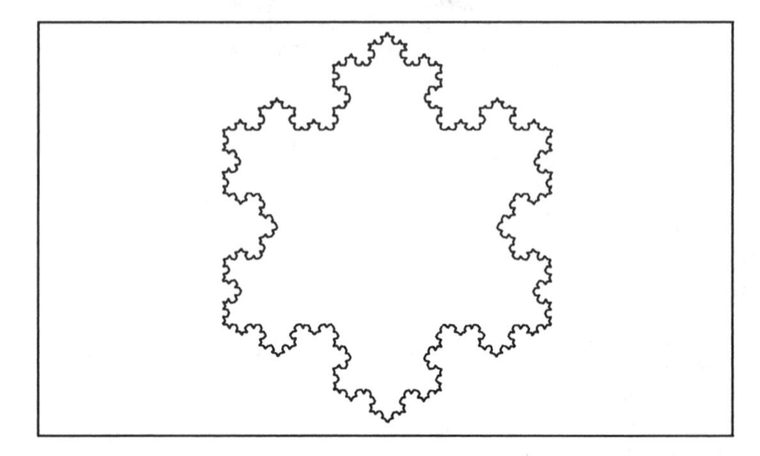

Figure 2.29 : The outline of the Koch snowflake is composed of three congruent parts, each of which is a Koch curve as shown in figures 2.31 and 2.33.

Some Natural Flakes

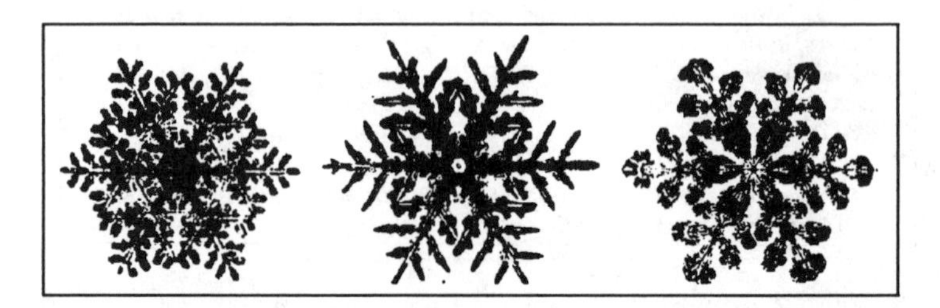

Figure 2.30 : The Koch snowflake obviously has some similarities with real flakes, some of which are pictured here.

Geometric Construction

Here is the simple geometric construction of the Koch curve. Begin with a straight line. This initial object is also called the *initiator*. Partition it into three equal parts. Then replace the middle third by an equilateral triangle and take away its base. This completes the basic construction step. A reduction of this figure, made of four parts, will be reused in the following stages. It is called the *generator*. Thus, we now repeat, taking each of the resulting line segments, partitioning them into three equal parts, and so on.

Figure 2.31 illustrates the first steps. Self-similarity is built into the construction process, i.e. each part of the 4 parts in the kth step is again a scaled down version — by a factor of 3 — of the entire curve in the previous $(k-1)^{\text{st}}$ step.

Koch Curve Construction

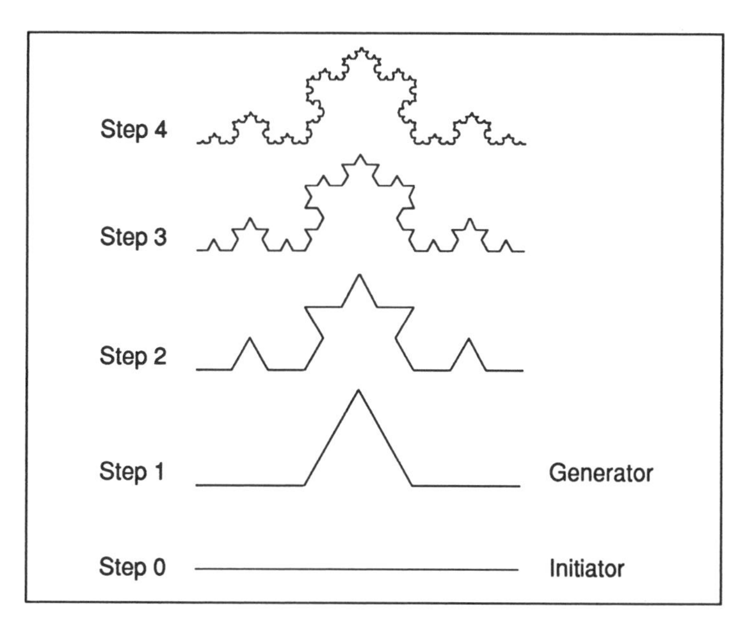

Figure 2.31 : The construction of the Koch curve proceeds in stages. In each stage the number of line segments increases by a factor of 4.

Actually Koch wanted to provide another example for a discovery first made by the German mathematician Karl Weierstraß, who in 1872 had precipitated a minor crisis in mathematics. He had described a curve that could not be differentiated, i.e. a curve which does not admit a tangent at any of its points. The ability to differentiate (i.e., to calculate the slope of a curve from point to point) is a central feature of calculus, that was invented independently by Newton and Leibniz some 200 years before Weierstraß. The idea of slope is a fairly intuitive one and goes hand in hand with the notion of a tangent, see figure 2.32.

If a curve has a corner then there is a problem. There is no way to fit a unique tangent. The Koch curve is an example of a curve which in a sense is made out of corners everywhere, i.e. there is no way to fit a tangent to any of its points.

Generalized Koch Constructions

It is almost obvious how one can generalize the construction to obtain a whole universe of self-similar structures. Such a Koch construction is defined by an initiator, which may be a collection of line segments, and a generator, which is a polygonal line, com-

Tangents of Curves

Figure 2.32 : At corners the tangent of a curve is not uniquely defined.

Comparing Koch Curve Construction Steps

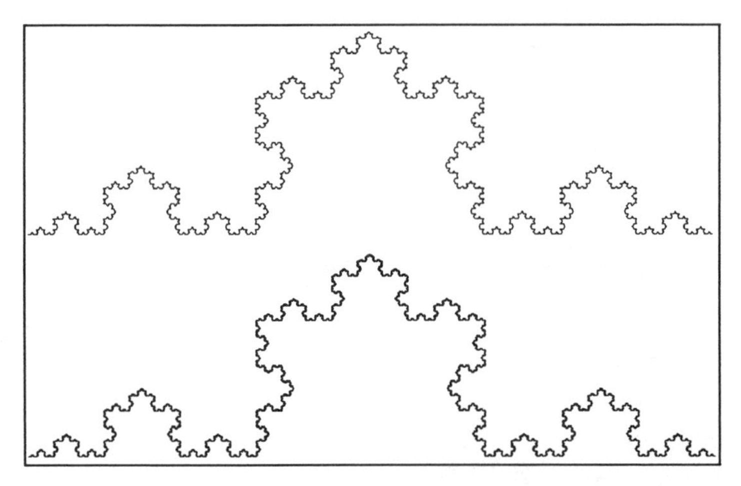

Figure 2.33 : Construction process of the Koch curve, step 5 (top) and step 20 (bottom).

posed of a number of connected line segments. Beginning with the initiator, one replaces each line segment by a properly scaled down copy of the generator curve. Here it is necessary to carefully match end points of the line segment and the generator. This procedure is repeated ad infinitum. In practice, of course, one stops as soon as the length of the longest line segment in the graph is below the resolution of the graphics device. Whether or not the Koch construction yields a converging sequence of images or even curves depends on the choice of initiator and generator. Figure 2.34 shows an example.

The Length of the Koch Curve

Let us return to the original Koch curve and discuss its length. In each stage we obtain a curve. After the first, we are left with a curve which is made up of four line segments of the same length, after the second step we have 4×4, and then $4 \times 4 \times 4$ line segments after the third step, and so on. If the original line had length L, then after the first step a line segment has length $L \times 1/3$, after the

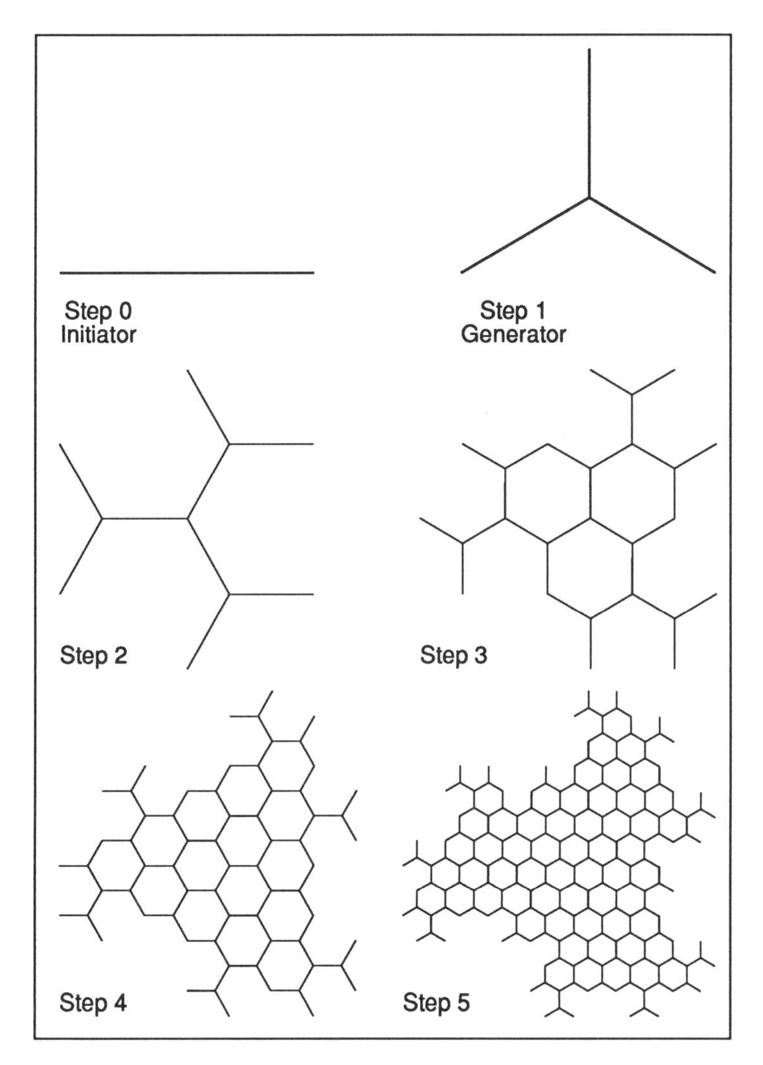

Figure 2.34 : A different choice of initiator and generator produces another fractal with self-similarities.

Another Koch Construction

second step we have $L \times 1/3^2$, then $L \times 1/3^3$, and so on. Since each of the steps produces a curve of line segments, there is no problem in measuring their respective lengths. After the first step it is $4 \times L \times 1/3$, then $4^2 \times L \times 1/3^2$, and so on. After the kth step, it is $L \times 4^k/3^k$. We observe that from step to step the length of the curves grows by a factor of 4/3.

Now there are several problems. First of all, the Koch curve is the object which one obtains if one repeats the construction steps infinitely often. But what does that mean? Next, even if we could answer this question, why is it a curve which comes out? Or, why is it that the curves in each step do not intersect themselves?

In figure 2.33 we see two curves which we can hardly distinguish. But they are different. The top one shows the result of the construction after 5 steps, while the other curve shows the result after 20 steps. In other words, since the length of the individual line segments is $1/3^k$, where k is the number of steps, it is clear that any of the changes in the construction are soon below visibility unless one works under a microscope. Thus, for practical purposes, one is tempted to be satisfied with a display of something like the 10th step, or whatever is appropriate to fool the eye. But such an object is not the Koch curve. It would have finite length and would still show its straight line construction segments under sufficient magnification. It is of crucial importance to distinguish between the objects which we obtain at any (single) step in the construction and the final object. We will pick up this difficulty, which of course is also present in the previous classical fractals, in the following chapters.

2.5 Space-Filling Curves

Talking about dimension in an intuitive way, we perceive lines to be typical for one-dimensional and planes as typical for two-dimensional objects. In 1890 Giuseppe Peano[13] (1858–1932) and immediately after that in 1891, David Hilbert[14] (1862–1943), discussed curves which live in a plane and which dramatically demonstrate that our naive idea about curves is very limited.[15] They discussed curves which 'fill' a plane, i.e. given some patch of the plane, there is a curve which meets every point in that patch. Figure 2.37 indicates the first steps of the iterative construction of Peano's original curve.

Space-Filling Structures in Nature

In Nature the organization of space-filling structures is one of the fundamental building blocks of living beings. An organism must be supplied with the necessary supporting substances such as water and oxygen. In many cases these substrates will be transported through vessel systems that must reach every point in the volume of the organism. For example, the kidney is an organ housing three interwoven tree-like vessel systems, the arterial, the venous, and the urinary systems (see the color plates). Each one of them has access to every part of the kidney. Fractals solve the problem of how to organize such a complicated structure in an efficient way. Of course, this was not what Peano and Hilbert were interested in almost 100 years ago. It is only now, after Mandelbrot's work, that the omnipresence of fractals becomes apparent.

Construction with Initiator and Generator

The Peano curve is obtained by another version of the Koch construction. We begin with a single line segment, the initiator, and then substitute the segment by the generator curve as shown in figure 2.37. Apparently the generator has two points of self-intersection. More precisely, the curve touches itself at two points. Observe that this generator curve fits nicely into a square, which is shown in dotted lines. It is this square whose points will be reached by the Peano curve.

Let us carefully describe the next step. Take each straight line piece of the curve in the first stage and replace it by the properly scaled down generator. Obviously, the scaling factor is 3. This constitutes stage 2. There are a total of 32 self-intersection points in the curve. Now we repeat, i.e. in each step, line segments are scaled down by a factor of 3. Thus, in the kth step, a line segment has length $1/3^k$, which is a very rapidly declining number. Since

[13] G. Peano, *Sur une courbe, qui remplit toute une aire plane*, Mathematische Annalen 36 (1890) 157–160.

[14] D. Hilbert, *Über die stetige Abbildung einer Linie auf ein Flächenstück*, Mathematische Annalen 38 (1891) 459–460.

[15] Hilbert introduced his example in Bremen, Germany during the annual meeting of the *Deutsche Gesellschaft für Naturforscher und Ärzte*, which was the meeting at which he and Cantor were instrumental in founding the *Deutsche Mathematiker Vereinigung*, the German mathematical society.

Hilbert's Paper — Page 1

Ueber die stetige Abbildung einer Linie auf ein Flächenstück.*)

Von

David Hilbert in Königsberg i. Pr.

Peano hat kürzlich in den Mathematischen Annalen**) durch eine arithmetische Betrachtung gezeigt, wie die Punkte einer Linie stetig auf die Punkte eines Flächenstückes abgebildet werden können. Die für eine solche Abbildung erforderlichen Functionen lassen sich in übersichtlicherer Weise herstellen, wenn man sich der folgenden geometrischen Anschauung bedient. Die abzubildende Linie — etwa eine Gerade von der Länge 1 — theilen wir zunächst in 4 gleiche Theile 1, 2, 3, 4 und das Flächenstück, welches wir in der Gestalt eines Quadrates von der Seitenlänge 1 annehmen, theilen wir durch zwei zu einander senkrechte Gerade in 4 gleiche Quadrate 1, 2, 3, 4 (Fig. 1). Zweitens theilen wir jede der Theilstrecken 1, 2, 3, 4 wiederum in 4 gleiche Theile, so dass wir auf der Geraden die 16 Theilstrecken 1, 2, 3, ..., 16 erhalten; gleichzeitig werde jedes der 4 Quadrate 1, 2, 3, 4 in 4 gleiche Quadrate getheilt und den so entstehenden 16 Quadraten

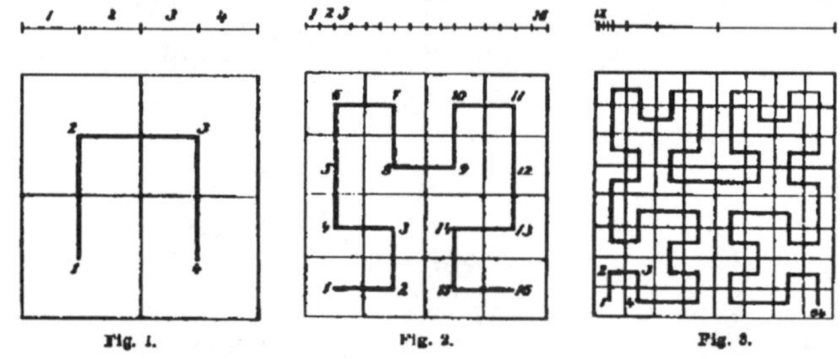

werden dann die Zahlen 1, 2 ... 16 eingeschrieben, wobei jedoch die Reihenfolge der Quadrate so zu wählen ist, dass jedes folgende Quadrat sich mit einer Seite an das vorhergehende anlehnt (Fig. 2). Denken wir uns dieses Verfahren fortgesetzt — Fig. 3 veranschaulicht den

*) Vergl. eine Mittheilung über denselben Gegenstand in den Verhandlungen der Gesellschaft deutscher Naturforscher und Aerzte. Bremen 1890.

**) Bd. 36, S. 157.

30*

Figure 2.35 : The first page of Hilbert's original 1890, 2-page paper with the first visualization of a fractal, his space-filling curve.

each line segment is replaced by 9 line segments of 1/3 the length of the previous line segments, we can easily calculate the length of the curves in each step. Assume that the length of the original line segments constituting the initiator was 1, then we obtain in stage 1: $9 \times 1/3 = 3$, and stage 2: $9 \times 9 \times 1/3^2 = 9$. Expressed as a general rule, in each step of the construction, the resulting curve increases in length by a factor of 3. In stage k, the length thus is 3^k.

Self-Similarity

The Peano curve construction , though as easy, or as difficult, as the construction of the Koch curve, bears within it several difficul-

Hilbert's Paper — Page 2

460 DAVID HILBERT. Stetige Abbildung einer Linie auf ein Flächenstück.

nächsten Schritt —, so ist leicht ersichtlich, wie man einem jeden gegebenen Punkte der Geraden einen einzigen bestimmten Punkt des Quadrates zuordnen kann. Man hat nur nöthig, diejenigen Theilstrecken der Geraden zu bestimmen, auf welche der gegebene Punkt fällt. Die mit den nämlichen Zahlen bezeichneten Quadrate liegen nothwendig in einander und schliessen in der Grenze einen bestimmten Punkt des Flächenstückes ein. Dies sei der dem gegebenen Punkte zugeordnete Punkt. *Die so gefundene Abbildung ist eindeutig und stetig und umgekehrt einem jeden Punkte des Quadrates entsprechen ein, zwei oder vier Punkte der Linie.* Es erscheint überdies bemerkenswerth, dass durch geeignete Abänderung der Theillinien in dem Quadrate sich leicht *eine eindeutige und stetige Abbildung finden lässt, deren Umkehrung eine nirgends mehr als dreideutige ist.*

Die oben gefundenen abbildenden Functionen sind zugleich einfache Beispiele für überall stetige und nirgends differentiirbare Functionen.

Die mechanische Bedeutung der erörterten Abbildung ist folgende: *Es kann sich ein Punkt stetig derart bewegen, dass er während einer endlichen Zeit sämmtliche Punkte eines Flächenstückes trifft.* Auch kann man — ebenfalls durch geeignete Abänderung der Theillinien im Quadrate — zugleich bewirken, *dass in unendlich vielen überall dichtvertheilten Punkten des Quadrates eine bestimmte Bewegungsrichtung sowohl nach vorwärts wie nach rückwärts existirt.*

Was die analytische Darstellung der abbildenden Functionen anbetrifft, so folgt aus ihrer Stetigkeit nach einem allgemeinen von K. Weierstrass bewiesenen Satze*) sofort, dass diese Functionen sich in unendliche nach ganzen rationalen Functionen fortschreitende Reihen entwickeln lassen, welche im ganzen Intervall absolut und gleichmässig convergiren.

Königsberg i. Pr., 4. März 1891.

*) Vergl. Sitzungsberichte der Akademie der Wissenschaften zu Berlin, 9. Juli 1885.

Figure 2.36 : The other page of Hilbert's paper.

ties which did not occur or were hidden in the latter (construction). For example, take the intuitive concept of self-similarity. For the construction of the Koch curve, it seemed that we could say that the final curve (i.e. the curve which you see on a graphics terminal after many steps) has similarity with each of the preceding steps. If you look at the Peano curve in the same intuitive way, each of the steps has similarity with the preceding steps; but if you look at the final curve (i.e. the curve which results after many steps on a graphics terminal), essentially we see a 'filled out' square which doesn't look at all similar to the early steps of the construction. In other words, either the Peano curve is not self-similar, or we have to be much more careful in describing what self-similarity means.

Peano Curve Construction

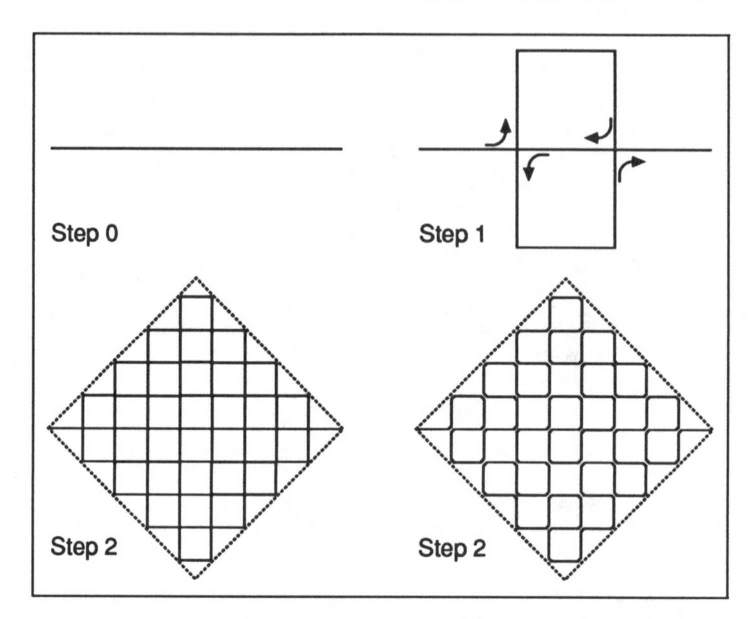

Figure 2.37 : Construction of a plane filling curve with initiator and generator. In each step one line segment is replaced by 9 line segments scaled down by a factor of 3. For reasons of clarity the corners in these polygonal lines, where the curve may intersect itself, have been slightly rounded.

In fact, we will see in chapter 4 that the Peano curve is perfectly self-similar. The problem is to 'see' the final object as a curve because, in any graphical representation, it 'looks' much more like a piece of the plane.

Parametrization of a Square by the Peano Curve

Let us explore the space-filling property a bit further. When you look at the displayed stages of the development of the curve, you notice that approximately the first $1/9^{\text{th}}$ of the curve stays within the left subsquare, and in fact, seems to fill just that area. Corresponding statements hold for the other subsquares. You will also notice that each subsquare can also be tiled into 9 sub-subsquares, each one reduced by 1/9 when compared to the whole one. The curve first traces out all tiles of a subsquare before it enters the next subsquare. This process goes on and on through all stages of the curves.

The implication of this is as follows: if we trace out a stage of the Peano curve up to a certain percentage, let us say to 10/27 of its total length, i.e. about 37%, then we end up at a certain point in the square. Now we go to the next stage and again trace out 37% of the new, longer curve (see figure 2.38). Again we end up at a point in the square, and this point is not far from the first one. When repeating this procedure for the following stages, we obtain a sequence of points. These points will converge to a unique point in

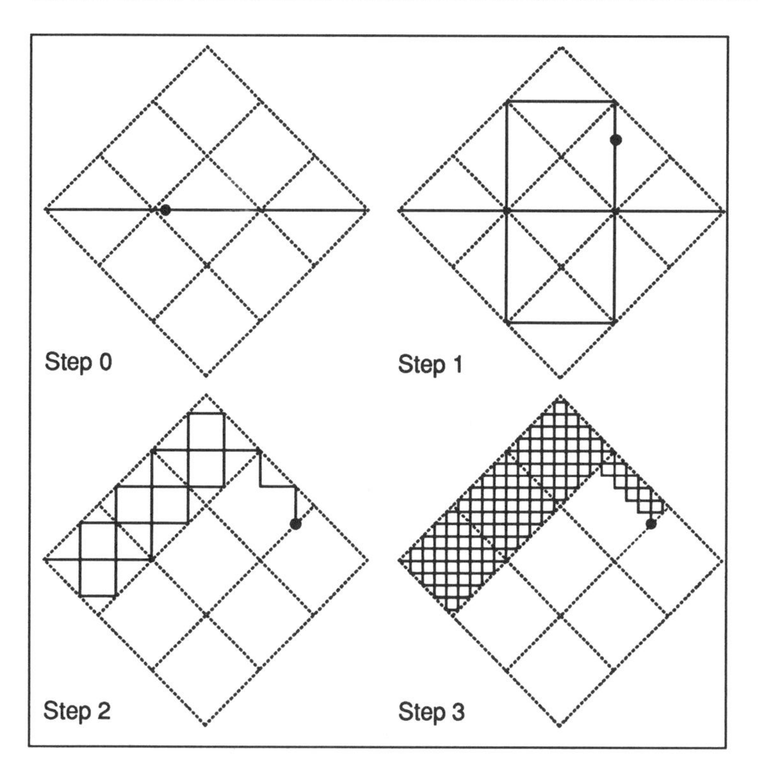

Figure 2.38 : The Peano curves of four different stages are traced out to $1/3+1/27 = 10/27$ of the total length. The rest of the curve is not shown in the bottom figures. The parameter $10/27$ identifies a point marked in each graph. These points converge to a unique point in the square as the number of steps increases.

the square. This point may be called the point with address 10/27. In this manner, we can define for all percentages — for all numbers between 0.0 and 1.0 — a point in the square. These points will form a curve that passes through every point in the square! Using mathematical terms, we say that 'the square can be parametrized by the unit interval'. Thus, a curve, which by nature is something one-dimensional, can fill something two-dimensional. It seems that the use of the intuitive notion of dimension here is rather slippery.

To make the argument precise, one would have to introduce an addressing system, which for the case of the Peano curve would be based on strings composed of 9 symbols, or digits. For each point in the square there is an address, which is an infinite string. This string also identifies points from each stage of the Peano curve construction. That sequence of points (one from each stage) will converge to the original point in the square.

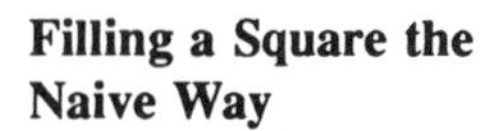

Filling a Square the Naive Way

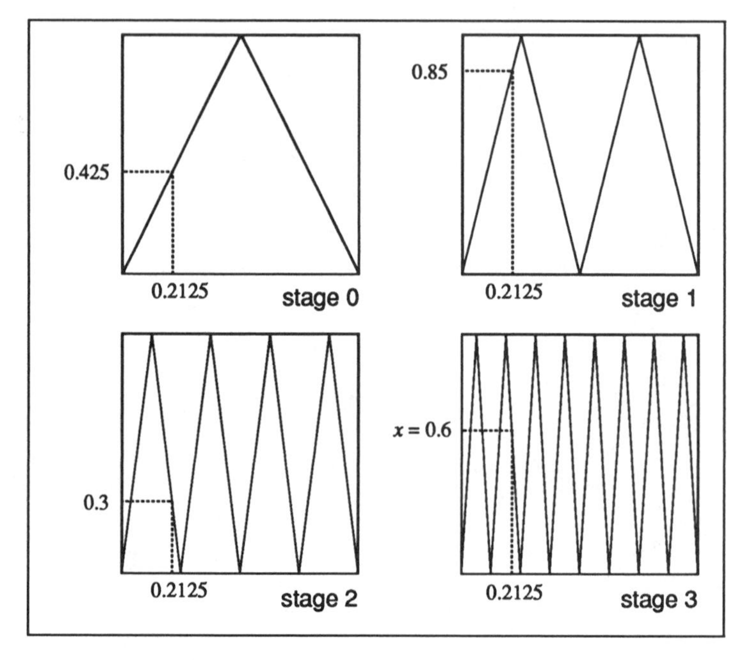

Figure 2.39 : The first four stages in an attempt to construct another space-filling curve using zigzag curves.

Is There a Better Way to Fill a Square by a Curve?

The space-filling Peano curve, or rather any finite stage of the construction, certainly is very awkward to draw by hand or even by a plotter under computer control. The number of small line segments that must be drawn to fill the square is just enormous. Moreover, there is a 90 degree turn after every segment. Therefore it is fair to ask whether there is a much simpler way to fill a square with a line. Think of how you would approach that problem with a pencil in your hand. It seems the easiest way would be to just draw a zigzagging line from one side of the square to the other, making sure that the turns are narrow enough in order to avoid any white specs on the paper.

The Naive Construction ...

Let us put this procedure into a more mathematical description analogous to the stages in the Peano curve construction. Stage 0 is a simple line from the lower left corner to the middle of the top side of the square and back to the base line ending at the lower right corner. For the next stage let us double the resolution in the sense that a horizontal line somewhere in the middle of the square will intersect the next curve at twice as many points. This is easy to achieve just by doing two zigzags at half the distance, see figure 2.39.

It is obvious how to continue the construction. For each stage we just double the number of zigzags. For any given resolution

$\varepsilon > 0$ we can certainly find a stage at which the generated curve passes by every point in the square with a distance less than ε, and we would say that the job is done. Moreover, we could even claim to have invented a space-filling curve, which is in some sense self-similar, because in each stage the curve is composed of two copies of the curve of the previous stage, properly scaled in the horizontal direction.[16] Any child will accomplish something like this at an early age. Certainly, the brilliant minds of Peano and Hilbert must have been aware of this. Then, what was it that drove them to invent such complicated constructions, which then were even accepted for publication in most renowned mathematical journals?

... Does Not Lead Anywhere

The answer seems contra-intuitive, but logical after a little bit of analysis. What comes out of the Peano curve construction in the limit is a curve as pointed out in the technical section starting on page 112. This curve has infinite length, self-similarities and it reaches every point in the square. By contrast, the naive construction from above will *not* lead to a curve, although every stage of it is a curve! Let us explore this astonishing fact. We label the horizontal and vertical axes of the square by x and y, which both range from 0 to 1. A curve from some stage, say the nth stage, of the construction then is given by the graph of a zigzag function, which we call y_n. Let us now fix some coordinate x between 0 and 1 and look at the corresponding values $y_n(x)$ as the stage n increases. If the construction really leads to a well defined limit curve, then one must expect that the sequence of points $y_1(x), y_2(x), \ldots$ also converges, namely to the y-value of the limit curve at position x. Clearly this is true for all points x which allow a finite binary representation such as 1/4 or 139/256 because by construction at such points all curves will have a y-value of 0, provided the stage is large enough. But then there are other points which clearly violate the crucial convergence property. For example, at $x = 1/7$ the y-values of the curves are 2/7, 4/7, 6/7, 2/7, 4/7, 6/7, ... and so on in a periodic fashion. Therefore there is no limit object, no space-filling curve, and no new insight. This naive way of filling the square is essentially the same as just filling a finite array of pixels in an image by assigning 'black' to every pixel. After a certain number of steps we are done, and continuing for higher resolutions would not make sense. There is no self-similarity and certainly not a fractal behind the picture. So this is the real ingenuity of Peano and Hilbert — they created a 'monster' with unforeseen properties which were never thought possible before.

[16] For cases like this one, where the scaling factor is different in different directions, the term *self-affine* is more appropriate. Affine transformations are discussed further in chapters 5 and 6.

Analysis of the Naive Approach to Space-Filling

It is not hard to analyse the sequence of curves from the naive space-filling construction. For this purpose let us introduce the periodically extended tent transformation

$$h(x) = \begin{cases} 2\text{frac}(x) & \text{if } \text{frac}(x) \leq 0.5 \\ 2(1 - \text{frac}(x)) & \text{if } \text{frac}(x) > 0.5 \end{cases}$$

where frac(x) denotes the fractional part of x, i.e.

$$\text{frac}(x) = x - \max\{k \mid k \leq x, k \text{ integer}\} .$$

With this notation we can write the curves in the construction simply as

$$\begin{aligned} y_0(x) &= h(x), \\ y_1(x) &= h(2x), \\ y_2(x) &= h(4x), \\ &\vdots \\ y_n(x) &= h(2^n x), \\ &\vdots \end{aligned}$$

where in all cases x ranges from 0 to 1. We see that only the fractional parts of $x, 2x, 4x, ..., 2^n x, ...$ determine the y-values of the curves at position x. Considering the example $x = 1/7$ from the text, we now see that

$$\begin{aligned} \text{frac}(1/7) &= 1/7 \\ \text{frac}(2/7) &= 2/7 \\ \text{frac}(4/7) &= 4/7 \\ \text{frac}(8/7) &= 1/7 \end{aligned}$$

Thus, the fractional part of $16/7$ again is $2/7$ and so on in a periodic fashion. Applying the tent transformation to these fractional parts $1/7, 2/7, 4/7, 1/7, ...$ finally yields the sequence of values $2/7, 4/7, 6/7, 2/7, 4/7, 6/7, ...$ as claimed in the text. Therefore the limit

$$\lim_{n \to \infty} y_n(1/7)$$

does not exist, and the sequence of curves $y_0, y_1, ...$ cannot have a limit curve.

To conclude we ask whether our choice of $x = 1/7$ for the convergence test is a rather artificial and rare case. The answer here is no. In fact, it is true for almost all positions x that the sequence $y_0(x), y_1(x), ...$ has no limit. Let us briefly elaborate. The fractional parts of $2^n x$ are most easily found, when the position x is given in a binary representation. The example from the text, $x = 1/7$, has a binary extension $0.001001...$:

$$x = \frac{1}{8} + \frac{1}{64} + \frac{1}{512} + \cdots = \frac{\frac{1}{8}}{1 - \frac{1}{8}} = \frac{1}{7} .$$

Multiplying a binary number by 2 is equivalent to just shifting all binary digits one position to the left. Taking the fractional part of the result means deleting any leading digits before the 'decimal' point. For example the repeated binary shifts of $1/7$ are 0.010010..., 0.100100..., and 0.001001..., which is equal to $1/7$ again. The complete operation is also known as the *binary shift*, and we remark in passing that it is central to the analysis of chaos in chapter 10. Applying the binary shift repeatedly to a number is thus the same as placing that number at the corresponding position on an infinitely precise ruler, and looking at it through a microscope, increasing the magnifying power continuously by a factor of two. If we take a random number between 0 and 1, with random binary digits, then the binary shift will produce a sequence of random numbers, which certainly will never settle down to a limiting value. Because 'most' numbers have random digits, the lack of convergence with regard to our naive curve construction is typical and not the exception.

An Application of Space-Filling Curves

It may seem that space-filling curves are mostly an academic curiosity — regarded as 'monsters' initially. However, they became important roots in Mandelbrot's development of fractals as models of nature. Moreover — and this may come as a real surprise — those early monsters are also good for down-to-earth technical applications 100 years after their discovery. Let us briefly describe an image processing application published at the prestigeous SIGGRAPH[17] convention in 1991. It introduces a novel digital halftoning technique useful to render a grey scale image on a bilevel graphic device such as a laser printer.[18] The problem lies in the fact that a printer renders a bitmap, an array of black and white pixels, while shades of grey cannot be represented at the pixel level. To cope with this difficulty, several so called dithering techniques have been used. They are based on scanning a given grey scale image line by line or in small square blocks. A black and white approximation of the image is produced with the objective to minimize an overall error. Usually, there are artefacts in the result which make it obvious that a dithering process was involved. How can space-filling curves help? Imagine a Hilbert curve that passes through all pixels of the given grey scale image. The curve offers an alternative to scanning the image line by line, namely to sample the image pixel by pixel along the Hilbert curve. Now a sequence of consecutive pixels along this convoluted path can be replaced by a black and white approximation. The advantage of the image scan along the Hilbert curve is that is is free of any directional features

[17] SIGGRAPH is the Special Interest Group Graphics of the Association for Computing Machinery (ACM). Their yearly conventions draw about 30,000 professionals from the field of computer graphics.

[18] Luiz Velho, Jonas de Miranda Gomes, *Digital Halftoning with Space-Filling Curves*, Computer Graphics 25,4 (1991) 81–90.

Two Dithering Methods

Figure 2.40 : Dithering with the Hilbert curve (left) versus traditional dithering. The methods are applied to a square which is continuously shaded from white (lower left corner) to black (upper right corner).

present in the traditional methods. It produces aperiodic patterns of clustered dots which are perceptually pleasant with characteristics similar to photographic grain structures. Figure 2.40 compares the traditional approach, called clustered-dot ordered dither, with the new method. Besides this dithering algorithm there are other earlier applications of space filling curves in image processing.[19]

Hilbert Curve Dithering Algorithm

Let us describe the details of a simplified version of the dithering algorithm with the Hilbert curve. We consider a square image with continuously varying grey shades which must be approximated by an image which may contain only black and white pixels. The resolution of the output image must be a power of 2. For example, we consider images with 4, 8, 16, and 32 pixels per row and per column in figure 2.41. As shown for the first few of these cases, we can fit a corresponding Hilbert curve exactly to such a tiling of the image. This introduces an ordering of the pixels. For the 4 by 4 pixel example, where we label columns by letters A, B, C, and D, and rows by 1, 2, 3, and 4, the pixels of the image are ordered as follows

$$A1, B1, B2, A2, ..., D2, C2, C1, D1 \ .$$

Let us denote by

$$I_1, I_2, ..., I_n$$

the intensity values of the corresponding pixels of the shaded input image (ranging from 0 for black to 1 for white). Here n is a power of

[19] R. J. Stevens, A. F. Lehar, F. H. Preston, *Manipulation and Presentation of Multidimensional Image Data Using the Peano Scan,* IEEE Transactions on Pattern Analysis and Machine Intelligence 5 (1983) 520–526.

Dithering with the Hilbert Curve

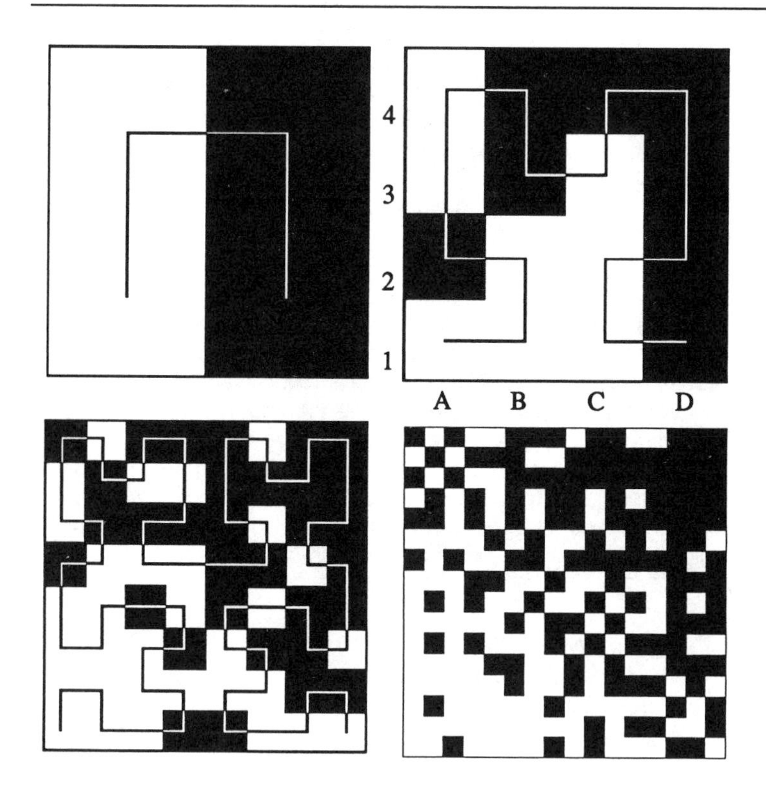

Figure 2.41 : The principle of the dithering algorithm based on four successive stages of the Hilbert scan of an image. The same shaded square is used as in figure 2.40.

2, the total number of pixels in the image. For the definition of the output image we have to compute corresponding intensity values

$$O_1, O_2, ..., O_n \in \{0, 1\} .$$

To begin we set

$$O_1 = \begin{cases} 0 \text{ if } I_1 \leq 0.5 \\ 1 \text{ if } I_1 > 0.5 \end{cases} .$$

This approximation carries an error

$$E_1 = I_1 - O_1 .$$

Instead of ignoring this error we can pass it along to the next pixel in the sequence. More precisely, we set

$$O_k = \begin{cases} 0 \text{ if } I_k + E_{k-1} \leq 0.5 \\ 1 \text{ if } I_k + E_{k-1} > 0.5 \end{cases}$$

$$E_k = I_k - O_k .$$

In other words, the error diffuses along the sequence of pixels. The goal of this error diffusion is to minimize the overall error in the intensities averaged over blocks of various sizes of the image. For example, we have that the errors, summed up over the complete image, are equal to

$$\sum_{k=1}^{n} E_k = E_n$$

which is expected to be relatively small. The crucial point of the algorithm is that the error diffuses along the Hilbert curve which traces out the image in a way that is conceived as very irregular to our sensory system. If we replace the Hilbert curve for example by a curve which scans the image row by row, the regular error diffusion will produce disturbing artifacts. The algorithm proposed by Stevens, Lehar, and Preston at SIGGRAPH is a generalization of this method. It considers blocks of consecutive pixels from the Hilbert scan at a time, instead of individual pixels.[20]

Conclusion

In conclusion, we have shown that the notion of self-similarity in a strict sense requires a discussion of the object which finally results from the constructions of the underlying feedback systems; and it can be dangerous to use the notion without care. One must carefully distinguish between a finite construction stage and the fractal itself. But if that is so, then how can we discuss the forms and patterns which we see in nature, as for example the cauliflower, from that point of view?

The cauliflower shows the same forms — clusters are composed of smaller clusters of essentially the same form — over a range of several, say five or six, magnification scales. This suggests that the cauliflower should be discussed in the framework of fractal geometry very much like our planets are suitably discussed for many purposes as perfect spheres within the framework of Euclidean geometry. But a planet is not a perfect sphere and the cauliflower is not perfectly self-similar. First, there are imperfections in self-similarity: a little cluster is not an exact scaled down version of a larger cluster. But more importantly, the range of magnification within which we see similar forms is finite. Therefore, fractals can only be used as models for natural shapes, and one must always be aware of the limitations.

[20] The simplified version presented here was first published in I. H. Witten and M. Neal, *Using Peano curves for bilevel display of continuous tone images*, IEEE Computer Graphics and Applications, May 1982, 47–52.

2.6 Fractals and the Problem of Dimension

The invention of space filling curves was a major event in the development of the concept of dimension. They questioned the intuitive perception of curves as one-dimensional objects, because they filled the plane (i.e., an object which is intuitively perceived as two-dimensional). This contradiction was part of a discussion which lasted several decades at the begining of this century. Talking about fractals we usally think of fractal dimension, Hausdorff dimension or boxcounting dimension (we will discuss this in detail in chapter 4), but the original concepts reside in the early development of topology.

A World Behaving Like Rubber

Topology is a branch of mathematics which has essentially been developed in this century. It deals with questions of form and shape from a qualitative point of view. Two of its basic notions are 'dimension' and 'homeomorphism'. Topology deals with the way shapes can be pulled and distorted in a space that behaves like rubber.

Circle, Square and Koch Island

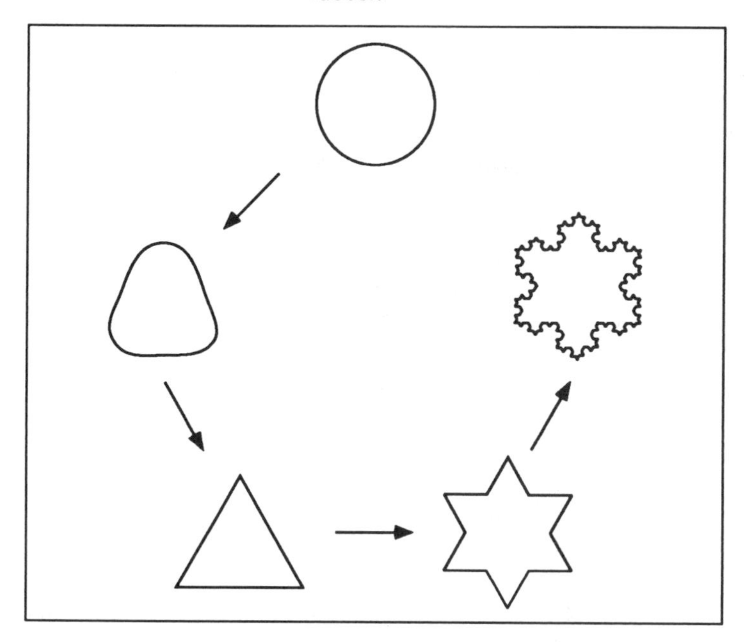

Figure 2.42 : A circle can be continuously deformed into a triangle. A triangle can be deformed into a Koch Island. Topologically they are all equivalent.

In topology straight lines can be bent into curves, circles pinched into triangles or pulled out as squares. For example, from the point of view of topology a straight line and the Koch curve cannot be distinguished. Or the coast of a Koch island is the same as a circle.

Or a plane sheet of paper is equivalent to one which is infinitely crumpled. However, not everything is topologically changeable. Intersections of lines, for example remain intersections. In mathematicians' language an intersection is invariant; it cannot be destroyed nor can new ones be born, no matter how much the lines are stretched and twisted. The number of holes in an object is also topologically invariant, meaning that a sphere may be transformed into the surface of a horseshoe, but never into a doughnut. The transformations which are allowed are called homeomorphisms,[21] and when applied, they must not change the invariant properties of the objects. Thus, a sphere and the surface of a cube are homeomorphic, but the sphere and a doughnut are not.

Topological Equivalence

We have mentioned already that a straight line and the Koch curve are topologically the same. Moreover, a straight line is the prototype of an object which is of dimension one. Thus, if the concept of dimension is a topological notion, we would expect that the Koch curve also has topological dimension one. This is, however, a rather delicate matter and troubled mathematicians around the turn of the century.

The history of the various notions of dimension involves the greatest mathematicians of that time: men like H. Poincaré, H. Lebesgue, L. E. J. Brouwer, G. Cantor, K. Menger, W. Hurewicz, P. Alexandroff, L. Pontrjagin, G. Peano, P. Urysohn, E. Čech, and D. Hilbert. That history is very closely related to the creation of the early fractals. Hausdorff remarked that the problem of creating the right notion of dimension is very complicated. People had an intuitive idea about dimension: the dimension of an object, say X, is the number of independent parameters (coordinates), which are required for the unique description of its points.

Poincaré's idea was inductive in nature and started with a point. A point has dimension 0. Then a line has dimension 1, because it can be split into two parts by a point (which has dimension 0). And a square has dimension 2, because it can be split into two parts by a line (which has dimension 1). A cube has dimension 3, because it can be split into two parts by a square (which has dimension 2).

Topological Dimension

In the development of topology, mathematicians looked for qualitative features which would not change when the objects were transformed properly (technically by a homeomorphism). The (topological) dimension of an object certainly should be preserved. But it turned out that there were severe difficulties in arriving at a proper and detailed notion of dimension which would behave that way. For example, in 1878 Cantor found a transformation f

[21] Two objects X and Y (topological spaces) are homeomorphic if there is a homeomorphism $h : X \to Y$ (i.e., a continuous one-to-one and onto mapping that has a continuous inverse h^{-1}).

from the unit interval $[0,1]$ to the unit square $[0,1]\times[0,1]$ which was one-to-one and onto.[22] Thus it seemed that we need only one parameter for the description of the points in a square. But Cantor's transformation is not a homeomorphism. It is not continuous, i.e. it does not yield a space-filling *curve*!

But then the plane-filling constructions of Peano and later Hilbert yielded transformations g from the unit interval $[0,1]$ to the unit square $[0,1]\times[0,1]$ which were even continuous. But they were not one-to-one (i.e., there are points, say x_1 and x_2, $x_1 \neq x_2$, in the unit interval which are mapped to the same point of the square $y = g(x_1) = g(x_2)$).

This raised the question — which so far seemed to have an obvious answer — whether or not there is a one-to-one and onto transformation between $I = [0,1]$ and $I^2 = [0,1]\times[0,1]$ which is continuous in both directions? Or more generally, is the n-dimensional unit cube $I^n = [0,1]^n$ homeomorphic to the m-dimensional one, $I^m = [0,1]^m, n \neq m$? If there were such a transformation mathematicians felt that they were in trouble: a one-dimensional object would be homeomorphic to a two-dimensional one. Thus, the idea of topological invariance would be wrong.

Line and Square are not Equivalent

Between 1890 and 1910 several 'proofs' appeared showing that I^n and I^m are not homeomorphic when $n \neq m$, but the arguments were not complete. It was the Dutch mathematician Brouwer who ended that crisis in 1911 by an elegant proof, which enriched the development of topology enormously. But the struggle for a suitable notion of dimension and a proof that obvious objects — like I^n — had obvious dimensions went on for two more decades. The work of the German mathematician Hausdorff (which led eventually to the fractal dimension) also falls in this time span.

During this century mathematicians came up with many different notions of dimension (small inductive dimension, large inductive dimension, covering dimension, homological dimension).[23] Several of them are topological in nature; their value is always a natural number (or 0 for points) and does not change for topologically equivalent objects. As an example of these notions we will discuss the covering dimension. Other notions of dimension capture properties which are not at all topologically invariant. The most prominent one is the Hausdorff dimension. The Hausdorff dimension for the straight line is 1, while for the Koch curve it is log4/log3. In other words, the Hausdorff dimension has changed, though from a topological point of view the Koch curve is just a straight line. Moreover, $\log 4/\log 3 = 1.2619...$ is not an integer.

---

[22] The notion 'onto' means here that for every point z of the unit square there is exactly one point x in the unit interval that is mapped to $z = f(x)$.

[23] C. Kuratowski, *Topologie II*, PWN, 1961. R. Engelking, *Dimension Theory*, North Holland, 1978.

Rather it is a fraction, which is typical for fractal objects. Other examples which are of (covering) dimension 1 are the coast of the Koch island, the Sierpinski gasket and also the Sierpinski carpet. Even the Menger sponge, whose basic construction steps are indicated in figure 2.43 is of (covering) dimension 1. Roughly, the gasket, carpet and sponge have (covering) dimension 1 because they contain line elements but no area, or volume elements. The Cantor set is of dimension 0 because it consists of disconnected points and it does not contain any line segment.

Construction of the Menger Sponge

Figure 2.43 : An object which is closely related to the Sierpinski carpet is the Menger sponge, after Karl Menger (1926). Take a cube, subdivide its faces into nine congruent squares and drill holes as shown from each central square to the opposite central square (the cross-section of the hole must be a square). Then subdivide the remaining eight little squares on each face into nine smaller squares and drill holes again from each of the central little squares to their opposite ones, and so on.

The Cover Dimension

Let us now discuss the topologically invariant cover dimension. The idea behind its concept — which is attributable to Lebesgue — is the following observation: let us take a curve in the plane, see figure 2.44, and try to cover it with disks of a small radius. The arrangement of disks on the left part of the curve is very different from the one in the middle, which in turn is very different from the one on the right part of the curve. What is the difference? In the right part we can only find pairs of disks which have non-empty intersection, while in the center part we can find triplets and in the left part even a quadruplet of disks which have non-empty intersection.

Covering a Curve

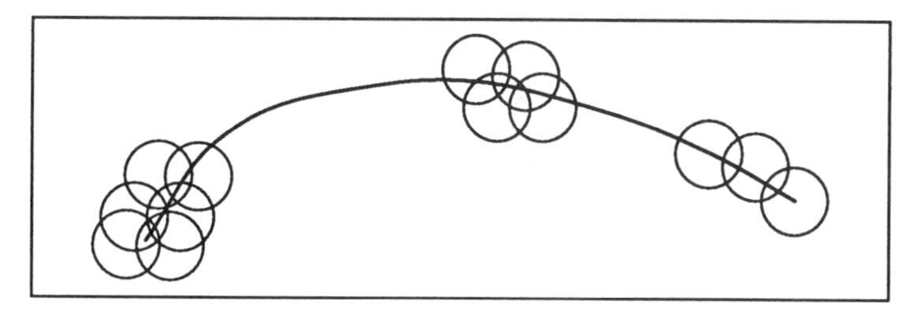

Figure 2.44 : Three different coverings of a curve by circles.

This is the crucial observation, which leads to a definition. We say that the curve has covering dimension 1 because we can arrange coverings of the curve with disks of small radius so that there are no triplets and quadruplets, but only pairs of disks with non-empty intersection, and moreover, there is no way to cover the curve with sufficiently small disks so that there are no pairs with non-empty intersection.

Covering a Plane

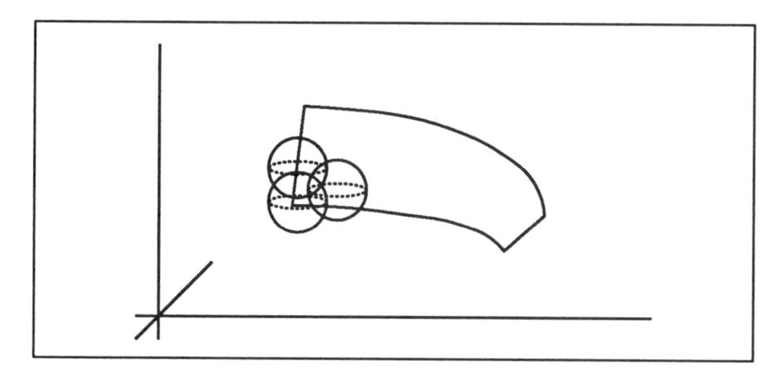

Figure 2.45 : The covering of a plane by balls.

This observation generalizes to objects in space (in fact also to objects in higher dimensions). For example, a surface in space, see figure 2.45 has covering dimension 2, because we can arrange coverings of the surface with balls of small radius so that there are no quadruplets but only triplets of balls with non-empty intersection, and there is no way to cover the surface with sufficiently small balls so that there are only pairs with non-empty intersection.

Refinement of Covers and Covering Dimension

We are accustomed to associating dimension 1 with a curve, or dimension 2 with a square, or dimension 3 with a cube. The notion of covering dimension is one way to make this intuition more precise. It is one of several notions in the domain of topological dimensions. Let us first discuss the covering dimension for two examples, a curve in the plane and a piece of a surface in space, in figure 2.44 and figure 2.45.

We see a curve covered with little disks and focus our attention on the maximal number of disks in the cover which have non-empty intersection. This number is called the order of the cover. Thus, at the left end of figure 2.44 the order is 4, while in the center it is 3, and at the right end it is 2. In figure 2.45 we see a piece of a surface in space covered with balls and the order of the cover is 3.

We have almost arrived at the definition. For that, we introduce the notion of a refinement of a cover. For us, covers of a set X in the plane (or in space) are just collections of finitely many open disks (or balls) of some radius,[24] say $A = \{D_1, ..., D_r\}$, such that their union covers X. More precisely we assume that we have a compact metric space X. A finite cover, then, is a finite collection of open sets, such that X is contained in the union of these open sets. An open cover $B = \{E_1, ..., E_r\}$ is called a refinement of $A = \{D_1, ..., D_r\}$ provided for each E_i there is D_k such that $E_i \subset D_k$. The order of an open cover A is the maximal integer k, such that there are disjoint indices $i_1, ..., i_k$ with $D_{i_1} \cap \cdots \cap D_{i_k} \neq \emptyset$.

Now we say that an object X has *covering dimension* n provided any open cover of X admits an open refinement of order $n + 1$ but not of order n. n can be any nonnegative integer.

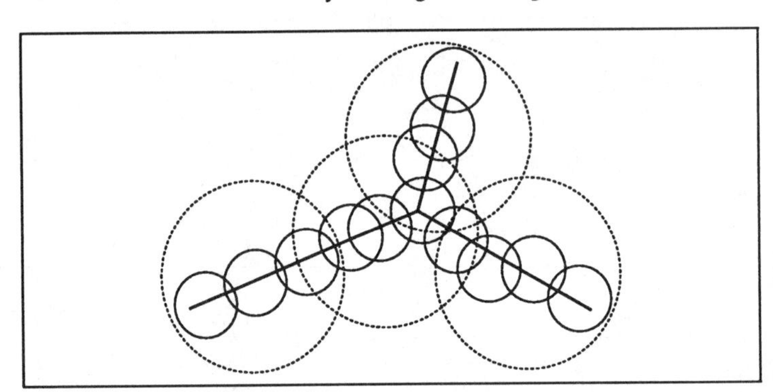

Figure 2.46 : The covering at a branching point and a refinement.

Figure 2.46 illustrates that the notion of refinement is really crucial. The large cover (dotted circles) covers the y-shaped object so that in one case 3 disks intersect. But there is a refinement with smaller disks (solid circles, each smaller open disk is contained in a large open disk), such that at most 2 disks intersect.

[24]'Open' means that we consider a disk (resp. ball) without the bounding circle (resp. sphere) or, more generally, unions of such disks.

Now it is intuitively clear that a finite number of points can be covered so that there is no intersection. Curves can be covered so that the order of the cover is 2 and there is no cover of sufficiently small disks or balls with order 1. Surfaces can be covered so that the order of the cover is 3 and there is no cover of sufficiently small disks or balls with order 2. Thus the covering dimension of points is 0, that of curves is 1, and that of surfaces is 2.

The same ideas generalize to higher dimensions. Moreover, it does not matter whether we consider a curve imbedded in the plane or in space and use disks or balls to cover it.[25]

---

[25]For more details about dimensions we refer to Gerald E. Edgar, *Measure, Topology and Fractal Geometry*, Springer-Verlag, New York, 1990.

2.7 The Universality of the Sierpinski Carpet

We have tried to obtain a feeling for what the topological notion of dimension is and we have learned that from this point of view not only a straight line, but also for example, the Koch curve are one-dimensional objects. Indeed, from the topological point of view the collection of one-dimensional objects is extremely rich and large, going far beyond objects like the one in figure 2.47, which come to mind at first.

A Tame One-Dimensional Object

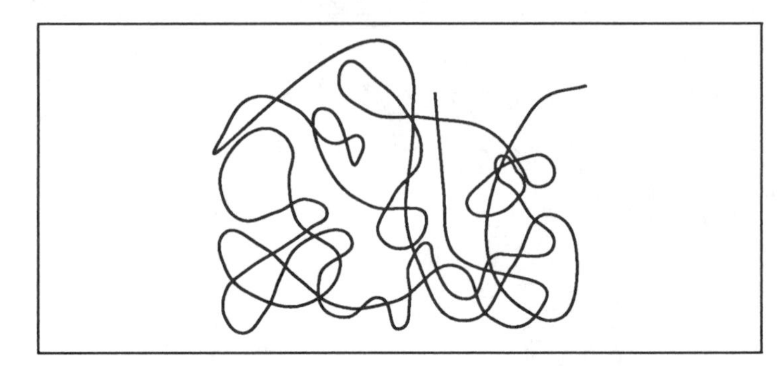

Figure 2.47 : This wild looking curve is far from a really complex one-dimensional object.

The House of One-Dimensional Objects

We are now prepared to get an idea of what Sierpinski was trying to accomplish when he invented the carpet. We want to build a house or hotel for all one-dimensional objects. This house will be a kind of *super* object which contains all possible one-dimensional objects in a topological sense. This means that a given object may be hidden in the super object not exactly as it appears independently, but rather as one of its topologically equivalent mutants. Just imagine that the object is made out of rubber and can adjust its form to fit into the super object. For example, the spider with five arms in figure 2.48 may appear in any one of the equivalent forms in the super object.

In which particular form a spider with five arms will be hidden in the super object is irrelevant from a topological point of view. In other words, if one of the arms were as wild as a Koch curve, that would be acceptable too.

Sierpinski's marvelous result[26] in 1916 says that the carpet is such a super object. In it we can hide any one-dimensional object whatsoever. In other words, any degree of (topological) complexity a one-dimensional object may have must also be present in the Sierpinski carpet. The exact result of Sierpinski's work is:

[26]W. Sierpinski, *Sur une courbe cantorienne qui content une image binnivoquet et continue detoute courbe donneé*, C. R. Acad. Paris 162 (1916) 629–632.

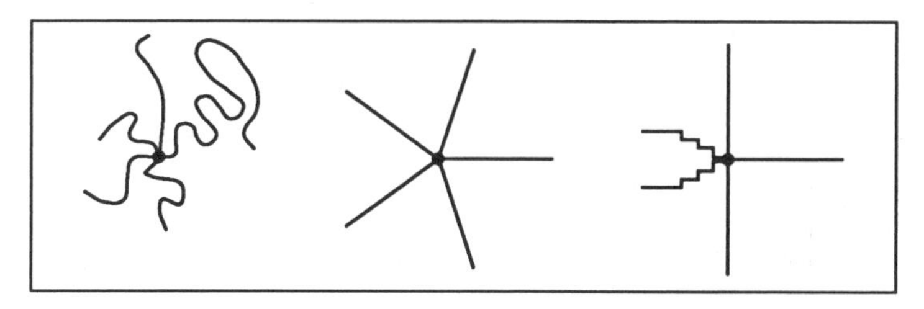

Topologically equivalent spiders

Figure 2.48 : All these spiders with five arms are topologically equivalent.

Fact. *The Sierpinski carpet is universal for all compact*[27] *one-dimensional objects in the plane.*

Let us get some initial idea about the meaning of the above statements. Take a piece of paper and draw a curve (i.e., a typical one-dimensional object) which fits (this makes it compact) on the paper. Try to draw a really complicated one, as complicated as you can think, with as many self-intersections as you may wish. Even draw several curves on top of each other. Whatever complication you may invent, the Sierpinski carpet is always ahead of you. That is, any complication which is in your curve is also in some subset (piece) of the Sierpinski carpet. More precisely, within the Sierpinski carpet we may find a subset which is topologically the same as the object which you have drawn on your paper. The Sierpinski carpet is really a super object. It looks orderly and tame, but its true nature goes far beyond what can be seen. In other words, what we can see with our eyes and what we can see with our mind are as disparate as they can be. We might say the Sierpinski carpet is a hotel which accommodates all possible (one-dimensional, compact) species living in flatland. But not everything can live in flatland.

Planar and Non Planar Curves

We can draw curves in a plane or in space. But can we draw all curves that we can draw in space also in the plane? At first glance yes, but there is a problem. Take the figure eight (a) in the plane in figure 2.49, for example.

Is it a real figure eight (with one self-intersection) or does it just look like a figure eight because it is a projection of the twisted circle which lies in space as in (b)? Without further explanation it could be both. However, from a qualitative point of view a figure eight is

[27] Compactness is a technical requirement which can be assumed to be true for any drawing on a sheet of paper. For instance, a disk in the plane without its boundary would not be compact, or a line going to infinity would also not be compact. Technically, compactness for a set X in the plane (or in space) means that it is bounded, i.e. it lies entirely within some sufficiently large disk in the plane (or ball in space) and that every convergent sequence of points from the set converges to a point from the set.

Circle vs. Figure Eight

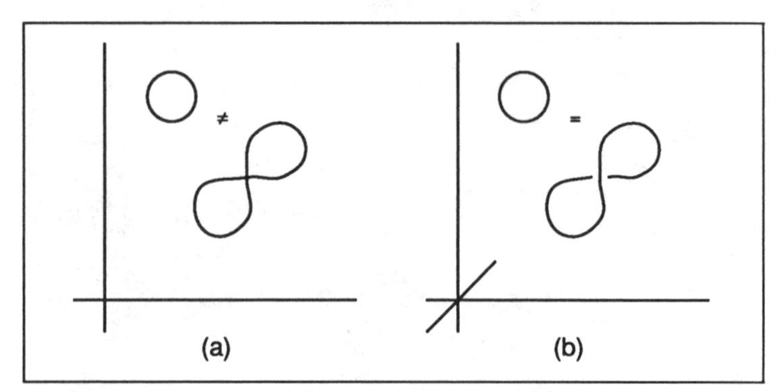

Figure 2.49 : The figure eight is not equivalent to the circle (a). The twisted circle is equivalent to a circle (b).

very different from a circle because it has a self-intersection and it separates the plane into three regions rather than only two as for a circle. Thus, topologically we have to distinguish them from each other. The curve in (b) is a circle from a topological point of view and can be embedded easily into a plane without self intersections once we untwist it.

This raises the question whether any curve in space can be embedded into a plane without changing its topological character. The answer is no. Here is a simple illustration, the WG&E example in figure 2.50. Imagine that we have three houses A, B, and C which have to be supplied with water, gas and electricity from W, G, and E so that the supply lines do not cross (if drawn in a plane). There is no way to bring a line from W to C without a crossing. The only way to escape a crossing is to go into space (i.e., run the supply lines at different levels).

Thus, if we are interested in maintaining the topological character of one-dimensional objects, we may have to go into space. In fact, every one-dimensional object can be embedded into 3-dimensional space. Generalizations of this question are at the heart of topology. It is this branch which goes beyond the intuitive understanding of why the skeleton in figure 2.50 cannot be embedded into the plane by providing very deep methods which generalize to higher dimensions. For instance, any two-dimensional object can be embedded into a five-dimensional space, and the five dimensions are actually needed in order to avoid obtaining effects similar to self-intersections, which would change the topological character.

Note that the graph in figure 2.50 cannot be drawn in the plane without self-intersections. Thus this graph cannot be represented in the Sierpinski carpet. This leads to the question, what is the universal object for one-dimensional objects in general (i.e., both in the plane and in space)?

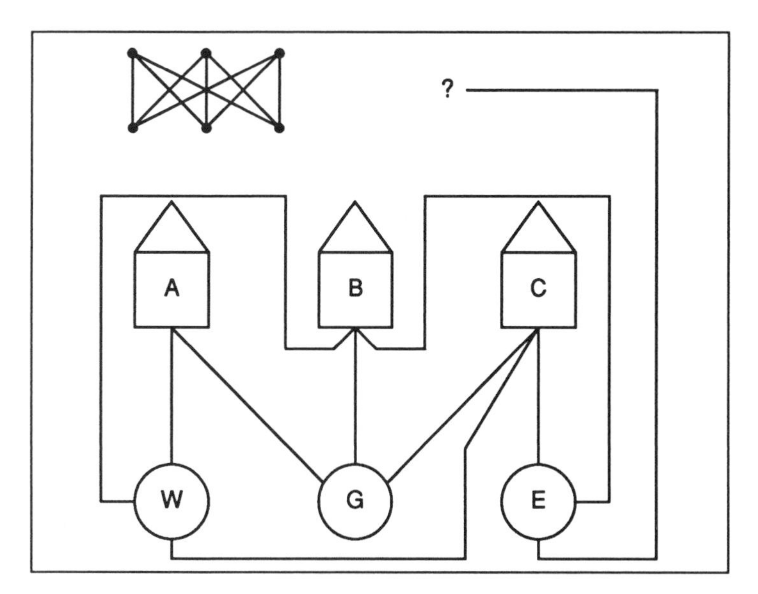

Figure 2.50 : A tough problem: can one get water, gas and electricity to all three houses without any intersection? The complete graph (with intersections!) is shown in the upper left corner of the figure.

Water, Gas & Electricity

The Universality of the Menger Sponge

About ten years after Sierpinski had found his result, the Austrian mathematician Karl Menger solved this problem and found a hotel for all one-dimensional objects. He proved around 1926 the following.[28]

Fact. *The Menger sponge is universal for all compact one-dimensional objects.*

Roughly this means that for any admissible object (compact, one-dimensional) there is a part in the Menger sponge which is topologically the same as the given object.[29] That is, imagine that the given object again is made of rubber. Then some deformed version of it will fit exactly into the Menger sponge.

We cannot demonstrate the proof of Menger or Sierpinski's amazing results; they are beyond the scope of this book. But we want to give an idea of the complexity of the one-dimensional objects we can find here. Let us discuss just one of many methods to measure this complexity. In particular this will allow us to distinguish between the Sierpinski gasket and the Sierpinski carpet.

[28]K. Menger, *Allgemeine Räume und charakteristische Räume, Zweite Mitteilung: „Über umfassenste n-dimensionale Mengen"*, Proc. Acad. Amsterdam 29, (1926) 1125–1128. See also K. Menger, *Dimensionstheorie*, Leipzig 1928.

[29]Formally, for any compact one-dimensional set A there is a compact subset B of the Menger sponge which is homeomorphic to A.

Since their basic construction steps are so similar (see section 2.2), we may ask whether the gasket is also universal? In other words, how complicated is the Sierpinski gasket? Is it as complicated as the carpet, or less? And if less, how much less complicated is it? Would you bet on your guess or visual intuition?

The answer is really striking: the Sierpinski gasket is absolutely tame when compared with the carpet, though visually there seems to be not much of a difference. The Sierpinski gasket is a hotel which can accommodate only a few (one-dimensional, compact) very simple species from flatland. Thus, there is, in fact, a whole world of a difference between these two fractals. Let us look at objects like the ones in figure 2.51

Order of Spiders

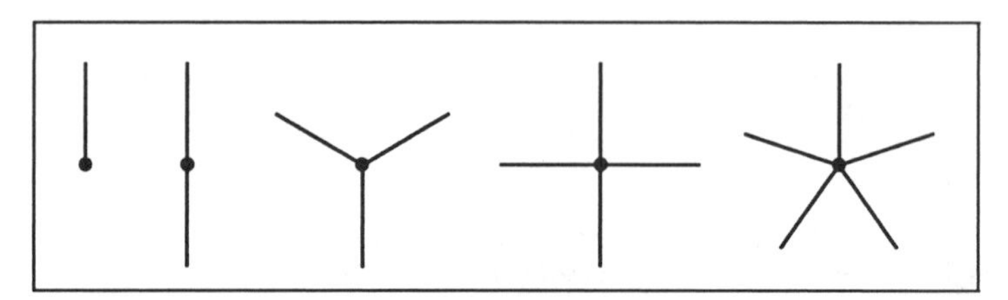

Figure 2.51 : Spiders with increasing branching.

What we see are line segments with crossings. Or, we could say we see a central point to which there are different numbers of arms attached. We like to count the number of arms by a quantity which we call the branching order of a point. This will be a topological invariant. That is, this number will not change when we pass from one object to a topologically equivalent one. We can easily come up with one-dimensional objects of any prescribed branching order.

Branching Order

There is a very instructive way to distinguish one aspect of many different (topological) complexity features for one-dimensional objects. This concerns their branching and is measured by the branching order,[30] which we introduce next. Figure 2.52 shows some different types of branched structures.

The branching order is a local concept. It measures the number of branches which meet in a point. Thus, for a point on a line we count two branches, while for an endpoint we count one branch. In example (d) of figure 2.52 we have one point — labelled ∞ — from which there are infinitely (countably) many line segments. Thus, this point would have branching order ∞, while points on the generating line segments (disregard the limit line) again would have branching order 2. Let us call the objects in figure 2.52 spiders. Thus (a) is a spider with 2, (b) with 3, (c) with 6, and (d) with ∞ many arms.

[30]See A. S. Parchomenko, *Was ist eine Kurve*, VEB Verlag, 1957.

Branching Order of Sets

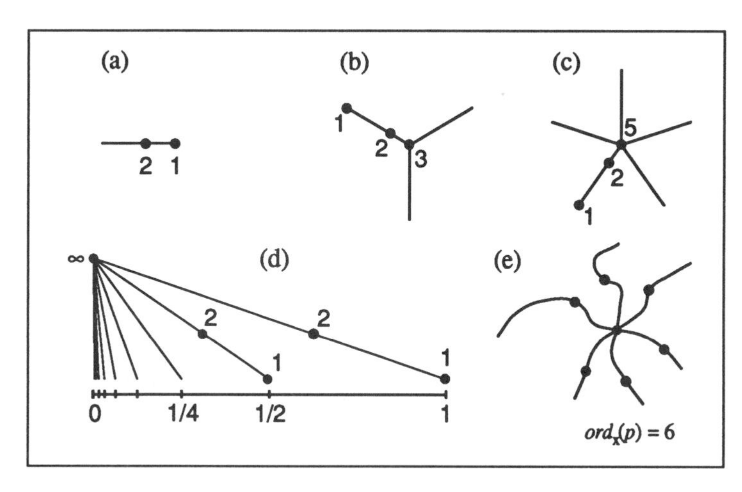

Figure 2.52 : Several examples of finite and (countably) infinite branching order. The numbers indicate the branching order of the corresponding points.

Let X be a set[31] and let $p \in X$ a point. Then we define the branching order of X at p to be[32]

$$\text{ord}_X(p) = \text{the number of branches of } X \text{ at } p.$$

One way to count these branches would be to take sufficiently small disks around the point of interest and count the number of intersection points of the boundary of the disks with X.

Let us now construct a monster spider whose branching order has the cardinality of the continuum, i.e. there are as many branches as there are numbers in the unit interval $[0,1]$.

We begin by taking a single point, say P, in the plane at $(1/2, 1)$ (see figure 2.53), together with the Cantor set C in $[0,1]$. Now from each point in C we draw a line segment to P. You will recall that the cardinality of the Cantor set is the same as the cardinality of [0,1]. Therefore the cardinality of the points in the boundary of a small disk around P will again have the same cardinality. We call this set a *Cantor brush.*

Any pictorial representation of the Cantor brush can be a bit misleading. It might suggest that there is a countable number of bristles, while in fact they are uncountable. It can be shown that the Cantor brush is, however, a set of (covering) dimension 1, as is, of course any spider with k arms.

[31] Formally, we need that X is a compact metric space.

[32] A formal definition goes like this. Let α be a cardinal number. Then one defines $\text{ord}_X(p) \leq \alpha$, provided for any $\varepsilon > 0$ there is a neighborhood U of p with a diameter $\text{diam}(U) < \varepsilon$ and such that the cardinality of the boundary ∂U of U in X is not greater than α, $\text{card}(\partial U) \leq \alpha$. Moreover, one defines $\text{ord}_X(p) = \alpha$, provided $\text{ord}_X(p) \leq \alpha$ and additionally there is $\varepsilon_0 > 0$, such that for all neighborhoods U of x with diameter less than ε_0 the cardinality of the boundary of U is greater or equal to α, $\text{card}(\partial U) \geq \alpha$.

The Cantor Brush

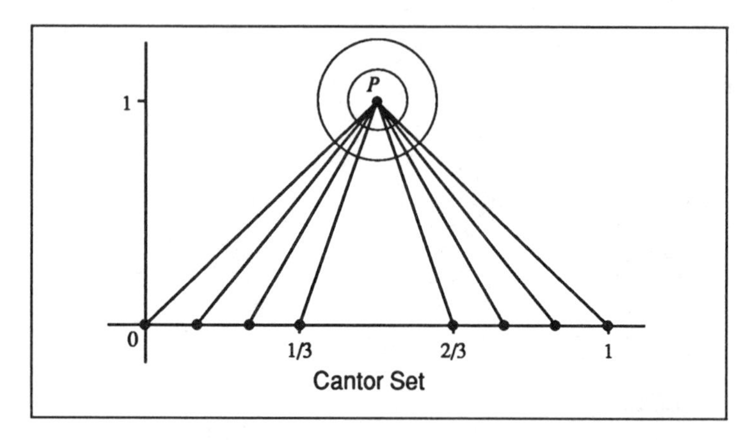

Figure 2.53 : An example of uncountably infinite branching order, the Cantor brush.

Branching Order of Sierpinski Gasket

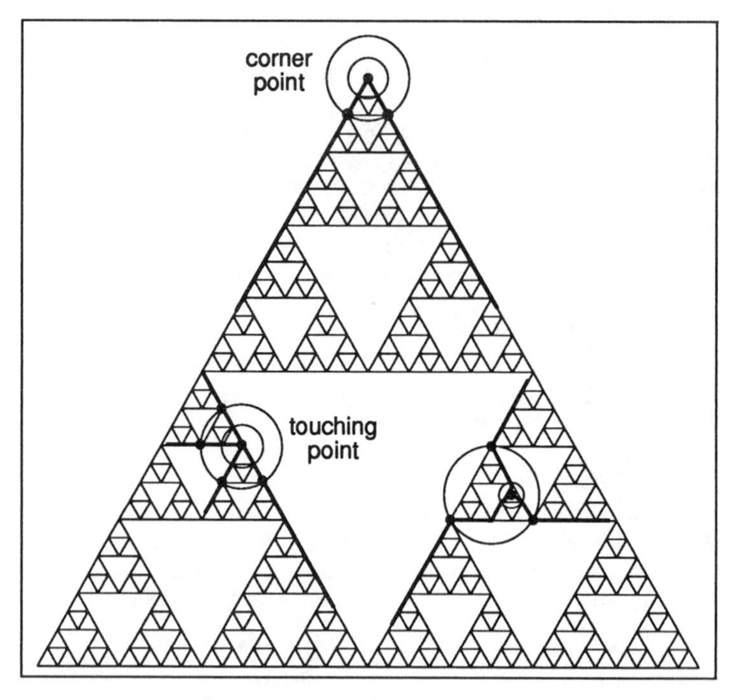

Figure 2.54 : The Sierpinski gasket allows only the branching orders 2, 3, and 4.

Now let us look at the Sierpinski gasket in terms of the branching order, see figure 2.54. Which spiders can be found in the gasket? It can be shown that if p is any point in the Sierpinski gasket S then

$$\text{ord}_S(p) = \begin{cases} 2, \text{ if } p \text{ is a corner of the initial triangle} \\ 4, \text{ if } p \text{ is a touching point} \\ 3, \text{ if } p \text{ is any other point} \end{cases} .$$

If p is a point, exactly two arms lead to this point. Observe, how the circles drawn around the corner point (they need not to be centered at p) intersect the Sierpinski gasket at just two points. If p is a touching point, we can trace 4 arms to this point. In this case one can see circles around p that intersect the gasket at exactly 4 points. Now if p is any other point it must be right within an infinite sequence of sub-triangles. These sub-triangles are connected to the rest of the Sierpinski gasket at just 3 points. Thus we can find smaller and smaller circles around p which intersect the Sierpinski gasket at exactly 3 points and we can construct 3 arms which pass though these points leading to p.

Topologically Equivalent Six-Armed Spiders

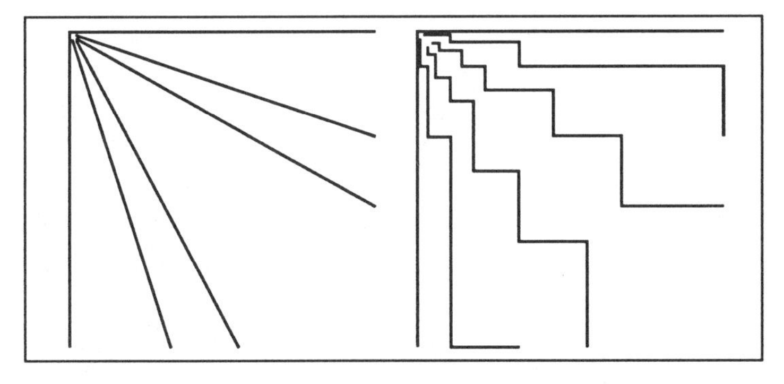

Figure 2.55 : These two spiders are topologically equivalent.

The Universality of the Sierpinski Carpet

We can observe that the Sierpinski gasket has points with branching order 2, 3 and 4, see figure 2.54. These are the only possibilities. In other words, a spider with five (or more) arms *cannot* be found in the Sierpinski gasket![33]

On the other hand, the Sierpinski carpet is universal. Therefore it must accommodate spiders with any branching order, and, in particular, it must even contain a (topological) version of the Sierpinski gasket. Let us try to construct, as a very instructive example, a spider with five or six arms in the Sierpinski carpet. This is demonstrated in figure 2.56 and figure 2.57. Figure 2.55

[33]This is rather remarkable and it is therefore very instructive to try to construct a spider with five arms and understand the obstruction!

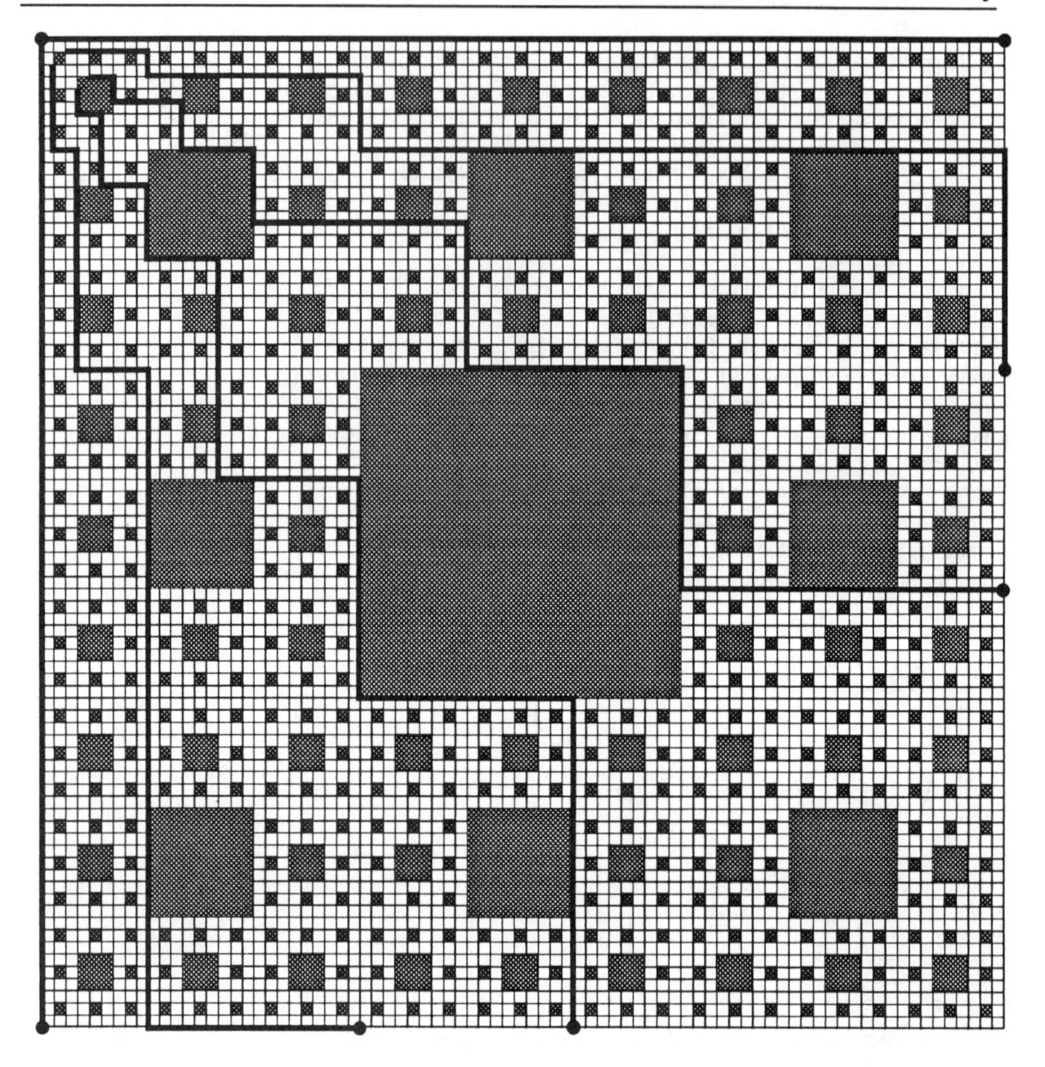

Figure 2.56 : Construction of a spider with six arms using symmetry and a recursive construction

shows the actual spider which we have found in the carpet (right) and a topologically equivalent spider (left).

Let us summarize. Our discussion of the universality of the Sierpinski carpet shows that fractals in fact have a very firm and deep root in a beautiful area of mathematics, and in varying an old chinese saying[34] we may say fractals are more than pretty images.

[34] A picture is worth a thousand words.

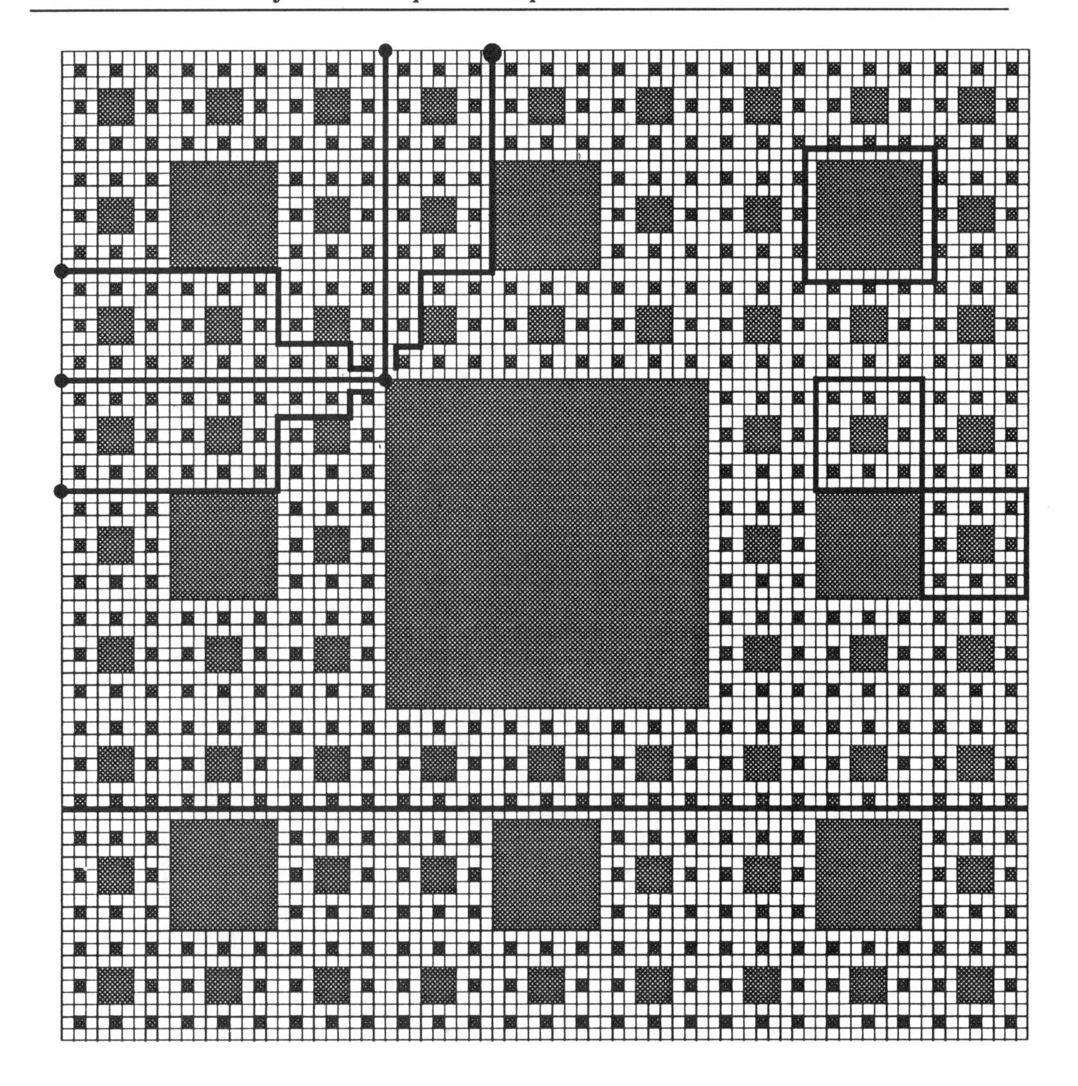

Figure 2.57 : The Sierpinski carpet houses any one-dimensional object: lines, squares, figure-eight like shapes, five-arm spiders or even deformed versions of the Sierpinski gasket (this is not shown — can you construct it?).

2.8 Julia Sets

Gaston Julia (1893–1978) was only 25 when he published his 199 page masterpiece[35] in 1918, which made him famous in the mathematics centers of his days. As a French soldier in the First World War, Julia had been severely wounded, as a result of which he lost his nose. Figure 2.58 shows him in the 1920's. Between several painful operations, he carried on his mathematical research in a hospital. Later he became a distinguished professor at the École Polytechnique in Paris.

Gaston Julia

Figure 2.58 : Gaston Julia, 1893–1978, one of the forefathers of modern dynamical systems theory.

Julia's Work Was Famous in the 1920's

Although Julia was a world famous mathematician in the 1920's, his work was essentially forgotten until Mandelbrot brought it back to light at the end of the seventies through his fundamental experiments. Mandelbrot was introduced to Julia's work through his uncle Szolem Mandelbrojt, who was a mathematics professor in Paris. In fact, he was the successor of Jacques Salomon Hadamard at the prestigious Collège de France.

Mandelbrot was born in Poland in 1924, and after his family had emigrated to France in 1936, his uncle felt responsible for his education. Around 1945, his uncle recommended Julia's paper to him as a masterpiece and source of good problems, but Mandelbrot didn't like it. Somehow he could not relate to the style and kind

[35] G. Julia, Mémoire sur l'iteration des fonctions rationnelles, Journal de Math. Pure et Appl. 8 (1918) 47–245.

of mathematics which he found in Julia's paper and chose his own very different course, which, however, brought him back to Julia's work around 1977 after a path through many sciences, which some characterize as highly individualistic or nomadic. With the aid of computer graphics, Mandelbrot showed us that Julia's work is a source of some of the most beautiful fractals known today. In this sense, we could say that this masterpiece is full of classical fractals which had to wait to be kissed awake by computers. In the first half of this century Julia was indeed world famous. To learn about his results, Hubert Cremer organized a seminar at the University of Berlin in 1925 under the auspices of Erhard Schmidt and Ludwig Bieberbach. The list of participants reads almost like an excerpt of a 'Who's Who' in mathematics of that time. Among them were Richard D. Brauer, Heinrich Hopf, and Kurt Reidemeister. Cremer also produced an essay on the topic which contains the first visualization of a Julia set (see figure 2.59).[36]

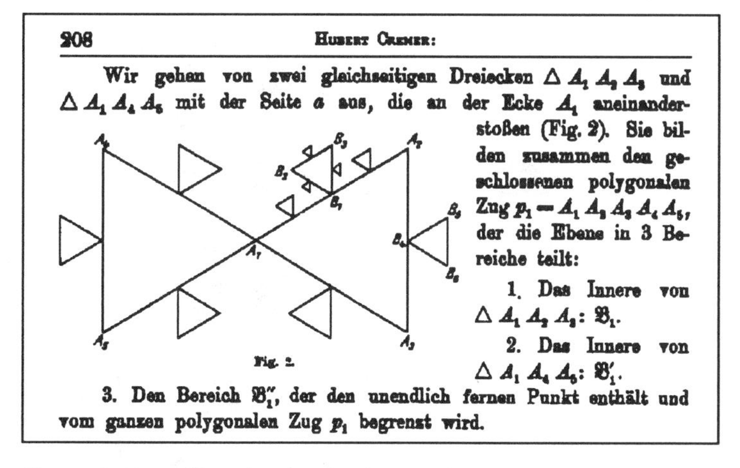
208 HUBERT CREMER:

Wir gehen von zwei gleichseitigen Dreiecken $\triangle A_1 A_2 A_3$ und $\triangle A_1 A_4 A_5$ mit der Seite a aus, die an der Ecke A_1 aneinanderstoßen (Fig. 2). Sie bilden zusammen den geschlossenen polygonalen Zug $p_1 = A_1 A_2 A_3 A_4 A_5$, der die Ebene in 3 Bereiche teilt:

1. Das Innere von $\triangle A_1 A_2 A_3$: $\mathfrak{B}_1$.

2. Das Innere von $\triangle A_1 A_4 A_5$: $\mathfrak{B}'_1$.

3. Den Bereich $\mathfrak{B}''_1$, der den unendlich fernen Punkt enthält und vom ganzen polygonalen Zug p_1 begrenzt wird.

Figure 2.59 : First drawing by Cremer in 1925 visualizing a Julia set.

The Quadratic Feedback System

Julia sets live in the complex plane. They are crucial for the understanding of iterations of polynomials like x^2+c, or x^3+c, etc. A detailed introduction will be given in chapter 12, but here we assume that you are familiar with the concept of complex numbers. If you aren't, we propose that for now you simply think of real numbers. Let us look at $x^2 + c$ as an example. Iterating means that we fix c and choose some value for x and obtain $x^2 + c$. Now we substitute this value for x and evaluate $x^2 + c$ again, and so on. In other words, for an arbitrary but fixed value of c we generate a

[36]H. Cremer, *Über die Iteration rationaler Funktionen*, Jahresberichte der Deutschen Mathematischen Vereinigung 33 (1925) 185–210.

Some Samples of Julia Sets

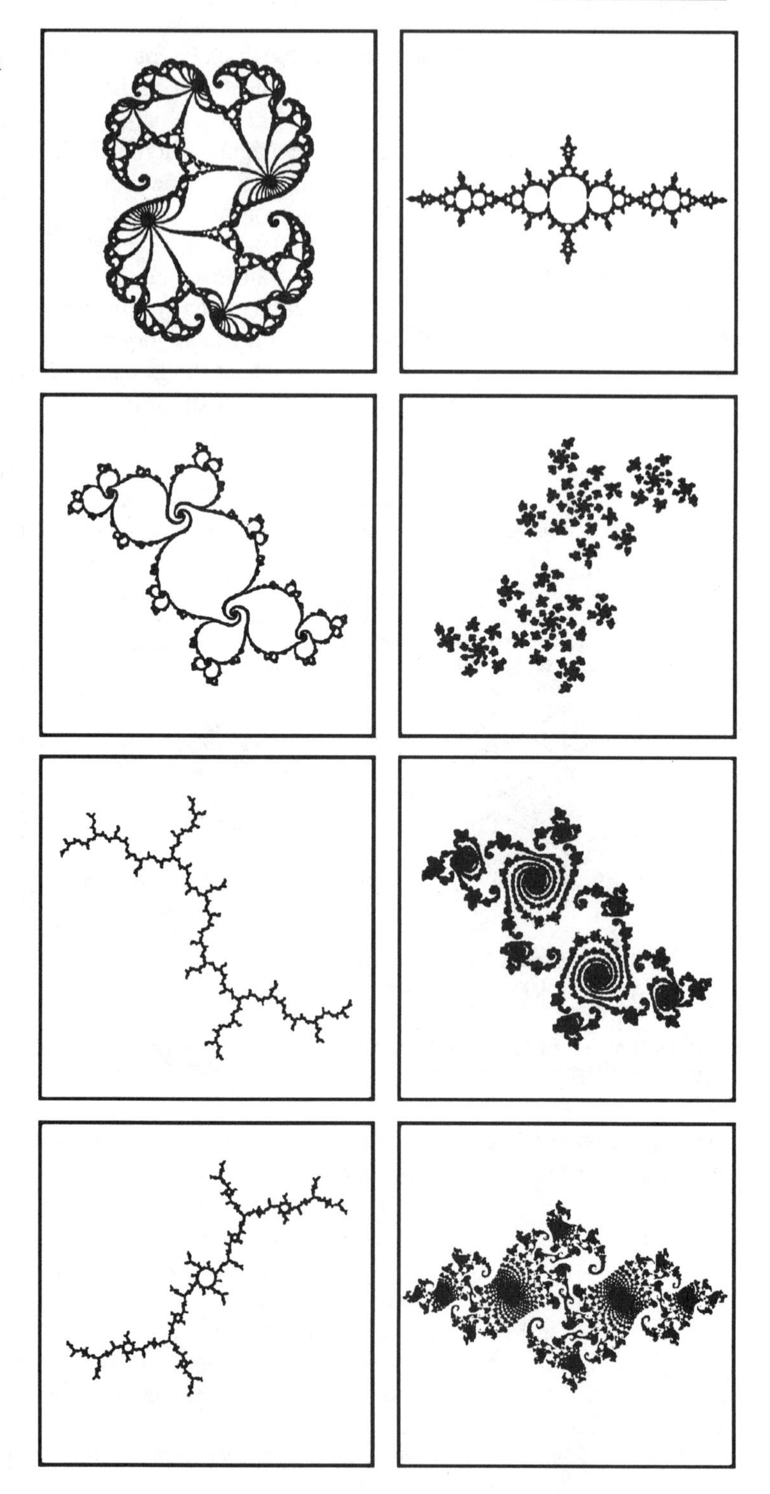

Figure 2.60 : Some first Julia sets.

sequence of complex numbers

$$x \to x^2 + c \to (x^2 + c)^2 + c \to ((x^2 + c)^2 + c)^2 + c \to \dots$$

This sequence must have one of two following properties:

The Julia Set Dichotomy

- either the sequence becomes unbounded: the elements of the sequence leave any circle around the origin,
- or the sequence remains bounded: there is a circle around the origin which is never left by the sequence.

The collection of points which lead to the first kind of behavior is called the *escape set* for c, while the collection of points which lead to the second kind of behavior is called the *prisoner set* for c. This terminology has already been used in the section on the Cantor set. Both sets are non-empty. For example, given c, then for x sufficiently large, $x^2 + c$ is even larger than x. Thus, the escape set contains all points x which are very large. On the other hand, if we choose x so that $x = x^2 + c$, then iteration remains stationary. Starting with such an x the sequence produced by the iteration will be constant $x, x, x, \dots$. In other words, neither can the prisoner set be empty.

The Julia Set

Both sets cover some part of the complex plane and complement each other. Thus, the boundary of the prisoner set is simultaneously the boundary of the escape set, and that is the Julia set for c (or rather $x^2 + c$). Figure 2.60 shows some Julia sets obtained through computer experiments.

Is there self-similarity in the Julia sets? Already from our first crude figure it seems obvious that there are structures that repeat at different scales. In fact, any Julia set may be covered by copies of itself, but these copies are obtained by a *nonlinear* transformation. Thus, the self-similarity of Julia sets is of a very different nature as compared to the Sierpinski gasket, which is composed of reduced but otherwise congruent copies of itself.

2.9 Pythagorean Trees

Pythagoras, who died at the beginning of the fifth century B.C., was known to his contemporaries, and later even to Aristotle, as the founder of a religious brotherhood in southern Italy, where Pythagoreans played a political role in the sixth century B.C. The linking of his name to the Pythagorean theorem is, however, rather recent and spurious. In fact, the theorem was known long before the life time of Pythagoras. An important discovery ascribed to Pythagoras, or in any case to his school, is that of the *incommensurability* of side and diagonal of the square; that is, the ratio of diagonal and side of the square is not equal to the ratio of two integers.

$a^2 + b^2 = c^2$

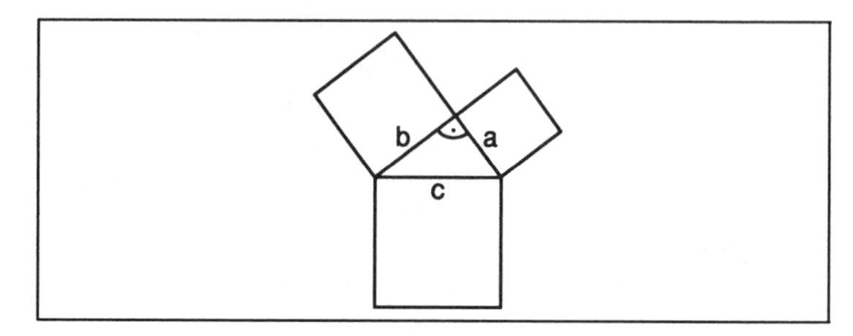

Figure 2.61 : The Pythagorean theorem $a^2 + b^2 = c^2$.

The Incommensurability of Side and Diagonal of the Square

The discovery that the ratio of diagonal and side of the square is not equal to the ratio of two integers produced the necessity to extend the number system to *irrational numbers*. $\sqrt{2}$, the length of the diagonal in the unit square, is irrational. Let us give the argument. Assume that $\sqrt{2} = p/q$. We may also assume that p and q have no common divisor. Then $p^2 = 2q^2$, i.e. p^2 must be an even number. But this implies that p itself must be even, because the square of an odd number is odd. Thus $p = 2r$. But then $p^2 = 2q^2$ means that $4r^2 = 2q^2$ or $2r^2 = q^2$, which means that q must be even as well. But this contradicts the assumption that p and q have no common divisor. Thus, $\sqrt{2}$ is irrational. This proof is found in the tenth book of Euclid around 300 B.C.

The computation of square roots is a related problem and has inspired mathematicians to discover some wonderful geometric constructions. One of them allows us to construct $\sqrt{n}$ for any integer n. It could be called the square root spiral, and it is a geometric feedback loop. Figure 2.62 explains the idea.

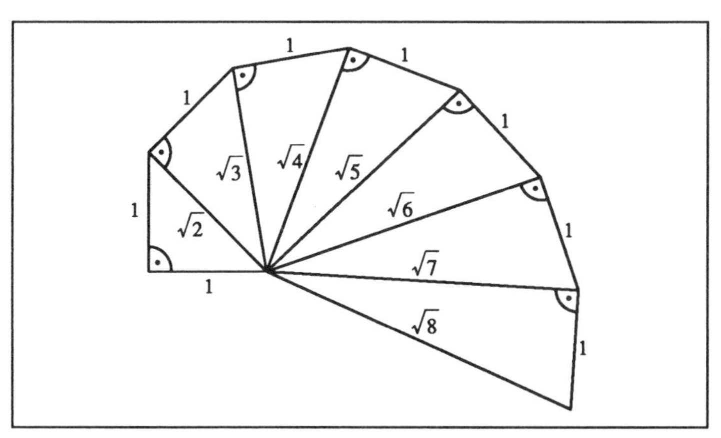

Figure 2.62 : Construction of a square root spiral. We begin with a right-angled triangle so that the sides forming the right angle are of length 1. Then the hypothenuse is of length $\sqrt{2}$. Now we continue by constructing another right triangle so that the sides adjacent to the right angle have length 1 and $\sqrt{2}$. The hypothenuse of that triangle has length $\sqrt{3}$, and so on.

The Square Root Spiral

The Construction of Pythagorean Trees

The construction which yields the family of Pythagorean trees and their relatives is very much related to the construction of the square root spiral. The construction proceeds along the following steps and is shown in figure 2.63.

Step 1: Draw a square.

Step 2: Attach a right triangle to one of its sides along its hypothenuse (here with two equal sides).

Step 3: Attach two squares along the free sides of the triangle.

Step 4: Attach two right triangles.

Step 5: Attach four squares.

Step 6: Attach four right triangles.

Step 7: Attach eight squares.

Once we have understood this basic construction we can modify it in various ways. For example, the right triangles which we attach in the process need not be isosceles triangles. They can be any right triangle. But once we allow such variations we have, in fact, an additional degree of freedom. The right triangles can always be attached in the same orientation, or we may flip their orientation after each step. Figure 2.64 shows the two possibilities.

Figure 2.65 shows the results of these constructions after some 50 steps. It is most remarkable that the only thing which we have changed is the orientation of the rectangles, not their size. The results, however, could not look more different. In the first case

Pythagorean Tree Construction

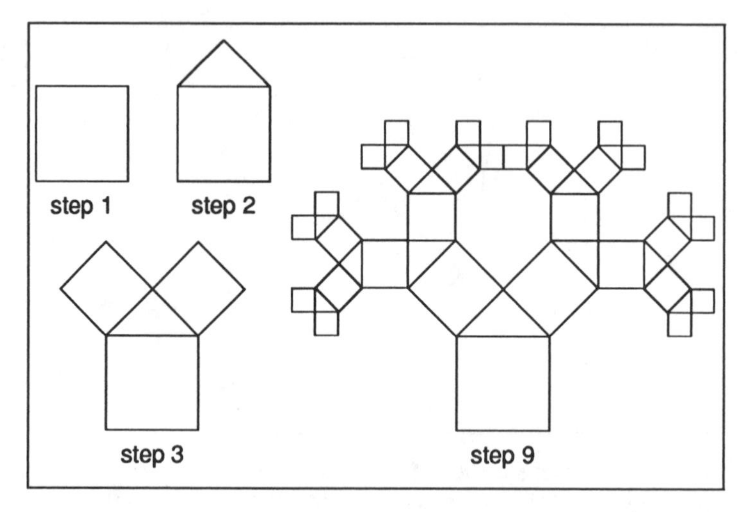

Figure 2.63 : Basic idea of a Pythagorean tree.

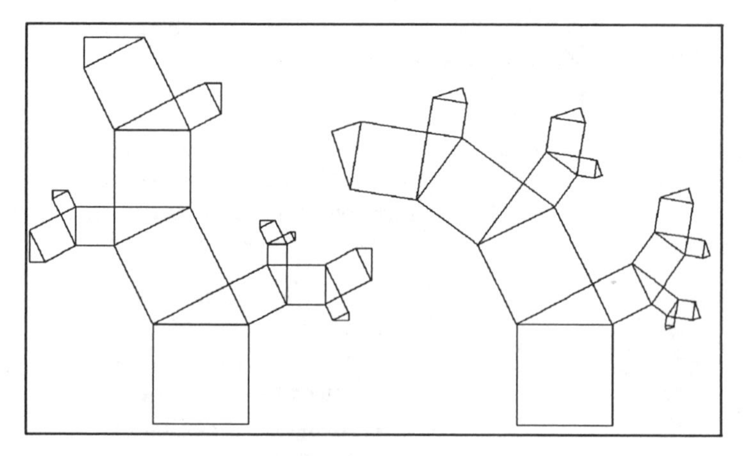

Figure 2.64 : Two constructions with non isosceles triangles.

we see some kind of spiraling leaf, while the second reminds us of a fern or pine tree. Note that in the bottom construction in figure 2.65 we see a major stem from which branches radiate off in a left, right, left, right, ... pattern. This seems to be quite different in the other construction. There we see a major stem which curls and from which we have only a branching away to the left. Would you have guessed that both 'plants' derive from the same feedback principle? Didn't they look at first as if they belonged to totally different families? They don't. They are, however, very close relatives, and this becomes apparent when analyzing the corresponding construction processes. This is one way that fractals

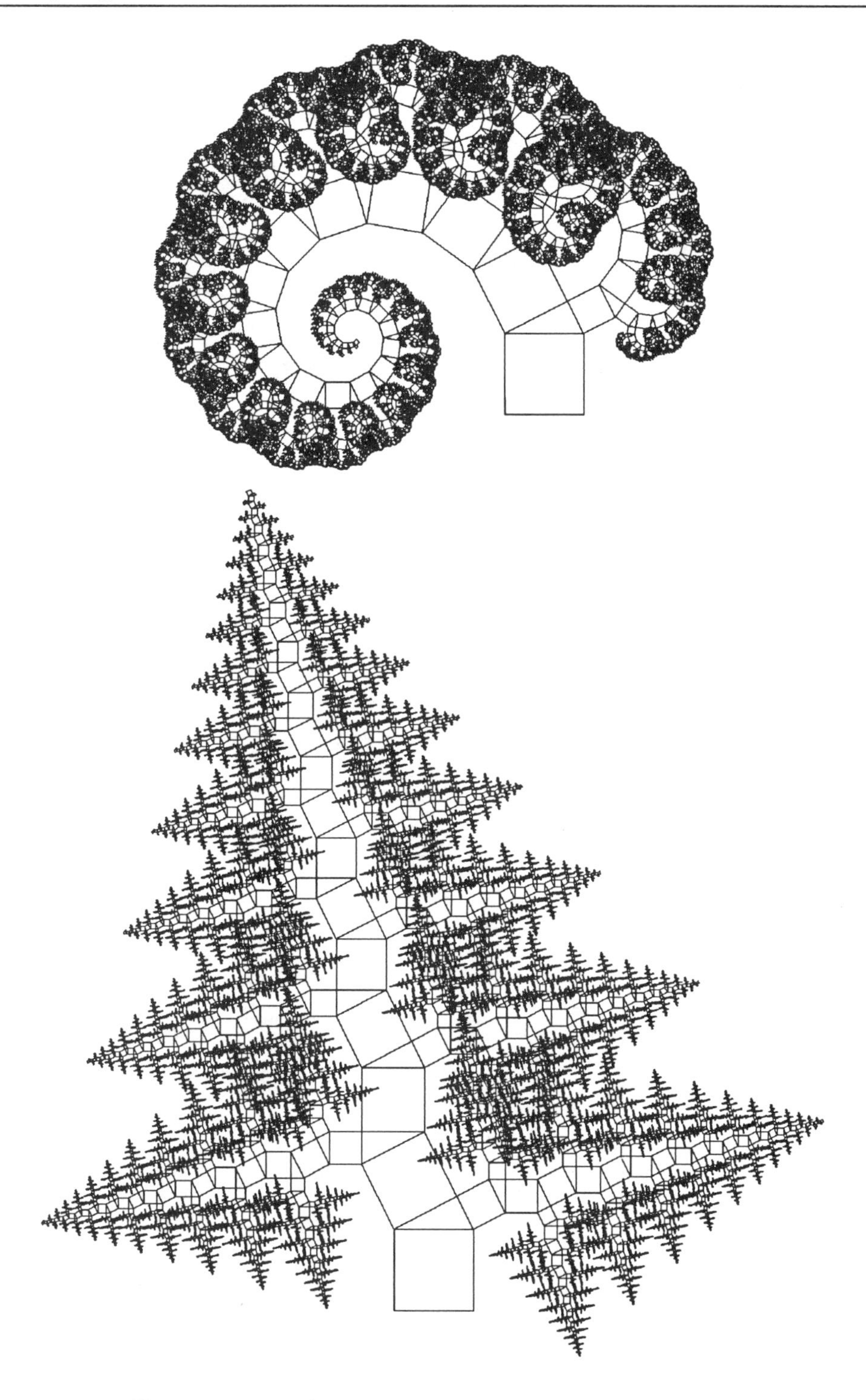

Figure 2.65 : The two constructions carried out some 50 times. Note that the size of the triangles is the same in both.

Periodic Tiling

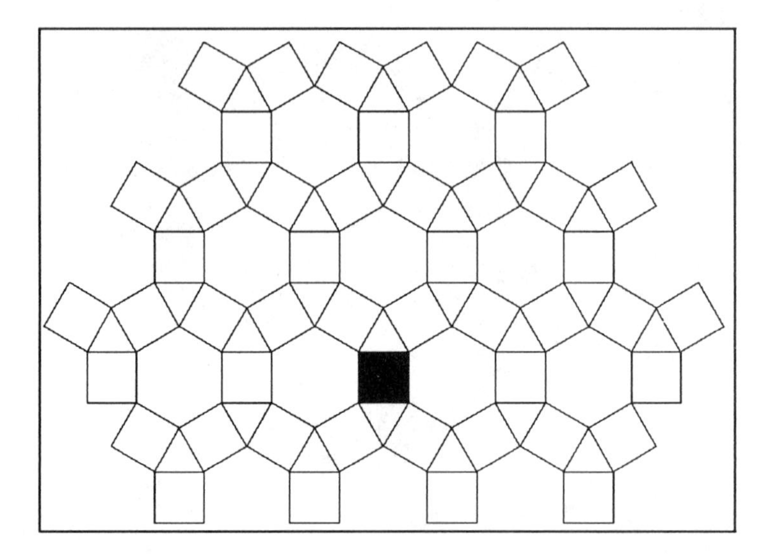

Figure 2.66 : Periodic tiling.

Broccoli-Like Phytagoras Tree

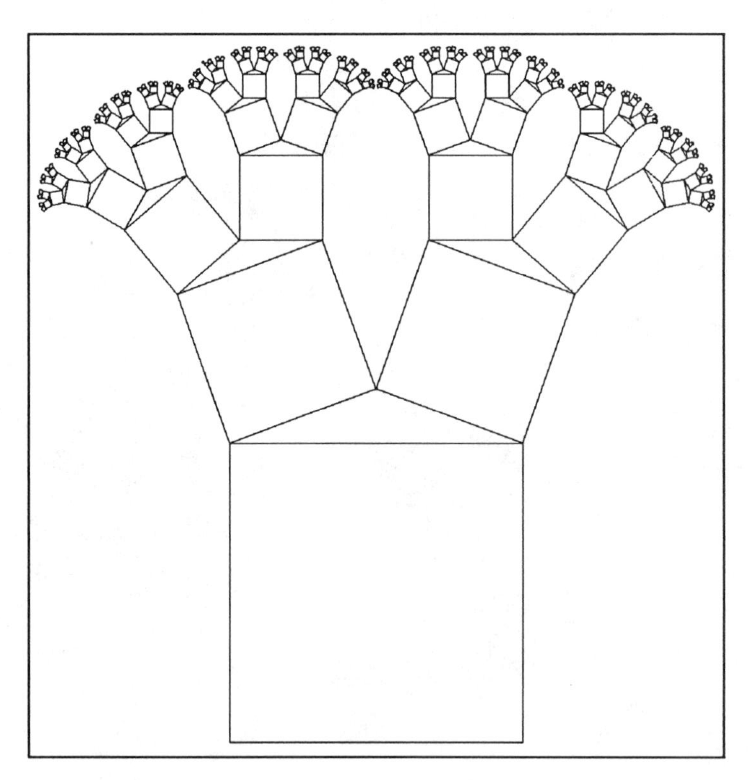

Figure 2.67 : Construction with isosceles triangles which have angle greater than 90°.

may help to introduce new tools into botany. The biologist Aristid Lindenmayer (1925–1989) introduced the concept of *L-Systems* along these lines, and we will discuss that approach in some detail in chapter 8.

Let us continue to look into our primitive, but amazing, constructions using some other modifications. Why not take just any kind of triangle? To keep some regularity we should take similar ones. This opens the door to a large variety of fascinating forms which range from plant-like ones to tilings to who knows what. In figure 2.66 we have attached equilateral triangles, and notice that the construction becomes periodic.

Passing from equilateral triangles to isosceles triangles with angles greater than 90° yields another surprise — a form which is broccoli-like (see figure 2.67). These constructions raise a number of interesting questions. When does the construction lead to an overlap? By what law do the lengths of the sides of the triangles or squares decrease in the process? Moreover, we have beautiful examples of structures which are self-similar, i.e. each structure subdivides during construction into two major branches, and these again into two major branches, and so on; and each of these branches is a scaled down version of the entire structure.

Our gallery of historical fractals ends here, though we have not discussed the contributions of Henry Poincaré, Karl Weierstraß, Felix Klein, L. F. Richardson, or A. S. Besicovitch. They all would deserve more space than we could give them here, but we refer the interested reader to Mandelbrot's book.[37]

[37] B. Mandelbrot, *The Fractal Geometry of Nature*, Freeman, New York, 1982.

2.10 Program of the Chapter: Sierpinski Gasket by Binary Addresses

The Sierpinski gasket is one of the major actors in this book. It plays an important role in several chapters. There are different ways to obtain this set. Perhaps the most surprising algorithm is the chaos game presented in chapter 1. It is discussed further in chapter 6. Here we show a method which is unbelievably short. Essentially all the information needed to generate it is hidden in one line of the code of our program, namely:

```
IF (x AND (y - x)) = 0 THEN
       PSET (x + 158 - 0.5 * y, y + 30)
```

Indeed, this is a very special line. It is not at all easy to understand and the arguments which we present in the following explains only roughly how it works. Later in chapter 9 we completely uncover the secret of this seemingly mysterious fact which lies in a beautiful relation to Pascal's triangle. There we also give extensions allowing you to generate even more complicated patterns.

Screen Image of Sierpinski Gasket

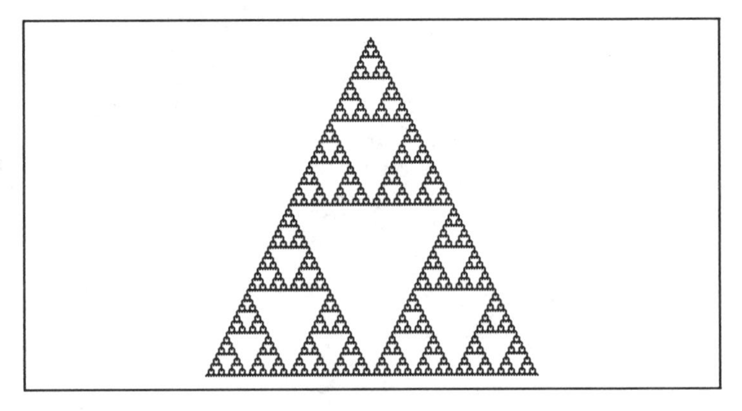

Figure 2.68 : Output of the program 'Sierpinski Gasket by Binary Addresses'.

Address Testing

Strictly speaking this program does not compute the Sierpinski gasket. Rather, it computes a color coding of the Pascal triangle. You will recall that we found the pattern of the Sierpinski gasket when color coding the divisibility of the numbers of the Pascal triangle: white cells for even numbers, black cells for odd numbers. Assume that we intend to compute the black or white color of an arbitrary cell. The question arises whether it is possible to avoid running the process of computing the numbers of the Pascal triangle through all the rows above it. The algorithm of this program provides an affirmative answer: the coding can be determined directly from the coordinates of the cell. However, this algorithm requires quite a bit of mathematical work. It is based on a binary coding of the coordinates. Each entry is determined by a pair (x, y) of integer coordinates with respect to a special coordinate system.

Binary Coordinate System for Pascal's Triangle

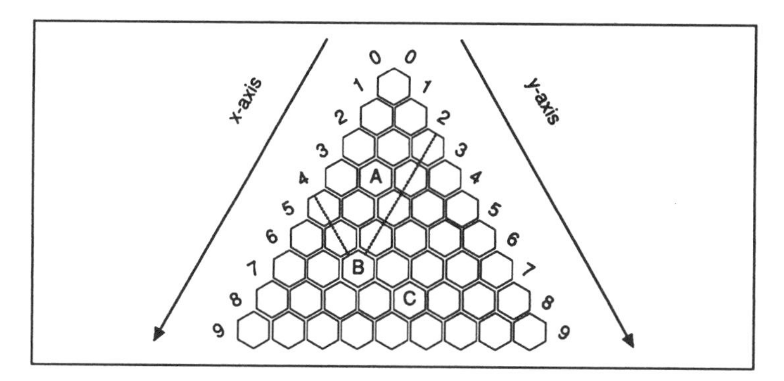

Figure 2.69 : Binary address system for Pascal's triangle. The labeled cells have coordinates A(2,1), B(4,2), and C(3,4).

Start with the origin $(0,0)$ as the top entry in a triangular array. Let the x-axis run diagonally to the left and the y-axis diagonally to the right (see figure 2.69). Then each integer pair of coordinates (x, y) corresponds to a specific location in the array.

To determine the color of a given cell, place the binary expansions of the two coordinates of the cell above each other and follow this rule: if two 1's appear above each other in any one of the columns, then the cell is left white. Otherwise, it is shaded in as black. For example, the cell labeled C in figure 2.69 has coordinates $(3,4)$. In binary representation these are $(011, 100)$, and since corresponding digits in the two coordinates are not both 1, the cell must be colored black in Pascal's triangle. The justification of this algorithm goes back to a result of Ernst Eduard Kummer (1810–1893).[38] Note that the criterion here is the same as in section 5.3. This gives some clues why we see the Sierpinski gasket in the mod 2 coloring of Pascal's triangle.

Now how is this done in our BASIC program? The comparison of the binary expansions is just an `AND` operation. Doing this the straightforward way as in the program 'Skewed Sierpinski Gasket' we obtain a skewed variation of the familiar pattern. Note also that in the program the triangle is drawn with the offset of 30 in x and y. But how can we generate the desired shape as in figure 2.68? First, we have to compare x and $y-x$. This is equivalent to having an x- and y-axis as in figure 2.69. Furthermore, we need to

[38] E. E. Kummer, *Über Ergänzungssätze zu den allgemeinen Reziprozitätsgesetzen,* Journal für die reine und angewandte Mathematik 44 (1852) 93–146. S. Wilson was probably the first who gave a rigorous explanation for the Sierpinski gasket appearing in Pascal's triangle, however, not using Kummer's result. See S. Wilson, *Cellular automata can generate fractals,* Discrete Appl. Math. 8 (1984) 91–99.

BASIC Program **Skewed Sierpinski Gasket**
Title The shortest possible program to do it skewed

```
DEFINT x, y
FOR y = 0 TO 255
    FOR x = 0 TO 255
        IF (x AND y) = 0 THEN PSET (x+30, y+30)
    NEXT x
NEXT y
END
```

BASIC Program **Sierpinski Gasket by Binary Addresses**
Title The shortest possible program to do it

```
DEFINT x, y
FOR y = 0 TO 255
    FOR x = 0 TO y
        IF (x AND (y-x)) = 0 THEN PSET (x+158-.5*y,y + 30)
    NEXT x
NEXT y
END
```

transform the point to be drawn, `(x+158-0.5*y,y+30)`. In other words, we shear the top of the triangle a bit to the right and center it at $x = 158$ (note, that $158 = 128 + 30$).

Can you draw the image with reversed colors (this requires substituting only one character)? Is it possible to modify the crucial line of the program in such a way that the IF statement compares `x AND y` but the resulting image is the same?

Note, that this program draws a figure which is only 256 pixels wide. If you want to change to a different size, do not forget to adapt the number 158 which is the width divided by 2 plus 30.

Hints for PC Users

If you are using an IBM-compatible PC and the resolution of your screen is only 320 x 200 pixels you can see only the first 200 lines of the Sierpinski gasket generated by this program (also, you should change `y+30` to `y` in the program). To see a 'complete' image change the number 255 in the second line of the code to 127. This produces a smaller version which fits completely on the screen.

Chapter 3

Limits and Self-Similarity

Now, as Mandelbrot points out [...] nature has played a joke on the mathematicians. The 19th-century mathematicians may have been lacking in imagination, but nature was not. The same pathological structures that the mathematicians invented to break loose from 19th-century naturalism turn out to be inherent in familiar objects all around us in Nature.

Freeman Dyson[1]

Dyson is referring to mathematicians, like G. Cantor, D. Hilbert, and W. Sierpinski, who have been justly credited with having helped to lead mathematics out of its crisis at the turn of the century by building marvelous abstract foundations on which modern mathematics can now flourish safely. Without question, mathematics has changed during this century. What we see is an ever increasing dominance of the algebraic approach over the geometric. In their striving for absolute truth, mathematicians have developed new standards for determining the validity of mathematical arguments. In the process, many of the previously accepted methods have been abandoned as inappropriate. Geometric or visual arguments were increasingly forced out. While Newton's *Principia Mathematica*, laying the fundamentals of modern mathematics, still made use of the strength of visual arguments, the *new objectivity* seems to require a dismissal of this approach. From this point of view, it is ironic that some of the constructions which Cantor, Hilbert, Sierpinski and others created to perfect their extremely abstract foundations simultaneously hold the clues to understanding the patterns of nature in a visual sense. The Cantor set, Hilbert curve, and Sierpinski gasket all give testimony to the delicacy and problems of modern set theory and at the same time, as Mandelbrot

[1] Freeman Dyson, *Characterizing Irregularity*, Science 200 (1978) 677–678.

Romanesco

Figure 3.1 : The new bread romanesco, a crossing between cauliflower and broccoli, exhibits striking self-similarity.

has taught us, are perfect models for the complexity of nature.

Finding the right abstract formulation for the old concept of a *limit* was part of the struggle to build a safer foundation for modern mathematics. As we know, the concept of limits is one of the most powerful and fundamental ideas in mathematics and the sciences. At the same time, it is one which troubles many non-mathematicians. This is very unfortunate, especially because of the fact that contemporary mathematics seems to tell us that the concept of limits is trivial. The truth is, of course, that building the right mathematical framework for the understanding of limits took the best mathematicians thousands of years. It is therefore very inappropriate for us to ignore the problems of non-mathematicians today, for they are sometimes of the same quality and depth as those which puzzled the great mathematicians in the past.

Self-similarity, by contrast seems to be a concept which can be understood without any trouble. The term self-similarity hardly needs an explanation. One would guess that the term has been around for centuries, but it has not. It is only some 25 years old. The new bread romanesco, a crossing between cauliflower and broccoli, illustrates the concept (see figure 3.1 the color plates). Macroscopically we see a form which is best described as a cluster. That cluster is composed of smaller clusters, which look almost identical to the entire cluster, however scaled down by some factor. Each of these smaller clusters again is composed of smaller ones, and those again of even smaller ones. Without difficulty we can identify three generations of clusters on clusters. The second, third, and all the following generations are essentially scaled down versions of the previous ones. In a rough sense, this is what we call *self-similarity*.

We will see that a rigorous discussion of the concept of self-similarity is intimately related to the concept of limit, and therefore it will require some care. The visual observation in nature, however, is simple and immediate. Once one has been introduced to this basic phenomenon, it is hard to walk through the fields and woods without constantly examining plants and other structures.

Fractals add a new dimension to the problems of dealing with limits; but also — and this is our point here — a refreshingly new perspective from which to understand the concept of limits. On one hand fractals may visualize the limit object in a feedback process, on the other hand some fractals demonstrate self-similarity in its purest form. In fact, many fractals can be completely characterized and defined by their self-similarity properties.

3.1 Similarity and Scaling

What is Similarity?

Self-similarity extends one of the most fruitful notions of elementary geometry: similarity. Two objects are similar if they have the same shape, regardless of their size. Corresponding angles, however, must be equal, and corresponding line segments must all have the same factor of proportionality. For example, when a photo is enlarged, it is enlarged by the same factor in both horizontal and vertical directions. Even an oblique, i.e. non-horizontal, non-vertical, line segment between two points on the original will be enlarged by the same factor. We call this enlargement factor *scaling factor*. The transformation between the objects is called similarity transformation.

Similarity Transformations

Similarity transformations are compositions involving a scaling, a rotation and a translation. A reflection may additionally be included, but we skip the details of that. Let us be more specific for similarity transformations in the plane. Here we denote points P by their coordinate pairs $P = (x, y)$. Let us apply scaling, rotation and translation to one point $P = (x, y)$ of a figure. First, a scaling operation, denoted by S, takes place yielding a new point $P' = (x', y')$. In formulas,

$$x' = sx,$$
$$y' = sy,$$

where $s > 0$ is the scale factor. A scale reduction occurs, if $s < 1$, and an enlargement of the object will be produced when $s > 1$. Next, a rotation R is applied to $P' = (x', y')$ yielding $P'' = (x'', y'')$.

$$x'' = \cos\theta \cdot x' - \sin\theta \cdot y',$$
$$y'' = \sin\theta \cdot x' + \cos\theta \cdot y'.$$

This describes a counterclockwise (mathematically positive) rotation of P' about the origin of the coordinate system by an angle of θ. Finally, a translation T of P'' by a displacement (T_x, T_y) is given by

$$x''' = x'' + T_x,$$
$$y''' = y'' + T_y$$

which yields the point $P''' = (x''', y''')$. Summarizing, we may write

$$P''' = T(P'') = T(R(P')) = T(R(S(P)))$$

or, using the notation

$$W(P) = T(R(S(P)))$$

we have $P''' = W(P)$. W is the similarity transformation. In formulas,

$$x''' = s\cos\theta \cdot x - s\sin\theta \cdot y + T_x ,$$
$$y''' = s\sin\theta \cdot x + s\cos\theta \cdot y + T_y .$$

Applying W to all points of an object in the plane produces a figure which is similar to the original.

Similarity Transformation in Two Dimensions

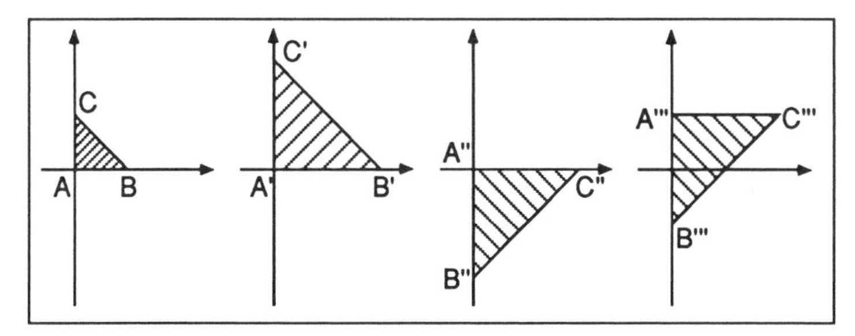

Figure 3.2 : A similarity transformation is applied to the triangle ABC. The scaling factor is $s = 2$, the rotation is by 270°, and the translation is given by $T_x = 0$ and $T_y = 1$.

The similarity transformations can also be formulated mathematically for objects in other dimensions, for example for shapes in three or only one dimension. In the latter case we have points x on the real line, and the similarity transformations can simply be written as $W(x) = sx + t$, $s \neq 0$.

Consider a photo which is enlarged by a factor of 3. Note that the area of the resulting image is $3 \cdot 3 = 3^2 = 9$ times the area of the original. More generally, if we have an object with area A and scaling factor s, then the resulting object will have an area which is $s \cdot s = s^2$ times the area A of the original. In other words, the area of the scaled-up object increases as the square of the scaling factor.

Scaling 3D-Objects

What about scaling three-dimensional objects? If we take a cube and enlarge it by a scaling factor of 3, it becomes 3 times as long, 3 times as deep, and 3 times as high as the original. We observe that the area of each face of the enlarged cube is $3^2 = 9$ times as large as the face of the original cube. Since this is true for all six faces of the cube, the total surface area of the enlargement is 9 times as much as the original. More generally, for objects of any shape whatever, the total surface area of a scaled-up object increases as the square of the scaling factor.

What about volume? The enlarged cube has 3 layers, each with $3 \cdot 3 = 9$ little cubes. Thus the total volume is $3 \cdot 3 \cdot 3 = 3^3 = 27$

DISCORSI
E
DIMOSTRAZIONI
MATEMATICHE,
intorno à due nuoue ſcienze
Attenenti alla
MECANICA & i MOVIMENTI LOCALI;
del Signor
GALILEO GALILEI LINCEO,
Filoſofo e Matematico primario del Sereniſſimo
Grand Duca di Toſcana.
Con vna Appendice del centro di grauità d'alcuni Solidi.

IN LEIDA,
Appreſſo gli Elſevirii. M. D. C. XXXVIII.

Figure 3.3 : Galileo's *Dialogues Concerning Two New Sciences* from 1638.

times as much as the original cube. In general, the volume of a scaled-up object increases as the cube of the scaling factor.

These elementary observations have remarkable consequences, which were the object of discussion by Galileo (1564–1642) in his 1638 publication *Dialogues Concerning Two New Sciences*. In fact Galileo[2] suggested 300 feet as the limiting height for a tree.

[2]We quote D'Arcy Thompson's account from his famous 1917 *On Growth and Form* (New Edition, Cambridge

Giant sequoias, which grow only in the Western United States and hence were unknown to Galileo, have been known to grow as high as 360 feet. However, Galileo's reasoning was correct; the tallest giant sequoias adapt their form in ways that evade the limits of his model.

What was his reasoning? The weight of a tree is proportional to its volume. Scaling up a tree by a factor s means that its weight will be scaled by s^3. At the same time the cross-section of its stem will only be scaled by s^2. Thus the pressure inside the stem would scale by $s^3/s^2 = s$. That means that if s increases beyond a certain limit, then the strength of wood will not be sufficient to bear the corresponding pressure.[3]

This tension between volume and area also explains why mountains do not exceed a height of 7 miles, or why different creatures respond differently to falling.[4] For example, a mouse may be unharmed by a ten-story fall, but a man may well be injured by just falling from his own height. Indeed, the energy which has to be absorbed is proportional to the weight, i.e. proportional to the volume of the falling object. This energy can only be absorbed over the surface area of the object. With scaling up volume, hence weight, hence falling energy, goes up much faster than area. As volume increases the hazards of falling from the same height increase.

Similarity and Growth of Ammonites

In chapter 4 we will continue to discuss scaling properties. In particular we will look at spirals, as for example the logarithmic spiral. We all have seen how a spiral drawn on a disk seems to grow continuously as it is turned around its center. In fact, the logarithmic spiral is special in that magnifying it is the same as rotating the spiral. Figure 3.4 illustrates this remarkable phenomenon, which as such is another example of a self-similar structure. Figure 3.5 shows an ammonite which is a good example of a logarithmic spiral in nature. In other words, an ammonite grows according to a law of similarity. It grows in such a way that its shape is preserved.

Babies Are Not Similar to Their Parents

Most living things, however, grow by a different law. An adult is not simply a baby scaled up. In other words, when we wonder about the similarity between a baby and its parents we are not talking about (the mathematical term of) geometric similarity! In

University Press, 1942, page 27): "[Galileo] said that if we tried building ships, palaces or temples of enormous size, yards, beams and bolts would cease to hold together; nor can Nature grow a tree nor construct an animal beyond a certain size, while retaining the proportions and employing the material which suffice in the case of a smaller structure. The thing will fall to pieces of its own weight unless we either change its relative proportions, which will at length cause it to become clumsy, monstrous and inefficient, or else we must find new material, harder and stronger than was used before. Both processes are familiar to us in Nature and in art, and practical applications, undreamed of by Galileo, meet us at every turn in this modern age of cement and steel."

[3]Here is a related problem. Suppose a nail in a wall supports some maximum weight w; how much weight would a nail support which is twice as big?

[4]See J. B. S. Haldane, *On Being the Right Size*, 1928, for a classic essay on the problem of scale.

Magnifying a Logarithmic Spiral

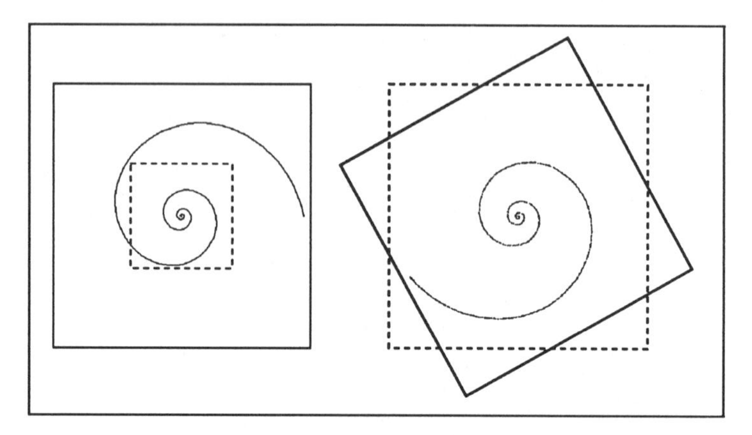

Figure 3.4 : The magnifying of a logarithmic spiral by a factor b shows the same spiral, however, rotated by an angle θ.

Ammonite

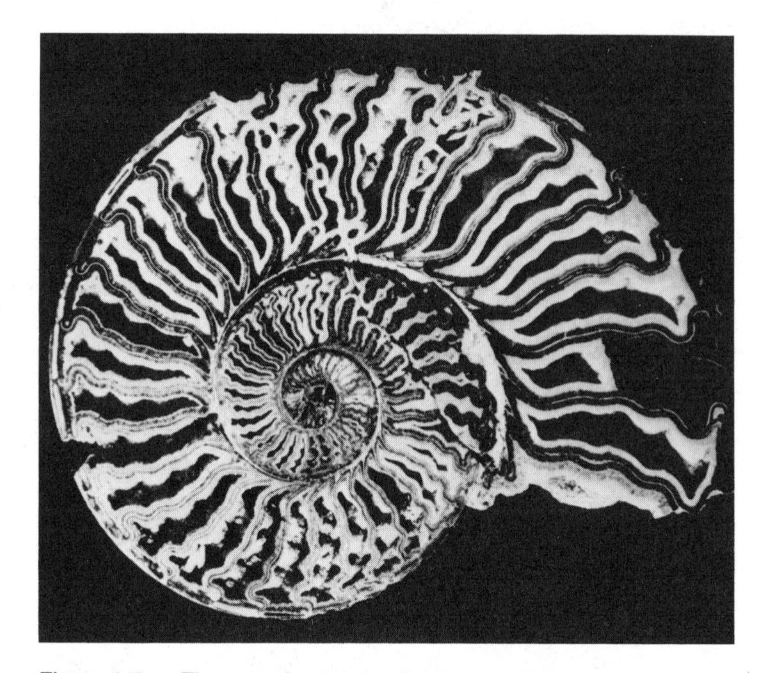

Figure 3.5 : The growth pattern of ammonite follows a logarithmic spiral.

the growth from baby to adult, different parts of the body scale up, each with a different scale factor. Two examples are:

- Relative to the size of the body, a baby's head is much larger than an adult's. Even the proportions of facial features are different: in a baby, the tip of the nose is about halfway down the face; in an adult, the nose is about two thirds of the way down. Figure 3.6 illustrates the deformation of a square grid necessary to measure the changes in shape of a human head from infancy to adulthood.

- If we measure the arm length or head size for humans of different ages and compare it with body height, we observe that humans do not grow in a way that maintains geometric similarity. The arm, which at birth is one-third as long as the body, is by adulthood closer to two-fifths as long. Figure 3.7 shows the changes in shape when we normalize the height.

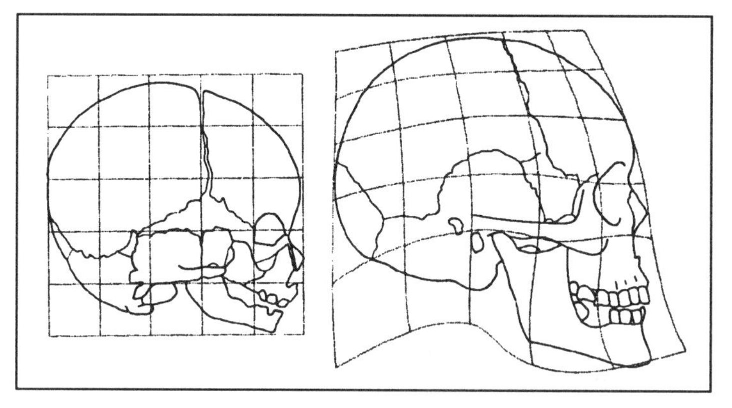

Figure 3.6 : The head of a baby and an adult are not similar, i.e. they do not transform by a simple scaling. Figure adapted from *For All Practical Purposes,* W. H. Freeman, New York, 1988.

Isometric and Allometric Growth

In summary, the growth law is far from being a similarity law. A way to get insight into the growth law of, for example, the head size versus the body height is by plotting the ratio of these two quantities versus age. In table 3.8, we list these data for a particular person.[5] Entering the ratio and the age in a diagram, and connecting the dots, we get a curve, see figure 3.9. If the growth were proportional, that is, according to similarity, the ratio would be constant throughout the life-time of the person, and we would have gotten a straight horizontal line. Graphing therefore provides a way to test for proportional growth. In our example data this is not the case. We can discern two different phases: one that fits early development, up to the age of about three years, and another that fits development after that time. In the first period we have proportional growth, sometimes called *isometric growth.* After the age of three years, however, the ratio drops significantly, indicating that body height is growing relatively faster than head size. This is called *allometric growth.* At about the age of 30

[5]The data in this table is taken from D'Arcy Thompson, *On Growth and Form*, New Edition, Cambridge University Press, 1942, page 190.

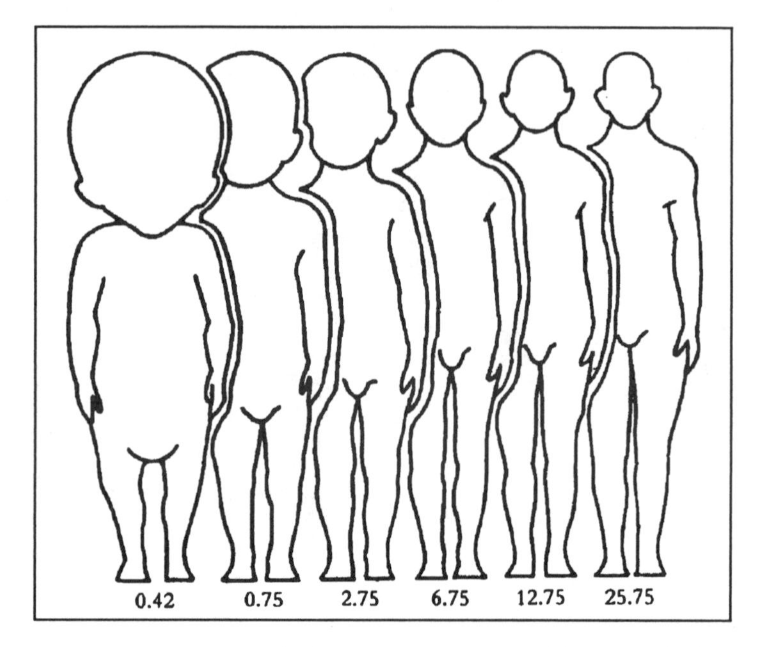

Figure 3.7 : Changes in shape between 0.5 and 25 years. Height is normalized to 1. Figure adapted from *For All Practical Purposes*, W. H. Freeman, New York, 1988.

Head Size Versus Body Height Data

Age years	Body Height cm	Head Size cm	Ratio
0	50	11	0.22
1	70	15	0.21
2	79	17	0.22
3	86	18	0.21
5	99	19	0.19
10	127	21	0.17
20	151	22	0.15
25	167	23	0.14
30	169	23	0.14
40	169	23	0.14

Table 3.8 : Body height and head size of a person. The last column lists the ratio of head size to body height. The first few years this ratio is about constant, while later it drops, indicating a change from isometric to allometric growth.

the growth process is completed and the ratio is constant again. A more sophisticated analysis of this data yielding mathematical growth laws will be presented in the next chapter. In fact, the well known phenomenon of nonproportional growth above is at the

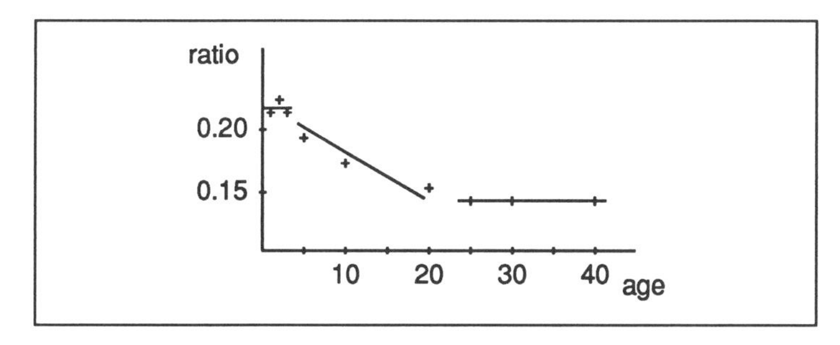

Graphing Growth

Figure 3.9 : Growth of head relative to height for the data from table 3.8. On the horizontal axis age is marked off, while the vertical axis specifies the head to body height ratio.

heart of fractal geometry as we will see shortly. Having discussed similarities and ways of scaling, let us now return to the central theme of this chapter: what is self-similarity?

What is Self-Similarity?

Intuitively, this seems clear; the word self-similar hardly needs a definition — it is self-explanatory. However, talking in precise mathematical terms about self-similarity really is a much more difficult undertaking. For example, in the romanesco, or for that matter, in any physically existing object, the self-similarity may hold only for a few orders of magnitude. Below a certain scale, matter decomposes into a collection of molecules, atoms, and, going a bit further, elementary particles. Having reached that stage, of course, it becomes ridiculous to consider miniature scaled-down copies of the complete object. Also, in a structure like a cauliflower the part can never be exactly equal to the whole. Some variation must be accounted for. Thus, it is already clear at this point that there are several variants of mathematical definitions of self-similarity. In any case, we like to think of mathematical fractals as objects which possess recognizable details at all microscopic scales — unlike real physical objects. When considering cases of fractals where the small copies, while looking like the whole, have variations, we have so called *statistical self-similarity*, a topic which we will get back to in chapter 7. Moreover, the miniature copies may be distorted in other ways, for example somewhat skewed. For this case there is the notion of *self-affinity*.

Self-Similarity of the Koch Curve

To exemplify the concept, we choose the Koch curve which is already familiar from the second chapter. Can we find similitudes (= similarity transformations) in the Koch curve? The Koch curve looks like it is made up of four identical parts. Let us look at one of these, say the one at the extreme left. We take a variable zoom lens and observe that at exactly $3\times$ magnifying power the little piece seems to be identical to the entire curve. Each one of the little pieces breaks into four identical pieces again, and each

Blowup of Koch Curve

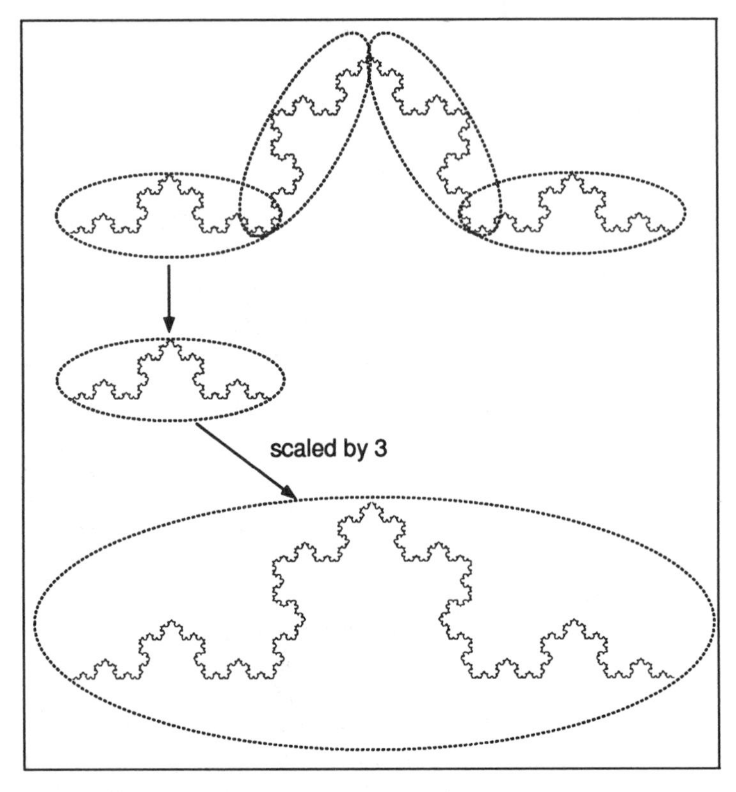

Figure 3.10 : One quarter of the Koch curve (top) is magnified by a factor of 3. Due to the self-similarity of the Koch curve the result is a copy of the whole curve.

of them seems to be identical to the entire Koch curve when we apply a magnifying lens of 9×, and so on ad infinitum. This is the self-similarity property, in its mathematically purest form.

Different Degrees of Self-Similarity

But even in this case, where copies of the whole appear at all stages and are exact and not distorted in any way, there are still various degrees of self-similarity possible. Consider for example a cover of a book that contains on it a picture of a hand holding that very book. Surprisingly, this somewhat innocent sounding description leads to a cover with a rather complex design. As we look deeper and deeper into the design, we see more and more of the rectangular covers. Contrast that with an idealized structure of a two-branch tree as shown in figure 3.11. Also pictured is the self-similar Sierpinski gasket. All three examples are self-similar structures: they contain small replicas of the whole. However, there is a significant difference. Let us try to find points which have the property that we can identify small replicas of the whole

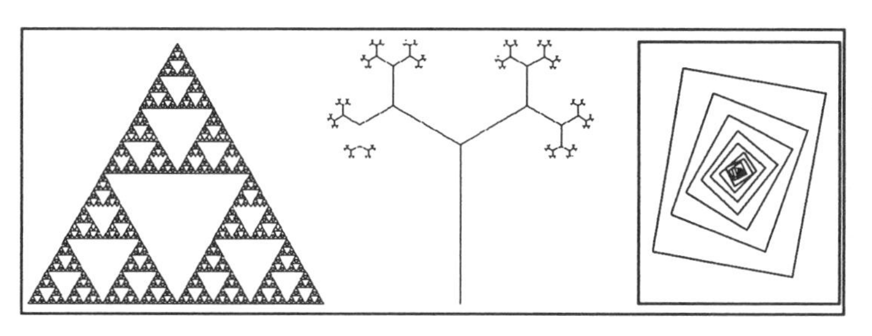

Three Different Self-Similar Structures

Figure 3.11 : A sketch of an image that contains on it a picture of itself is shown on the right. The two-branch tree is self-similar at the leaves, while the Sierpinski gasket is self-similar everywhere.

in their neighborhoods at any degree of magnification.

Self-Similarity at a Point

In the case of the book design, the copies are arranged in one nested sequence, and clearly the self-similarity property can be found only at *one particular point*. This is the limit point at which the size of the copies tends to zero. The book cover is self-similar at this point.[6]

Self-Affinity

The situation is much different in the two-branch tree. The complete tree is made up of the stem and *two* reduced copies of the whole. Thus, smaller and smaller copies accumulate near the leaves of the tree. In other words, the property of self-similarity condenses in the set of leaves. The whole tree is not strictly self-similar, but *self-affine*. The stem is not similar to the whole tree but we can interpret it as an affine copy which is compressed to a line.

Strict Self-Similarity

Finally, in the Sierpinski gasket, similar to the Koch curve above, we can find copies of the whole near *every* point of it, which we have already discussed. The gasket is composed from small but exact copies of itself. Considering these differences, we call all three objects self-similar, while only the Sierpinski gasket and the Koch curve are in addition called *strictly self-similar*. Also the set of leaves without the stem and all the branches is strictly self-similar. Now what would a cauliflower be in these categories? It would be a physical approximation of a self-similar, but not *strictly* self-similar, object akin to the two-branch tree.

[6]The notion of self-similarity at a point is central to the discussion of the self-similarity properties of the Mandelbrot set (see chapter 13).

3.2 Geometric Series and the Koch Curve

Fractals such as the Koch curve, the Sierpinski gasket and many others are obtained by a construction process. Ideally, however, this process should never terminate. Any finite stage of it produces an object, which may have a lot of fine structure, depending on how far the process has been allowed to proceed; but essentially it still is far from a true fractal. Thus, the fractal only exists as an idealization. It is what we would get if we let the process run 'indefinitely'. In other words, fractals really are limit objects, and their existence is not as natural as it may seem. This is important, and the mathematical foundation of such limits is one of the goals of this chapter and some others.

Limits often lead to new quantities, objects or qualities; this is true particularly for fractals (we will come back to that later on). However, given a sequence of objects, there are cases where it is not immediately obvious whether a limit exists at all. As for example, the first sum in

$$\sum_{k=1}^{\infty} \frac{1}{k} = \frac{1}{1} + \frac{1}{2} + \frac{1}{3} + \cdots$$

$$\sum_{k=1}^{\infty} \frac{1}{k^2} = \frac{1}{1} + \frac{1}{4} + \frac{1}{9} + \cdots$$

is divergent[7] (i.e. the sum is infinite) whereas the second one converges to $\pi^2/6$, as shown by Euler.

Let us recall for a moment the discussion of geometric series. For a given number $-1 < q < 1$ the question is, does

$$\sum_{k=0}^{\infty} q^k = 1 + q + q^2 + q^3 + \cdots$$

have a limit, and what is the limit? To this end one defines

$$S_n = 1 + q + q^2 + q^3 + \cdots + q^n \ .$$

Then on the one hand we have $S_n - qS_n = 1 - q^{n+1}$, and on the other hand $S_n - qS_n = S_n(1 - q)$. Putting these two identities together we obtain

$$S_n = \frac{1 - q^{n+1}}{1 - q} \ . \tag{3.1}$$

[7]The sum $1 + 1/2 + 1/3 + 1/4 + \cdots$ is infinite. An argument for this fact goes as follows. Assume that the sum has a finite value, say S. Then, clearly $1/2 + 1/4 + 1/6 + \cdots = S/2$. It follows that $1 + 1/3 + 1/5 + \cdots = S - (1/2 + 1/4 + 1/6 + \cdots) = S/2$. But also since $1 > 1/2$, $1/3 > 1/4$, $1/5 > 1/6$, ... we must have that $1 + 1/3 + 1/5 + \cdots > 1/2 + 1/4 + 1/6 + \cdots$. This is a contradiction, as both sums should equal $S/2$. Therefore our assumption, namely that the sum $1 + 1/2 + 1/3 + \cdots = S$ must be wrong. A finite limit of this sum cannot exist.

In other words, as n becomes larger, q^{n+1} becomes smaller, which means that S_n gets closer and closer to $1/(1-q)$. In short, we have justified writing

$$\sum_{k=0}^{\infty} q^k = \frac{1}{1-q} \,. \tag{3.2}$$

This is one of the elementary limit considerations which is useful,[8] even though the limit $1/(1-q)$ is not at all enlightening. Nevertheless, it will help us to understand a particular point about fractal constructions. In theory S_n will be different from the limit $1/(1-q)$, no matter how large we choose n. But practically, for example in a finite accuracy computer, both will be indistinguishable provided n is large enough.

The Construction Process of Geometric Series

The geometric series has an analogy in the construction of basic fractals. There is an initial object, here the number 1, and a scaling factor, here q. The important property of the scaling factor is that it be less than 1 in magnitude. Then there is a construction process.

Step 1: Start with 1.
Step 2: Scale down 1 by the scaling factor q and add.
Step 3: Scale down 1 by the scaling factor $q \cdot q$ and add.
Step 4: ...

The point is that this infinite construction leads to a new number, representing that process, the limit of the geometric series.

The Construction Process of the Koch Island

The Koch island, which we see in its basic construction in figure 3.12, is obtained in an analogous manner, except that rather than adding up numbers, we 'add up' geometrical objects. 'Addition', of course, is here interpreted as a union of sets; and the important point is that in each step we add a certain number of scaled down versions of the initial set.

Step 1: We choose an equilateral triangle T with sides of length a.
Step 2: We scale down T by a factor of 1/3 and paste on 3 copies of the resulting little triangle as shown. The resulting island is bounded by $3 \cdot 4$ straight line segments, each of length $a/3$.
Step 3: We scale down T by a factor of $1/3 \cdot 1/3$ and paste on $3 \cdot 4$ copies of the resulting little triangle as shown. The resulting island is bounded by $3 \cdot 4 \cdot 4$ straight line segments, each of length $1/3 \cdot 1/3 \cdot a$.
Step 4: ...

---

[8] Remember, for example, the problem of understanding infinite decimal expansions of the form 0.154399999... We know that it is just 0.1544, but why? Well, first $0.1543999... = 0.1543000...+9 \cdot 10^{-5}(1+10^{-1}+10^{-2}+10^{-3}+...)$. Then one can apply eqn. (3.2) with $q = 10^{-1}$ and obtain $1+10^{-1}+10^{-2}+10^{-3}+\cdots = 10/9$. Thus $0.1543000...+9 \cdot 10^{-5}(1+10^{-1}+10^{-2}+10^{-3}+...) = 9 \cdot 10^{-5} \cdot 10/9$, which is 10^{-4}. Finally, $0.1543999... = 0.1543000+10^{-4} = 0.1544$.

Koch Island Construction

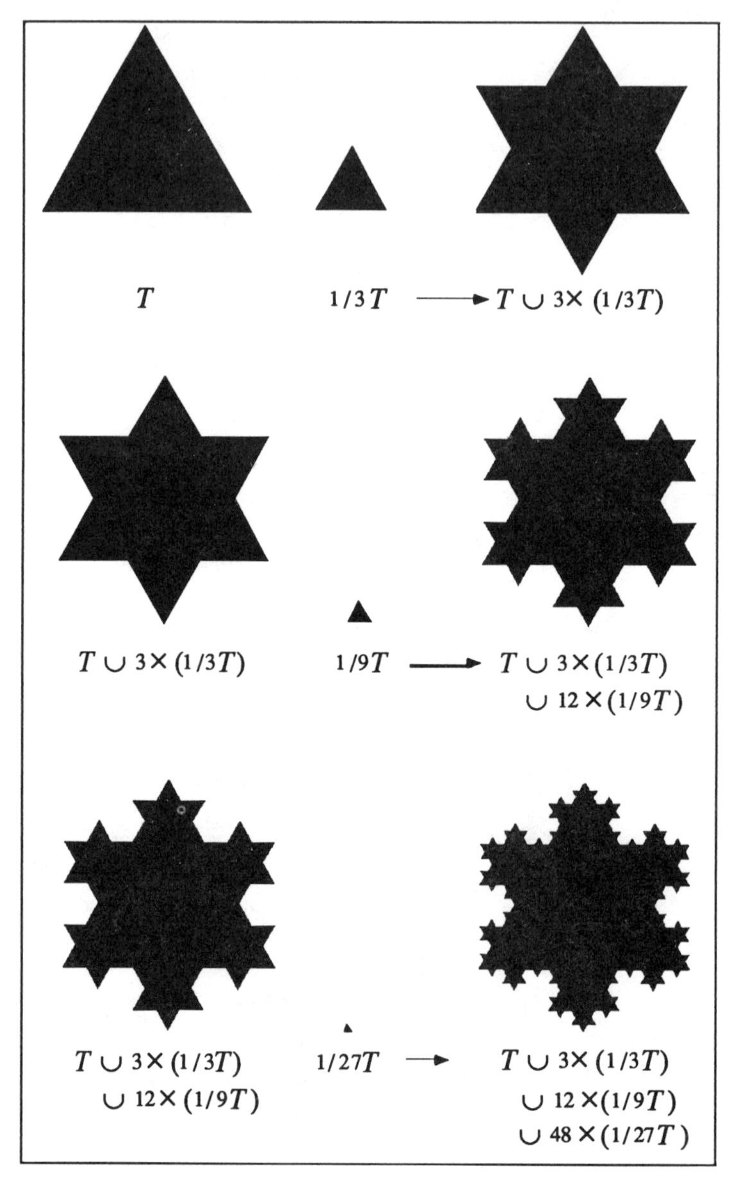

Figure 3.12 : The Koch island is the limit object of the construction and has area $\frac{2}{5}\sqrt{3}a^2$.

The point here is that this infinite construction leads to a *new* geometric object, the Koch island. In fact, the analogy between the geometric process and geometric series goes much further. Let us get a first impression. What is the area of the Koch island, the geometric object which we see as a limit of the above process?

The Area of the Koch Island

Well, let us try to figure out how much area we add in each step. At the beginning we have the area A_1 for the initial triangle T, and calculate $A_1 = \sqrt{3}/4 \cdot a^2$. In each step k, we have to add the area of n_k little equilateral triangles with sides s_k. Convince yourself that $n_1 = 3$, $n_2 = 3 \cdot 4$, $n_3 = 3 \cdot 4 \cdot 4$, ... In other words, $n_k = 3 \cdot 4^{k-1}$. The sides of the little triangles are obtained by successively scaling down the side of the original triangle by a factor 1/3. In other words, $s_k = (1/3)^n a$. Combining these results we get

$$\begin{aligned} A_{k+1} &= A_k + n_k \cdot \frac{\sqrt{3}}{4} \cdot s_k^2 \\ &= A_k + 3 \cdot 4^{k-1} \cdot \frac{\sqrt{3}}{4} \cdot \frac{1}{3^{2k}} a^2 \\ &= A_k + \frac{\sqrt{3}}{12} \cdot \left(\frac{4}{9}\right)^{k-1} a^2 . \end{aligned}$$

In other words, if we develop the terms step by step we have the series

$$A_{k+1} = A_1 + \frac{\sqrt{3}}{12} \left(1 + \frac{4}{9} + \frac{4^2}{9^2} + \cdots + \frac{4^{k-1}}{9^{k-1}}\right) a^2 .$$

The bracket in this formula contains a partial sum of the geometric series $1 + \frac{4}{9} + \frac{4^2}{9^2} + \frac{4^3}{9^3} + \cdots$ which has the limit $\frac{1}{1-4/9} = \frac{9}{5}$. That means that the Koch island, the geometric object in the limit, has area

$$A = A_1 + \frac{\sqrt{3}}{12} \cdot \frac{9}{5} a^2$$

and since $A_1 = \sqrt{3}/4 \cdot a^2$, we finally obtain

$$A = \frac{2}{5}\sqrt{3}a^2 .$$

This is quite a convincing argument that there is indeed a new geometric object resulting from the infinite process. But a rigorous argument would need much more.

It would need a language which would allow us to talk about the process of adding new and smaller shapes in the above construction exactly in the same way as is used to discuss adding smaller and smaller numbers in a series. In fact, this language already exists. One of the great achievements of what is called *point set topology* was to extend the idea of limits as known when dealing with numbers to far-reaching abstractness. This, together with a notion called *Hausdorff distance*, which is a generalization of the usual distance between points to the distance between two *point sets*, provides the right framework in which we can indeed find a perfect analogy between the infinite process of adding numbers in a geometric series and its limit behavior on the one hand, and the infinite adding of smaller and smaller triangles in the Koch island construction and its limit behavior on the other. In some sense, nothing new and exciting happens or has to be understood. Everything is just an appropriate translation of how we are used to thinking about geometric series. In that sense the Koch island is a visualization of the limit of a geometric series.

Limits Lead to New Qualities

Let us now look at properties of the limit, which are not shared by any of its finite stage approximations. The most important property is that of self-similarity. For example, the self-similarity of the Koch curve is reflected by the fact that the curve is made up of four identical parts. Can we actually verify the self-similarity with our images on paper? Of course not. There are two reasons: a technical one and a mathematical one.

The Technical Problem with Demonstrating Self-Similarity

The technical reason is obvious. Black ink on white paper comes in little dots which under a sufficiently high powered microscope look more or less like random specks and certainly not like a Koch curve. This effect could be called limited resolution and is very similar to the problem of representing numbers in a computer. Recall that $\sqrt{2}$ in a computer representation is never really $\sqrt{2}$, but rather some approximation such as 1.414215. Magnifying an image can be compared with multiplying a number by some factor greater than 1. For example, if we multiply $\sqrt{2}$ by $\sqrt{2}$ again and again we will get 2, $2\sqrt{2}$, 4, $4\sqrt{2}$, 8, $8\sqrt{2}$, ... In other words, we will get powers of 2, or powers of 2 multiplied by $\sqrt{2}$. If we multiply an approximation of $\sqrt{2}$ with itself again and again, then for a while we obtain results which approximate what we get in the case of multiplying $\sqrt{2}$ quite well. But sooner or later our numerical results will deviate more and more, and they will eventually totally disagree with our theoretical expectations.

The Mathematical Problem with Demonstrating Self-Similarity

The mathematical reason for the impossibility of running these experiments on paper is similar. Only the limit structure, but none of the intermediate construction steps, has the property of perfect self-similarity; and the limit structure cannot be obtained by any computer whatsoever. This is very much like the true and precise numerical value of $\sqrt{2}$ not being representable by any computer. It would need infinitely many digits. The only pictures of the Koch curve which are possible are approximating images. For example, if we draw an image of the 5th step and compare it with an image of the 10th step, we do not see a difference. But there is, of course, a dramatic difference. The change, however, is below the resolution of the device (printer or screen). No matter which step we may choose to represent our Koch curve, it will be indistinguishable from the true image of the Koch curve if the step is sufficiently

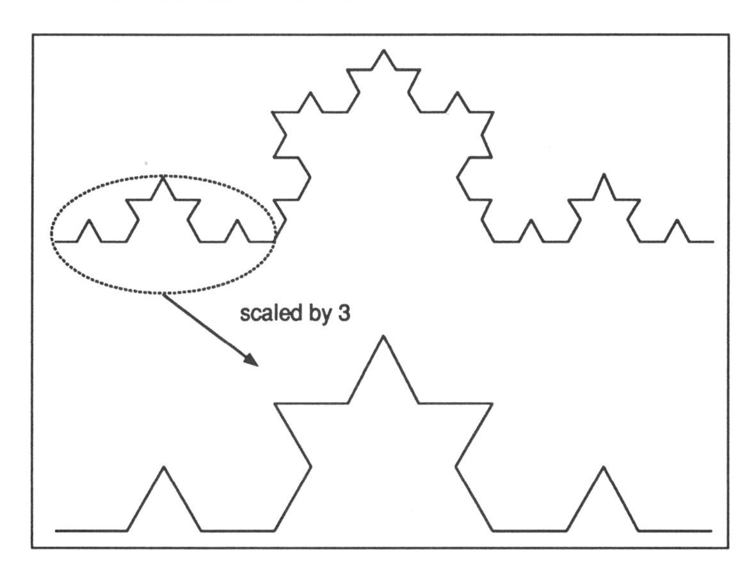

Figure 3.13 : Each single construction step of the Koch curve does not generate a self-similar curve. For example, scaling a part of the stage 3 approximation (top) by a factor of 3 does not yield a curve equal to the stage 3 curve (bottom).

No Self-Similarity at Finite Stage

advanced. But theoretically the two objects (i.e. some step in the process here and the Koch curve there) are dramatically different. For example, no matter which step we choose, the boundary of the respective object is made by tiny little straight line segments. Thus under sufficient magnification those will become macroscopically visible. In other words, if we look at one of the pieces in, say, the 10th step under the microscope, we will see a piece which is familiar, say, from the 2nd step while magnifications (with the correct magnifying factor) of the boundary in the limit structure will look exactly like the Koch curve. Also, an approximation of the Koch curve by any of its finite stage constructions cannot be self-similar, no matter how accurate the approximation is (see figure 3.13). The fact is, however, the Koch curve contains no straight line segment of any length.[9]

A Second New Quality of the Limit Object

Another property of the Koch curve, that is not shared by any of its finite stage approximations, is that its length is infinite (see section 2.4). As the Koch curve is one third of the boundary of the Koch island, we have that the boundary of the Koch island is also infinitely long. In contrast to this, the area of the Koch island is a finite and well defined number, as seen above. That, in

[9]Mathematically it is a continuous curve which is nowhere differentiable. It was invented by Helge von Koch just to provide an example for such a curve, see H. v. Koch, *Une méthode géometrique élémentaire par l'étude de certain questions de la théorie des courbes planes,* Acta. Mat. 30 (1906) 145–174.

fact, is the metaphoric message of Mandelbrot's 1967 article in the magazine *Science*, entitled *How long is the Coast of Britain?* We will discuss this in more detail in chapter 4.

Self-Similarity in Geometric Series

Looking back at the geometric series one may see a remarkable correspondence to the self-similarity of the Koch curve. If we formally scale the series

$$\sum_{k=0}^{\infty} q^k = 1 + q + q^2 + q^3 + \cdots$$

with the factor q, we obtain

$$q \sum_{k=0}^{\infty} q^k = q + q^2 + q^3 + q^4 + \cdots$$

Therefore,

$$\sum_{k=0}^{\infty} q^k = 1 + q \sum_{k=0}^{\infty} q^k \, . \tag{3.3}$$

This is the 'self-similarity' of the geometric series. The value of the sum is 1 plus the scaled down version of the whole series. As in the case of the Koch curve, the self-similarity only holds for the limit but not for any finite stage. For example, denote $S_2 = 1 + q + q^2$, then $1 + qS_2 = 1 + q + q^2 + q^3 \neq S_2$.

In summary, we have linked the Koch curve and island to the geometric series which provides an analogy and strong evidence for the existence of these fractals. Let us see in the next two sections how we can approach these objects from another direction, namely as solutions of appropriate equations.

3.3 Corner the New from Several Sides: Pi and the Square Root of Two

Limits have always had something mysterious about them, and it would be a great loss not to communicate that. Therefore, let us make an excursion and see how limits can reach out into the unknown. Limits create and characterize new quantities and new objects. The study of these unknowns was the pacemaker in early mathematics and has led to the creation of some of the most beautiful mathematical inventions. When Archimedes computed π by his approximation of the circle by a sequence of polygons, or when the Sumerians approximated $\sqrt{2}$ by an incredible numerical scheme, which was much later rediscovered by no one less than Newton, they were well aware of the fact that π and $\sqrt{2}$ were unusual numbers. The beautiful relation between the Fibonacci sequence $1, 1, 2, 3, 5, 8, 13, 21, 34, 55, \ldots$ and the golden mean
$\frac{1}{2}(1+\sqrt{5})$ has, over several centuries, inspired scientists and artists alike to wonderful speculations. It is almost ironic that mathematics and physics at the most advanced levels have recently taught us that some of these speculations, which motivated Kepler, among others, to speculate about the harmony of our cosmos, have an amazing parallel in modern science: it has been understood that in scenarios, which describe the breakdown of order and the transition to chaos, the golden mean number characterizes something like the last barrier of order before chaos sets in. Moreover, the Fibonacci numbers occur in a most natural way in the geometric patterns which can occur along those routes.

In this section we focus on two numbers, $\pi = 3.14\ldots$ and $\sqrt{2} = 1.41\ldots$, and their approximations from various directions. While the story of π is in some sense a diversion from fractals, the central theme of the book, the other example will be developed along lines which parallel the definition and approximation of fractals as worked out in the following sections.

Archimedes' Method for π

The method used by Archimedes for the calculation of π is based on inscribed and circumscribed regular polygon. In our presentation we use modern mathematical tools such as the sine and tangent functions which were not known to Archimedes, of course. We start with an inscribed hexagon. It has $n = 6$ sides. The angle covered by one half side is $\theta = \pi/6$ (see figure 3.14).

The length of the inscribed side is $2r \sin\theta$. The length of a side of the circumscribed hexagon is $2r \tan\theta$. Thus, for the length of the circle $U = 2\pi r$ we have

$$2rn \sin\theta < U < 2rn \tan\theta \ .$$

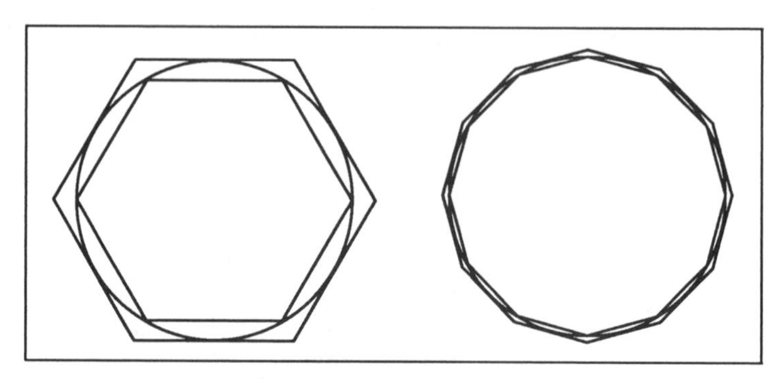

Figure 3.14 : Inscribed and circumscribed regular polygons.

Dividing by $2r$ we obtain a lower and an upper estimate for π,

$$n \sin\theta < \pi < n \tan\theta \ .$$

In numbers this is $3 < \pi < 3.464$, not a very accurate result. But we can easily improve the result simply by doubling the number n of sides and replacing θ by $\theta/2$ which yields

$$2n \sin\frac{\theta}{2} < \pi < 2n \tan\frac{\theta}{2} \ .$$

This is $3.106 < \pi < 3.215$. Further doubling, i.e. going from a regular polygon of 12 sides to one of 24 sides, and then to 48, 96, and so on, we can obtain an estimate that can be as sharp as we want to. After k such doubling steps the formula is

$$2^k n \sin\frac{\theta}{2^k} < \pi < 2^k n \tan\frac{\theta}{2^k} \ .$$

It is not clear, exactly what method Archimedes used to compute the sines and tangents. Probably he used an iteration method based on formulas similar to

$$\sin\frac{\theta}{2} = \sqrt{\frac{1-\cos\theta}{2}},$$
$$\tan\frac{\theta}{2} = \frac{\sin\theta}{1+\cos\theta}.$$

π and the Length of a Circle

The computation of the length of a circle, i.e. the computation of π, is a problem which challenged ancient mathematicians to a great extent. This problem has a history which is more than 4000 years old. The *Old Testament* uses $\pi = 3$ (see 1. Kings 7:23). The Babylonians used $\pi = 3.125$ and the Egyptians[10] (around

[10] In fact they proposed an algorithm for the computation of the area of a circle: take away $1/9$ of the diameter and square the remaining $8/9$ of the result.

1700 B. C.) proposed $\pi = 3.1604...$ Also in China philosophers and astronomers were very active in deriving approximations of π. One of the best goes back to Zu Chong-Zhi (430–501), who used the value 355/113, which has seven correct digits. At that time Chinese silk was sold as far as Rome. But it is not clear whether the fundamental work of Archimedes was also known to the Chinese. Archimedes was the first (around 260 B.C.) to provide a definite solution to the problem. He considered the circle with radius 1 and approximated half of its circumference by a sequence of regular polygons. In fact, he considered a sequence of approximating regular polygons which were inscribed and another sequence of regular polygons which were circumscribed. He carried out the approximation a few steps and obtained the numerical value 3.141031951 which has already four leading correct digits. He could have gone to even higher accuracy because his method was absolutely correct.

A more elegant method was discovered by the medieval scholar and philosopher Nicolaus Cusanus around 1450. It is another example for a feedback system and a forerunner of the very sophisticated methods used nowadays to compute π on mainframe computers up to millions of digits.

Cusanus' Method of Computation of π

Archimedes considered a fixed circle and approximated its circumference by a sequence of polygons. In a way Cusanus turned this approach around and employed a sequence of regular polygons with fixed circumference. More precisely, the regular polygons have 2^n, $n = 2, 3, 4, ...$ vertices such that the circumference always has length 2! He then computed the length of the corresponding circles which were inscribed and circumscribed (see figure 3.15).

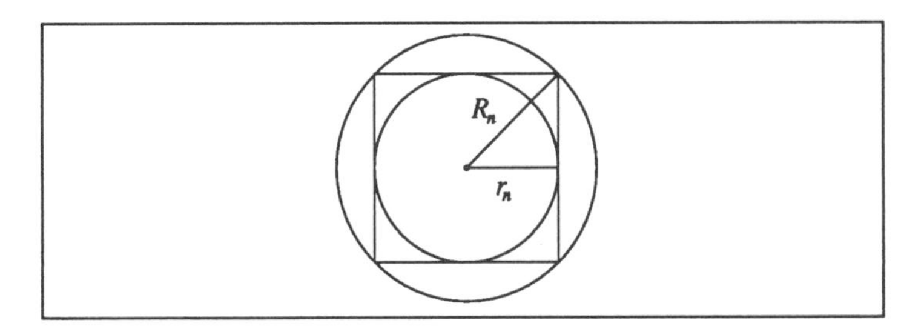

Figure 3.15 : Initial circle and square in Cusanus' method. For a given regular polygon with 2^n sides which sum up to a circumference of two units, the inscribed and circumscribed circles are considered.

Let R_n (or r_n) denote the radius of the circumscribed (or inscribed) circle of the nth polygon. Then we have

$$2\pi r_n < 2 < 2\pi R_n ,$$

The First Steps in Cusanus' Method

n	r_n	R_n	p_n	Error
2	0.250000	0.353553	3.313708	0.172116
3	0.301777	0.326641	3.182598	0.041005
4	0.314209	0.320364	3.151725	0.010132
5	0.317287	0.318822	3.144118	0.002526
6	0.318054	0.318438	3.142224	0.000631
7	0.318246	0.318342	3.141750	0.000158
8	0.318294	0.318318	3.141632	0.000039
9	0.318306	0.318312	3.141603	0.000010
10	0.318309	0.318310	3.141595	0.000002
11	0.318310	0.318310	3.141593	0.000001

Table 3.16 : The first few steps of Cusanus' method for the iterative calculation of π. The approximation p_n in the fourth column is computed by $p_n = 2/(r_n + R_n)$. The error $p_n - \pi$ decreases by about a factor of four in each step.

or equivalently

$$\frac{1}{R_n} < \pi < \frac{1}{r_n}. \tag{3.4}$$

For $n = 2$ we have a square with circumference 2, see figure 3.15, and thus we compute using the Pythagorean theorem $R_2 = \sqrt{2}/4$ and $r_2 = 1/4$. Then Cusanus continued to extract the following useful relations from geometric considerations which were already known to Archimedes, Pythagoras, and others.

$$\begin{aligned} r_{n+1} &= \frac{R_n + r_n}{2}, \\ R_{n+1} &= \sqrt{R_n r_{n+1}} \end{aligned}$$

for $n = 2, 3, \ldots$ It turns out that $r_n < R_n$ for all n and that r_n increases while R_n decreases as n grows. Thus, both sequences have limits, and these limits must be the same.[11] But then eqn. (3.4) implies that the limit must be $\frac{1}{\pi}$. It turns out that Cusanus' method yields π up to 10 correct decimal places if one computes the feedback system for up to $n = 18$. Table 3.16 lists the first eleven steps, the corresponding approximations of π, and their errors.

Other Approaches to π

F. Vieta (1540–1603):

$$\frac{2}{\pi} = \frac{\sqrt{2}}{2} \cdot \frac{\sqrt{2+\sqrt{2}}}{2} \cdot \frac{\sqrt{2+\sqrt{2+\sqrt{2}}}}{2} \cdots$$

[11] If they were not the same, say $R_n \to R$ and $r_n \to r$ with $r \neq R$, then $(r + R)/2 \neq R$, as it should.

J. Wallis (1616–1703):

$$\frac{\pi}{2} = \frac{2 \cdot 2}{1 \cdot 3} \cdot \frac{4 \cdot 4}{3 \cdot 5} \cdot \frac{6 \cdot 6}{5 \cdot 7} \cdot \frac{8 \cdot 8}{7 \cdot 9} \cdots$$

J. Gregory (1638–1675) and G. W. Leibniz (1646–1716):

$$\frac{\pi}{4} = 1 - \frac{1}{3} + \frac{1}{5} - \frac{1}{7} + \frac{1}{9} - \frac{1}{11} + \frac{1}{13} - \cdots \qquad (3.5)$$

L. Euler (1707–1783):

$$\frac{\pi^2}{6} = \frac{1}{1^2} + \frac{1}{2^2} + \frac{1}{3^2} + \frac{1}{4^2} + \frac{1}{5^2} + \cdots$$
$$\frac{\pi^4}{90} = \frac{1}{1^4} + \frac{1}{2^4} + \frac{1}{3^4} + \frac{1}{3^4} + \frac{1}{4^4} + \cdots$$

C. F. Gauss (1777–1855):

$$\pi = 48 \arctan \frac{1}{48} + 32 \arctan \frac{1}{57} - 20 \arctan \frac{1}{239} \qquad (3.6)$$

S. Ramanujan (1887–1920):

$$\frac{1}{\pi} = \frac{\sqrt{8}}{9801} \sum_{n=0}^{\infty} \frac{(4n!)(1103 + 26390n)}{(n!)^4 396^{4n}}$$

J. M. Borwein and P. M. Borwein (1984):

$$x_{n+1} = \frac{1 - \sqrt{1 - x_n^2}}{1 + \sqrt{1 - x_n^2}} \qquad x_0 = \frac{1}{\sqrt{2}}$$
$$y_{n+1} = (1 + x_{n+1})^2 y_n - 2^{n+1} x_{n+1} \quad y_0 = \frac{1}{2}$$

With these settings y_n converges quadratically to $1/\pi$.

The following is another interesting characterization.[12] An integer is called *square free* provided it is not divisible by the square of a prime number. For example, 15 is square free ($15 = 3 \cdot 5$), but 50 is not ($50 = 2 \cdot 5^2$). Now let $h(n)$ be the number of, and $q(n) = h(n)/n$ be the fraction of, square free numbers between 1 and n. Then

$$\lim_{n \to \infty} q(n) = \frac{6}{\pi^2} \, .$$

---

[12] C. R. Wall, *Selected Topics in Elementary Number Theory*, University of South Carolina Press, Columbia, 1974, page 153.

World Records in π

Like no other irrational number, π has fascinated the giants of science as well as amateurs around the world. For hundreds and even thousands of years more and more digits of π have been worked out using sometimes extremely tedious methods. This enormous effort stands in absolutely no relation to its use. It would be hard to find applications in scientific computing, where more than some 20 digits of π are necessary. Nonetheless, people have been pushing the number of known digits of π higher and higher as if it were a sport like the high jump, where athletes are driven to equal and surpass the standing world record. When asking mountaineers about their motivation to painfully climb a particularly high peak, they very well might answer, that they do it 'because it is there'. In this sense the number π is even better than Mount Everest because the number of digits in π is unlimited. Once arrived at a world record, there is already the challenge to also conquer the next ten, or hundred, or million digits.

Ludolph's Number

Let us give some examples of the craze that went on in the previous centuries and that is still continuing today with the help of computers. The Dutch mathematician Ludolph von Ceulen (1539–1610) dedicated a large portion of his work to the computation of π. In 1596 he reported 20 digits of π, and just before his death he succeeded in computing 32 and even 35 digits pushing the method of Archimedes to its extreme: he used inscribed and circumscribed polygons with $2^{62} \approx 10^{18}$ vertices. The last three digits are inscribed on his tombstone, and henceforth the number π was also known as *Ludolph's number*.

Approximation of π Using Series and Hand Calculation

Year	Name	Digits
1700	Sharp	72
1706	Machin	100
1717	Delaney	127
1794	Vega	140
1824	Rutherford	208
1844	Strassnitzky and Dase	200
1847	Clausen	248
1853	Rutherford	440
1855	Richter	500
1873	Shanks	707
1945	Ferguson	620

Table 3.17 : Partial list of the world records of the computation of π from 1700 until computing machines became available.

Machin's Formula for π

In 1706 John Machin discovered an elegant and computable way to represent π as a limit. Before, in 1671, Gregory had discovered that the area under the curve $1/(1+x^2)$ from 0 to x was $\arctan x$. The arctangent series

$$\arctan x = x - \frac{x^3}{3} + \frac{x^5}{5} - \frac{x^7}{7} + \cdots \tag{3.7}$$

was a direct conclusion of this. Substituting $x = 1$ then gives an easy formula for $\pi/4$, see eqn. (3.5). However, this series is very slowly convergent and, thus, not useful for actual computations. Machin devised a neat like trick to modify the Gregory series and improve its convergence dramatically. The derivation is easy using the trigonometric identities

$$\tan \alpha \pm \beta = \frac{\tan \alpha \pm \tan \beta}{1 \mp \tan \alpha \tan \beta} .$$

Let β be the unique angle less than $\pi/4$ such that

$$\tan \beta = \frac{1}{5} .$$

Using the above trigonometric formulas, we compute

$$\tan 2\beta = \frac{2 \tan \beta}{1 - \tan^2 \beta} = \frac{\frac{2}{5}}{1 - \frac{1}{25}} = \frac{5}{12}$$

and

$$\tan 4\beta = \frac{2 \tan 2\beta}{1 - \tan^2 2\beta} = \frac{\frac{5}{6}}{1 - \frac{25}{144}} = \frac{120}{119} .$$

From the last result we see that $\tan 4\beta \approx 1$, and therefore $4\beta \approx \pi/4$. Now the *tangent* of the difference between these two angles can again be computed

$$\tan(4\beta - \frac{\pi}{4}) = \frac{\tan 4\beta - 1}{1 + \tan 4\beta} = \frac{\frac{1}{120}}{\frac{239}{120}} = \frac{1}{239} .$$

In other terms,

$$4\beta - \frac{\pi}{4} = \arctan \frac{1}{239} ,$$

or, solved for $\pi/4$, we obtain the final result

$$\frac{\pi}{4} = 4 \arctan \frac{1}{5} - \arctan \frac{1}{239} .$$

In contrast to Gregory's formula there are two series to be computed here, but this drawback is more than compensated for by the fact that these series converge much more rapidly, especially the second one. Following Machin's idea many more similar formulas expressing π as a sum of arctangents have been developed, among them one from Gauss, see eqn. (3.6).

The Number Cruncher's Pain ...

After the discovery of differential calculus in the 17th century, new and better methods were devised for the computation of π. These methods used the series expansions of the arcsine and arctangent. The most convenient one for calculation with paper and pencil was provided by John Machin (1680–1752). Table 3.17 lists the progress made along these lines.[13] Computations typically took several months. Of course, some mistakes in such immense work were unavoidable. So when Vega computed his 140 digits in 1794, he discovered an error in the 113th place of Delaney's result. The 200 digits of Strassnitzky and Dase also were not in agreement with Rutherford's. Clausen then showed that the error was in Rutherford's calculation. Also Shanks' result was wrong from digit 527 on. Of all these, Strassnitzky deserves special mention. The actual calculations were carried out by Johann Martin Zacharias Dase (1824–1861), who was a calculating prodigy. His extraordinary calculating powers were verified by renowned mathematicians. He multiplied two 8-digit numbers in 54 seconds, two 20-digit numbers in 6 minutes, and two 100-digit numbers in under 9 hours, all of it in his head! There are at least two abilities that such prodigies must have: rapid execution of arithmetical operations, and something like a photographic memory to store the vast amount of information. On the other hand it seems that extraordinary intelligence is not necessary; on the contrary, this would be counter-productive. Dase, for example, was no exception. All who knew him agree, that except for calculating and numbers, he was quite dull. At the age of 20 Strassnitzky taught him an arctangent formula for π similar to Machin's formula, and in two months time Dase produced 200 correct digits. But that was not all. In three years he computed natural logarithms of the first million integers, each to seven decimal places, and continued to work on a table of hyperbolic functions. He was brought to the attention of Gauss, and upon Gauss' recommendation, he started to make a listing of the factors of all numbers from 7,000,000 to 10,000,000, a work sponsored by the Hamburg Academy of Sciences. However, Dase died in 1861, after finishing about half of them.

... and the Circle Squarer's Ease

In 1885 F. Lindemann succeeded in proving a fundamental theorem on transcendental numbers which also solved an age-old problem: π is a transcendental number,[14] which implies that squaring the circle is an impossible task. Nonetheless, people continued to find 'solutions' to the circle squaring problem, some more obscure than others. Here is just one example. In 1897, the house of representatives of Indiana, USA, passed a bill 'for an act intro-

[13] Our exposition here is based in part on the book *A History of Pi* by Petr Beckmann, Second Edition, The Golem Press, Boulder, 1971.

[14] A number x is called algebraic provided that it is the root of a polynomial equation with rational coefficients. A transcendental number is one that is not algebraic.

ducing a new mathematical truth', which defined two(!) values of π, namely 3.2 and 4. Fortunately, the senate of Indiana postponed further consideration of the law indefinitely.

Approaching π with Technology

In the 20th century it became more and more difficult to break the record in the computation of π — until computers came on the scene. It was a relatively simple task to program a computer to evaluate, for example, Machin's formula up to a thousand digits. Of course, as soon as it became possible, it was done. Table 3.18 lists the records in this second phase.

Approximation of π by Computer

Year	Name	Computer	Digits
1949	Reitwiesner	ENIAC	2037
1945	Nicholson et al	NORC	3,089
1958	Felton	Pegasus	10,000
1958	Genuys	IBM704	10,000
1959	Unpublished	IBM704	16,167
1961	Shanks, Wrench	IBM7090	100,000
1973	Guilloud, Bouyer	CDC7600	1,000,000
1983	Kanada et al	Hitachi S-810	16,000,000
1985	Gosper	Symbolics	17,000,000
1986	Bailey	Cray2	29,300,000
1987	Kanada	SX 2	134,000,000
1989	Kanada	HITAC S-820/80	1,073,740,000

Table 3.18 : World records of the computation of π in the computer age. The computing times are mostly on the order of 5 to 30 hours, the shortest one being 13 minutes (1945) and the longest run (100 hours) gave the 1989 record.

The computations up until the seventies were all based on arctangent series that had already been used by the pre-computer age pioneers. A complete listing of the first 100,000 digits of π was published by Shanks and Wrench in 1962.[15] In the last section of the paper the authors speculate about the possiblity of computing a million digits, concluding that "One would really want a computer 100 times as fast, 100 times as reliable, and with memory 10 times as large. No such machine now exists. [...] In 5 to 7 years such a computer [...] will, no doubt become a reality. At that time a computation of π to 1,000,000 digits will not be difficult." The authors were too optimistic; it took 12 more years until Jean Guilloud and Martine Bouyer checked off that millionth digit.

The simplicity of the method using, for example, the Gauss formula (3.6) in connection with the arctan series (3.7) is a temptation for any ambitious programmer. It provides an excellent exercise

[15] D. Shanks and J. W. Wrench, Jr., *Calculation of π to 100,000 Decimals*, Mathematics of Computation 16, 77 (1962) 76–99.

for a programming course. We have tried it out and succeeded in computing the first 200,000 digits.[16] However, the undertaking turned out to be not quite as easy as initially assumed. In the first run only the first 60,000 digits were correct. The mistake was due to insufficient treatment of overflow errors.

How Far Can We Go?

The question arises, how many digits one can possibly hope to be able to compute? The algorithms based on arctangent expansions have the property that doubling the number of digits in the result requires a computation which is *four* times as long. The 1973 computation of a million decimals took 23 hours. For example, to get from one million to 128 million digits, one must double the number of digits seven times ($128 = 2^7$). On the same computer, the time of 23 hours, quadrupled seven times, would have yielded a computing time requirement of about 43 years ... Even though computers were becoming faster and faster, it was clear that there would be an end to that development sooner or later. Thus, a couple more millions of digits seemed possible, but certainly not hundreds of millions of digits. So the record of a million decimals stood for 10 years. But the grounds had already been prepared for yet another escalation.

Another Breakthrough: New Algorithms

A major breakthrough occurred in 1976 when algorithms which yield a quadratically convergent iteration procedure were discovered independently by Brent and Salamin.[17] This means that in each iteration step of these methods the number of correct digits is doubled. More recently the Borwein brothers have worked out a family of even more efficient methods.[18] All the new algorithms are more efficient than the good old arctangent series, however, only because of another breakthrough in a different area — arithmetic. Addition of two n-digit numbers costs about n operations (add all corresponding pairs of digits and add up). However, the direct, naive multiplication of two n-digit numbers would essentially have to be carried out in n^2 operations (multiply each digit with all other digits and add up). Thus, when the number of digits n is of the order of a million or more, the difference between addition and multiplication is dramatic. Thus, the discovery, that the complexity of the multiplication of numbers is effectively not significantly larger than that of the addition of numbers is almost unbelievable: a multiplication can be made almost as fast as an addition.[19] Practical implementations make use of a form of Fast

---

[16] The program ran about 15 hours on a Macintosh FX.

[17] R. P. Brent, *Fast multiple-precision evaluation of elementary functions,* Journal Assoc. Comput. Mach. 23 (1776) 242–251. E. Salamin, *Computation of π Using Arithmetic-Geometric Mean,* Mathematics of Computation 30, 135 (1976) 565–570.

[18] See the book J. M. Borwein, P. B. Borwein, *Pi and the AGM — A Study in Analytic Number Theory,* Wiley, New York, 1987.

[19] More precisely, the way the computing requirements grow as the number of digits in the factors of the multi-

Fourier Transformation techniques. The combination of the new feedback methods for π and the fast multiplication algorithms for very long numbers facilitated computations of π with millions of digits of precision. The record at the time of writing this book stands at one billion digits,[20] and the prospects are good to get even 2 billion digits in the very near future.[21] Of course, the first million digits are just as useless as all other millions of digits that may follow.

Two Reasons to Compute π

However, there are two new reasons for this excessive digit hunting. The first one is related to a longstanding conjecture which states that the digits in π as well as the pairs of digits, the triplets of digits, and so on are uniformly distributed. In mathematical terms, π is believed to be a normal number. By extensive computer study, one may be able to find signs about the truth or falsity of this conjecture. At least up to the 29.3 million digits that Bailey has computed, all statistical tests indicate that π is, in fact, normal. Of course, this is far from a proof. The other reason, why programs for the calculation of π should be written, is that they can be used to effectively test the reliability of computer hardware. It is claimed that some computer manufacturers indeed perform such tests.[22] Even the smallest error at any operation in the calculation will invariably produce wrong digits from some place on, and these errors are very obviously detectable.

Is There a Message in π?

The advanced and more recent efforts to compute π may have inspired Carl Sagan to a part of his novel *Contact*[23] where he presents speculation about a hidden pattern or message God may have provided in the digits of π. In the story a super computer makes a discovery after countless hours of number crunching: the sequence of digits of π, located very far from the beginning, interpreted bitwise and displayed as a rectangular picture, shows a well known figure — a circle. The novel concludes:

"In whatever galaxy you happen to find yourself, you take the circumference of a circle, divide by its diameter, measure closely enough, and uncover a miracle — another circle, drawn kilometers downstream of the decimal point. There would be richer messages

plication is increased is not much worse than the corresponding (linear) growth of computing time for the addition of long numbers. The interested reader is referred to the survey in D. Knuth, *The Art of Computer Programming, Volume Two, Seminumerical Algorithms*, Addison Wesley, 1981, pages 278–299.

[20] Besides Yasumasa Kanada from the University of Tokyo, also David and Gregory Chudnovsky from Columbia University, New York succeeded in computing one billion digits. Their results agree.

[21] For a recent update on techniques and algorithms see J. M. Borwein, P. B. Borwein, and D. H. Bailey: *Ramanujan, modular equations, and approximations to pi, or how to compute one billion digits of pi*, American Mathematical Monthly 96 (1989) 201–219.

[22] In fact, in the 1961 paper by Shanks and Wrench, one instance of such hardware failure was reported, and an auxiliary run of the program was made to correct for the error. Thus, at least in the time about 30 years ago, reliability of the arithmetic was an important practical issue even for the 'end user'.

[23] Carl Sagan, *Contact*, Pocket Books, Simon & Schuster, New York, 1985.

further in. It doesn't matter what you look like, or what you're made of, or where you come from. As long as you live in this universe, and have a modest talent for mathematics, sooner or later you'll find it. It's already here. It's inside everything. You don't have to leave your planet to find it. In the fabric of space and in the nature of matter, as in a great work of art, there is, written in small, the artist's signature. Standing over humans, gods, and demons, [...] there is an intelligence that antedates the universe."

We now return to more worldly issues of numbers. Although limits are very useful for numerical computation of irrational numbers such as π, e or square roots, it is more satisfying from a theoretical point of view to have a more direct definition of the numbers. This could be an implicit definition in the form of a suitable equation that simultaneously prescribes an approximation by a feedback process, namely just by iterating the equation. Let us look at this issue in the remainder of this section.

$\sqrt{2}$ and Incommensurability

We recall the problem of the *incommensurability* of the side and the diagonal of a square: the ratio of the diagonal and side of a square is not equal to the ratio of two integers.[24] In other words, $\sqrt{2} \neq p/q$ for any integer p and q. No doubt the diagonal is real, but does that mean that $\sqrt{2}$ exists as a number in some sense? This was a great question; and though it sounds naive from todays point of view, it was not and still is not. Just ask yourself how you would *convince* somebody (of the existence of such a number). Certainly you could not expect much aid from the decimal expansion, which goes on and on in a seemingly totally disorganized fashion: the first 100 digits in the decimal expansion are

$$\begin{aligned}\sqrt{2} = 1.&41421\ 35623\ 73095\ 04880\ 16887\\ &24209\ 69807\ 85696\ 71875\ 37694\\ &80731\ 76679\ 73799\ 07324\ 78462\\ &10703\ 88503\ 87534\ 32764\ 15727\ldots\end{aligned}$$

But there is a different way to expand $\sqrt{2}$. Namely, to represent it as a special kind of limit, and then $\sqrt{2}$ looks almost as 'natural' as an integer does. This and some of the other most beautiful and mysterious limits are related to continued fraction expansions.

Continued Fractions

Let us begin with a seemingly strange way of writing rational numbers. Here is an example:

$$\frac{57}{17} = 3 + \cfrac{1}{2 + \cfrac{1}{1 + \cfrac{1}{5}}}\ .$$

[24] Compare chapter 2, page 142.

Let us see how this representation is obtained step by step:

$$\frac{57}{17} = 3 + \frac{6}{17} = 3 + \frac{1}{\frac{17}{6}} = 3 + \frac{1}{2 + \frac{5}{6}}$$
$$= 3 + \frac{1}{2 + \frac{1}{\frac{6}{5}}} = 3 + \frac{1}{2 + \frac{1}{1 + \frac{1}{5}}}$$

In this way any rational number can be written as a *continued fraction expansion.* The point is that a rational number has a finite expansion (i.e. the process terminates after some definite number of steps). In our example we write for short

$$\frac{57}{17} = [3, 2, 1, 5] \ .$$

The same algorithm applies to irrational numbers. However, in this case the algorithm never stops. It produces an infinite continued fraction representation.

Continued Fraction Expansion of $\sqrt{2}$

Let us look into a slightly more general situation which brings us back to $\sqrt{2}$. We begin with the equation:

$$x^2 + 2x - 1 = 0 \ .$$

The positive root of this equation is $x = \sqrt{2} - 1 < 1$. Note that $x^2 + 2x - 1 = 0$ can be rewritten as $x^2 + 2x = 1$, or $x(2 + x) = 1$, or

$$x = \frac{1}{2 + x} \ .$$

Moreover, after replacing x by $\frac{1}{2+x}$ on the right side,

$$x = \frac{1}{2 + \frac{1}{2 + x}}$$

and then, doing it again,

$$x = \frac{1}{2 + \frac{1}{2 + \frac{1}{2 + x}}}$$

etc. In other words, there will be an infinite repetition of 2's in the continued fraction expansion of $\sqrt{2} - 1$. Naturally, this implies that $\sqrt{2}$ has the expansion

$$x = 1 + \frac{1}{2 + \frac{1}{2 + \frac{1}{2 + \frac{1}{2 + \cdots}}}} \ .$$

This remarkable identity relates $\sqrt{2}$ with the sequence of numbers $[1,2,2,2,2,...]$, the digits of the continued fraction expansion of $\sqrt{2}$. We write $\sqrt{2} = [1,2,2,2,2,...]$ and mean that $1,2,2,2,...$ are placed into the fractions as above. In other words, $\sqrt{2}$ is the limit of the sequence $[1] = 1$, $[1,2] = 1.5$, $[1,2,2] = 1.4$, $[1,2,2,2] = 1.416$, ... Thus, $\sqrt{2}$ has a perfectly regular and periodic continued fraction expansion, while in an expansion with respect to some base like 10 the expansion looks like a big mess. It will never be periodic, because otherwise $\sqrt{2}$ would be a rational number.

Continued Fraction Expansion of the Golden Mean

The process which we discussed in detail for the equation $x^2+2x = 1$ works the same in a slightly different case,

$$x^2 = ax + 1$$

where a is an integer. After dividing by x and substituting for x twice we obtain

$$x = a + \frac{1}{x} = a + \cfrac{1}{a + \cfrac{1}{x}} = a + \cfrac{1}{a + \cfrac{1}{a + \cfrac{1}{x}}}$$

and so on. Thus, the continued fraction expansion will be

$$x = [a, a, a, ...] \; .$$

Specifically, if $a = 1$, then the positive root of $x^2 - x - 1 = 0$ is the *golden mean* $x = (1 + \sqrt{5})/2$ and we obtain

$$x = \frac{1+\sqrt{5}}{2} = [1,1,1,...] = 1 + \cfrac{1}{1 + \cfrac{1}{1 + \cdots}} \; .$$

Therefore the golden mean has the simplest possible continued fraction expansion. All roots of quadratic equations with integer coefficients have continued fraction expansions, which are eventually periodic, like $[2,2,2,3,2,3,2,3,...]$ or $[2,1,1,4,1,1,4,1,1,4,...]$. Rational numbers are characterized by a finite continued fraction expansions.

The Characterization by Equations

Let us summarize what our main point about irrational numbers is so far. If we only had a limit representation such as the decimal expansion of $\sqrt{2}$, we would feel quite uncomfortable. Comfort comes from some other characterization:

1. $\sqrt{2}$ has an elementary continued fraction expansion, [1,2,2,2,...].
2. $\sqrt{2}$ solves an equation, $x^2 - 2 = 0$.

But we can do even better. Consider the function

$$N(x) = \frac{1}{2}\left(x + \frac{2}{x}\right)$$

and its fixed points $x = N(x)$. Compute

$$x = \frac{1}{2}\left(x + \frac{2}{x}\right) = \frac{x}{2} + \frac{1}{x}, \quad \frac{x}{2} = \frac{1}{x}, \quad x^2 = 2\,.$$

Thus, the fixed points of the function $N(x)$ are just the square roots of two, and we may replace $x^2 - 2 = 0$ in our list above by

$$x = N(x) = \frac{1}{2}\left(x + \frac{2}{x}\right)\,.$$

There is an important reason for favoring this fixed point formulation over $x^2 - 2 = 0$: we can use $N(x)$ as the governing of the feedback process,

$$x_{n+1} = N(x_n), \quad n = 0, 1, 2, \ldots \tag{3.8}$$

This iteration will surely converge to the positive root of two, provided we start with a positive initial number $x_0 > 0$. We have discussed this already in chapter 1, page 33, and just give an example here, choosing $x_0 = 100$, see table 3.19.

Approximation of the Square Root of 2

n	x_n	Correct Digits
0	100.0000000000000000	0
1	50.0100000000000000	0
2	25.0249960007998400	0
3	12.5524580467459030	0
4	6.3558946949311400	0
5	3.3352816092804338	0
6	1.9674655622311490	1
7	1.4920008896897231	2
8	1.4162413320389438	3
9	1.4142150140500532	6
10	1.4142135623738401	13
11	1.4142135623730950	all

Table 3.19 : Approximation of square root of two using the iteration $x_{n+1} = (x_n + 2/x_n)/2$. The initial guess is $x_0 = 100$. Once the method is about the same magnitude as the true value 1.4142135623730950..., the iterates converge very rapidly, the number of correct digits double in each step.

We see that the iteration converges to $\sqrt{2}$ very rapidly after some initial iterations have brought the number x_n into a region close to the root. The number of correct leading digits roughly doubles in each step. Of course, this is no coincidence, but, in fact, the predominantly used method for the calculation of square roots, called *Newton's method.* Let us summarize our findings:

1. There is a well defined approximation procedure for $\sqrt{2}$, the feedback process

$$x_{n+1} = \frac{1}{2}\left(x_n + \frac{2}{x_n}\right), \quad x_0 > 0$$

   with a rapid convergence.
2. There is a corresponding fixed point equation

$$x = \frac{1}{2}\left(x + \frac{2}{x}\right) .$$

   which characterizes the limit, $\sqrt{2}$.

The fixed point equation should be seen in connection with symmetries, e.g. a regular hexagon is rotationally symmetric by a rotation of 60°, it also has a reflectional symmetry. In other words, one has an object, applies some operation (transformation) to it and obtains the same object. Our goal will be to corner fractals in the same way as one does irrational numbers, i.e. by an elementary limit process stemming from a fixed point equation, which characterizes the fractal by an invariance property.

3.4 Fractals as Solution of Equations

Let us return to fractals and find out how we can carry over the concepts we have learned from dealing with the square root of two. Summarizing the main point about the Koch curve we have that the curve is a limit of a process, a limit which has special properties, and which we can characterize in a similar way as $\sqrt{2}$ is characterized by its beautiful continued fraction expansion. But does it exist? Well, this question is very much of the same nature as the question of the existence of irrational numbers. Recall that in that case we take comfort from the fact that we believe in the validity of some closely related and characterizing concept. For example, for $\sqrt{2}$ we argue that this is the number which solves the equation $x^2 - 2 = 0$ or $x = (x + 2/x)/2$. Or for 2π we argue that this is the number which gives a length to the unit circle. Observe that here neither number is characterized as a limit of a sequence, and this really helps us to accept these numbers! The hypothesis that π might still not be known in mathematics if it did not relate to a circle so beautifully is speculative. Nevertheless, would Euler have discovered that $1 + 1/2^2 + 1/3^2 + 1/4^2 + \cdots$ is some very special number ($\pi^2/6$) worth being investigated if π was not somehow a reality?

In other words, we need some further reasons to accept the existence of the Koch curve, as well as characterizations which relate it to different ideas and concepts or principles. This is a major desire in mathematics. If an object or result suddenly becomes interpretable from a new point of view, mathematicians usually feel that they have made progress and are satisfied.

Is There an Invariance Property for the Koch Curve?

We may ask: is there an invariance property for the Koch curve? Can we find a characterization which is similar to that of $\sqrt{2}$? One type of invariance transformation is apparent. The Koch curve has an obvious reflectional symmetry. But this is not characterizing in the sense that it singles out the Koch curve. Ideally, we would like to find a transformation or a set of transformations which leave the Koch curve invariant. Such a transformation then could be viewed as some kind of symmetry. Recall the discussion of the self-similarity of the Koch curve at the end of section 3.2. Let us now be a little bit more formal and precise. Figure 3.20 illustrates the similarity transformation of the Koch curve. First, we reduce the Koch curve by a factor of 1/3. We put it onto a copier with reduction features and produce four copies. Then we paste the four identical copies along the edges of the broken line polygon in figure 3.20 (bottom) and obtain a curve which looks like the original one. The Koch curve is a collage of the four copies.

The Koch Collage

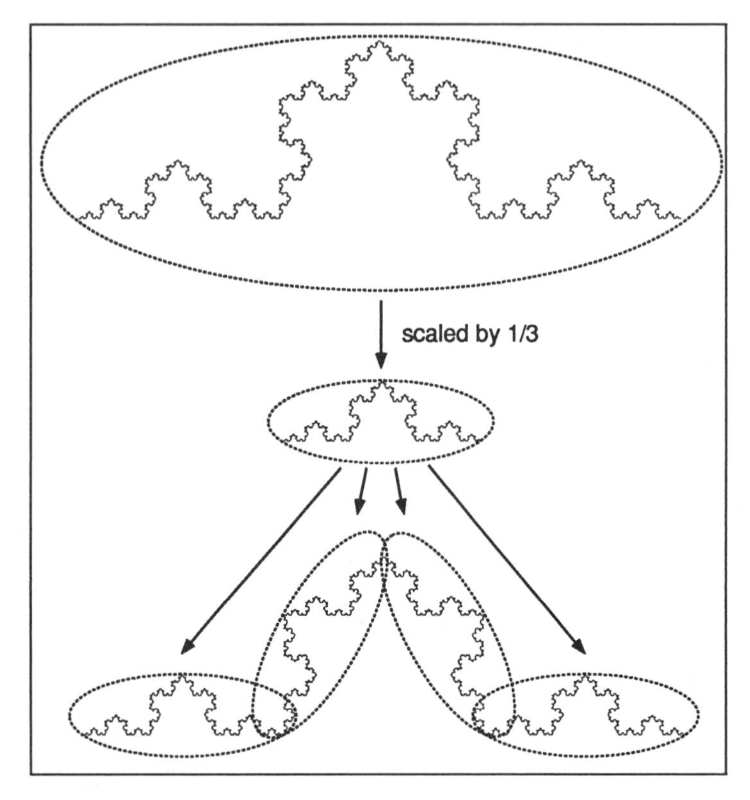

Figure 3.20 : The Koch curve is invariant under the transformations w_1 to w_4.

The Similarity Transformations of the Koch Curve

The following table lists the details of the similarity transformations w_1 to w_4 of the Koch curve as shown in figure 3.20.

number	scale	rotation	translation	
k	s	θ	T_x	T_y
1	1/3	0°	0	0
2	1/3	60°	1/3	0
3	1/3	−60°	1/2	$\sqrt{3}/12$
4	1/3	0°	2/3	0

Table 3.21 : Similarity transformations of the Koch curve collage. The transformations are carried out first by applying the scaling, then the rotation, and finally the translation (see section 3.1).

transformation	x-part	y-part
$w_1(x,y)$	$\frac{1}{3}x$	$\frac{1}{3}y$
$w_2(x,y)$	$\frac{1}{6}x - \frac{\sqrt{3}}{12}y + \frac{1}{3}$	$\frac{\sqrt{3}}{12}x - \frac{1}{6}y$
$w_3(x,y)$	$\frac{1}{6}x + \frac{\sqrt{3}}{12}y + \frac{1}{2}$	$-\frac{\sqrt{3}}{12}x - \frac{1}{6}y + \frac{\sqrt{3}}{12}$
$w_4(x,y)$	$\frac{1}{3}x + \frac{2}{3}$	$\frac{1}{3}y$

Table 3.22 : Explicit formulas for the similarity transformations of the Koch curve collage.

When we take into account that

$$\cos 60° = \cos(-60°) = \tfrac{1}{2} \ ,$$
$$\sin 60° = -\sin(-60°) = \tfrac{\sqrt{3}}{4} \ ,$$

we obtain explicit formulas for the transformations as given in table 3.22

Characterization by an Equation for the Self-Similarity

This collage-like operation can be described by a single mathematical transformation. We let w_k, $k = 1, ..., 4$, be the four similarity transformations given by a reduction with factor 1/3 composed with a positioning (rotation and translation) at piece k along the polygon as shown in figure 3.20(b). Then, if A is any image, let $W(A)$ denote the collection (union) of all four transformed copies

$$W(A) = w_1(A) \cup w_2(A) \cup w_3(A) \cup w_4(A) \ . \qquad (3.9)$$

This is a transformation of images, or more precisely, subsets of the plane. Figure 3.23 shows the result of this transformation when applied to an arbitrary image, for example, when A is the logo NCTM. When comparing the results in figure 3.20 and figure 3.23, we make a fundamental observation. In the case where we apply the transformation W from eqn. (3.9) to the image of the Koch curve, we obtain the Koch curve back again. That is, if we formally introduce a symbol K for the Koch curve, we have the important identity

$$W(K) = K \ ,$$

which is the desired invariance (or fixed point) property. In other words, if we pose the problem of finding a solution X to the equation $W(X) = X$, then the Koch curve K solves the problem. Moreover, this equation also shows the self-similarity of K since

$$K = w_1(K) \cup w_2(K) \cup w_3(K) \cup w_4(K)$$

states that K is composed of four similar copies of itself. In other words, we have characterized K by its self-similarity. If we further

The NCTM Collage

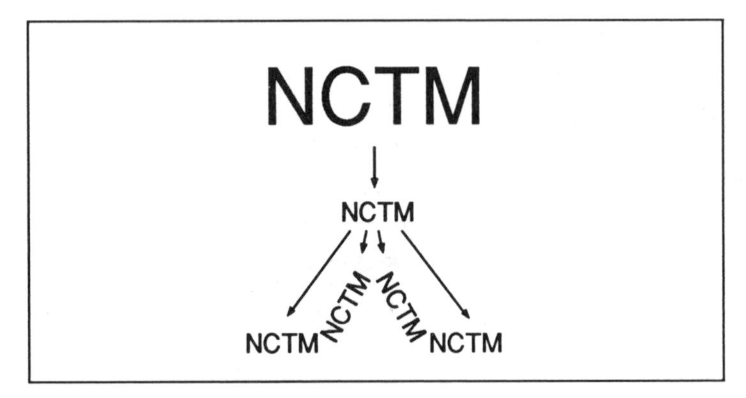

Figure 3.23 : The NCTM logo is not invariant under W.

substitute for K the collection of the four copies on the right hand side of the equation, then it becomes clear that K is made of 16 similar copies of itself, and so on. We will come back to this interpretation of self-similarity later in this section.

When we apply the same transformation W to the logo NCTM (i.e. X is the image 'NCTM'), we do not get back the logo NCTM at all. Rather, we see some strange collage.

Only the Koch Curve is Invariant Under W

We are led to conjecture that maybe the only image which is left invariant under the collage transformation W is the Koch curve. Indeed, that is a theorem which has far reaching consequences which will be discussed in chapter 5. A collage transformation like W above is called a *Hutchinson operator*, after J. Hutchinson, who was the first to discuss its properties.[25]

Koch Curve as a Limit Object

Having characterized the Koch curve as a fixed point of the Hutchinson operator, we now conclude the analogy to the computation of $\sqrt{2}$ (see eqn. (3.8)). It remains to show that mere iteration of the operator W applied to a starting configuration A_0 yields a sequence

$$A_{k+1} = W(A_k), \quad k = 0, 1, 2, ...$$

which converges to the limit object, the Koch curve. This is indeed the case, and figure 3.24 visualizes the limit process providing pictorial evidence that there is such a self-similar object. Let us summarize.

1. There is a well defined approximation procedure for the Koch curve, the feedback process

$$A_{k+1} = W(A_k), \quad k = 0, 1, 2, ...$$

[25] J. Hutchinson, *Fractals and self-similarity*, Indiana University Journal of Mathematics 30 (1981) 713–747.

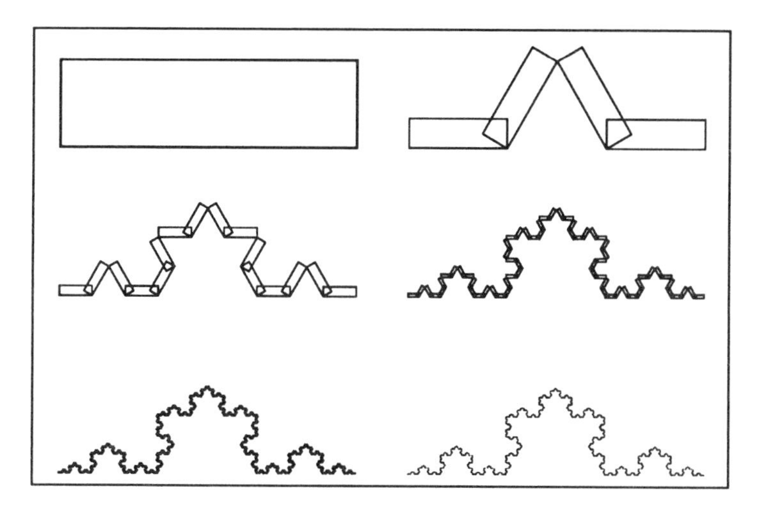

Figure 3.24 : Starting with an arbitrary shape, a rectangle, iteration of the Hutchinson operator produces a sequence of images, which converge to the Koch curve.

Limit Object Koch Curve

where A_0 can be any initial image and W denotes the Hutchinson operator

$$W(A) = w_1(A) \cup w_2(A) \cup w_3(A) \cup w_4(A)$$

for the Koch curve.

2. There is a corresponding fixed point equation

$$A = W(A)$$

which uniquely characterizes the limit, the Koch curve.

How can we make sure that what we see — W applied to the Koch curve yields the Koch curve again — is actually true? Can we really trust an image, or better, a graphic experiment? The answer is that we should take it as some supporting evidence, but not more than that. After all, it might be that in some invisibly small detail there is a difference between $W(K)$ and K itself. In other words, we must go on and convince ourselves that this remarkable self-similarity property is actually a fact and not just an experimental artifact. This will be our next goal. However, we will first discuss this property in two simpler examples, the *Cantor set* and *Sierpinski gasket*.[26]

Equation for the Cantor Set

In chapter 2 the Cantor set was introduced as a limit in a geometric feedback process (begin with the unit interval, remove the open interval of length 1/3 centered at 1/2, then remove the

[26]The mathematical discussion must be postponed to chapter 5 where we will look at the convergence of images and the characterization of fractals by Hutchinson operators in detail.

The Cantor Set Construction

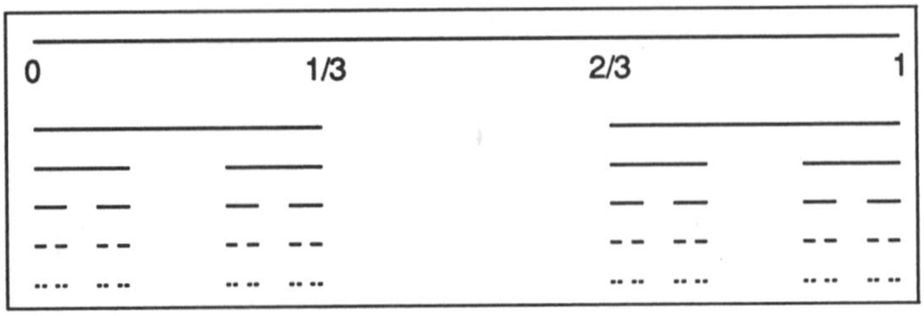

Figure 3.25 : The geometric feedback construction of the Cantor set.

The Cantor Set Transformations

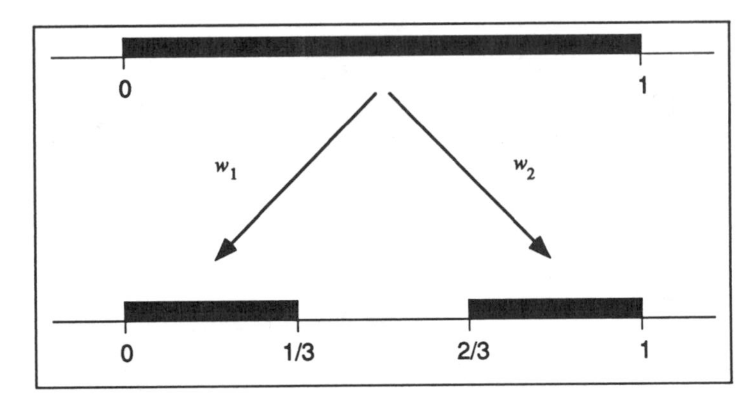

Figure 3.26 : The similarity transformations w_1 and w_2 for the Cantor set.

middle thirds of the remaining intervals, and so on). Moreover, it has been described as the set of numbers between 0 and 1, for which there exists a triadic expansion that does not contain the digit 1. This last characterization allows us to verify that the Cantor set is the fixed point of the appropriate Hutchinson operator W given by the two transformations

$$
\begin{aligned}
w_1(x) &= \frac{x}{3}\,, \\
w_2(x) &= \frac{x}{3} + \frac{2}{3}\,.
\end{aligned}
$$

Thus, for a given set A, $W(A) = w_1(A) \cup w_2(A)$. Figure 3.26 shows how the transformations act when applied to the unit interval.

Our claim is that the Cantor set is a solution to the equation

$$W(X) = X$$

i.e. the Cantor set C is invariant under W, and $W(C) = C$.

The Invariance of C under W

Cantor himself gave a characterization of the set named after him in terms of numbers expanded with respect to base 3, triadic numbers. Recall that any number x, $0 \leq x \leq 1$, can be expanded in

$$x = a_1 \cdot 3^{-1} + a_2 \cdot 3^{-2} + a_3 \cdot 3^{-3} + a_4 \cdot 3^{-4} + \ldots ,$$

where the digits a_k are from $\{0, 1, 2\}$. Then x is written in the form $x = 0.a_1a_2a_3\ldots$, i.e. we take the coefficients $a_1, a_2, a_3, \ldots$ as triadic digits. Then the Cantor set is determined by

$$C = \{x \mid x = 0.a_1a_2a_3\ldots, a_k \in \{0, 2\}\} ,$$

i.e. all numbers which admit a triadic expansion that misses the triadic digit 1. Using this characterization we can, in fact, convince ourselves that the invariance property, which characterizes self-similarity, is true: first, we have to understand how w_1 and w_2 act on triadic numbers, but that is really easy to explain: if $x = 0.a_1a_2a_3\ldots$, then $w_1(x) = 0.0a_1a_2a_3\ldots$, and $w_2(x) = 0.2a_1a_2\ldots$ Thus, if $a_k \in \{0, 2\}$, then the triadic digits of $w_1(x)$ and $w_2(x)$ will also have that property, i.e. $w_k(C)$, $k = 1, 2$, is contained in C again. But can we get all points of C in this way? Indeed, if $y \in C$, i.e. $y = 0.a_1a_2a_3\ldots$ and $a_k \in \{0, 2\}$, then there is x in C and exactly one of the two transformations w_k, $k = 1, 2$, will have the property $w_k(x) = y$. Simply take $x = 0.a_2a_3\ldots$ Then if $a_1 = 0$, choose w_1; otherwise choose w_2. This establishes that $W(C) = C$ holds.

The Invariance Property and Self-Similarity

The invariance property explains self-similarity. We start with

$$C = w_1(C) \cup w_2(C)$$

i.e. C is a collage of two similar copies of itself — scaled down by a factor of 1/3. Then we obtain

$$C = w_1(w_1(C) \cup w_2(C)) \cup w_2(w_1(C) \cup w_2(C)) ,$$

which leads to

$$C = w_1(w_1(C)) \cup w_1(w_2(C)) \cup w_2(w_1(C)) \cup w_2(w_2(C)),$$

i.e. C is a collage of four similar copies of itself — scaled down by a factor of 1/9, and so on. That is to say, we can identify smaller and smaller pieces in C which are just scaled down versions of C.

Let us discuss in a similar fashion the Sierpinski gasket. Again we begin by a limit characterization which is actually the one given by Sierpinski in 1916.

The Sierpinski Gasket as a Limit

Start with a triangle. It can be any kind, but for reasons which will soon become apparent, we will let T be a right triangle with two of its sides having length one. Now pick the midpoints of the sides. These define a center triangle, which we remove. We are

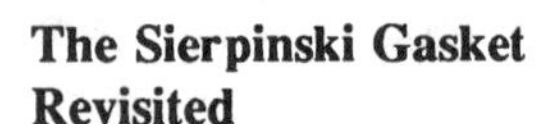

The Sierpinski Gasket Revisited

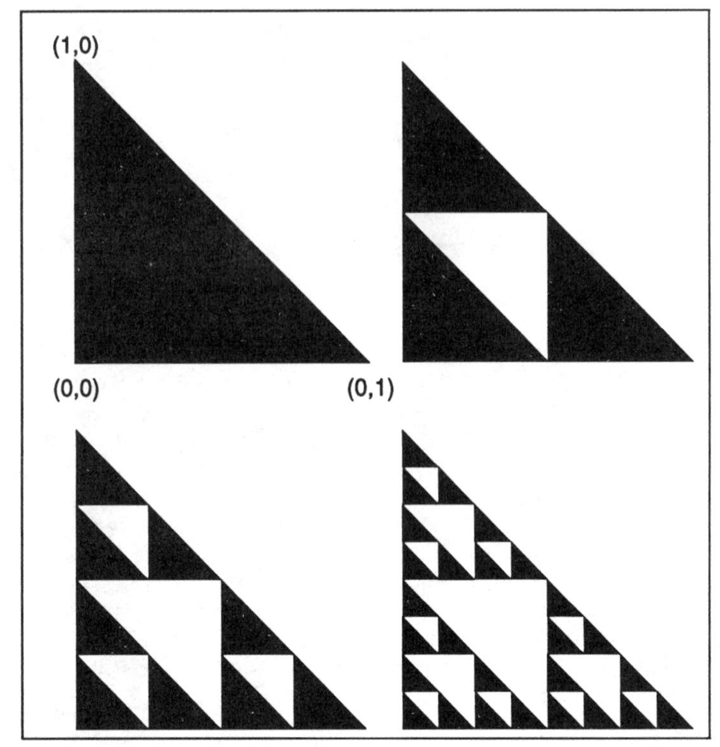

Figure 3.27 : Construction of Sierpinski gasket as limit. Stages 0 to 3 are shown.

left with three similar triangles, and for each we pick the midpoints of their sides, take away the center triangles, and are left with nine smaller triangles, and so on, see figure 3.27.

Also the Sierpinski gasket is self-similar. To discuss this feature, we think of it in a plane, so that the vertices are at coordinates (0,0), (1,0), and (0,1). Then we introduce three similarity transformations w_1, w_2, and w_3. Each of these transformations can be interpreted as a scaling by a factor of 1/2, together with a positioning such that

$$\begin{aligned} w_1(0,0) &= (0,0) \\ w_2(1,0) &= (1,0) \\ w_3(0,1) &= (0,1) \ . \end{aligned}$$

We claim that if S denotes the Sierpinski gasket then

$$S = w_1(S) \cup w_2(S) \cup w_3(S) \ . \tag{3.10}$$

The Invariance of the Sierpinski Gasket

In other words, if we introduce the Hutchinson operator

$$W(A) = w_1(A) \cup w_2(A) \cup w_3(A)$$

where A is any image in the plane, then

$$W(S) = S ,$$

i.e. the Sierpinski gasket is invariant under W, or it solves the equation $W(X) = X$.

This means that the Sierpinski gasket can be broken down into 3, or 9, or 27 (abstractly 3^k) triangles which are scaled down versions of the entire Sierpinski gasket S by a factor of 1/2, or $(1/2)^2$, or $(1/2)^3$ (abstractly $(1/2)^k$). In other words, once we have given an argument for eqn. (3.10), we have completely understood the self-similarity of the Sierpinski gasket.

The Binary Characterization of the Sierpinski Gasket and the Invariance of S under W

Though it seems to be obvious from the geometric construction that S should satisfy $S = w_1(S) \cup w_2(S) \cup w_3(S)$, we prefer to give a solid argument. The geometric removal process in figure 3.27 is equivalent to looking at certain points in the plane and taking away a certain subset in a systematic fashion. If (x, y) is a point in the plane with non-negative coordinates and $x + y \leq 1$, then (x, y) is in the triangle with vertices $(0,0), (1,0), (0,1)$. Given any point (x, y) from this triangle, it can be tested for membership in the Sierpinski gasket in the following way.

Write down a binary expansion of both coordinates

$$x = 0.a_1a_2a_3..., \text{ where } a_k \in \{0,1\} ,$$
$$y = 0.b_1b_2b_3..., \text{ where } b_k \in \{0,1\} .$$

The point (x, y) belongs to the Sierpinski gasket if and only if corresponding digits a_k and b_k are never both equal to 1. In other words, $a_k = 1$ must imply $b_k = 0$ and $b_k = 1$ must imply $a_k = 0$, and this holds for all $k = 1, 2, 3...$ We will derive this characterization in chapter 5, section 5.3.

Thus, a point z is disregarded if a dual expansion of its coordinates x and y have a pair of coefficients $a_k = 1$, $b_k = 1$. At first there seems to be a problem with some points like $(x, y) = (0.5, 0.5)$ for example. This is clearly a point in the Sierpinski gasket although it seems obvious from the equality of x and y that one can always find corresponding binary digits a_k and b_k of x and y which are both equal to 1. But note that 0.5 has two dual representations, one is $0.5 = 0.1000...$ and the other one is $0.5 = 0.0111...$ Choosing the first for x and the second for y we see that the point belongs to S also according to the binary characterization of the Sierpinski gasket.

Using the binary characterization of S we can now argue that Hutchinson's formula $S = w_1(S) \cup w_2(S) \cup w_3(S)$ is correct. All we have to do is understand how w_k acts upon a point (x, y) in S.

transformation	x-part	y-part
$w_1(x,y)$	$0.0a_1a_2a_3...$	$0.0b_1b_2b_3...$
$w_2(x,y)$	$0.1a_1a_2a_3...$	$0.0b_1b_2b_3...$
$w_3(x,y)$	$0.0a_1a_2a_3...$	$0.1b_1b_2b_3...$

Table 3.28 : Explicit formulas in dual expansions for the similarity transformations of the Sierpinski gasket. Here the point $z = (x, y)$ is given in the dual expansion, $x = 0.a_1a_2a_3...$ and $y = 0.b_1b_2b_3...$

The details are a bit tedious, but they are of the same nature as with the ternary characterization of the Cantor set from chapter 2. Table 3.28 lists the three points to which (x, y) is transformed under w_1, w_2 and w_3. Note that the points of the Sierpinski gasket can be grouped into three sets depending on the first binary digits of x and y. The first set collects points with $a_1 = b_1 = 0$, the second points with $a_1 = 1$ and $b_1 = 0$, and in the third set we find all points with $a_1 = 0$ and $b_1 = 1$. There are just three points which are contained in two of the above categories simultaneously, namely $(0.5, 0)$, $(0, 0.5)$, and $(0.5, 0.5)$. But this does not pose any problem for the following conclusion. Using the above table it becomes clear that $w_1(S)$ is equal to the first subset, $w_2(S)$ is the second and $w_3(S)$ is the third. Thus, indeed, we have that

$$W(S) = w_1(S) \cup w_2(S) \cup w_3(S) = S \; .$$

In the discussion of the Koch curve, Cantor set, and Sierpinski gasket we have learned that each of these basic fractals can be obtained by a limit process. But simultaneously there is a fixed point characterization by a Hutchinson operator which is a composition of appropriate similarity transformations. This is a very far reaching insight. For one thing, it explains the meaning of self-similarity. But in fact, the Hutchinson operator gives us much more. It also provides us an alternative way to talk about the existence of the Koch curve, or Cantor set, or Sierpinski gasket.

Tilings of Square and Triangle

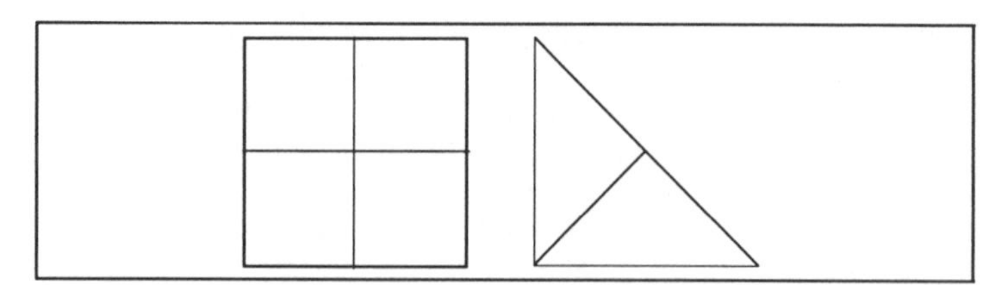

Figure 3.29 : Break-down of square into four scaled down squares, and of triangle into two scaled down and similar triangles.

A Unique Identification of Objects

It can be shown that each of the Hutchinson operators which we introduced earlier identifies a unique object in a plane (for the Koch curve and Sierpinski gasket) and on a line (for the Cantor set), which it leaves fixed. Or in other words, if W is the appropriate Hutchinson operator, then a solution to the equation $W(X) = X$ will automatically be either the Koch curve, the Cantor set, or the Sierpinski gasket. Thus we have a characteristic equation for each of these fractals. Naturally, these equations are not unique. Also for $\sqrt{2}$ there are several possible characterizations by equations, and the same is true here. This is a topic with very interesting variations, which we will pick up again in chapter 8. There are also characterizations of traditional geometric objects in terms of similarity invariance properties. Take, for example, a square or simple triangle. The break-down in figure 3.29 shows how these objects can be split up in a self-similar way. Thus, we can see fractals like the Sierpinski gasket in the same family as traditional geometrical objects. In fact, they solve the same kind of equations. Or in other words, from this point of view, fractals can be seen as extensions of traditional geometry, very much like irrational numbers can be seen as extensions of rational numbers by solving appropriate equations.

Self-Similarity in the Series of Hutchinson Operators

Using the Hutchinson operator W we can complete the analogy to the geometric series. Let us start out with with a triangle T of the coast of the Koch island, see figure 3.30.

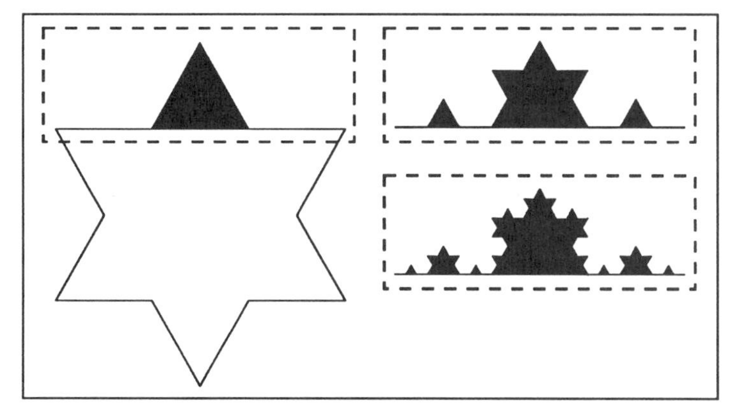

Figure 3.30 : The starting configuration (left) and the first two steps in the construction of a part of the Koch island in analogy to the geometric series.

We now apply the Hutchinson operator to T and add the result. Correspondingly, in a geometric series we would start with the number 1, and the first step would consist of a scaling of the number with a factor q with succeeding addition. Here, after the first application we have

$$T \cup W(T) = T \cup w_1(T) \cup w_2(T) \cup w_3(T) \cup w_4(T) \ .$$

Thus, we have added four triangles. In the next step, we again apply the Hutchinson operator W to our current configuration $T \cup W(T)$ and add the result:

$$T \cup W(T) \cup W^2(T) \ .$$

Here $W^2(T)$ denotes the repeated application of W, i.e. $W(W(T))$, and this is the collection of 16 triangles given by

$$w_1(w_1(T)), w_1(w_2(T)), w_1(w_3(T)), \ldots, w_4(w_3(T)), w_4(w_4(T)).$$

The next step yields

$$T \cup W(T) \cup W^2(T) \cup W^3(T) \ .$$

In analogy to the geometric series we may even write down the limit object of this construction as

$$\bigcup_{k=0}^{\infty} W^k(T)$$

where we imply the convention $W^0(T) = T$.

3.5 Box Self-Similarity: Grasping the Limit

We have discussed fractals as objects which are obtained from a limit process, and we have seen that it is the limit object (and not the final stages of the process) which exhibits self-similarity. In fact, we have learned that the limit object can be characterized by its self-similarity properties. Although the term self-similarity is at first very intuitive, in a sense, we have seen that it is in fact very abstract if one looks at it in a more rigorous way. So let us now try to finally get to a more operational meaning. This is done using a method which we call *box self-similarity*.

Box self-similarity offers not a new version or modification of the concept of self-similarity, but rather it is a method to grasp the self-similarity property of the limit object. Let us explain the method using the Sierpinski gasket as an example.

The single construction steps of the Sierpinski gasket (e.g. look at figure 3.27) are not self-similar objects. Just compare the number of pieces in the whole and a section of the gasket. On the other hand if we look at an image of the Sierpinski gasket we are led to believe that we see a self-similar object. But, in fact it is not. Any visualization of the Sierpinski gasket, printed on paper or displayed on a computer screen, is just a finite approximation and therefore cannot be self-similar, if we use the strict meaning of the term. Or, turned the other way around, what we see is self-similar only up to a certain precision. Box self-similarity makes this observation into a concept. We systematically look at objects in different resolutions and try to determine self-similarity only up to this resolution.

Pixel Approximation

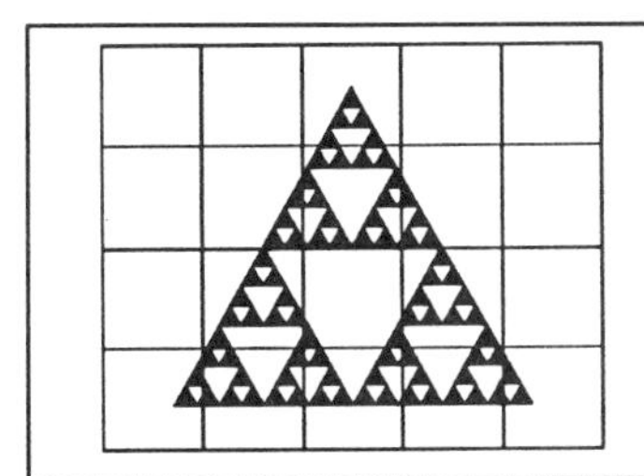

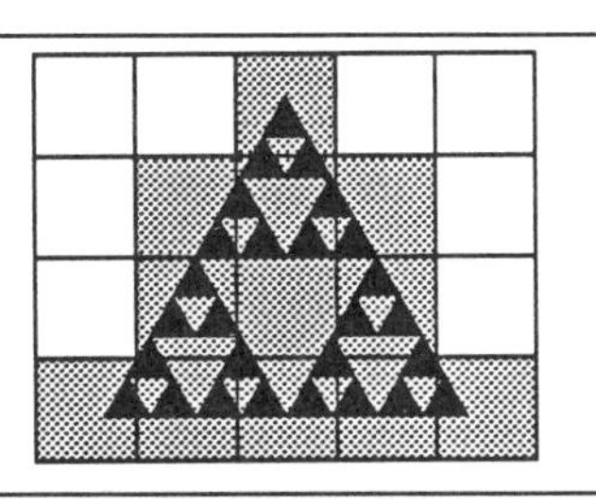

Figure 3.31 : Visualisation of the Sierpinski gasket in finite resolution. Imagine the grid represents the pixels of a computer screen. Pixels are lit if they are hit by the gasket. The right image shows the resulting picture which would be displayed.

Figure 3.31 shows a Sierpinski gasket drawn on top of a rather coarse grid. Imagine this grid would represent the pixels of a (very) low resolution computer screen. Displaying the Sierpinski gasket on this screen would be done by lighting the pixels which are hit by the points of the 'real' Sierpinski gasket as indicated in the figure.

The right image shows what we would see if we would display the gasket on such a low resolution display. We call this the pixel (or box) approximation with respect to the given grid. Indeed, it is hard to recognize the 'real' object. But in this way the principle of the approximation becomes visible and it is obviously the same for any finite resolution display.

Let us now look at the pixel approximation of the finite construction steps of the Sierpinski gasket. Figure 3.32 shows the stages of the construction drawn on top of grids of the different sizes. The pixel approximation of these objects is given by those boxes of the grid which are shaded grey. Let us first look at the coarse grid. When comparing the approximation of the different stages we observe that this grid is much too coarse to detect the differences between the construction steps. All pixel approximations are exactly the same.

Now look at the medium size grid. We observe that this grid detects a difference between stage 1 and stage 2 (there is one pixel which is lit in stage 1 but not in stage 2). But all further approximations with respect to this grid look the same. Finally, if we look at the grid with the highest resolution we note that this one stops detecting differences from stage 3 on. Obviously, this can be made to a general rule: any given grid eventually stops detecting differences at higher and higher stages. In fact, this is just what characterizes a limit process. It converges to a limit if a grid of any resolution will eventually see no differences.

Testing Self-Similarity

Let us now try to test for self-similarity in this sense. Exact self-similarity of the Sierpinski gasket means that if we take one of its subtriangles as a section, for example, the lower left main subtriangle, and scale it up by a factor of 2, we again see exactly the Sierpinski gasket. Let us now look once more at the finite stages of the construction process and compare the whole gasket with the lower left section scaled up by 2. Figure 3.33 shows this comparision on a coarse grid. Again the pixel approximation with respect to this grid is given by those boxes of the grid which are shaded grey. If we compare the pixel approximation of the whole figure and the section of stage 1, we observe no difference. The same is true for all the stages of the construction process. In other words, with respect to this coarse grid the whole and the scaled up section are identical.

We repeat this experiment with a grid of higher resolution (see figure 3.34). Now the grid detects a difference between the whole figure and the section of stage 1. But for the higher stages the pixel approximation of the whole figure and the section again are the same. Now we move on to an even higher resolution (see figure 3.35). As for the previous grid, we immediately see that the

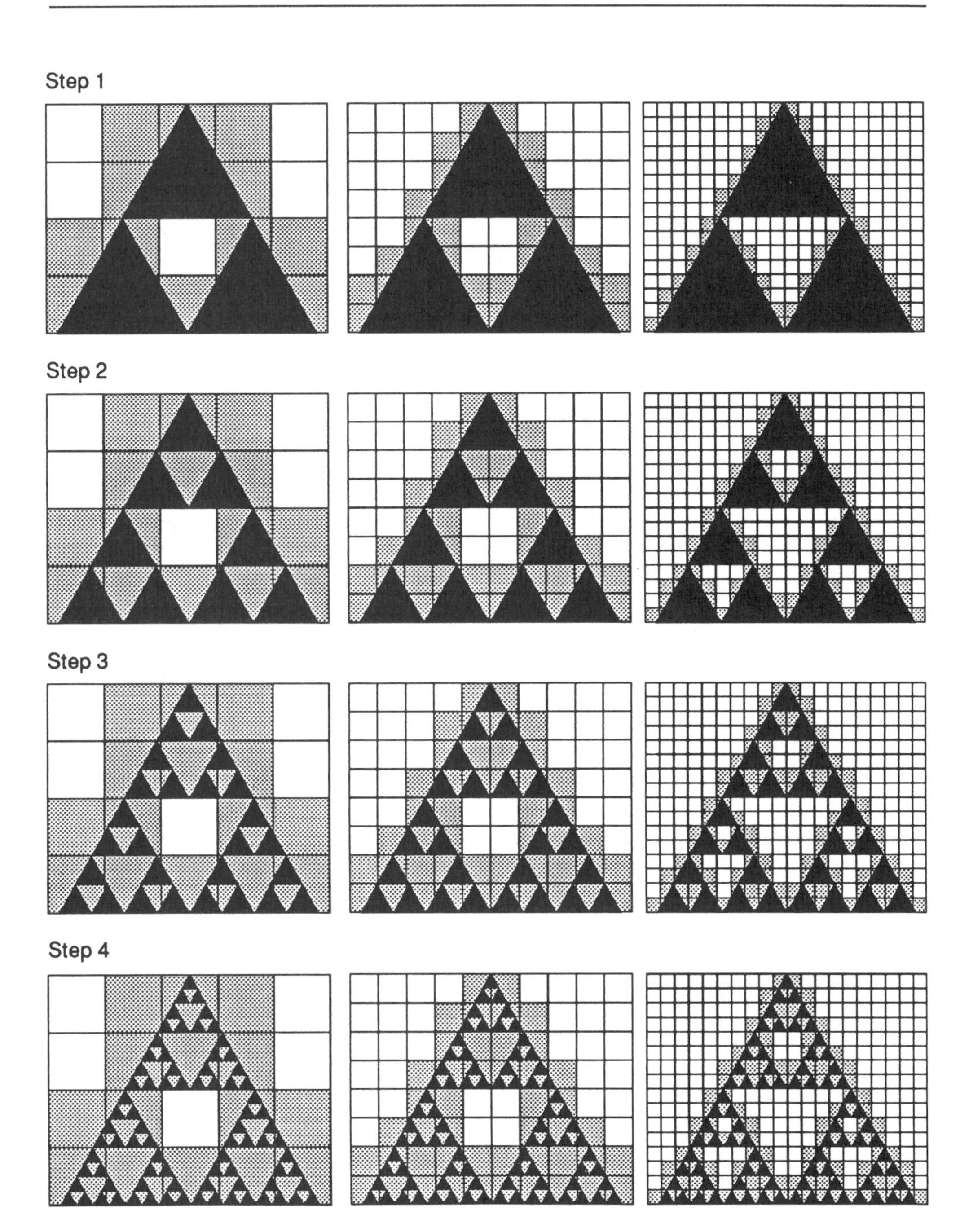

Figure 3.32 : The construction steps of the Sierpinski gasket and their pixel approximation. Try to observe where the different grids stop detecting differences between stages of the construction.

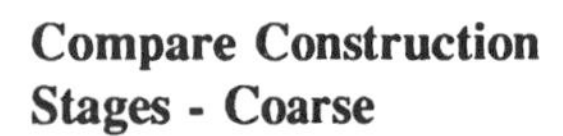
Compare Construction Stages - Coarse

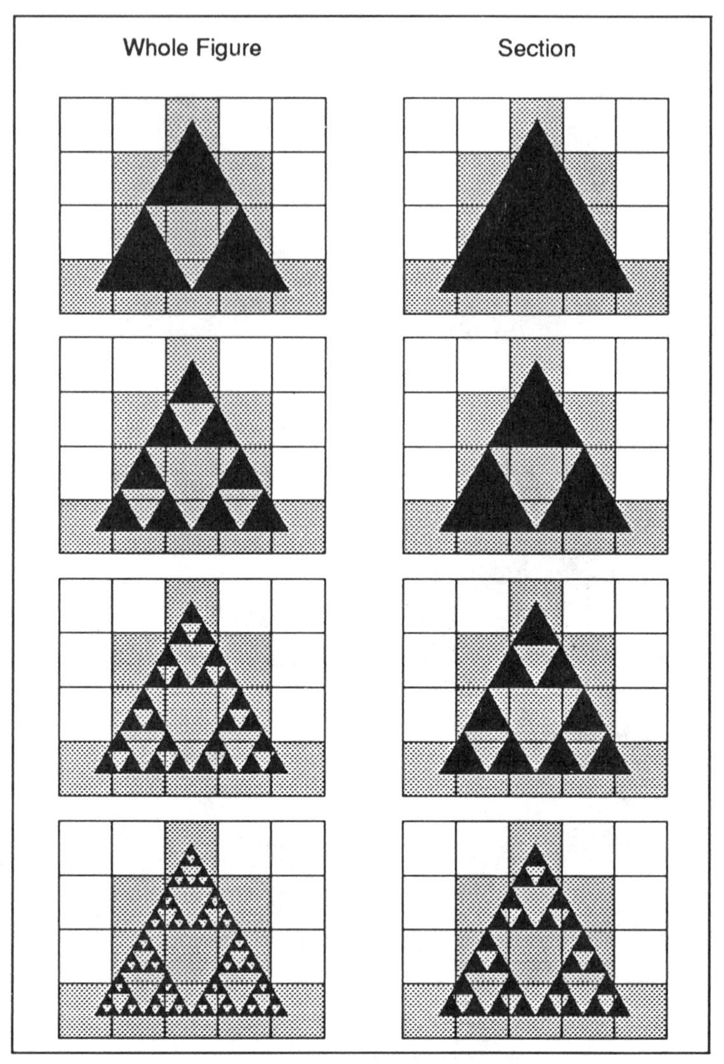

Figure 3.33 : Compare the pixel approximation of the construction steps of the Sierpinski gasket (left) and a section of the same stage (right). With respect to this grid they are the same.

pixel approximation of the whole figure and the scaled up section of stage 1 are different. Furthermore, this grid detects a difference in stage 2. But then again at all higher stages the approximation of the whole figure and the scaled up section are exactly the same with respect to the resolution of this grid.

What is the general rule? For any finite construction step of the Sierpinski gasket the whole figure and the sections are different

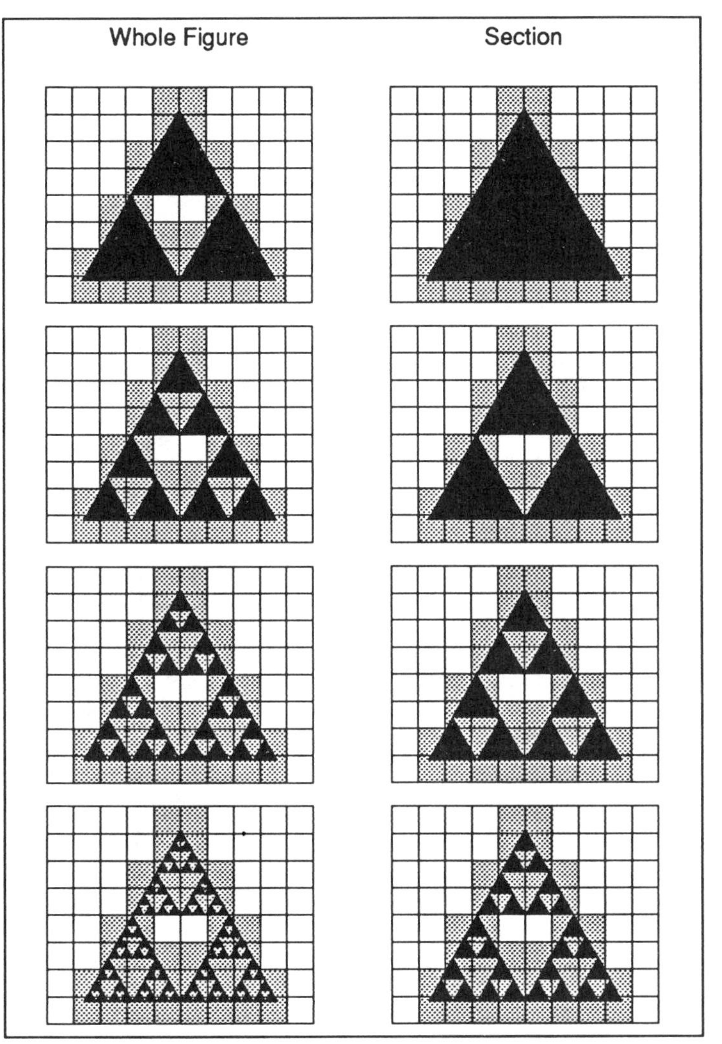

Figure 3.34 : Compare the pixel approximation of the construction steps of the Sierpinski gasket (left) and a section of the same stage (right). This grid detects a difference at stage 1.

Compare Construction Stages - Medium

and this difference becomes visible in the pixel approximation if the resolution of the grid is just fine enough. But on the other hand, however fine the resolution of a given grid might be, this grid will stop detecting differences between the whole figure and the scaled sections at a certain construction step of the gasket: all higher construction steps will appear to be the same.

Compare Construction Stages - Medium

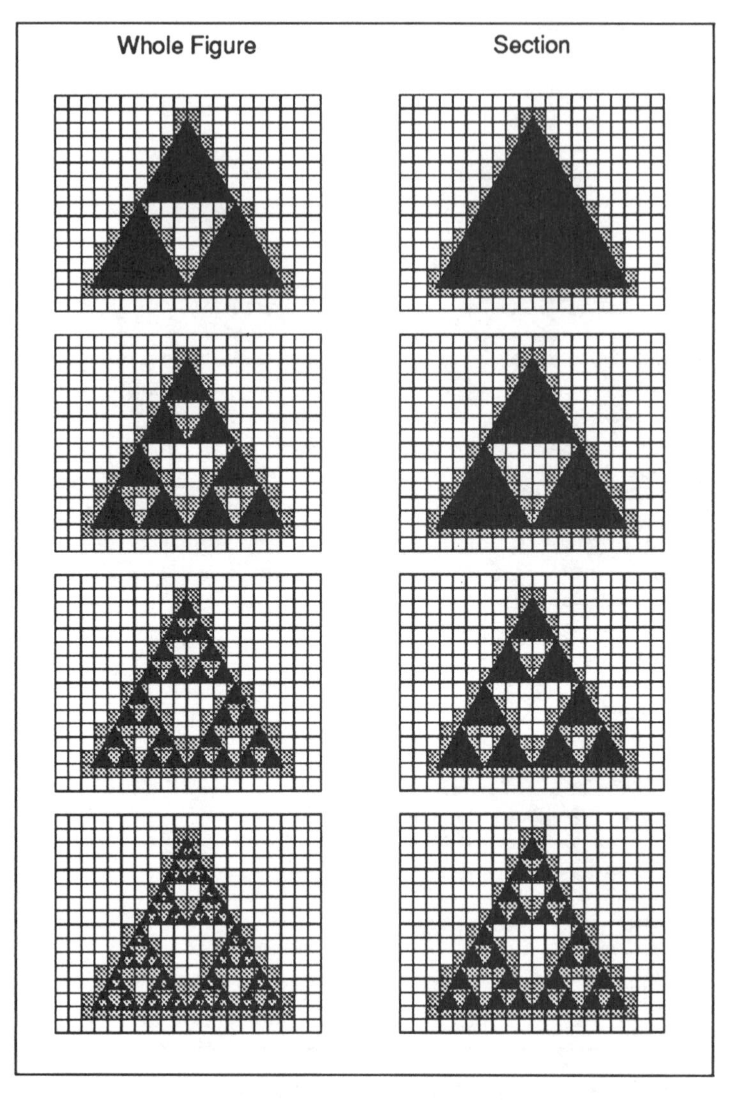

Figure 3.35 : Compare the pixel approximation of the construction steps of the Sierpinski gasket (left) and a section of the same stage (right). This grid detects a difference at stage 1 and 2.

Operational Characterization of Self-Similarity

In other words, we have arrived at an operational characterisation of strict self-similarity. We can test this property with respect to grids that we can choose as fine as we like. If an object is self-similar, then for any grid and sufficiently advanced stages of the construction process of the limit object the pixel approximation of the whole figure and the scaled up sections will appear to be exactly the same.

Let us conclude with a remark about the choice of the grids in our example. Apparently the grid and the triangles were very carefully aligned. In particular the blowup of the section is positioned on the grid exactly as the whole gasket. Imagine that this would not be the case (e.g., the figure would be shifted a bit). Then different boxes of the grid would be hit and the pixel approximation would be slightly different. Thus, we would not be able to compare the pixel approximation of the whole figure and the scaled up sections directly. In other words, the concept of box self-similarity is very sensitive to the proper choice (alignment) of grids. If all choices are made sufficiently carefully, then our concept of box self-similarity nicely allows us to grasp what is involved in approaching the limit. But in general (i.e., to really determine whether a given construction leads to a self-similar object or not), it is not practical. In chapter 4 we will discuss the problem of determining the fractal dimension of an object. There we will learn about a related concept, the box counting dimension. This concept, in contrast, will turn out to be very practical and should not be confused with box self-similarity.

3.6 Program of the Chapter: The Koch Curve

We discussed the Koch curve as the main example of this chapter. The program of the chapter follows the recursive definition of its construction. We start with a straight line. In the first step of the construction we replace this line by four appropriately positioned line segments. In the second step each of these segments is again replaced by four new segments, and so on. Thus, in the first stage we have 4 line segments and in the second there are 16 (then 64, 256, 1024, 4096, ...). In order to avoid keeping all these lines (or their end points) in the memory of the computer we organize the recursion of the procedure appropriately.

Screen Image of Koch Curve

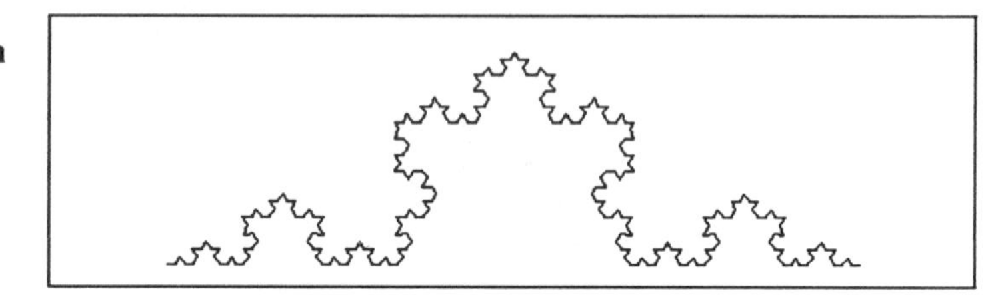

Figure 3.36 : Output of the program 'Koch Curve'.

Assume we want to display stage 2 of the construction process. Does this mean that we first have to compute all 4 segments of stage 1 and from those all 16 segments of stage 2? Fortunately not. Let us look at the first stage of the construction. The 4 segments, which we obtain from the initial one, are computed one after the other: left, middle left, middle right and right. Now as soon as we have computed the left one, we immediately take it and again apply the replacement procedure to obtain line segments of stage 2. Again we compute one segment after the other, and in this case we draw them immediately. When these first 4 segments of stage 2 are done, we go back and look at the next one of the segments of stage 1 (middle left). Now, the replacement procedure is applied to this one, and so on. Observe, that in this way we have to memorize at most one line segment from each stage of the recursion.

How can we do this in BASIC? We store the line segments for each stage using an array of point coordinates that mark the end points of the lines: `xleft(), yleft()` for the coordinates of the left end point and `xright(), yright()` for the right one. The index of these arrays indicate the stage of the recursion. Note that for the convenience of a compact program we count the stages of the construction as levels backwards.[27]

[27]This means, that we do not count 1, 2, 3, ... and then show stage number N, rather we start at `level` = N with the initial line segment and then count backwards down to `level = 1` and show the computed segments.

BASIC Program **Koch Curve**
Title A recursive program to draw the Koch curve

```
DIM xleft(10), xright(10),yleft(10), yright(10)
INPUT "Peak offset (0.29):",r
level = 5
xleft(level) = 30
xright(level) = 30+300
yleft(level) = 190
yright(level) = 190
GOSUB 100
END
REM DRAW LINE AT LOWEST LEVEL OF RECURSION
100 IF level > 1 GOTO 200
    LINE (xleft(1),yleft(1)) - (xright(1),yright(1))
    GOTO 300
REM BRANCH INTO LOWER LEVELS
200 level = level - 1
REM LEFT BRANCH
    xleft(level) = xleft(level+1)
    yleft(level) = yleft(level+1)
    xright(level) = .333*xright(level+1) + .667*xleft(level+1)
    yright(level) = .333*yright(level+1) + .667*yleft(level+1)
    GOSUB 100
REM MIDDLE LEFT BRANCH
    xleft(level) = xright(level)
    yleft(level) = yright(level)
    xright(level) = .5*xright(level+1) + .5*xleft(level+1)
        -r*(yleft(level+1)-yright(level+1))
    yright(level) = .5*yright(level+1) + .5*yleft(level+1)
        +r*(xleft(level+1)-xright(level+1))
    GOSUB 100
REM MIDDLE RIGHT BRANCH
    xleft(level) = xright(level)
    yleft(level) = yright(level)
    xright(level) = .667*xright(level+1) + .333*xleft(level+1)
    yright(level) = .667*yright(level+1) + .333*yleft(level+1)
    GOSUB 100
REM RIGHT BRANCH
    xleft(level) = xright(level)
    yleft(level) = yright(level)
    xright(level) = xright(level+1)
    yright(level) = yright(level+1)
    GOSUB 100
level = level + 1
300 RETURN
```

Let us now look at the details of the program. Before it starts any computation, it asks you for a parameter `r`. For now you should simply type 0.29. Now the initial line segment is set up 300 pixels wide (as in most of our other programs there is an offset by 30). Observe that we have chosen `level = 5`. Thus, we will see the 4th stage of the construction (with `level = 1` we would obtain the initial line, i.e. stage 0). You can change this variable to obtain the different stages of the Koch construction. Now we start the recursion (we jump to label 100).

In the recursive part we first check if we are at `level = 1`. If we are, a line segment is drawn. Otherwise, we prepare another level of the recursion (at label 200) and divide the current line segment. First we compute the left part and go to the next level of the recursion (this is done by the line `GOSUB 100`). Then the middle left, the middle right and finally the right part is computed and recursively subdivided. At the end of the recursive part we return to the previous level of the recursion, which will terminate the program if we are back at the initial level (the `END` that follows the first `GOSUB 100`).

If you like, you can try to modify the shape of the curve. The simplest way to do this is to change the input parameter `r` to different values (between 0 and 1). This will shift the point which represents the peak of the curve. But you can also change the computation in the program itself. Why not try to compute the 3/2-curve? This would require including the computation of 4 additional line segments in the recursive part of the program.

Chapter 4

Length, Area and Dimension: Measuring Complexity and Scaling Properties

Nature exhibits not simply a higher degree but an altogether different level of complexity. The number of distinct scales of length of natural patterns is for all practical purposes infinite.

Benoit B. Mandelbrot[1]

Geometry has always had two sides, and both together have played very important roles. There has been the analysis of patterns and forms on the one hand; and on the other, the measurement of patterns and forms. The incommensurability of the diagonal of a square was initially a problem of measuring length but soon moved to the very theoretical level of introducing irrational numbers. Attempts to compute the length of the circumference of the circle led to the discovery of the mysterious number π. Measuring the area enclosed between curves has, to a great extent, inspired the development of calculus.

Today measuring length, area and volume appears to be no problem. If at all, it is a technical one. In principle, we usually think these problems have long since been solved. We are used to thinking that what we see can be measured if we really want to do so. Or we look up an appropriate table. Mandelbrot tells the story that the length of the border between Spain and Portugal has two very different measurements: an encyclopedia in Spain claims 616 miles, while a Portuguese encyclopedia quotes 758 miles. Who is

[1] Benoit B. Mandelbrot, *The Fractal Geometry of Nature*, Freeman, 1982.

right? If you look up the length of the coast of Britain in various sources, again you will find that the results vary anywhere between 4500 and 5000 miles.[2] What is happening here? There seems to be a problem. That is the theme of Mandelbrot's 1967 article[3] *How long is the coast of Britain?* For a moment we are led to believe that somebody has done a sloppy job. We have all seen those people surveying in the field with their high precision optical gear. Is it possible that they made a mistake? And who made it, who is right and who is wrong? How do we find out?[4] And today with satellite surveying and laser precision, do we have more reliable results? The answer is no, and the fact is, we never will.

We will demonstrate that for all practical purposes, typical coastlines do not have a meaningful length! This statement seems to be ridiculous or at least counter-intuitive. An object like an island with some definitive area should also have some definitive length to its boundary.

We know that if we measure the circumference of a circular object, we will not obtain πd, d the diameter, but rather something close to it. We know we are inaccurate, but we don't worry because if we need a more accurate result, we just increase the level of precision in our measurement. Measurement requires units such as miles, yards, inches, etc.: all idealized straight line segments. If we have a curved object such as a circle, then there is no doubt that the object has a definitive length and that it can be measured as accurately as necessary. Somehow our experience is that objects which fit on a piece of paper have finite length. But this is a misleading intuition. We usually measure the length only of objects, for which the result in fact does make sense and is of some practical value. But coastlines (and fractals) are not the only exceptions.

[2]The *Encyclopedia Americana*, New York, 1958 states "Britain has coasts totaling 4650 miles = 7440 km". *Collier's Encyclopedia*, London 1986 states "The total mileage of the coastline is slightly under 5000 miles = 8000 km."

[3]B. B. Mandelbrot, *How long is the coast of Britain? Statistical self-similarity and fractional dimension*, Science 155 (1967) 636–638.

[4]Here are several methods of getting an answer: (1) Ask all the people in Britain and take the average of their answers. (2) Check encyclopedias. (3) Take a very detailed map of Britain and measure the coast using a compass. (4) Take a very detailed map of Britain and a thin thread, fit it on the coast, and then measure the length of the thread. (5) Walk the coast of Britain.

Plate 1: Stilben (used in some detergents) dendrites in polarized light, © Manfred Kage, Institut für wissenschaftliche Fotografie.

Plate 2: Cast of a child's kidney, venous and arterial system, © Manfred Kage, Institut für wissenschaftliche Fotografie.

Plate 3: Broccoli Romanesco.

Plate 4: Wadi Hadramaut, Gemini IV image, © Dr. Vehrenberg KG.

Plate 5: Broccoli Romanesco, detail.

Plate 6: Fractal forgery of a mountain range with Mandelbrot sky, © R.F. Voss.

Plate 7: Fractal forgery of mountain range (top left), inverted mountain range, showing valleys as mountains and mountains as valleys (bottom left), inverted mountain range rendered as cloud pattern (top right) used in Plate 6, © R.F. Voss.

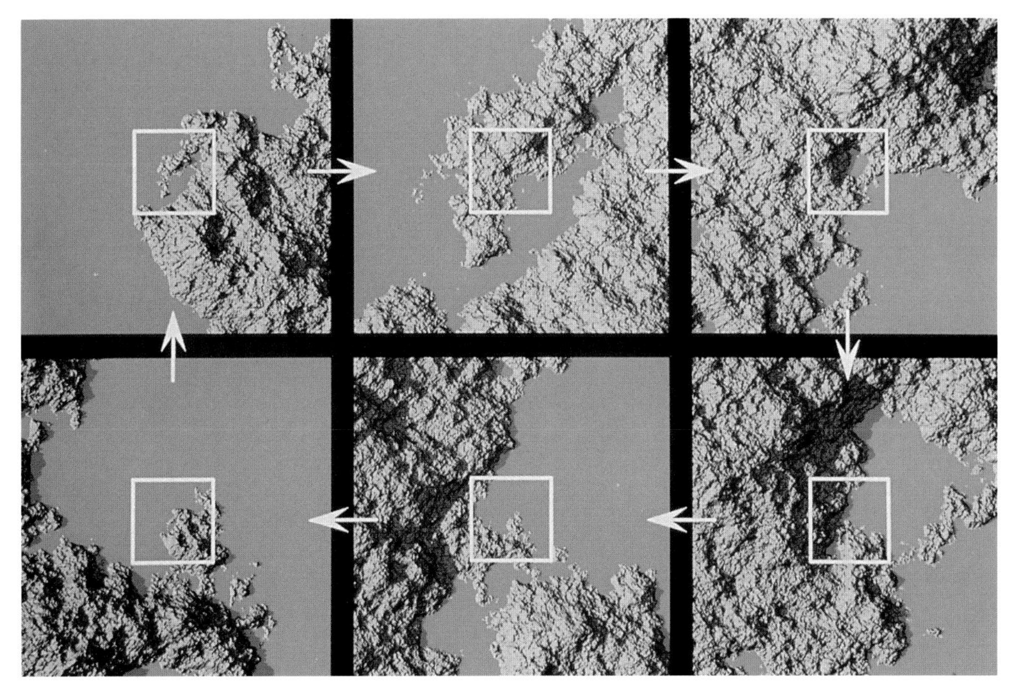

Plate 8: Fractal coast, repeating after 6 magnifications, © R.F. Voss.

Plate 9: Fractal Moon Craters, © R.F. Voss.

Plate 10: "Zabriski Point", fractal forgery of a mirage, © K. Musgrave, C. Kolb, B.B. Mandelbrot.

Plate 11: "Carolina", fractal forgery, © K. Musgrave.

Plate 12: Fractal forgery of planet rise, © K. Musgrave.

Plate 13: “Ein kleines Nachtlicht”, fractal forgery, stereoscopic image. View the left image with your right eye and the right image with your left eye. © K. Musgrave, C. Kolb, B.B. Mandelbrot.

Plate 14: Dawn over the Himalayas, Gemini IV image, © Dr. Vehrenberg KG.

4.1 Finite and Infinite Length of Spirals

One possible class of objects which defies length measurement are spirals, it seems. Spirals fit on a piece of paper, and obviously do have infinite length. Well, do spirals really have infinite length? This is a very delicate question. Some have, and others don't.

Spirals have fascinated mathematicians throughout the ages. Archimedes (287–212 B.C.) wrote a treatise on spirals, and one of them is even named after him. The Archimedean spiral is a good model for the track on a record, or the windings of a rolled carpet. The characteristic feature of an Archimedean spiral is that the distance between its windings is constant everywhere. The mathematical model for such a spiral is very easy to obtain once we introduce polar coordinates: a point in the plane is described by a pair (r, ϕ), where r is the distance to the origin of a coordinate system (the radius) and ϕ is the angle of the radius to the positive x-axis, measured in radians, i.e. $0 \leq \phi \leq 2\pi$.

Polar Coordinates

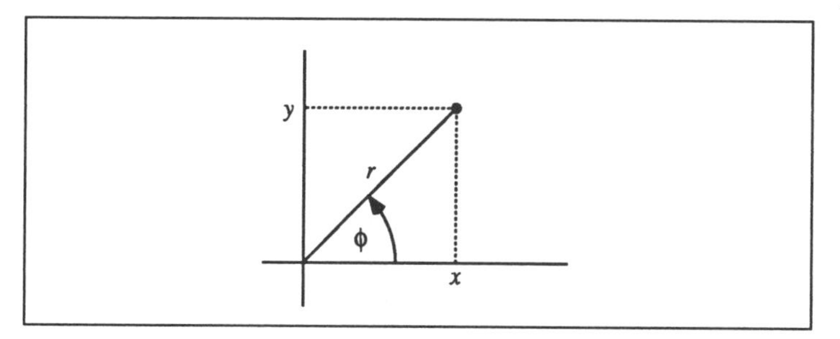

Figure 4.1 : The polar coordinates of the point at coordinates (x, y) are (r, ϕ), where $r = \sqrt{x^2 + y^2}$ is the distance to the origin and ϕ is the angle with the positive x-axis. Thus, $x = r\cos\phi$, and $y = r\sin\phi$.

In this frame of reference, an Archimedean spiral (seen from its center) can be modelled by the equation

$$r = q\phi$$

where we now allow ϕ to be any nonnegative number, i.e. $\phi = 2\pi$ is one turn, $\phi = 4\pi$ corresponds to two turns, and so on. When drawing this spiral we start in the center. As ϕ makes one complete turn (i.e. increases by 2π) the radius will increase by $2\pi q$, which is the constant distance between two successive windings.

If we replace r by the natural logarithm $\log r$, we obtain a model for the logarithmic spiral: $\log r = q\phi$, or equivalently

$$r = e^{q\phi} .$$

Archimedean Spiral

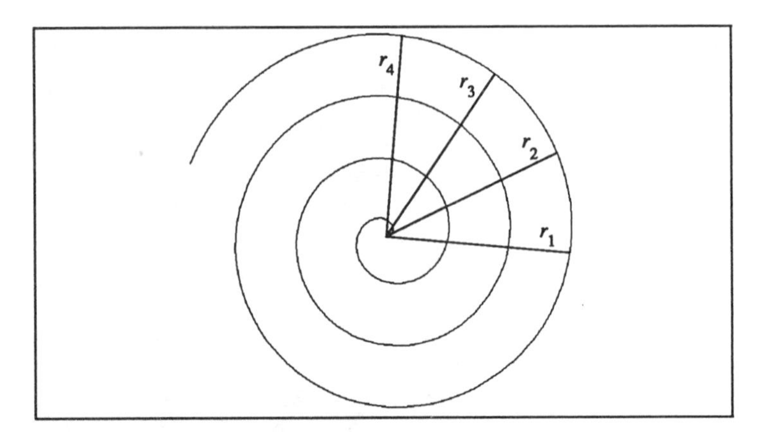

Figure 4.2 : Archimedean spiral. Stepping along the spiral in steps of a constant angle α yields an arithmetic sequence of radii $r_1, r_2, \ldots$

Logarithmic Spiral

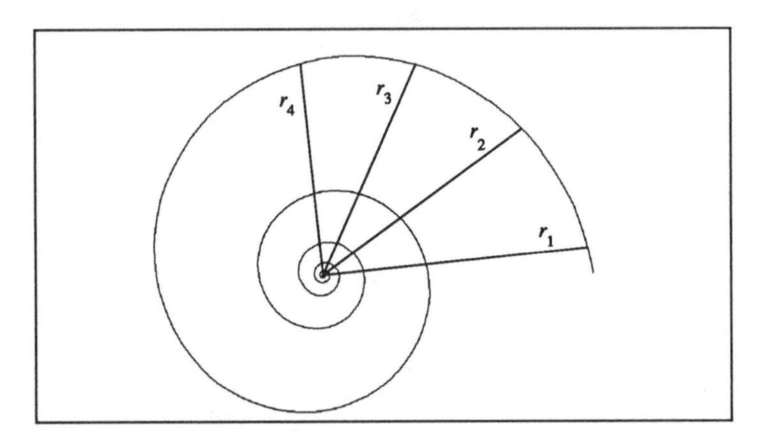

Figure 4.3 : Logarithmic spiral. Stepping along the spiral in steps of a constant angle α yields a geometric sequence of radii $r_1, r_2, \ldots$

When $q > 0$ and ϕ grows beyond all bounds, the spiral goes to infinity. When $q = 0$, we obtain a circle. And when $q < 0$, we obtain a spiral which winds into the center of the coordinate system as ϕ goes to infinity. This spiral is related to geometric series and has a remarkable property which is related to fractals. It is self-similar in a way which has equally inspired mathematicians, scientists and artists.

The great Swiss mathematician Jacob Bernoulli (1654–1705) devoted a treatise, entitled *Spira Mirabilis* (Wonderful Spiral), to the logarithmic spiral. He was so impressed by its self-similarity that he chose the inscription *Eadem Mutata Resurgo* (In spite of changes — resurrection of the same) for his tombstone in the Cathedral of Basel.

Spirals, Arithmetic and Geometric Mean

An Archimedean spiral is related to arithmetic series in the following way: we choose an arbitrary angle, say α, and points on the spiral, whose radii $r_1, r_2, r_3, \ldots$ have angle α to each other. Then the numbers r_i constitute an arithmetic sequence, i.e. the differences between consecutive numbers are the same. Thus, $r_3 - r_2 = r_2 - r_1$, and so on. Indeed, if $r_1 = a\phi_1$, then $r_2 = a(\phi_1 + \alpha)$, and $r_3 = a(\phi_1 + 2\alpha)$. In other words, $r_2 = (r_1 + r_3)/2$. That means any radius is the arithmetic mean of its two neighboring radii.

If we replace the arithmetic mean $(r_1 + r_3)/2$ by the geometric mean $\sqrt{r_1 r_3}$, we obtain the other classical spiral, the famous logarithmic spiral. Let us see why. Squaring the equation for the geometric mean gives $r_2^2 = r_1 r_3$, or, equivalently

$$\frac{r_1}{r_2} = \frac{r_2}{r_3} .$$

Taking logarithms this identity reads

$$\log r_3 - \log r_2 = \log r_2 - \log r_1 .$$

That means, the logarithms of successive radii form an arithmetic series. Thus, we obtain $\log r = q\phi$, the formula for the logarithmic spiral.

The radii r_i of the logarithmic spiral form a geometric sequence. We have

$$\frac{r_1}{r_2} = \frac{r_2}{r_3} = \frac{r_3}{r_4} = \frac{r_4}{r_5} = \cdots$$

is a constant, say a. Then, for any index n

$$\frac{r_n}{r_{n+1}} = a$$

and

$$r_{n+1} = \frac{r_n}{a} = \frac{r_{n-1}}{a^2} = \cdots = \frac{r_1}{a^n} .$$

Self-Similarity of the Logarithmic Spiral

What is the amazing property which he admired so much? Bernoulli observed that a scaling of the spiral with respect to its center has the same effect as simply rotating the spiral by some angle. Indeed, if we rotate the logarithmic spiral $r = e^{q\phi}$ by an angle of ψ clockwise, then the new spiral will be

$$r = r(\phi) = e^{q(\phi+\psi)} .$$

Since

$$e^{q(\phi+\psi)} = e^{q\psi} \cdot e^{q\phi} ,$$

rotating by ψ is the same as scaling by $s = e^{q\psi}$.

Spiral or not Spiral?

Figure 4.4 : A 'spiral' by Nicholas Wade. Reproduced with kind permission by the artist. From: Nicholas Wade, *The Art and Science of Visual Illusions,* Routledge & Kegan Paul, London, 1982.

Now what is the length of the spiral? Let us look at an example of a spiral where the construction process makes calculation easy. It is, by the way just another example of a geometric feedback system.

The Construction of Polygonal Spirals

We generate an infinite polygon. First choose a decreasing sequence $a_1, a_2, a_3, \ldots$ of positive numbers. Now a_1 is the length of our initial piece. We construct the polygon in the following way: draw a_1 vertically from bottom to top. At the end make a right turn and draw a_1 again (from left to right). At the end of that line start to draw a_2 (first, continue in the same direction, from left to right). At the end make another right turn and draw again (now from top to bottom). At the end of that line take a_3. Continue on using the same principles. Figure 4.5 shows the first steps of this construction.

How long is this polygonal spiral? Well, each segment a_k appears twice in the construction, and thus the length is twice the sum of all a_i, i.e. $2(a_1+a_2+a_3+\cdots)$. Let us now choose particular values of a_k. Let q be a positive number. If we take $a_k = q^{k-1}$,

Polygonal Spiral

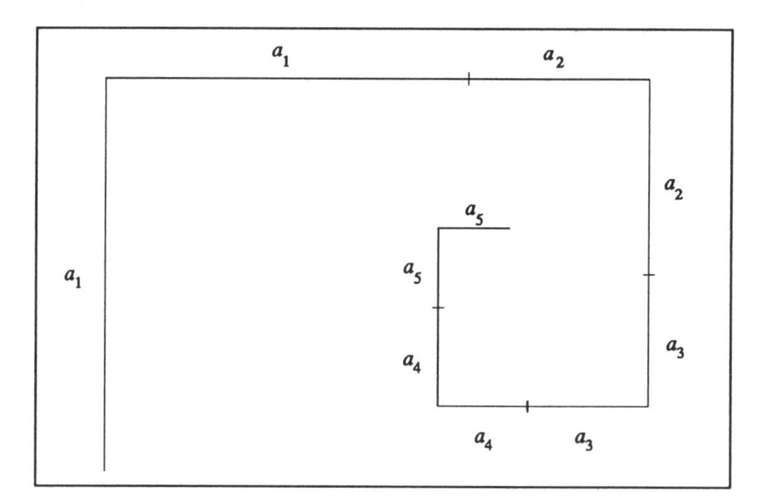

Figure 4.5 : The first construction steps of a polygonal spiral.

Infinite and Finite Polygonal Spiral

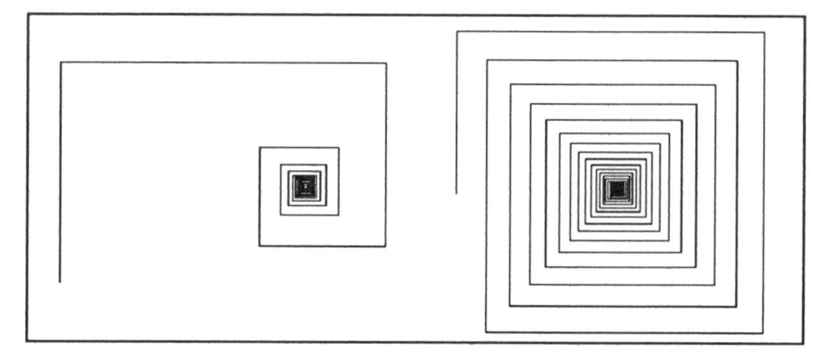

Figure 4.6 : Infinite and finite polygonal spiral. The spiral on the left is the one for $a_k = 1/k$ (i.e. the length is infinite). The spiral on the right is the one for $a_k = q^{k-1}$ with $q = 0.95$, a value slightly below 1 (i.e. it has a finite length).

we obtain as total length $2\sum_{k=0}^{\infty} q^k$, which is a geometric series. Provided that $q < 1$, the limiting length[5] is equal to $2/(1-q)$. Thus, this polygonal spiral has finite length.

On Finite Area an Infinitely Long Spiral

If we take, however, $a_k = 1/k$ we obtain a series which is known not to have a limit.[6] In other words, the associated spiral is infinitely long, although it fits onto a finite area! Figure 4.6 shows both cases. Can you *see* which of the two spirals is finite and which is infinite?

The above polygonal spiral constructions can easily be used to support a smooth spiral construction. Observe that the polygons

[5]Recall that the limit of the geometric series $1 + q + q^2 + q^3 + q^4 + \ldots$ is $1/(1-q)$, when $|q| < 1$.

[6]The sum $1 + 1/2 + 1/3 + 1/4 + \cdots$ is infinite (see the footnote on page 164).

Smooth Polygonal Spiral

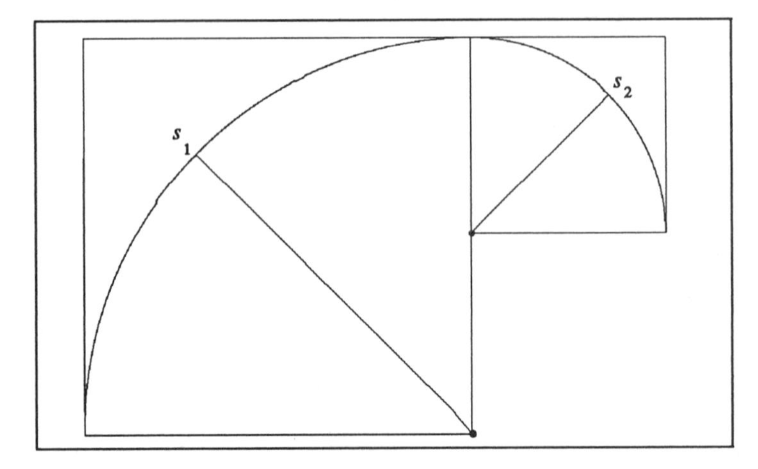

Figure 4.7 : Construction of a smooth polygonal spiral.

Infinite and Finite Smooth Spiral

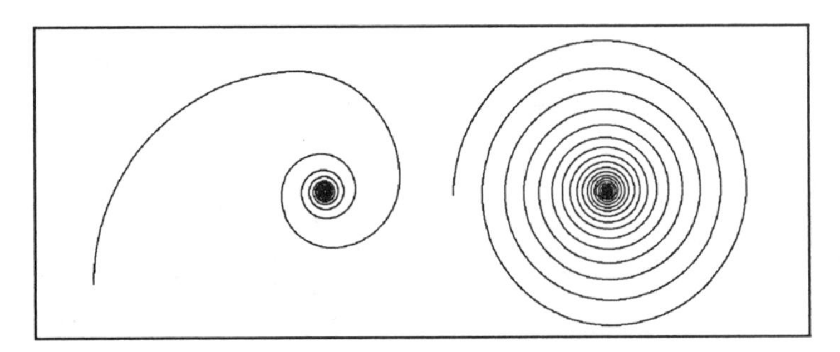

Figure 4.8 : The smooth spiral construction from figure 4.7 is carried out for the polygonal spirals from figure 4.6. $a_k = 1/k$ (left) and $a_k = q^{k-1}$ with $q = 0.95$ (right). Again the left spiral has infinite length while the right hand one has a finite length.

are composed of right angles with equal sides a_k. Each of them encompases a segment of a circle — in fact exactly a quarter of a circle with radius a_k. Putting together these segments appropriately produces a smooth spiral.

What is the length of these smooth spirals? Observe that the radii of the circle segments are of length a_k, while the circle segments then are of length $s_k = 2\pi a_k/4 = (\pi/2)a_k$. In other words, we have

$$\text{length} = \sum_{k=1}^{\infty} s_k = \frac{\pi}{2} \sum_{k=1}^{\infty} a_k,$$

which is finite for $a_k = q^{k-1}$ (where $q < 1$) but infinite for $a_k = 1/k$. Figure 4.8 shows both spirals.

Again, it is amazing how little our visual intuition helps us to 'see' finite or infinite length. In other words, the fact that a curve fits on a piece of paper does not tell us whether its length is finite or not. Fractals add a new dimension to that problem.

The Golden Spiral

If we take for our polygonal spiral construction $a_k = 1/g^{k-1}$, where $g = (1+\sqrt{5})/2$ is the golden mean, we obtain the famous golden spiral. For the length of this spiral we compute

$$\frac{2}{1-\frac{1}{g}} = \frac{2g}{g-1} = 2g^2 = 3+\sqrt{5}\ .$$

Here we have used that g satisfies $g^2-g-1=0$ (i.e. $g-1=1/g$).

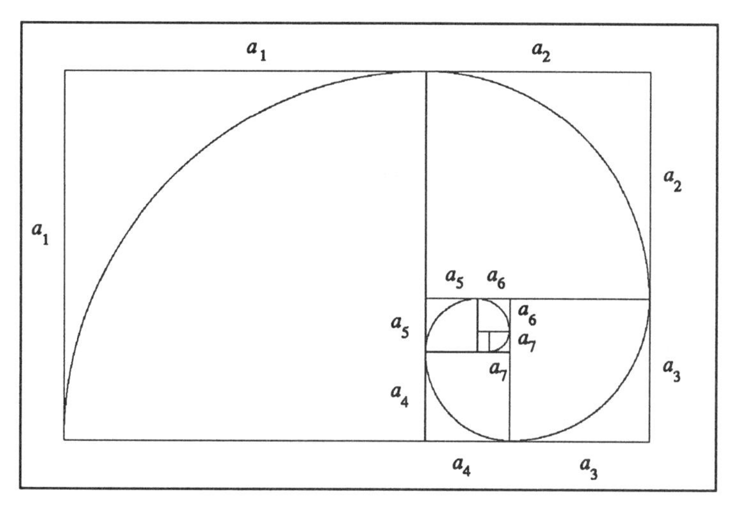

Figure 4.9 : The golden spiral.

The golden spiral can also be obtained in another beautiful construction: start with a rectangle with sides a_1 and $a_1 + a_2$, where $a_1 = 1$ and $a_2 = 1/g$ (i.e. , $a_1/a_2 = g$). The rectangle again breaks down into a square with sides a_2 and a smaller rectangle with sides a_3 and a_2, and so on (see Figure 4.9). Note that

$$\frac{a_2}{a_3} = \frac{\frac{1}{g}}{a_1-a_2} = \frac{\frac{1}{g}}{1-\frac{1}{g}} = \frac{1}{g-1} = \frac{1}{\frac{1}{g}} = g\ .$$

In general we have that $a_k/a_{k+1} = g$, or equivalently $a_k = 1/g^{k-1}$. The length of the inscribed smooth spiral is equal to $\frac{\pi}{2}g^2 = \frac{\pi}{4}(3+\sqrt{5})$.

4.2 Measuring Fractal Curves and Power Laws

The computation of the length of the various spirals — finite or infinite — is based on the corresponding mathematical formulas. The result on the infinite length of the Koch curve and the coast of the Koch island in chapter 2 is derived from the precise construction process of these fractals. Both of these methods for length computation of course fail when we consider fractals in nature such as real coastlines. There is no formula for the coastline of Great Britain, and also there is no defined construction process. The shape of the island is the result of countless years of tectonic activities of the earth on the one hand and the never stopping erosion and sedimentation processes on the other hand. The only way to get a handle on the length of the coastline is to measure. In practise we measure the coast on a geographical map of Britain rather than the real coast. We take a compass and set it at a certain distance. For example, if the scale of the map is 1:1,000,000 and the compass width is 5 cm, then the corresponding true distance is 5,000,000 cm or 50 km (approximately 30 miles). Now we carefully walk the compass along the coast counting the number of steps. Figure 4.11 shows a polygonal representation of the coast of Britain. The vertices of the polygons are assumed to be on the coast. The straight line segments have constant length and represent the setting of the compass. We have carried out this measurement using four different compass settings.[7]

Measuring the Coastline of Britain

Compass Setting	Length
500 km	2600 km
100 km	3800 km
54 km	5770 km
17 km	8640 km

Table 4.10 : Length of coast of Britain measured from maps of various scales and with a compass set to cover various distances.

Smaller Scales Give Longer Results

This elaborate experiment reveals a surprise. The smaller the setting of the compass, the more detailed the polygon and — the surprising result — the longer the resulting measurement will be. In particular, up in Scotland the coast has a very large number of bays of many different scales. With one compass setting many of the smaller bays are still not accounted for, while in the next smaller one they are, while still smaller bays are still ignored at that setting, and so on.

[7]In: H.-O. Peitgen, H. Jürgens, D. Saupe, C. Zahlten, *Fractals — An Animated Discussion,* Video film, Freeman, New York, 1990. Also appeared in German as *Fraktale in Filmen und Gesprächen,* Spektrum der Wissenschaften Videothek, Heidelberg, 1990.

Approximations of Britain

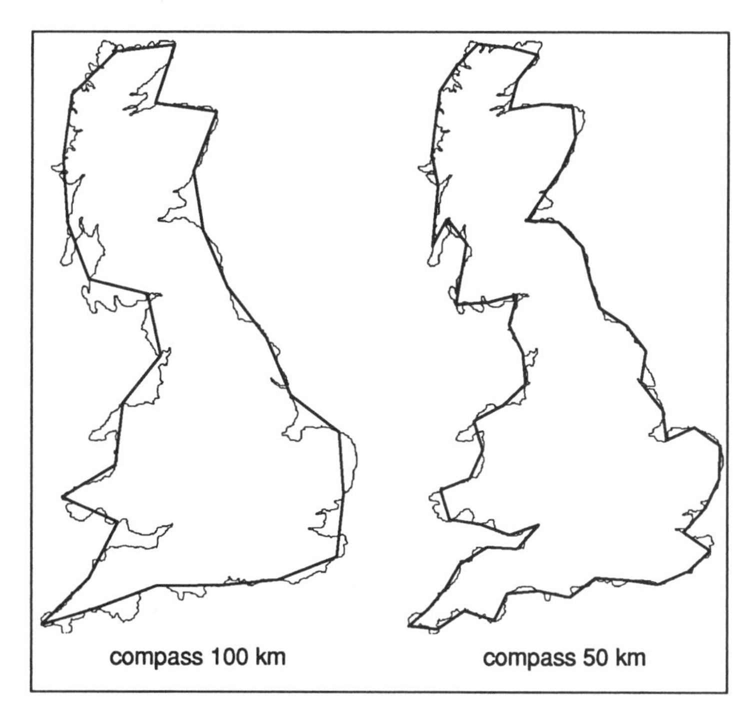

Figure 4.11 : Polygonal approximation of the coast of Britain.

Measuring the Circle

Number of Sides	Compass Setting	Length
6	500.0 km	2600 km
12	258.8 km	3106 km
24	130.5 km	3133 km
48	65.4 km	3139 km
96	32.7 km	3141 km
192	16.4 km	3141 km

Table 4.12 : Length of a circle of diameter 1000 km approximated using inscribed regular polygons. The entries are computed from the formula of Archimedes, see page 171.

Measuring a Circle

Let us compare this phenomenon with an experimental measurement of the length of a circle. We use a circle of diameter 1000 km, so that the length is of the same order of magnitude as the coast of Britain. We do not have to go through the process of actually walking a compass around the circle. Rather, we make use of the classical approach of Archimedes who had worked out a procedure to calculate what these measurements would be (see page 171 and table 4.12). In order to compare the results we enter the measurements in a graph. However, because the size of our

Log/log Diagram for Coast of Britain and Circle

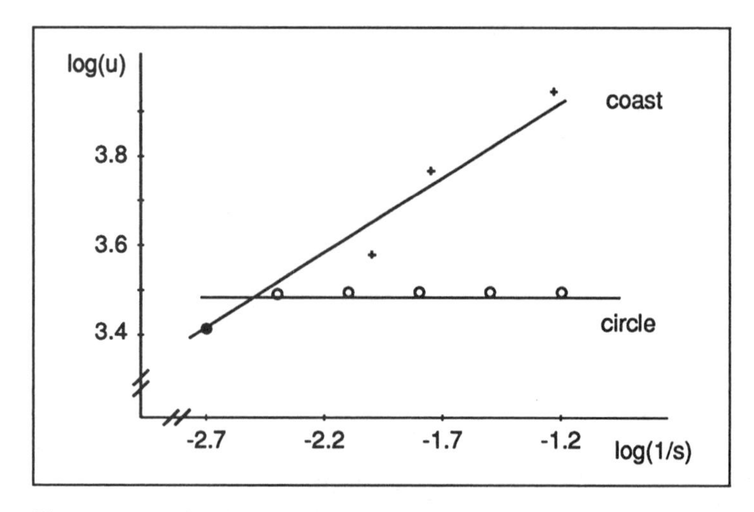

Figure 4.13 : Log/log diagram for measurements of coast of Britain and circle of diameter 1000 km. $u =$ length in units of 1000 km, $s =$ setting of compass in units of 1000 km. Rather than looking at $\log(s)$, we prefer to look at $\log(1/s)$ as a measure of the precision of the length.

compass setting varies over a broad spectrum from a few kilometers to several hundred, a length-versus-setting diagram is difficult to draw. In such a situation one usually passes to a log-log diagram. On the horizontal axis the logarithm of the inverse compass setting (1/setting) is marked. This quantity can be interpreted as the precision of the measurement. The smaller the compass setting is, the more precise is the measurement. The vertical axis is for the logarithms of the length. We take logarithms with respect to base 10, but that doesn't really matter. We need to introduce units for the length measurement u and compass setting s. Let us choose here

$$u = 0.951 \simeq 1000 \text{ km}$$
$$s = 1 \simeq 1000 \text{ km}.$$

In other words, $s = 0.1 \simeq 100$ km, and so on. Moreover, we like to interpret $1/s$ as a *measure of precision*, i.e. when s is small then the precision $1/s$ is large. Our log/log plots will always show how the total length ($\log(u)$) changes with an increase in precision ($\log(1/s)$). Figure 4.13 shows the result for the coastline of Britain.

Fitting a Straight Line to a Series of Points

We make a remarkable observation. Our points in the diagram roughly fall on straight lines. It is a topic of mathematical statistics how to define a line that approximates the points in such a diagram. Obviously, we cannot expect that the points fall exactly on a line, because of the nature of the measurements. However, a measure of the deviation of the line from the collection of points can be minimized. This leads to the widely used *method of least squares.* In our case we obtain a horizontal line for the circle and a line with some slope $d \approx 0.3$ for the coast of Britain.

Assume that we take these data and use them to make a forecast of the changes when passing to more precise measurements, i.e. when we use a smaller compass setting s. For this purpose we would simply extrapolate the lines to the right. This would yield about the same result for the circle since the line is horizontal. In other words, the circle has a finite length. However, the measured length of the coast would increase at smaller scales of measurement.

Let us denote by b the intercept of the fitting line with the vertical axis. Thus, b corresponds to the measured length at scale $s = 1$ corresponding to 1000 km. The relationship between the length u and the scale or size s covered by the compass can be expressed[8] by

$$\log u = d \cdot \log \frac{1}{s} + b \; . \tag{4.1}$$

Equation 4.1 expresses how the length changes when the setting of the compass is changed, assuming that in a log/log plot the measurements fall on a straight line. In that case the two constants, d and b, characteristize the growth law. The slope d of the fitting line is the key to the fractal dimension of the underlying object. We will discuss this in the next section.

Power Laws

We would not like to take it for granted that the reader is familiar with log/log diagrams. To explain the main idea, let us take some data from an experiment in physics. To investigate the laws governing free fall we may drop an object from different levels of a tall tower or building (of course with proper precautions taken). With a stopwatch we measure the time necessary for the object to reach the bottom. With height differences between levels being 4 meters, we get the following table of data.

Figure 4.15 shows the data graphically. Clearly, the plotted points are not on a straight line (left curve). Thus, the relation between height and drop time is not linear. The same plot on double logarithmic paper on the right, however, reveals that there is a law describing the relationship between height and drop time. This law is a *power law* of the form

$$t = c \cdot h^d \; . \tag{4.2}$$

Such a law is called a power law because t changes as if it were a power of h. The problem, then, is to verify the conjecture and

[8] Recall that a straight line in a x-y-diagram can be written as $y = dx + b$, where b is the y-intercept and d is the slope of the line (i.e. $d = (y_2 - y_1)/(x_2 - x_1)$), for any pair of points (x_1, y_1) and (x_2, y_2) on the line.

height h	drop time t	$\log h$	$\log t$
4 m	0.9 sec	0.60	−0.05
8 m	1.3 sec	0.90	0.11
12 m	1.6 sec	1.08	0.20
16 m	1.8 sec	1.20	0.26
20 m	2.0 sec	1.30	0.30
24 m	2.2 sec	1.38	0.34
28 m	2.4 sec	1.45	0.38
32 m	2.6 sec	1.51	0.41

Table 4.14 : Drop time versus height of free fall. The last two columns list the logarithms (base 10) for the data. The original and the logarithmic data is plotted in figure 4.15.

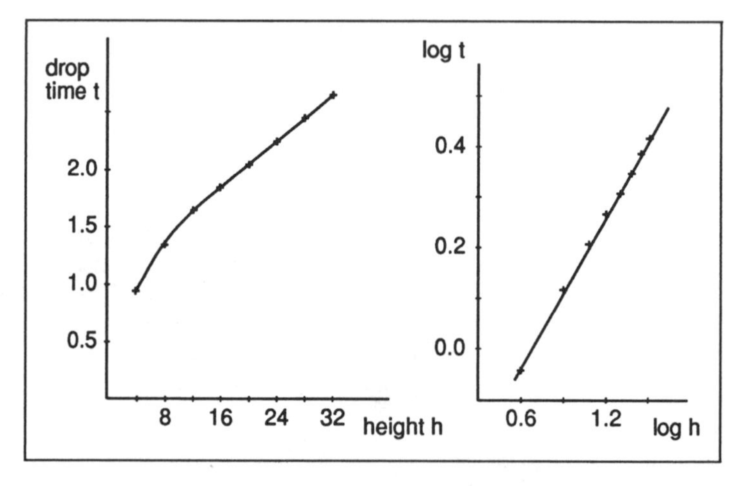

Figure 4.15 : The data from table 4.14 shows graphically the dependence of the drop time on the height of the fall. The data is displayed on the left using linear scales resulting in a parabola-like curve. On the right a double logarithmic representation of the same data is given. The data points appear to lie on a line.

determine c and d. To begin, let us assume that in fact eqn. (4.2) holds. Now we take the (base 10) logarithm[9] on both sides and obtain

$$\log t = d \cdot \log h + \log c \; .$$

In other words, if one plots $\log t$ versus $\log h$, rather than t versus h, one should see a straight line with slope d and y-intercept $b = \log c$, or $c = 10^b$. This is done in figure 4.15 on the right.

Thus, when measurements in a log/log-plot essentially fall onto a straight line, then it is reasonable to work with a power law which

[9] We may, of course, take the logarithm to any convenient base.

governs the relationship between the variables; and moreover, the log/log plot allows us to read off the power d in that power law as the slope of the straight line. In our example we can draw the fitting line in the double logarithmic plot and read off the slope and the y-intercept:

$$d = 0.48$$
$$\log c = -0.33‘.$$

Thus, $c = 10^{-0.33}$, and the power law determined from the measurements is

$$t = 0.47h^{0.48} . \qquad (4.3)$$

By the way, this is in good agreement with the Newtonian laws of motion, which state that the distance fallen is proportional to the square of the drop time. More precisely,

$$h = \frac{g}{2}t^2$$

where $g \approx 9.81$m/sec^2 is the gravitational constant. Solving this equation for t yields

$$t = \sqrt{\frac{2h}{g}} \approx 0.452 \cdot h^{0.5} ,$$

which is to be compared with our empirical result in eqn. (4.3).

Power Law for Allometric Growth

When we discussed allometric growth in chapter 3, we saw an interesting example of a power law. Let us remember that we compared measured head sizes with the body height as a baby developed into a child and then grew to adulthood. We learned that there were two phases. One up until the age of three, and the second after that until the growth process terminates. Using the approach of power laws with the tools of double logarithmic graphs we now try to model the allometric phase of growth by an appropriate power law. To this end we reconsider the original data from table 3.8 and extend it by corresponding logarithms (see table 4.16).

The plot in figure 4.17 on log/log scales reconfirms the two-stage growth of the measured person. We can fit two lines to the data. The first one reaching until age three and the second one for the rest of the data. The first line has a slope of about one. This corresponds to an equal growth of head size and body height; the two quantities are proportional and the growth is called isometric. The second line has a much lower slope, about $1/3$. This yields a power law stating that the head size should be proportional to the cube root of the body size. Or — turned around — we have that the body size scales like the cube of the head size,

$$\text{body height} \propto (\text{head size})^3 .$$

Head Size Versus Body Height Data

Age years	Body Height cm	Head Size cm	Body Height logarithm	Head Size logarithm
0	50	11	1.70	1.04
1	70	15	1.85	1.18
2	79	17	1.90	1.23
3	86	18	1.93	1.26
5	99	19	2.00	1.28
10	127	21	2.10	1.32
20	151	22	2.18	1.34
25	167	23	2.22	1.36
30	169	23	2.23	1.36
40	169	23	2.23	1.36

Table 4.16 : Body height and head size of a person with logarithms of the same data.

Head Size Versus Body Height Graph

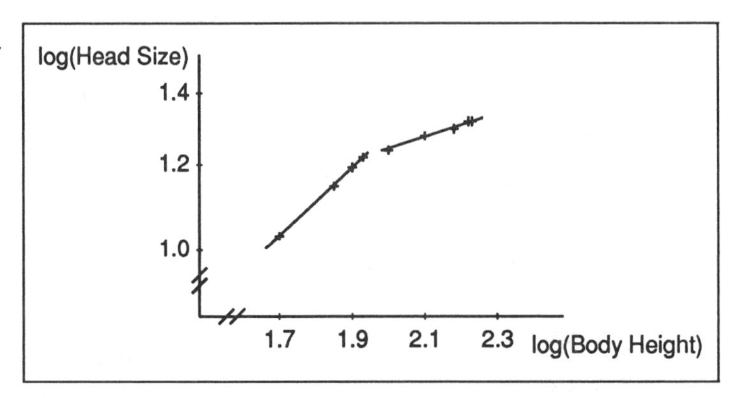

Figure 4.17 : Double logarithmic plot of head size versus body height data.

The body grows much faster than the head; here we speak of allometric growth. Of course our little analysis should not be mistaken for a serious research result. The measurements were taken from only one person and only at large time intervals. Moreover, the test person was born in the 19th century. Thus, the cubic growth law above is probably neither exact nor representative.

Let us summarize. If the x and y data of an experiment range over very large numerical scales, then it is possible that there is a power law which expresses y in terms of x. To test the power law conjecture we plot the data in a log/log plot. If, then, the measurements fit a straight line, we can read off the power of the law as the slope of that line.

Figure 4.13 supports that there is a power law (i.e., that eqn. (4.1) is true). Or equivalently, we may then conclude that ($u =$ length, $s =$ compass setting)

$$u = c \cdot \left(\frac{1}{s}\right)^d . \tag{4.4}$$

For the coast of Britain, we would then measure that $d \approx 0.3$. The result of this graphical analysis is, thus, that the measured length u of the coast grows in proportion to an increase of the precision $1/s$ raised to the power 0.3,

$$u \propto \frac{1}{s^{0.3}} .$$

Maps With More and More Detail

At this point we have to discuss several aspects of relation (4.4). One immediate consequence is that the length goes to infinity like $1/s^d$ as $s \to 0$. But can we really let the compass setting s go to zero? Of course we can, but there is some danger. If we let the size of the compass setting go to zero on some particular map of Britain, then the law (4.4) would be invalid due to the finite resolution of the map. In fact, in this case the measured length would tend to a limit. The power law and its consequences are only valid in a measured range of compass settings based on simultaneously picking maps with more and more detail. In other words, the power law characterizes the complexity of the coast of Britain over some range of scales by expressing how quickly the length increases if we measure with ever finer accuracy. Eventually, such measurements do not make much sense anymore because we would run out of maps and would have to begin measuring the coast in reality and face all the problems of identifying where a coast begins and ends, when to measure (at low or high tide), how to deal with river deltas and so on. In other words, the problem becomes somewhat ridiculous. But nevertheless, we can say that in any practical terms the coast of Britain has no length. The only meaningful thing we can say about its length is that it behaves like the above power law over a range of scales to be specified and that this behavior will be characteristic.

Characteristic Power Laws

What do we mean when we say 'characteristic'? Well, we mean that these numbers are likely to be different when we compare the coasts of Britain with those of Norway or California. The same will be true if we carry out an analogous experiment for the length of borders, e.g., the border of Portugal and Spain. Now we understand why the Portuguese encyclopedia came out with a larger value than the one in Spain. Since Portugal is very small in comparison with Spain, it is very likely that the map used in Portugal for the measurement of the common border had much more detail — was of much smaller scale — than the one in Spain.

Figure 4.18 : The Western States of the US.

The same reasoning explains the differences for the measurements of the coast of Britain.[10]

Measuring Utah

Let us look at the border of the State of Utah, one of the 50 states in the U.S.A. Figure 4.18 shows a rough map of Utah. Obviously (if you did not already know it) the border of Utah is very straight.[11] Table 4.20 collects a few measurements based on maps of various scales. If we represent these measurements in a log/log diagram, we obtain insight into the power law behavior. Apparently, the best way to fit a straight line to the points is by using a horizontal line. That is to say, the border of Utah has a power law with exponent $d = 0$ exactly like that of a circle, and that means that the border has, for all practical purposes, a finite length.

[10]The first measurements of this kind go back to the British scientist R. L. Richardson and his paper *The problem of contiguity: an appendix of statistics of deadly quarrels*, General Systems Yearbook 6 (1961) 139–187.

[11]We like Utah for many reasons. One of them is that we were introduced to fractals during a sabbatical in Salt Lake City during the 1982/83 academic year. And it was there where we did our first computer graphical experiments on fractals in the Mathematics and Computer Science Departments of the University of Utah.

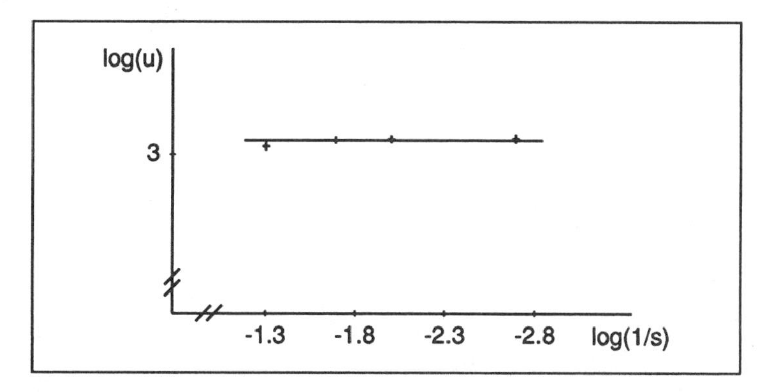

Figure 4.19 : Log/log representation of measurements of the boundary of Utah. $u =$ length measured in units of 100 km; $s =$ compass setting measured in units of 100 km.

Measuring the Border of Utah

setting	length
500 km	1450 km
100 km	1780 km
50 km	1860 km
20 km	1890 km

Table 4.20 : Length of border of Utah measured from maps of various scales and with a compass set to cover various distances.

Measuring the Koch Curve

Let us now try to understand the importance and meaning of the power law behavior in a pure mathematical situation. Recall the Koch island from chapter 3. The Koch island has a coast which is formed by three identical Koch curves. Now remember that each Koch curve can be divided into four self-similar parts, which are similar to the entire curve via a similarity transformation which reduces by a factor of 3.

Therefore, it is natural to choose compass settings covering sizes of the form $1/3$, $1/3^2$, $1/3^3$, ..., $1/3^k$. Of course there are two ways to work with these compass settings: an impossible one and the obvious one. It would be technically impossible to set a compass precisely to say $1/3^4 = 0.0123456789012...$ The thing to do would be to keep the compass setting constant and look at magnifications by a factor of 1, 3, 3^2, 3^3, ... Even that would be a waste of time because, from the construction of the Koch curve, we know exactly what the measurements would be, namely 4/3 for compass setting $s = 1/3$, 16/9 for $s = 1/9$, ..., $(4/3)^k$ for $s = (4/3)^k$.

Let us now represent these measurements in a log/log diagram (figure 4.22). Since we are free to choose a logarithm with respect

Measuring the Koch Curve

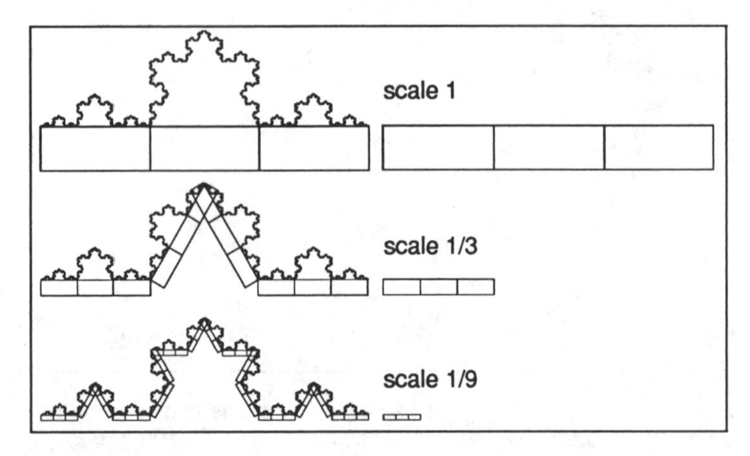

Figure 4.21 : Measuring the length of the Koch curve with different scales.

Log/log Plot for the Koch Curve

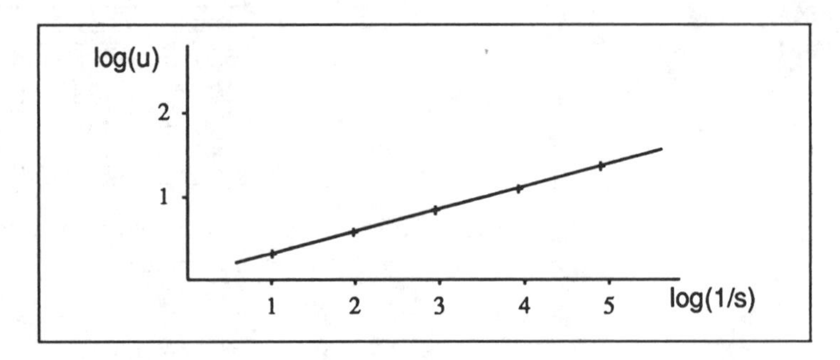

Figure 4.22 : $\log_3(u)/\log_3(1/s)$-diagram for the Koch curve.

to a convenient base, we take $\log_3$ so that for compass setting $s = 1/3^k$ we obtain length $u = (4/3)^k$ so that

$$\log_3 \frac{1}{s} = k \text{ and } \log_3 u = k \log_3 \frac{4}{3} .$$

Combining the two equations we obtain for the desired growth law

$$\log_3 u = d \log_3 \frac{1}{s} ,$$

that

$$d = \log_3 \frac{4}{3} \approx 0.2619 .$$

This numerical value for the slope d is remarkably close to the value which we measured for the coast of Britain.

4.3 Fractal Dimension

In our attempts to measure the length of the coast of Britain, we learned that the question of length — and likewise in other cases, of area or volume — can be ill-posed. Curves, surfaces, and volumes can be so complex that these ordinary measurements become meaningless. However, there is a way to measure the degree of complexity by evaluating how fast length, or surface, or volume increases if we measure with respect to smaller and smaller scales. The fundamental idea is to assume that the two quantities — length, or surface, or volume on the one hand, and scale on the other — do not vary arbitrarily but rather are related by a law, a law which allows us to compute one quantity from the other. The kind of law which seems to be relevant, as we explained previously, is a power law of the form $y \propto x^d$.

The Notion of Dimension

Such a law also turns out to be very useful for the discussion of *dimension*. Dimension is not easy to understand. At the turn of the century it was one of the major problems in mathematics to determine what dimension means and which properties it has (see chapter 2). And since then the situation has become somewhat worse because mathematicians have come up with some ten different notions of dimension: topological dimension, Hausdorff dimension, fractal dimension, self-similarity dimension, box-counting dimension, capacity dimension, information dimension, Euclidean dimension, and more. They are all related. Some of them, however, make sense in certain situations, but not at all in others, where alternative definitions are more helpful. Sometimes they all make sense and are the same. Sometimes several make sense but do not agree. The details can be confusing even for a research mathematician.[12] Thus, we will restrict ourselves to an elementary discussion of three of these dimensions:

- self-similarity dimension
- compass dimension (also called divider dimension)
- box-counting dimension

All are special forms of Mandelbrot's *fractal dimension*,[13] which in turn, was motivated by Hausdorff's[14] fundamental work from 1919. Of these three notions of dimension the box-counting dimension has the most applications in science. It is treated in the next section.

[12]Two good sources for those who want to pursue the subject are: K. Falconer, *Fractal Geometry, Mathematical Foundations and Applications*, Wiley, New York, 1990 and J. D. Farmer, E. Ott, J. A. Yorke, *The dimension of chaotic attractors*, Physica 7D (1983) 153–180.

[13]Fractal is derived from the Latin word *frangere*, which means 'to break'.

[14]Hausdorff (1868–1942) was a mathematician at the University of Bonn. He was a Jew, and he and his wife committed suicide in February of 1942, after he had learned that his deportation to a concentration camp was only one week away.

Felix Hausdorff

Figure 4.23 : Felix Hausdorff, 1868–1942.

Self-Similar Structures

We discussed the concept of self-similarity in the last chapter. Let us recall the essential points. A structure is said to be (strictly) self-similar if it can be broken down into arbitrarily small pieces, each of which is a small replica of the entire structure. Here it is important that the small piece can in fact be obtained from the entire structure by a *similarity transformation*. The best way to think of such a transformation is what we obtain from a photocopier with a reduction feature. For example, if we take a Koch curve and put it on a copying machine, set the reduction to 1/3 and produce four copies, then the four copies can be pasted together to give back the Koch curve. It then follows that if we copy each of the four reduced copies by a reduction factor of 1/3 four times (i.e., produce 16 copies which are reduced by a factor of 1/9 compared to the original), then these 16 copies can also be pasted together to reproduce the original. With an ideal copier, this process could be repeated infinitely often. Again, it is important that the reductions are similarities.

It would be a mistake to believe that if a structure is self-similar, then it is fractal. Take, for example, a line segment, or a square, or a cube. Each can be broken into small copies which are obtained by similarity transformations (see figure 4.24). These structures, however, are not fractals.

Scaling Factors Can Be Characteristic

Here we see that the reduction factor is 1/3, which is, of course, arbitrary. We could as well have chosen 1/2, or 1/7 or 1/356. But

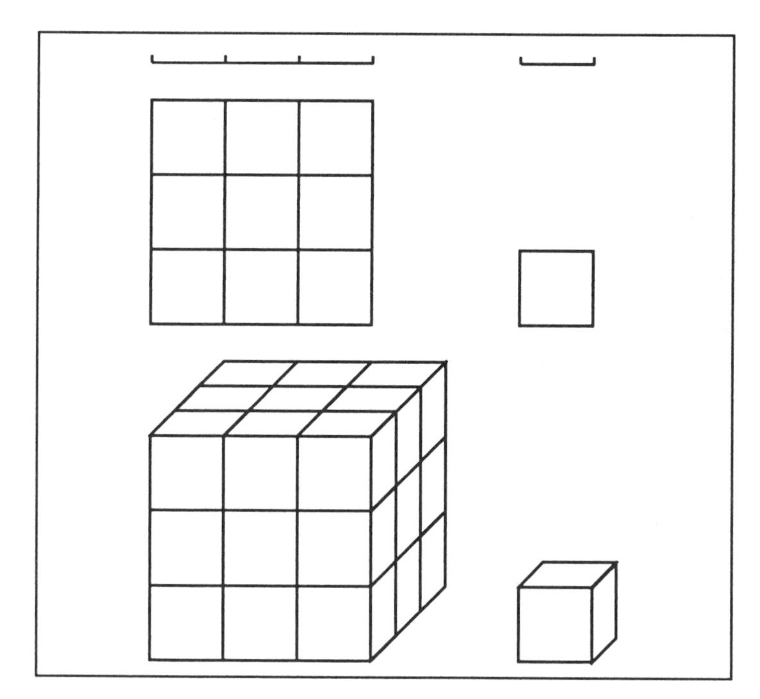

Figure 4.24 : Self-similarity of line, square, cube.

precisely in this fact lies the difference between these figures and fractal structures. In the latter the reduction factors — if they exist — are characteristic. For example, the Koch curve only admits 1/3, 1/9, 1/27, etc. The point, however, which is common to all strictly self-similar structures — fractal or not — is that there is a relation between the reduction factor (scaling factor) and the number of scaled down pieces into which the structure is divided (see table 4.25).

Apparently, for the line, square, and cube there is a nice power law relation between the number of pieces a and the reduction factor s. This is the law

$$a = \frac{1}{s^D} \tag{4.5}$$

where $D = 1$ for the line, $D = 2$ for the square, and $D = 3$ for the cube. In other words, the exponent in the power law agrees exactly with those numbers which are familiar as (topological) dimensions of the line, square, and cube. If we look at the Koch curve, however, the relationship of $a = 4$ to $s = 1/3$ and $a = 16$ to $s = 1/9$ is not so obvious.

But being guided by the relation for the line, square, and cube, we try a little bit harder. We postulate that eqn. (4.5) holds anyway.

object	number of pieces	reduction factor
line	3	1/3
line	6	1/6
line	173	1/173
square	$9 = 3^2$	1/3
square	$36 = 6^2$	1/6
square	$29929 = 173^2$	1/173
cube	$27 = 3^3$	1/3
cube	$216 = 6^3$	1/6
cube	$5177717 = 173^3$	1/173
Koch curve	4	1/3
Koch curve	16	1/9
Koch curve	4^k	$1/3^k$

Table 4.25 : Number of pieces versus scale factor for four objects.

In other words, $4 = 3^D$. Taking logarithms on both sides, we get

$$\log 4 = D \cdot \log 3 ,$$

or equivalently

$$D = \frac{\log 4}{\log 3} \approx 1.2619 .$$

But do we get the same if we take smaller pieces, as with a reduction factor of 1/9? To check this out, we would postulate that $16 = 9^D$, or $\log 16 = D \cdot \log 9$, or $D = \log 16/\log 9$, from which we compute

$$D = \frac{\log 4^2}{\log 3^2} = \frac{2\log 4}{2\log 3} = \frac{\log 4}{\log 3} \approx 1.2619 .$$

And as a general rule,

$$D = \frac{\log 4^k}{\log 3^k}$$

implies that $D = \log 4/\log 3$. Hence the power law relation between the number of pieces and the reduction factor gives the same number D, regardless of the scale we use for the evaluation. It is this number D, a number between 1 and 2, that we call the self-similarity dimension of the Koch curve.

Self-Similarity Dimension

More generally, given a self-similar structure, there is a relation between the reduction factor s and the number of pieces a into which the structure can be divided; and that is

$$a = \frac{1}{s^D}$$

or equivalently

$$D = \frac{\log a}{\log 1/s} \; .$$

D is called the *self-similarity dimension.* In cases where it is important to be precise, we use the symbol D_s for the self-similarity dimension in order to avoid confusion with the other versions of fractal dimension. For the line, the square and the cube we obtain the expected self-similarity dimensions 1, 2, and 3. For the Koch curve we get $D \approx 1.2619$, a number whose fractional part is familiar from measuring the length of the Koch curve in the last section. The fractional part 0.2619... is exactly equal to the exponent of the power law describing the measured length in terms of the compass setting used! Before we discuss this in more detail let us try a few more self-similar objects and compute their self-similarity dimensions. Figure 4.27 shows the Sierpinski gasket, carpet, and the Cantor set. Table 4.26 compares the number of self-similar parts with the correspopnding scaling dactors.

Other Dimensions

object	scale s	pieces a	dimension D_s
Cantor set	$1/3^k$	2^k	$\log 2 / \log 3 \approx 0.6309$
Sierpinski gasket	$1/2^k$	3^k	$\log 3 / \log 2 \approx 1.5850$
Sierpinski carpet	$1/3^k$	8^k	$\log 8 / \log 3 \approx 1.8927$

Table 4.26 : Self-similarity dimensions for other fractal objects.

Self-Similarity Dimension and Length Measurement

What is the relation between the power law of the length measurement using different compass settings and the self-similarity dimension of a fractal curve? It turns out that the answer is very simple, namely

$$D_s = 1 + d$$

where d, as before, denotes the slope in the log/log diagram of length u versus precision $1/s$, i.e. $u = c/s^d$. Let us see why. First, we simplify by choosing appropriate units of length measurements such that the factor c in the power law becomes unity

$$u = \frac{1}{s^d} \; . \tag{4.6}$$

Taking logarithms, we obtain

$$\log u = d \cdot \log \frac{1}{s} \; , \tag{4.7}$$

where u is the length with respect to compass setting s. On the other hand we have the power law $a = 1/s^D$, where a denotes the

Three More Fractals

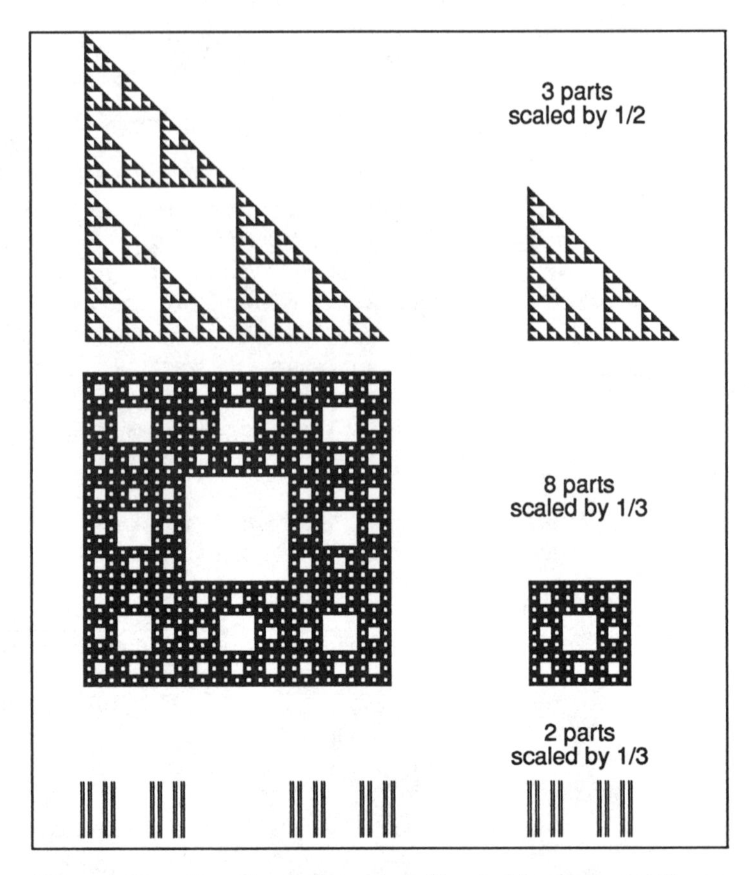

Figure 4.27 : The Sierpinski gasket, Sierpinski carpet, and Cantor set are shown with their building blocks, scaled down copies of the whole.

number of pieces in a replacement step of the self-similar fractal with scale factor s. In logarithmic form, this is

$$\log a = D_s \cdot \log \frac{1}{s} \ . \tag{4.8}$$

Now we can note the connection between length u and number of pieces a. At scale factor $s = 1$ we measure a length $u = 1$. This is true by construction: above in equation(4.6) we have set up units so that $u = 1$ when $s = 1$. Thus, when measuring at some other scale s, where the whole object is composed of a small copies each of size s, then we measure a total length of a times s,

$$u = a \cdot s \ .$$

This is the key to the following conclusion. Taking logarithms again we get

$$\log u = \log a + \log s \ .$$

In this equation we can substitute the logarithms $\log u$ and $\log a$ from equations (4.7) and (4.8). This yields

$$d \cdot \log \frac{1}{s} = D_s \cdot \log \frac{1}{s} + \log s \ .$$

Since $\log 1/s = -\log s$ we get

$$-d \cdot \log s = -D_s \cdot \log s + \log s$$

and dividing by $\log s$ and sorting terms we finally arrive at

$$D_s = 1 + d \ .$$

The result is that the self-similarity dimension can be computed in two equivalent ways:

- Based on the self-similarity of geometric forms find the power law describing the number of pieces a versus $1/s$, where s is the scale factor which characterizes the parts as copies of the whole. The exponent D_s in this law is the self-similarity dimension.
- Carry out the compass-type length measurement, find the power law relating the length with $1/s$, where s is the compass setting. The exponent d in this law, incremented by 1, is the self-similarity dimension, $D_s = 1 + d$.

Motivated by this result we may now generalize the dimension found in the second alternative also to shapes that are not self-similar curves such as coastlines and the like. Thus, we define the *compass dimension* (sometimes also called divider or ruler dimension) by

$$D_c = 1 + d$$

where d is the slope in the log/log-diagram of the measured length u versus precision $1/s$. Thus, since $d \approx 0.3$ for the coast of Britain, we can say that the coast has a fractal (compass) dimension of about 1.3. The fractal dimension of the state border of Utah of course is equal to 1.0, the fractal dimension of the straight line.

Measuring the 3/2-curve

We conclude this section with another example of a self-similar curve, the 3/2-curve. The construction process starts from a line segment of length 1. In the first step we replace the line segment by the *generator* curve, a polygonal line of 8 segments, each of length 1/4 (see figure 4.28). That is to say, the polygon has length 8/4, the length has doubled. In the next step, we scale down the polygonal line by a factor of 1/4 and replace each line segment of length 1/4 in step 1 by that scaled down polygon.

3/2-Curve: Two Steps

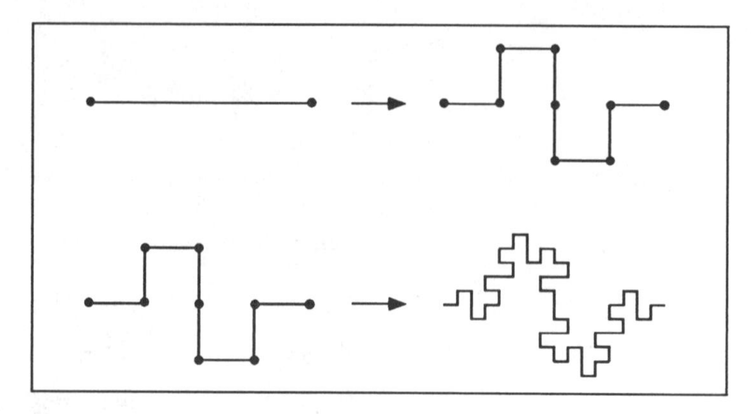

Figure 4.28 : The first two replacement steps in the construction of the 3/2-curve.

After the second step, we have 8^2 line segments, each of length $1/4^2$, so that the total length is now $8^2/4^2 = 2^2$. In the next step, we scale down the generator by a factor of $1/4^2$ and replace each line segment of length $1/4^2$ in step 2 by that scaled down generator, and so on. Apparently, the length of the resulting curve is doubled in each step (i.e., in step k the length is 2^k). The number of line segments grows by a factor of 8 in each step (i.e., in the k-th step we have 8^k line segments of length $1/4^k$). Entering these data in a log/log-diagram (preferably working with $\log_4$), we obtain figure 4.29.

Measuring the 3/2-Curve

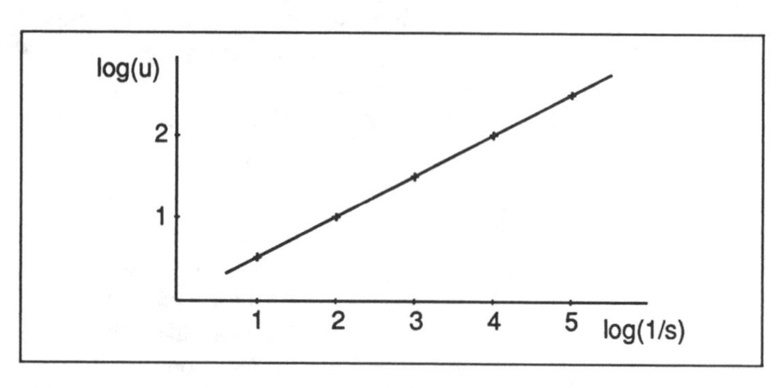

Figure 4.29 : Length versus 1/scale in the 3/2-curve. The graph displays logarithms with base 4. The result is a line with slope $1/2$.

Measuring the slope of the interpolating line, we obtain $d = 0.5$. More directly, the length computed at scale $1/4^k$ is 2^k, and this is reflected in the power law

$$u = \sqrt{\frac{1}{s}} = \frac{1}{s^{0.5}}$$

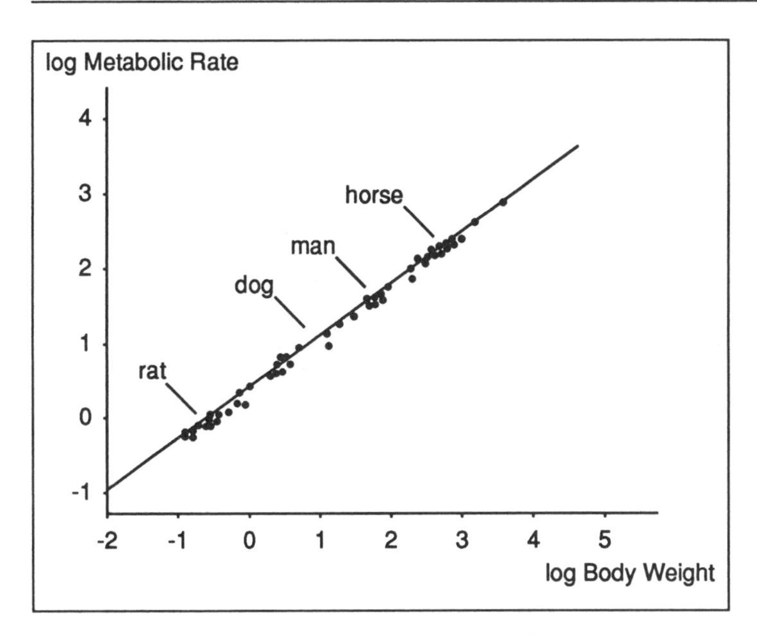

Figure 4.30 : The reduction law of metabolism, demonstrated in logarithmic coordinates, showing basal metabolic rate as power functions of body weight.

Metabolic Rate as Power Law

with exponent $d = 1/2$. Thus, the compass dimension and the self-similarity dimension are equal to $D = 1 + d = 1.5$, which justifies the name 3/2-curve.

The Fractal Nature of Organisms

We conclude this section with some fascinating speculations, which go back to a 1985 paper by M. Sernetz and others,[15] concerning the fractal nature of organisms. This paper discusses the metabolic rates of various animals (e.g. rats, dogs and horses) and relates them to their respective body weights. The metabolic rate is measured in Joules per second and the weight in kg. Since body weight is proportional to volume and since volume scales as r^3, when r is the scaling factor, a first guess would be that the metabolic rate should be proportional to the body weight (i.e., proportional to r^3). Figure 4.30 reveals, however, that the power law is significantly different from this expected value.

The slope a for the interpolating line is approximately 0.75. In other words, if m denotes the metabolic rate and w the body weight, then

$$\log m = a \log w + \log c \ ,$$

[15]From M. Sernetz, B. Gelléri, F. Hofman, *The Organism as a Bioreactor, Interpretation of the Reduction Law of Metabolism in terms of Heterogeneous Catalysis and Fractal Structure*, Journal Theoretical Biology 117 (1985) 209–230.

Figure 4.31 : Arterial and venous casts of a kidney of a horse as an example of fractal structures in organisms. Both systems in the natural situation fit entirely into each other and yet represent only the negative of the kidney. The remaining interspace between the vessels corresponds to the actual tissue of the organ (see also the color plates).

where $\log c$ is the m-intercept. Thus, $m = cw^\alpha$. Using $w \propto r^3$, this is equivalent to $m \propto r^{3\alpha}$, where $3\alpha \approx 2.25$.

This means that our guess, according to which the metabolic rate should be proportional to the weight/volume is wrong. It merely scales according to a fractal surface of dimension $D_F = 2.25$. How can that be explained? One of the speculations is that the above power law for the metabolic rate in organisms is a reflection of the fact that an organism is, in some sense, more like a highly convoluted surface than a solid body. In carrying this idea a little further — maybe too far — we could say that animals, including humans, look like three dimensional objects, but they are much more like fractal surfaces. Indeed, if we look beneath the skin, we find all kinds of systems (e.g. the arterial and venous systems of a kidney) which are good examples of fractal surfaces in their incredible vascular branching (see the color plates). From a physiology point of view, it is almost self-evident that the exchange functions of a kidney are intimately related to the size of the surfaces of its urinary and blood vessel systems. It is obvious that the volume of such a system is finite; it fits into the kidney! At the same time, the surface is in all practical terms infinite! And the relevant measuring task, quite like the ones for coastlines, would be to determine how the surface measurement grows as we measure it with higher and higher accuracy. This leads to the fractal dimension, which characterizes some aspects of the complexity of the bifurcation structure in such a system. This numerical evaluation of characteristic features of vessel systems can potentially become an important new tool in physiology. For example, questions like the following have been asked: What are the differences between systems of various animals? Or, is there a significant change in D_f when measured for systems with certain malfunctions?

4.4 The Box-Counting Dimension

In this section we discuss our third and final version of Mandelbrot's fractal dimension: the *box-counting dimension.* This concept is related to the self-similarity dimension, it gives the same numbers in many cases, but also different numbers in some others.

Non-Self-Similar Structures

So far, we have seen that we can characterize structures which have some very special properties such as self-similarity, or structures like coastlines, where we can work with a compass of various settings. But what can be done if a structure is not at all self-similar and as wild as figure 4.32, for example?

A Wild Fractal

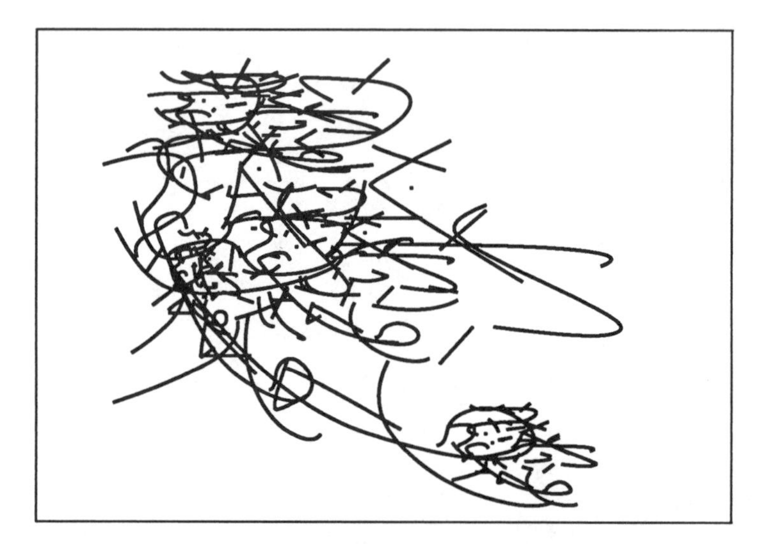

Figure 4.32 : A wild structure with some scaling properties.

In such a case, there is no curve which can be measured with a compass; and there is no self-similarity, though there are some scaling properties. For example, the 'cloud' in the lower right corner looks somewhat similar to the large 'cloud' in the upper portion. The box-counting dimension proposes a systematic measurement, which applies to any structure in the plane and can be readily adapted for structures in space. The idea is very much related to the coastline measurements.

We put the structure onto a regular grid with mesh size s, and simply count the number of grid boxes which contain some of the structure. This gives a number, say N. Of course, this number will depend on our choice of s. Therefore we write $N(s)$. Now change s to progressively smaller sizes and count the corresponding number $N(s)$. Next we make a log/log-diagram (i.e., more precisely, plot the measurements in a $\log(N(s))/\log(1/s)$-diagram).

Box-Count

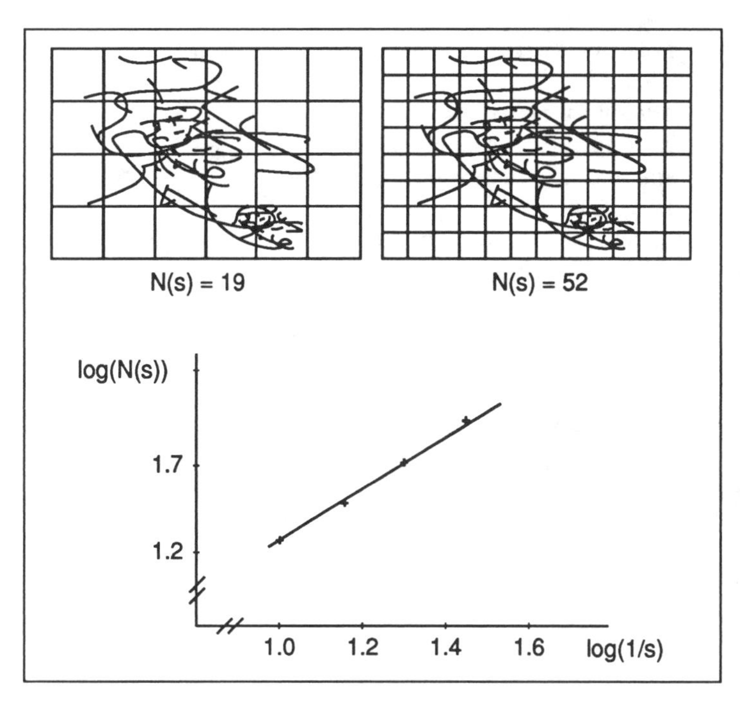

Figure 4.33 : The wild structure is box-counted.

The Box-Counting Dimension

We then try to fit a straight line to the plotted points of the diagram and measure its slope D_b. This number is the *box-counting dimension*, another special form of Mandelbrot's fractal dimension. Figure 4.33 illustrates this procedure. We measure a slope of approximately $D_b = 1.6$.

For practical purposes it is often convenient to consider a sequence of grids where the grid size is reduced by a factor of 2 from one grid to the next. In this approach each box from a grid is subdivided into four boxes each of half the size in the next grid. When box-counting a fractal using such grids we arrive at a sequence of counts $N(2^{-k}), k = 0, 1, 2, \ldots$ The slope of the line from one data to the next in the corresponding log/log diagram is

$$\frac{\log N(2^{-k-1}) - \log N(2^{-k})}{\log 2^{k+1} - \log 2^k} = \log \frac{N(2^{-k-1})}{N(2^{-k})}$$

where we have used logarithms with base 2. The result thus is the base 2 logarithm of the factor by which the box-count increases from one grid to the next. This slope is an estimate for the box-counting dimension of the fractal. In other words, if the number of boxes counted increases by a factor of 2^D when the box size is halved, then the fractal dimension is equal to D.

Self-Similarity and Box-Counting Dimension are Different

It is a nice exercise to experimentally verify the fact that the box-counting dimensions D_b of the Koch curve and the 3/2-curve are the same as the respective self-similarity and compass dimensions. Note, however, that in the plane a box-counting dimension D_b will never exceed 2. The self-similarity dimension D_s, however, can easily exceed 2 for a curve in the plane. To convince ourselves, we need only construct an example where the reduction factor is $s = 1/3$ and the number of pieces in a replacement step is $a > 9$ (see figure 4.34). Then

$$D_s = \frac{\log a}{\log 1/s} > 2 \; .$$

The reason for this discrepancy is that the curve generated in figure 4.34 has overlapping parts, which are ignored by the self-similarity dimension but not by the box-counting dimension. Here $s = 1/3$ and $a = 13$ (i.e., $D_s = \log 13/\log 3 = 2.33...$)

$$D_s = \frac{\log 13}{\log 3} \approx 2.335 \; .$$

Self-Intersection

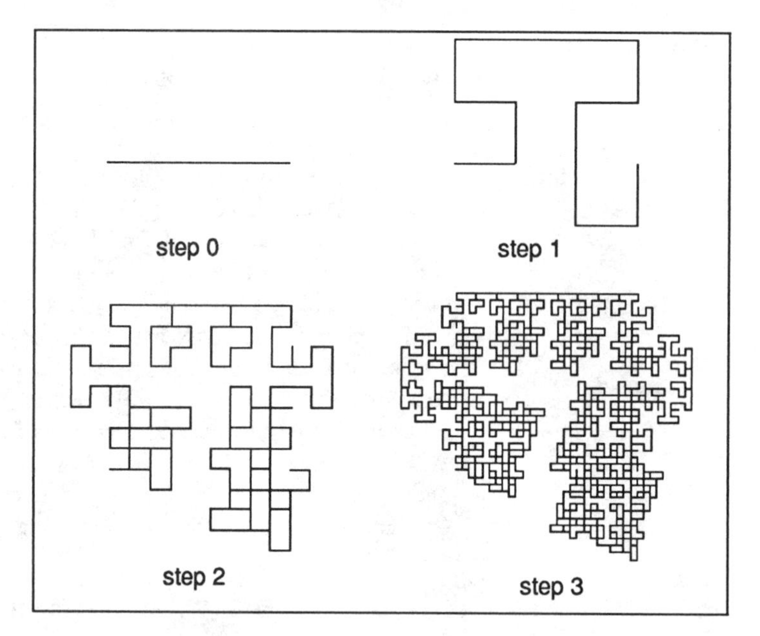

Figure 4.34 : First two steps of a curve generation with self-intersections.

Advantages of Box-Counting Dimensions

The box-counting dimension is the one most used in measurements in all the sciences. The reason for this dominance lies in the easy and automatic computability by machine. It is straightforward to count boxes and to maintain statistics allowing dimension calculation. The program can be carried out for shapes with and without self-similarity. Moreover, the objects may be embedded in higher dimensional spaces. For example, when considering objects in common three-dimensional space, the boxes are not flat but real three-dimensional boxes with height, width, and depth. But the concept also applies to fractals such as the Cantor set which is a subset of the nit interval, in which case the boxes are small intervals.

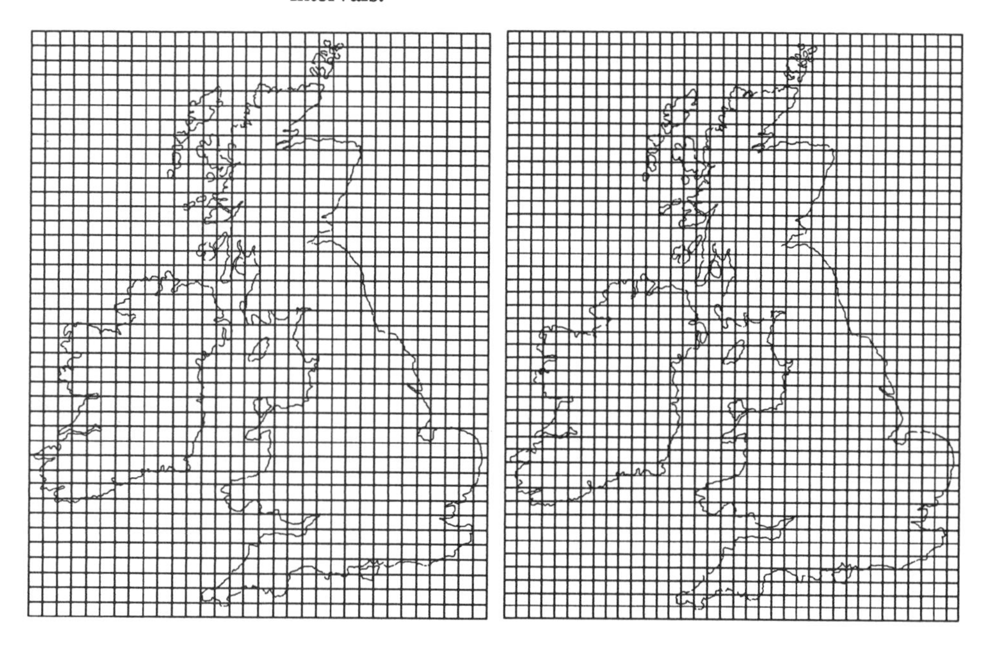

Figure 4.35 : Count all boxes that intersect (or even touch) the coastline of Great Britain, including Ireland.

Box-Counting Dimension of the Coast of Great Britain

As an example let us reconsider the classic example, the coastline of Great Britain. Figure 4.35 shows an outline of the coast with two underlying grids. Having normalized the width of the object, the grid sizes are 1/24 and 1/32. The box-count yields 194 and 283 boxes in each grid that intersect the coastline (check this carefully, if you have the time). From these data it is now easy to derive the box-counting dimension. When entering the data into a log/log-diagram, the slope of the line that connects the two points is

$$d = \frac{\log 283 - \log 194}{\log 32 - \log 24} \approx \frac{2.45 - 2.29}{1.51 - 1.38} \approx 1.31 \ .$$

This is in nice agreement with our previous result from the compass dimension.

Fractal Dimensions and Their Limitations

The concept of fractal dimension has inspired scientists to a host of interesting new work and fascinating speculations. Indeed, for a while it seemed as if the fractal dimensions would allow us to discover a new order in the world of complex phenomena and structures. This hope, however, has been dampened by some severe limitations. For one thing, there are several different dimensions which give different answers. We can also imagine that a structure is a mixture of different fractals, each one with a different value of box-counting dimension. In such a case, the conglomerate will have a dimension which is simply the dimension of the component(s) with the largest dimension. That means the resulting number cannot be characteristic for the mixture. What we would really like to have is something more like a spectrum of numbers which gives information about the distribution of fractal dimensions in a structure. This program has, in fact, been carried out and runs under the theme *multifractals*.[16]

[16]See B. B. Mandelbrot, *An introduction to multifractal distribution functions*, in: Fluctuations and Pattern Formation, H. E. Stanley and N. Ostrowsky (eds.), Kluwer Academic, Dordrecht, 1988.
J. Feder, *Fractals*, Plenum Press, New York 1988.
K. Falconer, *Fractal Geometry, Mathematical Foundations and Applications*, Wiley, New York, 1990.

4.5 Borderline Fractals: Devil's Staircase and Peano Curve

The fractals discussed in this chapter so far have a noninteger fractal dimension, but not all fractals are of this type. Thus, we want to expand our knowledge with two examples of fascinating fractals which represent very extreme cases: the first is the so-called devil's staircase, which implies a fractal curve of dimension 1.0. The second is a Peano curve of dimension equal to 2.0.

Devil's Staircase: Construction

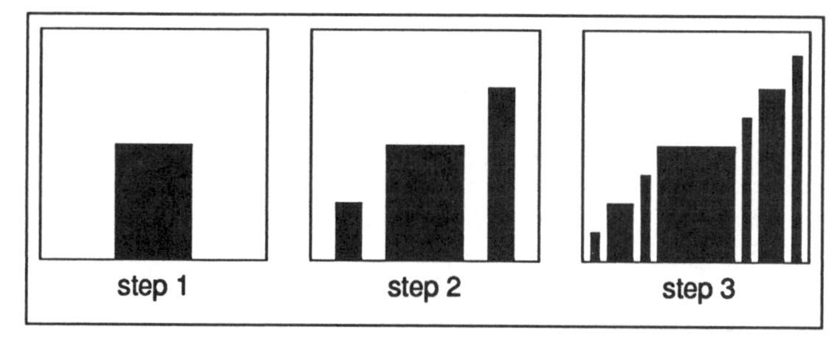

Figure 4.36 : The column construction of the devil's staircase.

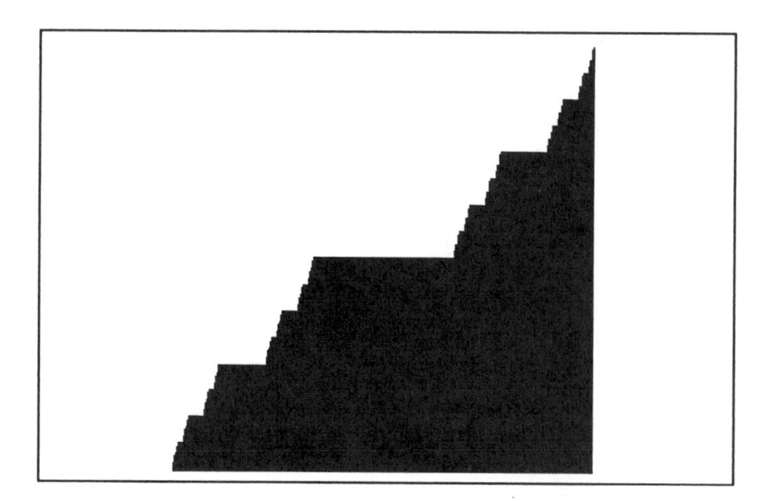

Figure 4.37 : The complete devil's staircase.

Devil's Staircase

The first one of these objects, the devil's staircase, is intimately related to the Cantor set and its construction. We take a square with sides of length 1. Then we start to construct the Cantor set on the base side (i.e., we take away successively middle thirds in the usual way). For each middle third of length $1/3^k$ which we take away, we paste in a rectangular column with base $1/3^k$ and a certain height. Let us see how this is done in figure 4.36.

In the first step, a column is erected over the middle third of the base side — the interval $[1/3, 2/3]$ — of the square with height $1/2$. In the second step, we erect two columns, one of height $1/4$ over the interval $[1/9, 2/9]$ and the other of height $3/4$ over the interval $[7/9, 8/9]$. In the third step, we erect four columns of heights $1/8, 3/8, 5/8, 7/8$, and in the k-th step, we erect 2^{k-1} columns of heights $1/2^{k-1}, 3/2^{k-1}, ..., (2^{k-1} - 1)/2^{k-1}$. In the limit, we obtain an object which is called the *devil's staircase*. Figure 4.37 shows an image obtained in a computer rendering. We see a staircase ascending from left to right, a staircase with infinitely many steps whose step heights become infinitely small. As the process continues, the square in figure 4.36 becomes filled with an upper white part and a lower black part. In the limit, there will be a perfect symmetry. The white part will be an exact copy of the black part. Put another way, the white part is obtained from the black part by a rotation of 180°. In this sense the boundary of the black staircase divides the square in two halves fractally. One immediate consequence of this argument is that the area used by the black staircase is exactly 1/2 of the initial square in the limit.

Area of the Devil's Staircase

We will look again at the columns in figure 4.36 and observe that the two narrow columns in step 2 make one column of height 1, likewise the four columns with base 1/27 make two columns of height 1, and so on. In other words, if we move the columns on the right over to the left side and cut the center column with base 1/3 into two equal parts, we obtain a figure which eventually fills half the square, see figure 4.38.

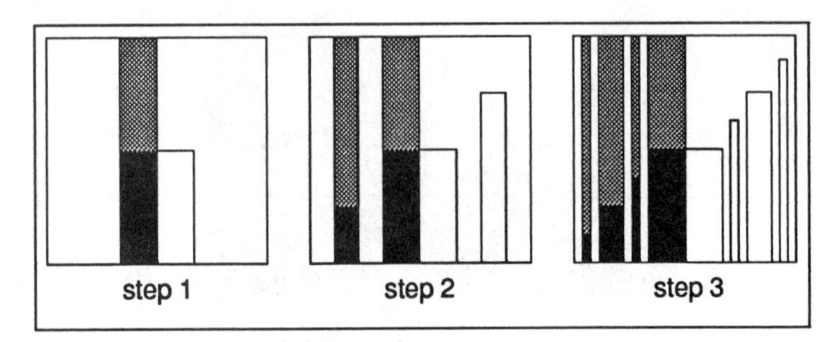

Figure 4.38 : The area under the devil's staircase is 1/2.

With the devil's staircase, we can also check an explicit argument. If we group the areas of the columns according to figure 4.38, using a geometric series, we obtain area A for the total area of the staircase as follows:

$$A = \frac{1}{3} \cdot \frac{1}{2} + \frac{1}{9} \cdot \left(\frac{1}{4} + \frac{3}{4}\right) + \frac{1}{27} \cdot \left(\frac{1}{8} + \frac{3}{8} + \frac{5}{8} + \frac{7}{8}\right) + \cdots$$

or,

$$A = \frac{1}{6} + \frac{1}{9} \cdot \left(1 + \frac{2}{3} + \frac{4}{9} + \cdots + \frac{2^k}{3^k} + \cdots\right) .$$

The sum of the geometric series in the bracket is 3. Thus, the result is

$$A = \frac{1}{6} + \frac{3}{9} = \frac{1}{2} .$$

Now to move on to our next questions: what is the nature of the boundary of the staircase, and how long is it? A polygonal approximation of the boundary of the staircase makes it obvious that

- the boundary is a curve, which has no gaps, and
- the length of that curve is exactly 2!

In other words, we have arrived at a very surprising result. We have obviously constructed a curve which is fractal, but it has a finite length. In other words, the slope d in the log/log-diagram of length versus 1/scale is $d = 0$, and thus the fractal dimension would be $D = 1 + d = 1$! This result is important because it teaches us that there are curves of finite length which we would like to call fractal nevertheless. Moreover, the devil's staircase looks self-similar at first glance, but is not. One may ask, of course, why those curves are called fractal in the first place? An argument in support of spending the characterization 'fractal' in this case is the fact that the devil's staircase is the graph of a very strange function, a function, that is constant everywhere except for those points that are in the Cantor set.

Polygon Construction of Devil's Staircase

It will be helpful in following the construction if you compare figure 4.36 and the following figure 4.39. We construct a polygon for each step in figure 4.36 by walking in horizontal and vertical directions only. We always start in the lower left corner and walk horizontally until we hit a column. At this point, we play fly and walk up the column until we reach the top. There we again walk horizontally until we hit the next column which we again surmount. At the top, we again walk on horizontally and then vertically, continuing on in the same pattern as often as necessary until we reach the right upper corner. In each step, the polygon constructed in this way has length 2 because, when summing up all horizontal lines the result is 1 and the sum of all vertical lines is also 1.

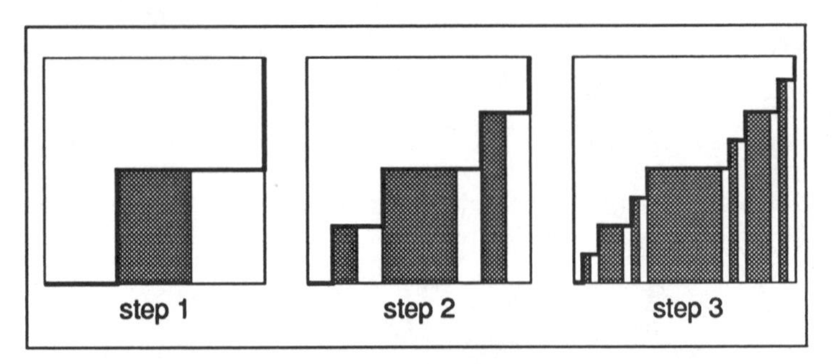

Figure 4.39 : Polygonal construction of devil's staircase.

Self-Affinity

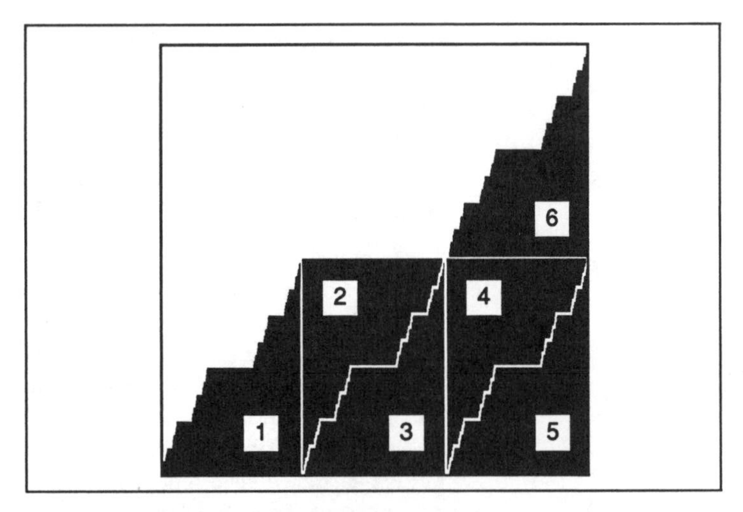

Figure 4.40 : The self-affinity of the devil's staircase.

Devil's Staircase Is Self-Affine

The devil's staircase is not self-similar. Let us explain (see figure 4.40). The devil's staircase can be broken down into six identical building blocks. Block 1 is obtained from the entire staircase by contracting the image horizontally by a factor of 1/3 and vertically by a factor of 1/2 (i.e., two different factors). This is why the object is not self-similar. For a self-similarity transformation, the two factors would have to be identical. Block 6 is the same as block 1. Moreover, a rectangle with sides of length 1/3 and 1/2 can house exactly one copy of block 1 together with a copy obtained by rotation by 180 degrees. This explains blocks 2 and 3, and 4 and 5. A contraction which reduces an image by different factors horizontally and vertically is a special case of a so called *affine transformation*. Objects that have parts that are affine copies of the whole are called *self-affine*. The devil's staircase is an example.

Devil's Staircase by Curdling

The devil's staircase may look like an odd mathematical invention. It is, indeed, a mathematical invention; but it isn't really so odd, for it has great importance in physics.[17] We want to discuss a problem — not really from physics, though it points in that direction — where the staircase comes out naturally.

Let us modify the Cantor set, see figure 4.41. Our initial object is no longer a line segment but rather a bar with density $r_0 = 1$. We suppose that we can compress and stretch the bar arbitrarily. The initial bar has length $l_0 = 1$ and therefore the mass $m_0 = 1$. Now we cut the bar in the middle obtaining two identical pieces of equal mass $m_1 = m_2 = 1/2$. Next we hammer them so that the length of each reduces to $l_1 = 1/3$ without changing the cross-section. Since mass is conserved, the density in each piece must increase to $r_1 = m_1/l_1 = 3/2$. Repeating this process, we find that in the n-th generation we have $N = 2^n$ bars, each with a length $l_n = 1/3^n$ and a mass $m_n = 1/2^n$. Mandelbrot calls this process *curdling* since an originally uniform mass distribution by this process clumps together into many small regions with a high density. The density of each of the small pieces is $r_n = m_n/l_n$. Figure 4.41 shows the density as height of the bars in each generation.

Assume now that the curdling process has been applied infinitely often, and we think of the resulting structure as put on an axis of length 1, measured from left to right. Then we can ask: what is the mass $M(x)$ of the structure in the segment from 0 to x?[18] The mass does not change in the gaps, but it increases by infinitesimal jumps at the points of the Cantor set. The graph of the function $M(x)$ turns out to be none other than the devil's staircase.

The Peano Curve

Having the fractal dimension $D = 1$, and yet not being an ordinary curve, the boundary curve of devil's staircase is one extreme case. Let us now look at extreme cases on the other end of the scale, curves which have fractal dimension $D = 2$. The first curve of this kind was discovered by G. Peano in 1890. His example created quite a bit of uncertainty about possible or impossible notions of curves, and for that reason also for dimension. We have already introduced the Peano curve in chapter 2, see figure 2.36. Recall that in its construction, line segments are replaced by a generator curve consisting of 9 segments each one being one third as long.

Based on the scaling factor 1/3 under which we observe self-similarity, we measure the curve with $s = 1/3^k, k = 0, 1, 2, \ldots$ as the size of the compass setting. This yields a total length of $u = (9/3)^k = 3^k$. Assuming the power law $u = c \cdot 1/s^d$, we first note that $c = 1$ because for $s = 1$ we have $u = 1$. Moreover, working with base 3 logarithms we conclude from the equation

[17] P. Bak, *The devil's staircase*, Phys. Today 39 (1986) 38–45.

[18] This can be written formally as $M(x) = \int_0^x dm(t)$.

Curdling

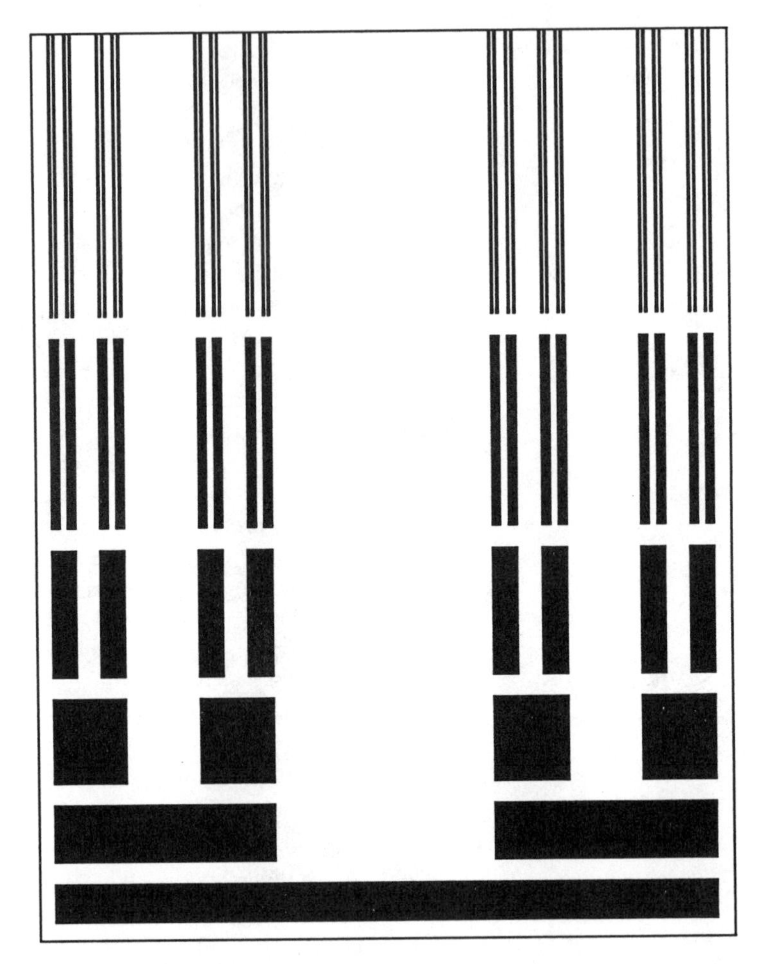

Figure 4.41 : Density shown as height in the successive generations of Cantor bars.

$\log_3 u = d \cdot \log_3 1/s$ that

$$d = \frac{\log_3 u}{\log_3 1/s} = \frac{k}{k} = 1 .$$

In other words, $D = 1 + d = 2$ (i.e., the Peano curve has fractal dimension 2). This reflects on the area-filling property of the Peano curve, which we discussed in chapter 2.

4.6 Program of the Chapter: The Cantor Set and Devil's Staircase

The devil's staircase is a borderline fractal. On one hand it is a curve having fractal dimension 1; on the other hand it is closely related to the Cantor set. This also becomes evident from this program which can compute both. Setting the variable `devil = 1` the staircase is produced, and the Cantor set is drawn if you set `devil = 0`.

You will recall that the classical Cantor set is obtained from a very simple rule: given a line, take away the middle third, then take away the middle thirds of the remaining two line segments, and so on. This program allows you to specify the relative size of the part to be removed. It asks the user at the beginning to specify this size. If you enter 0 nothing will we removed, for 1 everything will be removed, and for 0.333... you get the classical Cantor set. For the staircase construction this number specifies the size of the middle part for which the curve is at a constant value. In this case, 0.333... gives the classical staircase, 0 gives a diagonal and 1 a horizontal straight line. You should try these (and your own) settings.

Screen Image of Devil's Staircase

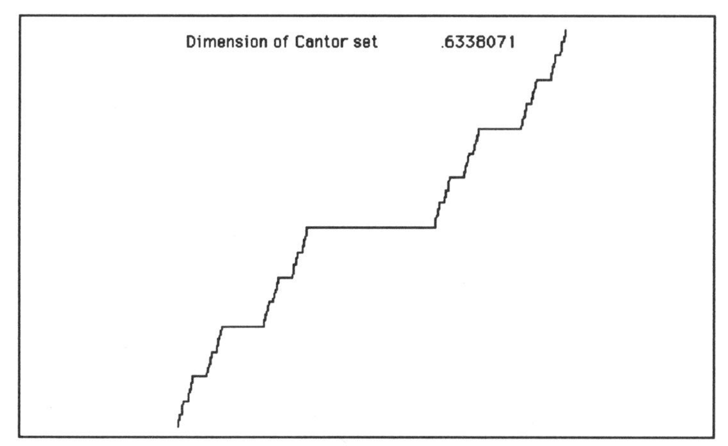

Figure 4.42 : Output of the program 'Cantor Set and Devil's Staircase'.

Now let us look at the program. The computation of the different stages of the Cantor set (or staircase) is very similar to the program for chapter 3 (we use the same recursive strategy of replacement of lines). Again the variable `level` specifies the depth of the recursion. If you change the initial value of the variable, you will see another stage of the construction process. Now assume we set `devil = 0` in order to compute the Cantor set. In

BASIC Program **Cantor Set and Devil's Staircase**
Title Drawing of Cantor set and devil's staircase

```
DIM xleft(10), yleft(10), xright(10), yright(10)
INPUT "part to remove (0 - 1):",r
level = 7
devil = 0
left = 30
w = 300
xleft(level) = left
xright(level) = left + w
yleft(level) = left + .5*(1+devil)*w
yright(level) = left + .5*(1-devil)*w
REM COMPUTE DIMENSION
IF r < 1 THEN d = LOG(2)/LOG(2/(1-r)) ELSE d = 0
PRINT "Dimension of Cantor set", d
GOSUB 100
END

REM DRAW LINE AT LOWEST LEVEL OF RECURSION
100 IF level > 1 GOTO 200
    LINE (xleft(1),yleft(1)) - (xright(1),yright(1))
    GOTO 300
REM BRANCH INTO LOWER LEVELS
200 level = level - 1
REM LEFT BRANCH
    xleft(level) = xleft(level+1)
    yleft(level) = yleft(level+1)
    xright(level) = .5*((1-r)*xright(level+1) + (1+r)*xleft(level+1))
    yright(level) = .5*(yright(level+1) + yleft(level+1))
    GOSUB 100
REM RIGHT BRANCH
    xleft(level) = .5*((1+r)*xright(level+1) + (1-r)*xleft(level+1))
    yleft(level) = .5*(yright(level+1) + yleft(level+1))
    IF devil THEN LINE (xleft(level),yleft(level))
        - (xright(level),yright(level))
    xright(level) = xright(level+1)
    yright(level) = yright(level+1)
    GOSUB 100
level = level + 1
300 RETURN
```

this case the program starts by setting up a horizontal initial line. Next it computes the fractal dimension (the similarity dimension) of the Cantor set and prints the result. Then the recursive line replacement starts (`GOSUB 100`).

In the recursive part we first check if we are at the lowest level. In this case we simply draw a line. Otherwise we compute the left branch of the replacement and go to the next level of the recursion (`GOSUB 100`). When we have computed and drawn the left branch, we compute the right branch of the replacement. Note, that since `devil = 0`, all the y-coordinates (`yleft` and `yright`) get the same value namely `left + 0.5*w`. Thus, we obtain a horizontal arrangement of points (or line segments) which represent the stage of the Cantor set construction.

Now let us discuss the case `devil = 1`. Here the initial line segment is a diagonal line. In the first step of the recursive replacement this line is replaced by two diagonal line segments connected by a horizontal one. In the following steps the short diagonal line segments are replaced recursively in the same way. In the program a horizontal connection is drawn whenever the further recursive replacement of the right branch starts (`IF devil THEN LINE ...`). In other words, in the case of the construction of the Cantor set we obtain for each stage horizontal line segments which converge to the Cantor set as the number of stages increases. In the case of the devil's staircase construction, the horizontal line segments from the Cantor set approximation become diagonal line segments and the gaps become horizontal line segments connecting the diagonal ones. You can make this visible if you alternate for lower stages (set `level = 1, 2, 3, ...`) between the computation of the Cantor set and the staircase.

Note that as in the previous programs you can adapt the position and size of the drawing by changing the variables `left` and `w`.

Chapter 5

Encoding Images by Simple Transformations

Fractal geometry will make you see everything differently. There is a danger in reading further. You risk the loss of your childhood vision of clouds, forests, galaxies, leaves, feathers, rocks, mountains, torrents of water, carpets, bricks, and much else besides. Never again will your interpretation of these things be quite the same.
Michael F. Barnsley[1]

So far, we have discussed two extreme ends of fractal geometry. We have explored fractal monsters, such as the Cantor set, the Koch curve, and the Sierpinski gasket; and we have argued that there are many fractals in natural structures and patterns, such as coastlines, blood vessel systems, and cauliflowers. We have discussed the common features, such as self-similarity, scaling properties, and fractal dimensions shared by those natural structures and the monsters; but we have not yet seen that they are close relatives in the sense that maybe a cauliflower is just a 'mutant' of a Sierpinski gasket, and a fern is just a Koch curve 'let loose'. Or phrased as a question, is there a framework in which a natural structure, such as a cauliflower, and an artificial structure, such as a Sierpinski gasket, are just examples of one unifying approach; and if so, what is it? Believe it or not, there is such a theory, and this chapter is devoted to it. It goes back to Mandelbrot's book, *The Fractal Geometry of Nature,* and a beautiful paper by the Australian mathematician Hutchinson.[2] Barnsley and Berger

[1]Michael F. Barnsley, *Fractals Everywhere,* Academic Press, 1988.

[2]J. Hutchinson, *Fractals and self-similarity,* Indiana Journal of Mathematics 30 (1981) 713–747. Some of the

have extended these ideas and advocated the point of view that they are very promising for the encoding of images.[3]

Fractal Geometry as a Language

We may regard fractal geometry as a new language in mathematics. As the English language can be broken down into letters and the Chinese language into characters, fractal geometry promises to provide a means to break down the patterns and forms of nature into primitive elements, which then can be composed into 'words' and 'sentences' describing these forms efficiently.

The word 'fern' has four letters and communicates a meaning in very compact form. Imagine two people talking over the telephone. One reports about a walk through a botanical garden admiring beautiful ferns. The person on the other end understands perfectly. As the word fern passes through the lines, a very complex amount of information is transmitted in very compact form. Note that 'fern' stands for an abstract idea of a fern and not exactly the one which was admired in the garden. To describe the individual plant adequately enough that the admiration can be shared on the other end, one word is not sufficient. We should be constantly aware of the problem that language is extremely abstract. Moreover, there are different hierarchical levels of abstractness, for example in the sequence: tree, oak tree, California oak tree, ...

Here we will discuss one of the major dialects of fractal geometry as if it were a language. Its elements are primitive transformations, and its words are primitive algorithms. For these transformations together with the algorithms, in section 1.2 we introduced the metaphor of the Multiple Reduction Copy Machine (MRCM),[4] which will be the center of interest in this chapter.

---

ideas can already be found in R. F. Williams, *Compositions of contractions*, Bol. Soc. Brasil. Mat. 2 (1971) 55–59.

[3]M. F. Barnsley, V. Ervin, D. Hardin, and J. Lancaster, *Solution of an inverse problem for fractals and other sets*, Proceedings of the National Academy of Sciences 83 (1986) 1975–1977.

M. Berger, *Encoding images through transition probablities*, Math. Comp. Modelling 11 (1988) 575–577.

A survey article is: E. R. Vrscay, *Iterated function systems: Theory, applications and the inverse problem*, in: Proceedings of the NATO Advanced Study Institute on Fractal Geometry, July 1989. Kluwer Academic Publishers, 1991.

A very promising approach seems to be presented in the recent paper A. E. Jacquin, *Image coding based on a fractal theory of iterated contractive image transformations*, to appear in: IEEE Transactions on Signal Processing, March 1992.

[4]A similar metaphor has been ussed by Barnsley in his popularizations of iterated function systems (IFS), which is the mathematical notation for MRCMs.

5.1 The Multiple Reduction Copy Machine Metaphor

MRCM = IFS

Let us briefly review the main idea of the MRCM, the multiple reduction copy machine. This machine provides a good metaphor for what is known as *deterministic iterated function systems* (IFS) in mathematics. From here on we use both terminologies interchangibly; sometimes it is more convenient to work with the machine metaphor, while in more mathematical discussions we tend to prefer the IFS notion. The reader may wish to skip back to the first chapter to take a look at figures 1.11 and 1.12. The copy machine takes an image as input. It has several independent lens systems, each of which reduces the input image and places it somewhere in the output image. The assembly of all reduced copies in some pattern is finally produced as output. Here are the dials of the machine

Dial 1: number of lens systems,
Dial 2: setting of reduction factor for each lens system individually,
Dial 3: configuration of lens systems for the assembly of copies.

The crucial idea is that the machine runs in a feedback loop; its own output is fed back as its new input again and again. While the result of this process is rather silly when there is only one reduction lens in the machine (only one point remains as shown in figure 1.11), this banal experiment turns into something extremely powerful and exciting when several lens systems are used. Moreover, we allow other transformations besides ordinary reductions (i.e., transformations more general than similarity transformations).

Imagine that such a machine has been built and someone wants to steal its secret — its construction plan. How much time and effort is necessary to get all the necessary information? Not very much at all. Our spy just has to run the machine once on an arbitrary image.[5] One copy reveals all the geometric features of the machine which we now start to operate in feedback mode.

An MRCM for Sierpinski Gaskets

Consider an MRCM with three lens systems, each of which is set to reduce by a factor of 1/2. The resulting copies are assembled in the configuration of an equilateral triangle. Figure 5.1 shows the effect of the machine run three times beginning with different initial images. In (a) we take a disk and use different shadings to keep track of the effect of the individual lens systems. In (b) we try a truly 'arbitrary' image. In just a few iterations the machine, or abstractly speaking the process, throws out images which look more and more like a Sierpinski gasket. In (c) we start with a Sierpinski gasket and observe that the machine has no effect on

[5] Almost any image can be used for this purpose. Images with certain symmetries provide some exceptions. We will study these in detail further below.

MRCM for the Sierpinski Gasket

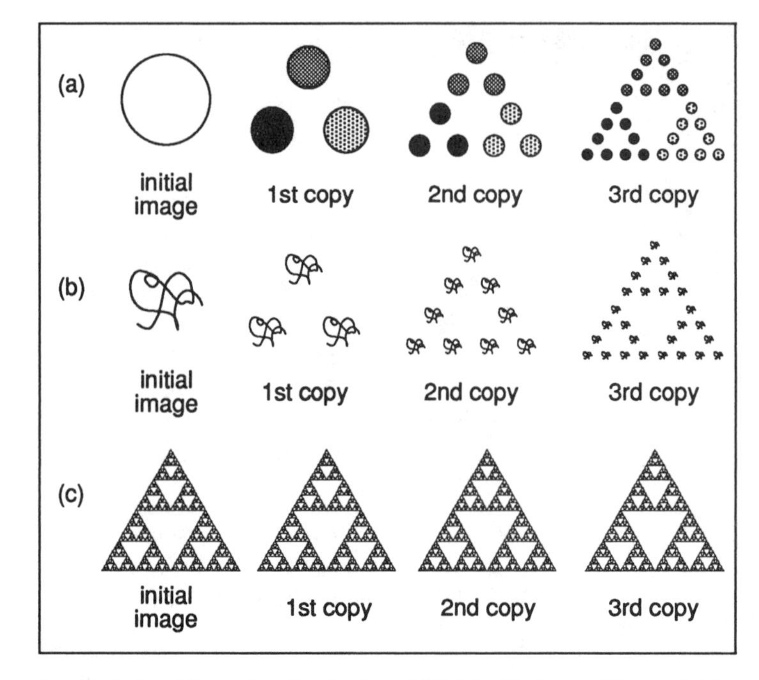

Figure 5.1 : Three iterations of an MRCM with three different initial images.

the image. The assembled reduced copies are the same as the initial image. That is, of course, because of the self-similarity property of the Sierpinski gasket.

The Attractor of the MRCM

Let us summarize this first experiment. No matter which initial image we take and run the MRCM with, we obtain a sequence of images which always tend towards one and the same final image. We call it the *attractor* of the machine or process. Moreover, when we start the machine with the attractor, then nothing happens, one says the attractor is left *invariant* or *fixed*. Perhaps it will help to explain this result if we compare our experiment with a physical one in which we have a bowl (figure 5.2, left) and observe how a little iron ball put into different initial positions and then let loose always comes to rest at the bottom, the rest point. But if we put the ball right at the bottom to begin with, nothing happens.

The bowl here corresponds to our machine. Initial positions of the ball here correspond to initial images in the machine. Observing the path of the ball in time corresponds to running the machine repeatedly, and the rest point of the ball corresponds to the final image. The fact that the ball moves continuously with time, while our machine operates in discrete steps, is not an essential difference. The point is that the ball in the bowl provides a metaphor for

Bowls

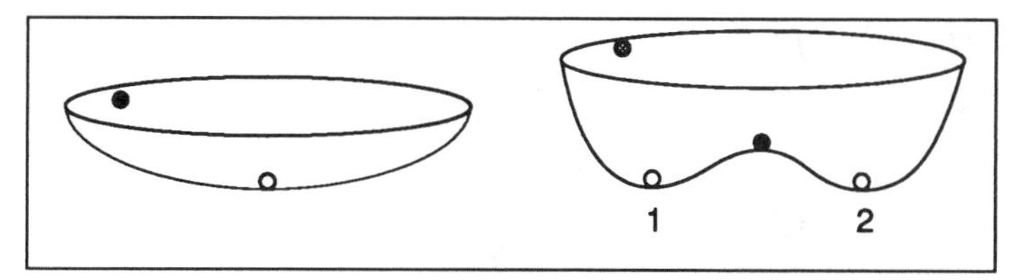

Figure 5.2 : Bowls with one and two dishes (attractors).

a *dynamical system* with only one attractor. The right hand image in figure 5.2 shows a situation with two different attractors. There the final development depends on where we start.

Is the MRCM more like a bowl with one dish or like a bowl with two or more dishes? And, how does the answer depend on the setting of the control dials? In other words, can it be that with one setting of the dials, the MRCM has one attractor, while there are several attractors with another setting? These are typical questions for modern mathematics, questions which are typical for a field called *dynamical systems theory*, which is a field which provides the framework for discussing chaos as well as the generation of fractals.

There are two ways to answer such a question. If we are lucky, we will be able to find a general principle in mathematics which is applicable and gives an answer. If that is not the case, we can either try to find a new theory, or if that turns out to be too hard at the moment, we can try to gain insight into the situation by carefully controlled experiments. It is quite clear that experiments alone will not be satisfactory in many cases. Often we do not know how the bowl is shaped. Then, if we find, for example, that for all tested initial positions, we always arrive at the same rest point, what does that tell us? Not much. We still could be in a situation with several rest points. That is to say, that quite by accident, the tested initial positions were not taken sufficiently arbitrarily.

Experiments Need Theoretical Support

In other words, finding that our MRCM seems to always run towards the same final image is a wonderful experimental discovery, but it needs theoretical support. It turns out that using some general mathematical principles and results developed by Felix Hausdorff and Stefan Banach, we can in fact show that any MRCM always has a unique final image as an attractor, and that this final image is invariant under the iteration of the MRCM. This is Hutchinson's beautiful and fundamental contribution to fractal theory. When we say 'any MRCM' here, we mean that the number and design of the lens systems may change in the MRCM. The only property which must be satisfied to have Hutchinson's result is that each lens system contracts images.

5.2 Composing Simple Transformations

The Multiple Reduction Copy Machine is based on a collection of contractions. The term contraction means, roughly speaking, that points are moved closer together when one contraction is applied. Of course, similarity transformations describing reduction by lenses are contractions. But we may also use transformations which reduce by different factors in different directions. For example, a transformation which reduces by one factor, say 1/3, horizontally and by a different factor, say 1/2, vertically is also allowed (see, for example, devil's staircase in section 4.5). Note that a *similarity transformation* maintains angles unchanged, while more general contractions may not.

Transformations of the MRCM

We may also take transformations of the latter kind combined with a shearing and/or rotation, and/or reflection. Figure 5.3 illustrates some admissible 'lens systems' for our MRCM. Mathematically, these are described as *affine linear transformations* of the plane.

Admissible Transformations

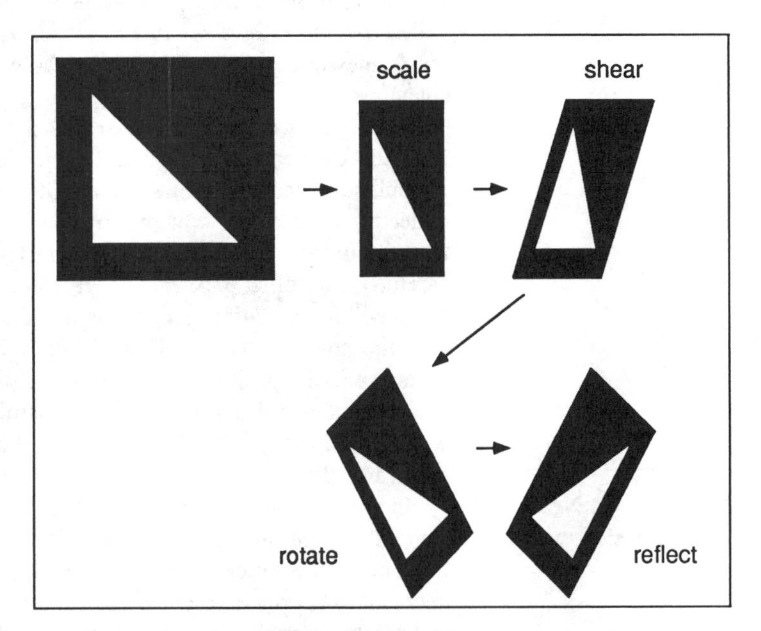

Figure 5.3 : Transformations with scaling, shearing, reflection, rotation and translation (not shown) are admissible in an MRCM.

Affine-Linear Transformations

The lens systems of our MRCMs can be described by *affine linear transformations* of the plane. Talking about a plane means that we fix a coordinate system, an x-axis, and a y-axis. Relative to that coordinate system every point P in the plane can be written as a pair (x, y). Sometimes we write $P = (x, y)$. In this way, points can be added together and can be multiplied by real numbers: if $P_1 = (x_1, y_1)$ and $P_2 = (x_2, y_2)$, then

$$P_1 + P_2 = (x_1 + x_2, y_1 + y_2)$$

and

$$sP = (sx, sy) .$$

Sum and Multiplication with Scalar

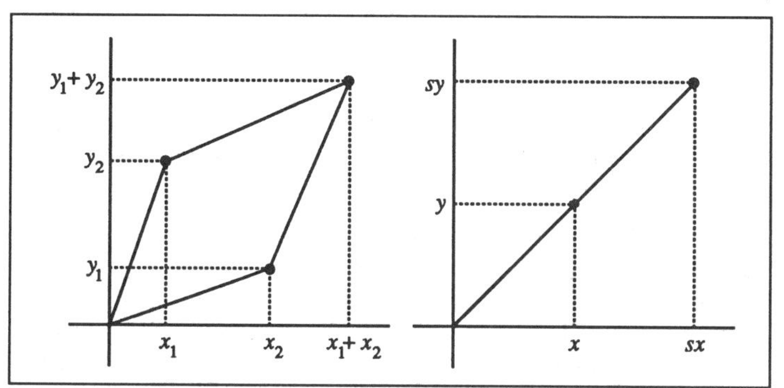

Figure 5.4 : (Left) Two points are added: $(x_1, y_1) + (x_2, y_2) = (x_1 + x_2, y_1 + y_2)$. (Right) A point is multiplied by a number: $s \cdot (x, y) = (sx, sy)$.

A *linear mapping* F is a transformation which associates with every point P in the plane a point $F(P)$ such that

$$F(P_1 + P_2) = F(P_1) + F(P_2)$$

for all points P_1 and P_2 and

$$F(sP) = sF(P)$$

for any real number s and all points P. A linear transformation F, or mapping, can be represented with respect to the given coordinate frame by a matrix

$$\begin{pmatrix} a & b \\ c & d \end{pmatrix}$$

where, if $P = (x, y)$ and $F(P) = (u, v)$, then

$$u = ax + by ,$$
$$v = cx + dy .$$

In other words, a linear transformation is determined by four coefficients a, b, c, and d. There are special representations which help us to discuss contractions more conveniently. To this end we write the four *coefficients* in our matrices as

$$\begin{pmatrix} r\cos\phi & -s\sin\psi \\ r\sin\phi & s\cos\psi \end{pmatrix}.$$

Such a representation is always possible. Just set

$$r = \sqrt{a^2 + c^2}$$

and

$$\phi = \arccos \frac{a}{\sqrt{a^2 + c^2}}$$

to obtain r and ϕ. Similar formulas hold for s and ψ. In this way it is easier to discuss reductions, rotations and reflections. For example:

- $s = r$, $0 \le r < 1$, and $\psi = \phi$ fixes a mapping which reduces by a factor of r and simultaneously rotates by the angle ϕ counterclockwise (the mapping is just a reduction, if $\phi = 0$).
- $s = r$, $0 \le r < 1$, $\phi = \pi$ and $\psi = 0$ fixes a mapping which reduces by a factor of r and simultaneously reflects with respect to the y-axis.
- $r = a$, and $s = b$, $0 \le a < 1$, $0 \le b < 1$, and $\phi = \psi = 0$ fixes a mapping which reduces by a factor of a in x-direction and by a factor of b in y-direction.
- $r = s > 0$ and $\phi = \psi$ defines a similarity transformation given by a rotation by an angle of ϕ and a scaling by a factor of r.

Affine linear mappings are simply the composition of a linear mapping together with a translation. In other words, if F is linear and Q is a point, then the new mapping $w(P) = F(P) + Q$, where P is any point in the plane, is said to be affine linear. Affine linear mappings allow us to describe contractions which involve positioning in the plane (i.e., the translation by Q). Since F is given by a matrix and Q is given by a pair of coordinates, say (e, f), an affine linear mapping w is given by six numbers,

$$\left(\begin{array}{cc|c} a & b & e \\ c & d & f \end{array}\right)$$

and if $P = (x, y)$ and $w(P) = (u, v)$ then

$$u = ax + by + e\ ,$$
$$v = cx + dy + f\ .$$

Another notation for the same equations is also sometimes used in this text,

$$w(x, y) = (ax + by + e, cx + dy + f)\ .$$

Affine Transformation

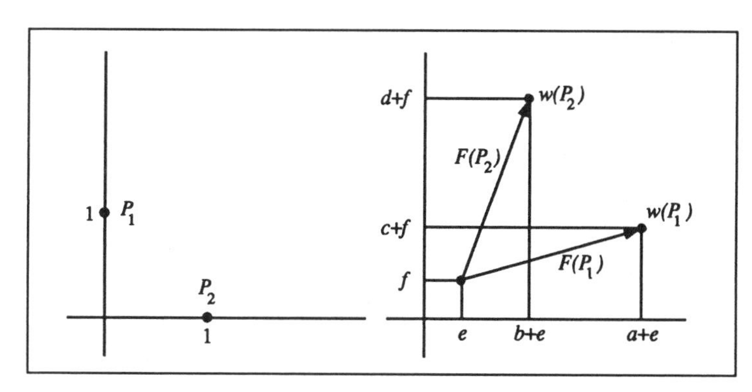

Figure 5.5 : The affine transformation described by six numbers a, b, c, d, e, f is applied to two points $P_1 = (1,0)$ and $P_2 = (0,1)$.

In the discussion of iterated function systems, it is crucial to study the objects which are left invariant under iteration of an IFS. Now given an affine linear mapping w, one can ask which points are left invariant under w? This is an exercise with a system of linear equations. Indeed, $w(P) = P$ means

$$x = ax + by + e \ ,$$
$$y = cx + dy + f \ .$$

Solving that system of equations yields exactly one solution, as long as the determinant $(a-1)(d-1) - bc \neq 0$. This point $P = (x, y)$ is called the fixed point of w. Its coordinates are

$$x = \frac{-e(d-1) + bf}{(a-1)(d-1) - bc} \ ,$$
$$y = \frac{-f(a-1) + ce}{(a-1)(d-1) - bc} \ .$$

The First Step: Blueprint of MRCM

Already the first application of the MRCM to a given image will usually reveal its internal affine linear contractions. This could be called the *blueprint* of the machine. Note, that it is necessary to select an initial image with sufficient structure in order to uniquely identify the transformations. Otherwise one cannot safely detect some of the possible rotations and reflections. Figure 5.6 illustrates this problem with three transformations. The first two images obviously are not suitable to fully unfold the blueprint of the machine. In the following images in this chapter we typically use a unit square $[0,1] \times [0,1]$ with an inscribed letter 'L' in the top left corner as an initial image to unfold the blueprint.

The lens systems of an MRCM are described by a set of affine transformations $w_1, w_2, ..., w_N$. For a given initial image A small

Unfolding the Blueprint

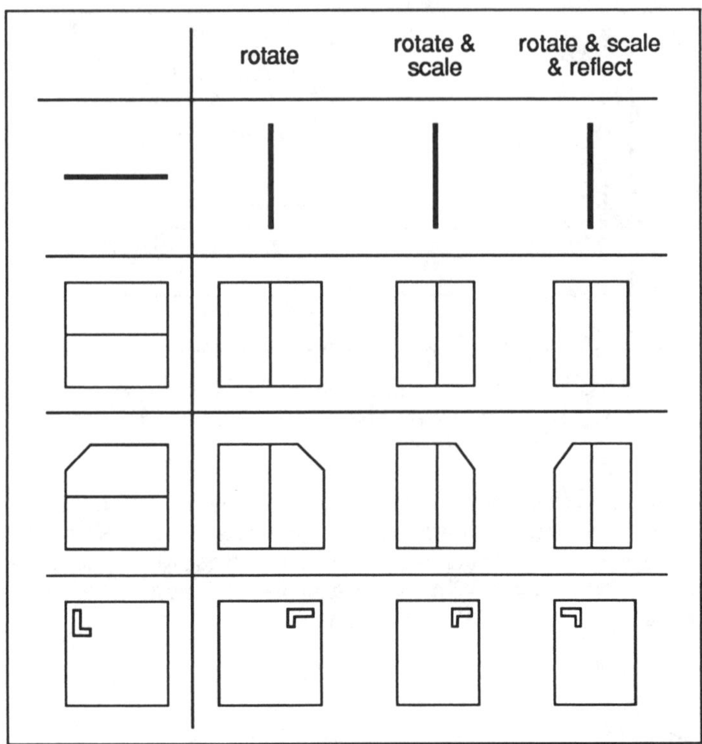

Figure 5.6 : The images of the first two rows do not allow the precise determination of the transformation.

affine copies $w_1(A), w_2(A), ..., w_N(A)$ are produced. Finally, the machine overlays all these copies into one new image, the output $W(A)$ of the machine:

$$W(A) = w_1(A) \cup w_2(A) \cup \cdots \cup w_N(A) \ .$$

Iterated Function System

W is called the *Hutchinson operator*. Running the MRCM in feedback mode thus corresponds to iterating the operator W. This is the essence of a deterministic iterated function system (IFS). Starting with some initial image A_0, we obtain $A_1 = W(A_0)$, $A_2 = W(A_1)$, and so on. Figures 5.7 and 5.8 summarize this setup. Shown are the MRCM as a feedback system and the blueprint of the machine for the Sierpinski gasket described by three transformations.

IFS and the Hutchinson Operator

Let $w_1, w_2, ..., w_N$ be N contractions of the plane (we will carefully discuss this term a little bit later). Now we define a new mapping — the Hutchinson operator — as follows: let A be any subset of the

MRCM Feedback

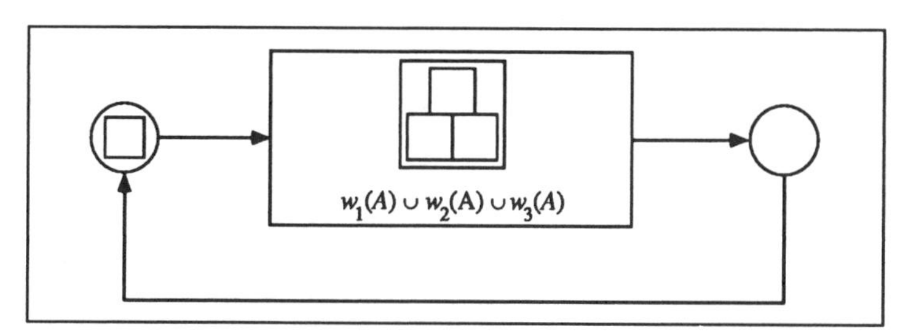

Figure 5.7 : The operation of an MRCM as a feedback system.

First Blueprint of an MRCM

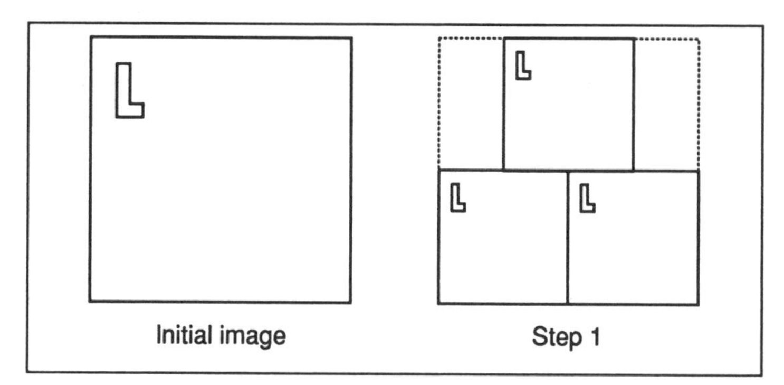

Figure 5.8 : Blueprint of an MRCM using a unit square with an inscribed letter 'L' in the top left corner as an initial image. The purpose of the outline of the initial image on the output on the right is to allow the identification of the relative positioning of the images.

plane.[6] Here we think of A as an image. Then the collage obtained by applying the N contractions to A and assembling the results can be expressed by the collage mapping:

$$W(A) = w_1(A) \cup w_2(A) \cup ... \cup w_N(A) \ . \qquad (5.1)$$

The Hutchinson operator turns the repeated application of the metaphoric MRCM into a dynamical system: an IFS. Let A_0 be an initial set (image). Then we obtain

$$A_{k+1} = W(A_k), k = 0, 1, 2, ...,$$

a sequence of sets (images), by repeatedly applying W. An IFS generates a sequence which tends towards a final image A_∞, which we call the attractor of the IFS (or MRCM), and which is left invariant by the IFS. In terms of W this means that

$$W(A_\infty) = A_\infty \ .$$

[6]Being more mathematically technical, we allow A to be any compact set in the plane. Compactness means, that A is bounded and that A contains all its limit points, i.e. for any sequence of points from A with a cluster point, we have that the cluster point also belongs to A. The open unit disk of all points in the plane with a distance less than 1 from the origin is not a compact set, but the closed unit disk of all points with a distance not exceeding 1 is compact.

Sierpinski Gasket Variation

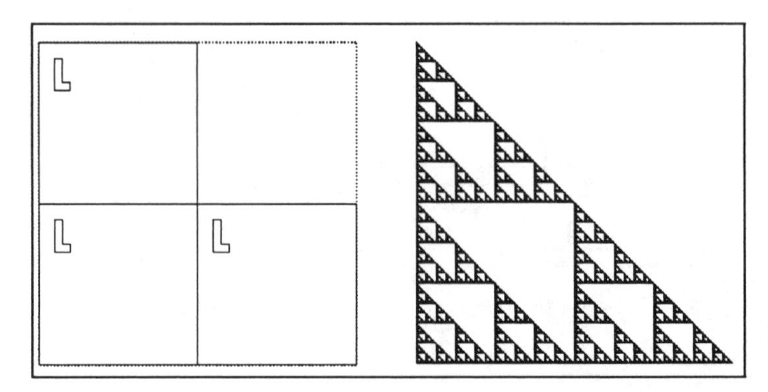

Figure 5.9 : IFS with three similarity transformations with scaling factor 1/2.

The Twin Christmas Tree

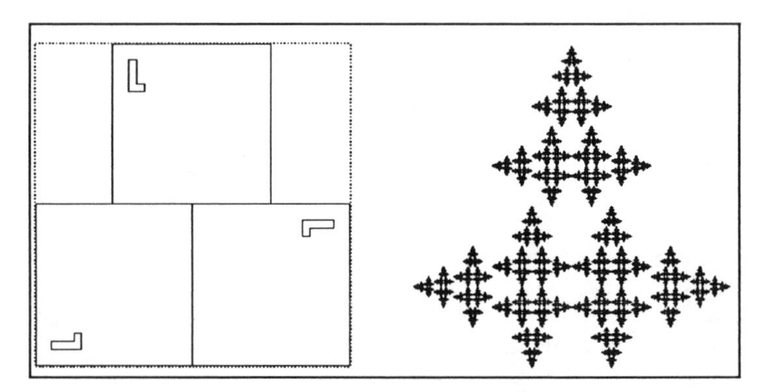

Figure 5.10 : Another IFS with three similarity transformations with scaling factor 1/2.

We say that A_∞ is a fixed point of W. Now how do we express that A_n tends towards A_∞? How can we make the term contraction precise? Is A_∞ a unique attractor? We will find answers to these questions in this chapter.

What happens if we change the transformations, or in other words if we play with the dials of the machine (i.e., if one changes the number of lenses, or changes their contraction properties, or assembles the individually contracted images in a different configuration)? In the following figures we show the results of some IFSs with different settings: the blueprint and the attractor. The blueprint is represented in a single drawing: the dotted square is for the initial image, and the solid line polygons represent the contractions.

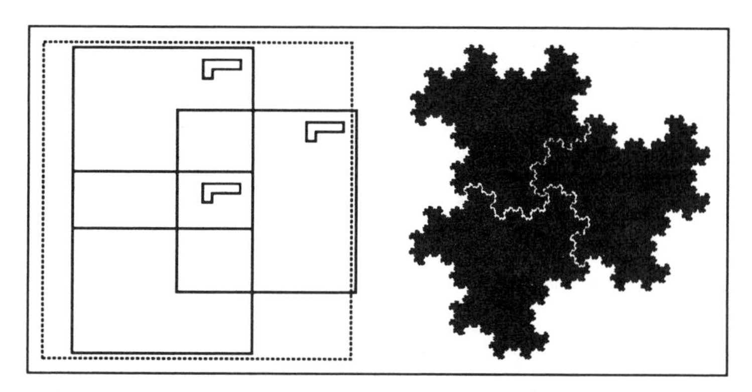

Figure 5.11 : The white line is inserted only to show that the figure can be made up from three parts similar to the whole.

A Dragon With Threefold Symmetry

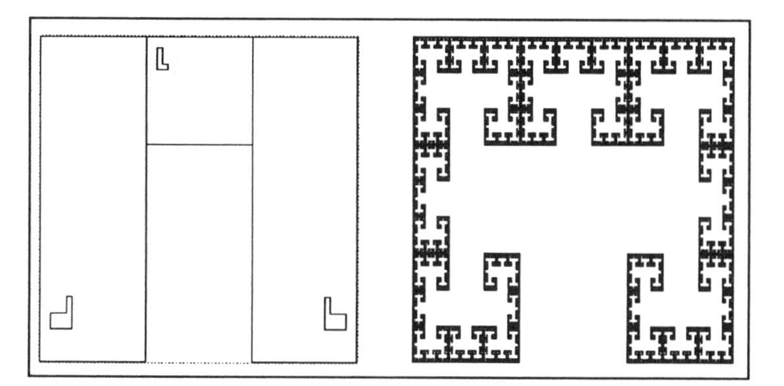

Figure 5.12 : IFS with three transformations, one of which is a similarity. The attrcator is related to the Cantor set.

The Cantor Maze

Table 5.48 lists the parameters of the corresponding affine transformations. They can be used with the program of this chapter to reproduce the figures displayed here.

Our first example is a small modification of the IFS which generates the Sierpinski gasket (see figure 5.9). It consists of three transformations each of which scales by a factor of 1/2 and translates as shown in the blueprint.

We are tempted to conjecture that all IFSs of three transformations that scale by 1/2 produce something very similar to the Sierpinski gasket. But this is far from the truth. In figure 5.10 we try another such IFS which differs from the original one for the Sierpinski gasket only by the addition of rotations. The lower right transformation rotates 90 degrees clockwise, while the lower left rotates by 90 degrees counter-clockwise. The result, called *twin christmas tree*, is clearly different from the Sierpinski gasket.

IFS for a Twig

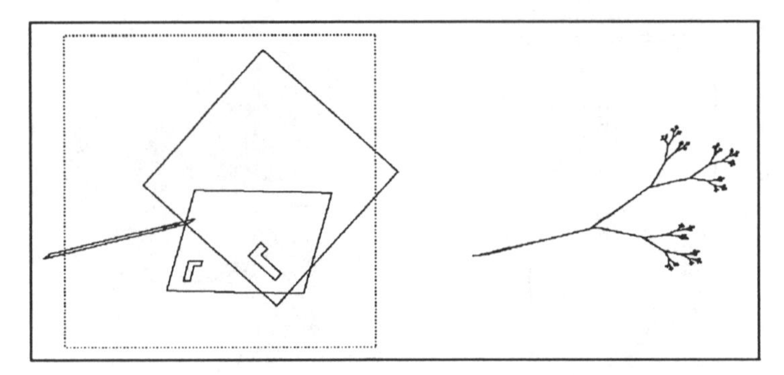

Figure 5.13 : IFS with three affine transformations (no similarities).

Crystal with Four Transformations

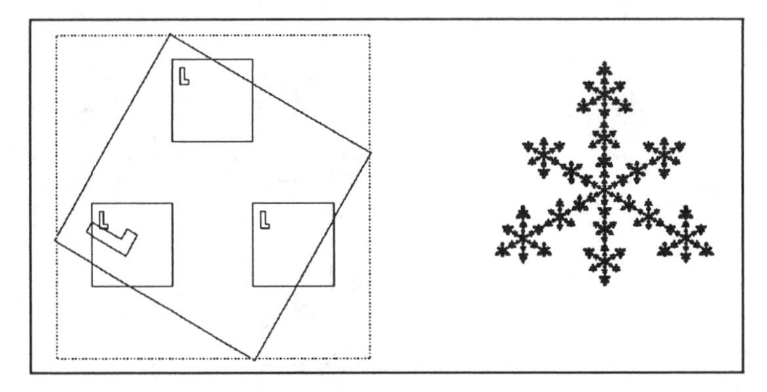

Figure 5.14 : IFS with four similarity transformations.

Now we start to also change the scaling factors of the transformations. In figure 5.11 we have chosen the factor of $s = 1/\sqrt{3}$ for all three transformations. Moreover, a clockwise rotation by 90 degrees is also included in each transformation. The result is a two-dimensional object with a fractal boundary: a type of dragon with threefold symmetry. It is invariant under a rotation of 120 degrees. It would be a good exercise at this point to compute the self-similarity dimension of the attractor using the techniques from the last chapter.

So far we have made use of only similarity transformations. In figure 5.12 there is only one similarity (which scales by 1/3) and two other transformations, which are rotations followed by a horizontal scaling by 1/3 and a reflection in one of the two cases. The result is a sort of maze, for which we have reason to introduce the name *Cantor maze*. The Cantor set is woven into the construction in all its details; all points of the cross product of two Cantor sets are connected in a systematic fashion.

Crystal with Five Transformations

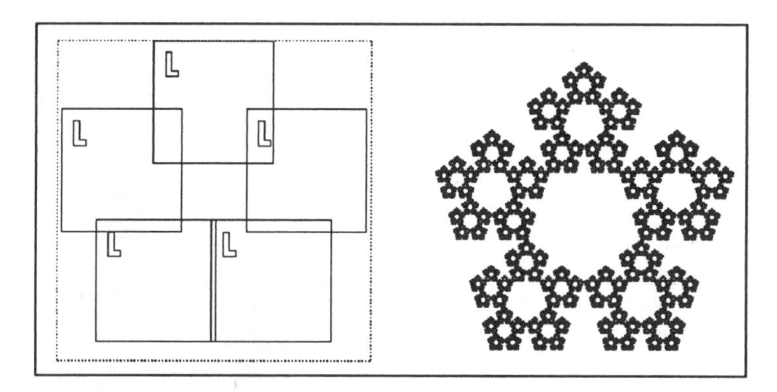

Figure 5.15 : IFS with five similarity transformations. Can you see Koch curves in the attractor?

A Tree

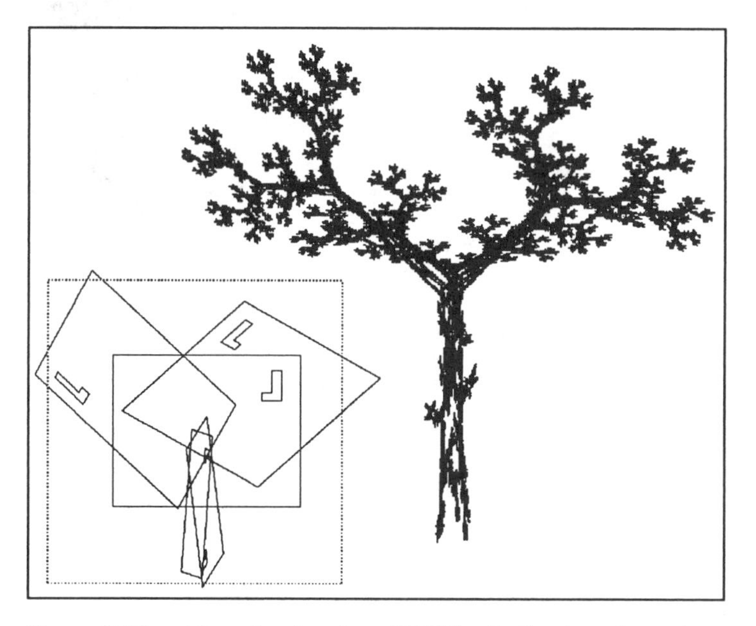

Figure 5.16 : The attractor of an MRCM with five transformations can even resemble the image of a tree (the attractor is shown twice as large as the blue-print indicates).

Here is our last example of an MRCM with only three transformations (see figure 5.13). The transformations have different horizontal and vertical scaling factors, two involve rotations, and one even a shear. What we get is very familiar: a nice twig.

We continue with two examples with more than three transformations (figures 5.14 and 5.15). All transformations are similarities with only scaling and translation. Only one transformation in figure 5.14 includes an additional rotation. These amazingly simple

Triangle, Square, and Circle

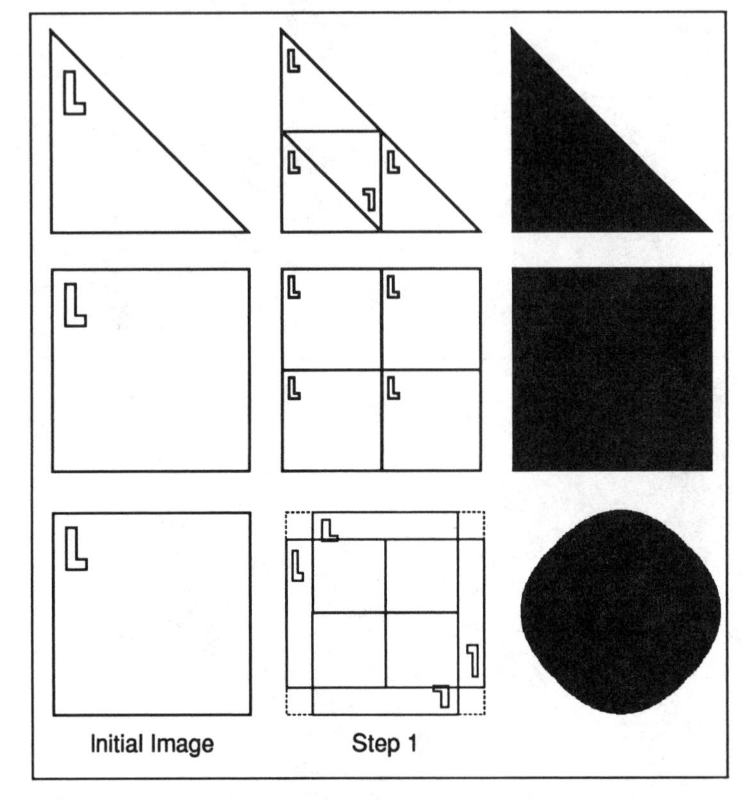

Figure 5.17 : The encoding of a triangle, a square and a circle by IFSs.

constructions already reveal quite complex and beautiful structures reminiscent of ice crystals.

Finally, let us close our little gallery by a surprisingly realistic drawing of a tree. Can you believe that even this image is a simple IFS attractor? In fact, it is encoded by just five affine transformations. In this case only one is a similarity. This image convincingly demonstrates the capabilities of IFSs in drawing fractal images.

Fractal Geometry Extends Classical Geometry

Given an arbitrarily designed MRCM, what is the final image (its attractor) which it will generate? Will it always be a fractal? Certainly not. Many objects of classical geometry can be obtained as attractors of IFSs as well. But often this way of representation is neither enlightening nor simpler than the classical description. We illustrate in figure 5.17 how the area of a square and a triangle can be obtained as IFS attractors. Representations of a plain circle, however, remain somewhat unsatisfactory using IFSs; only approximations are possible.

5.3 Classical Fractals by IFSs

The concept of an IFS allows us to make the construction of classical fractals much more transparent. They can be obtained as attractors of appropriate IFSs. In other words, the question of their existence as discussed in chapter 3 (we discussed the problem in detail for the Koch curve) can finally be settled by showing that for a given IFS there is a unique attractor. This will be done in the course of this chapter. But IFSs also allow us to better understand the number theoretical characterizations of some classical fractals like the Cantor set or the Sierpinski gasket.

Cantor Set

You will recall the characterization of the Cantor set by ternaries: it is the set of points of the unit interval which have a triadic expansion that does not contain the digit 1 (see chapter 2). Now we look at an IFS with

$$w_0(x) = \frac{1}{3}x, \quad w_1(x) = \frac{1}{3}x + \frac{1}{3}, \quad w_2(x) = \frac{1}{3}x + \frac{2}{3}\,.$$

Note, that this machine operates only on one variable (i.e., not in the plane).

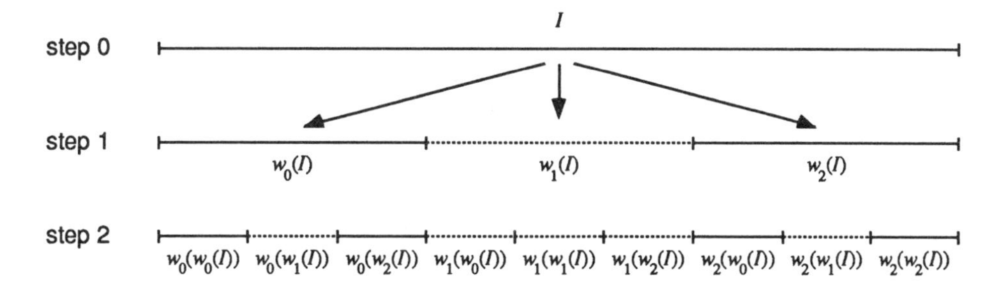

Figure 5.18 : First iteration steps of the triadic IFS. If w_1 is left out, the Cantor set is generated as the attractor.

Figure 5.18 shows the first steps of its iteration (using the unit interval as initial image). The attractor of this machine is clearly the unit interval (again and again, the unit interval is simply transformed into the unit interval). But what would happen if we used only the two transformations w_0 and w_2? In this case we obviously would obtain the Cantor set as the attractor (the iteration would correspond to the classical construction steps of the Cantor set: again and again middle thirds would be left out).

Now observe that w_1 transforms the unit interval to the interval $[\frac{1}{3}, \frac{2}{3}]$, i.e. points with triadic expansion from 0.1 to 0.1222... = $0.1\bar{2}$. In fact, whenever w_1 is involved in the iteration of the IFS this leads to points with an expansion that contains the digit 1.

Four Contractions

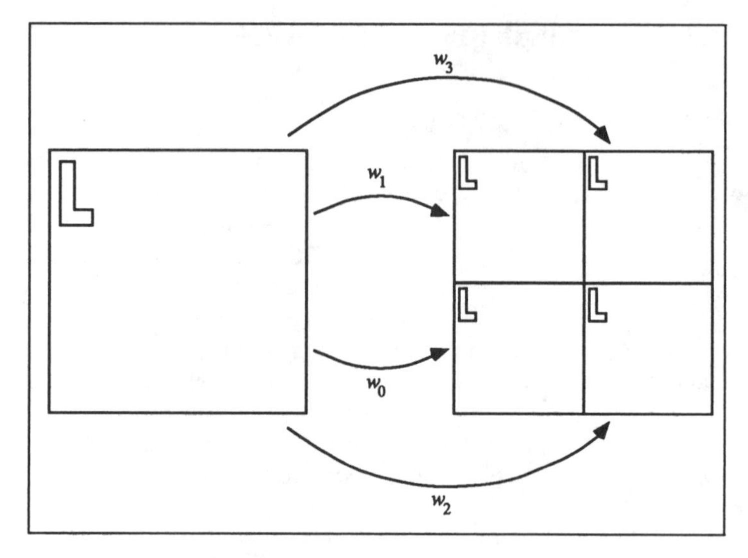

Figure 5.19 : Contractions transforming the unit square into its 4 congruent subquadrants.

In other words, leaving out everything which comes from w_1 just amounts to the ternary description of the Cantor set.

Sierpinski Gasket

Let us now turn to the Sierpinski gasket (or to its variation as already shown in figure 5.9). Now we look at the IFS that is given by four similarity transformations transforming the unit square Q into its four congruent quadrants (see figure 5.19).

It is convenient to label these transformations in binary form (i.e., 00, 01, 10 and 11 instead of 0, 1, 2, 3):

$$w_{00}(x,y) = (\tfrac{1}{2}x, \tfrac{1}{2}y), \qquad w_{01}(x,y) = (\tfrac{1}{2}x, \tfrac{1}{2}y + \tfrac{1}{2}),$$
$$w_{10}(x,y) = (\tfrac{1}{2}x + \tfrac{1}{2}, \tfrac{1}{2}y), \quad w_{11}(x,y) = (\tfrac{1}{2}x + \tfrac{1}{2}, \tfrac{1}{2}y + \tfrac{1}{2}).$$

Using all the four similarity transformations in an IFS will generate the unit square as an attractor. Figure 5.20 shows the first steps of the iteration of this machine. Note, that we use a binary coordinate system to identify the subsquares which are generated in each step. In each step, the IFS produces four times as many smaller squares. The binary coordinate system provides a very convenient way to do bookkeeping.

For example, in the first step, w_{01} has transformed the unit square Q into the subsquare $w_{01}(Q)$ at (0,1), w_{11} into the subsquare at (1,1), and so on. In the second step we find, for example, the square at $(10, 11)$ is $w_{11}(w_{01}(Q))$ (i.e., first apply w_{01} to Q and then w_{11} to the result). Here is another example, $w_{10}(w_{00}(w_{11}(Q)))$ would produce the square in the third step at (101,001). Do you see the labelling system? In the composition $w_{10}(w_{00}(w_{11}(Q)))$, take

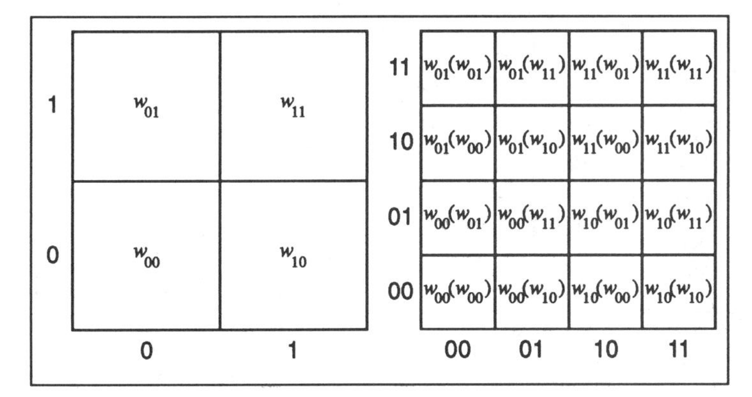

Figure 5.20 : The first two steps of the IFS. Observe that the generated subsquares can be identified by a binary coordinate system.

First Steps

the first digits from left to right, i.e. 101. This gives the binary x-coordinate of the subsquare. Then take the second digits in the composition from left to right, i.e. 001; this gives the y-coordinate.

We know already that the attractor of the IFS given by w_{00}, w_{01} and w_{10} will be the Sierpinski gasket. In other words, if we leave out everything in the unit square IFS which comes from w_{11}, we will also get the Sierpinski gasket. Now the binary bookkeeping pays off. Given any step k the 4^k little subsquares are identified by pairs of binary coordinates (with k digits). How can we recover whether w_{11} was involved in the production of a subsquare by the IFS? We just take the two binary coordinates which identify the little square and write them on top of each other, for example (100111, 010000) and (100111, 001100):

100111	100111
010000	001100
NO	YES

If we find the digit 1 simultaneously in corresponding places, then w_{11} was involved, otherwise not. Thus, omitting all these squares step by step will generate the Sierpinski gasket from the unit square.[7] This is in the same spirit as in the ternary description of the Cantor set. Moreover, we note that we have just built the interface to the geometrical patterns in Pascal's triangle, because our omission criterion here is exactly Kummer's number theoretical criterion for even binomial coefficients as used in the program of the chapter 2. We will explore this marvellous relation more in chapter 9.

[7]This explains the binary characterization of the Sierpinski gasket which we have used in chapter 3, page 195, for the discussion of self-similarity.

Nine Contractions

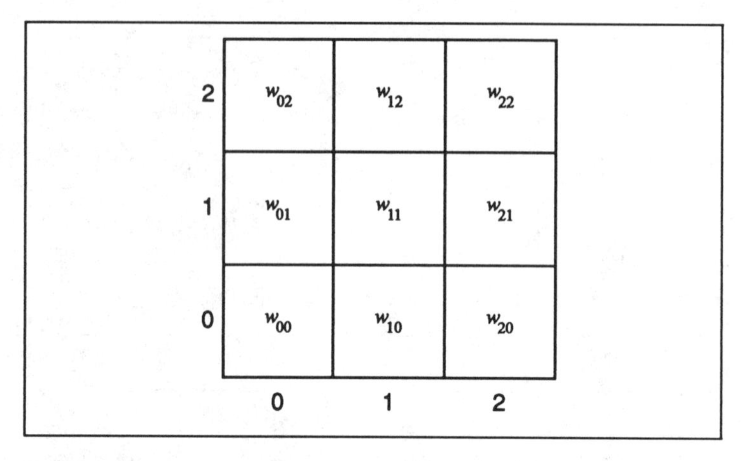

Figure 5.21 : The contractions transform the unit square into nine congruent subsquares that can be conveniently identified by a triadic coordinate system.

Sierpinski Carpet

The Sierpinski carpet has a very similar number theoretical description. Just start with the unit square and subdivide into nine congruent squares. For the appropriate IFS we use the transformations which transform the unit square into these subsquares, see figure 5.21 (again no rotations or reflections are allowed).

This time we label the transformations using ternary numbers like $w_{00}, w_{01}, w_{02}, w_{10}, \ldots, w_{22}$. Accordingly, each square in the kth step is identified by a pair of ternary coordinates (with k digits). In the limit, each point in the unit square is described by a pair of infinite ternary number strings like

(011201..., 210201...).

Now the Sierpinski carpet is obtained by omitting everything which comes from the transformation w_{11}. This means that we keep only those points in the unit square which admit a description by a pair of ternary numbers without the digit 1, or if the digit 1 appears in one of the coordinates, it must not appear at the same place in the other coordinate. For example, we keep $(11\overline{0}, 00\overline{1})$. Also $(201\overline{2}, 101\overline{0})$ belongs to the carpet, because it is equal to $(202\overline{0}, 101\overline{0})$. But we omit $(201\overline{0}, 101\overline{0})$, and so on. We remark that in this precise sense the Sierpinski carpet is the logical extension of the Cantor set into the plane.

In this book we have presented a gallery of classical fractals. This gallery has had no essential addition until very recently. B. Mandelbrot opened the doors wide to many new rooms in the gallery and added some potentially eternal masterpieces — like the Mandelbrot set — to it. But there are also two other creations or

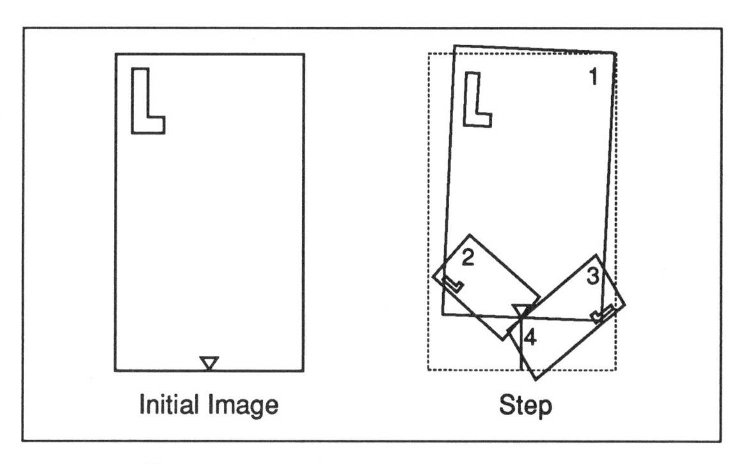

Blueprint of Barnsley's Fern

Figure 5.22 : Blueprint of Barnsley's fern.

Barnsley Fern Transformations

	Translations		Rotations		Scalings	
	e	f	θ	ψ	r	s
1	0.0	1.6	-2.5	-2.5	0.85	0.85
2	0.0	1.6	49	49	0.3	0.34
3	0.0	0.44	120	-50	0.3	0.37
4	0.0	0.0	0	0	0.0	0.16

Table 5.23 : Transformations for the Barnsley fern. The angles are given in degrees.

discoveries which have given current research significant momentum. One is the first *strange attractor* discovered by E. Lorenz at MIT in 1962, and the second is what we would like to call *Barns-*

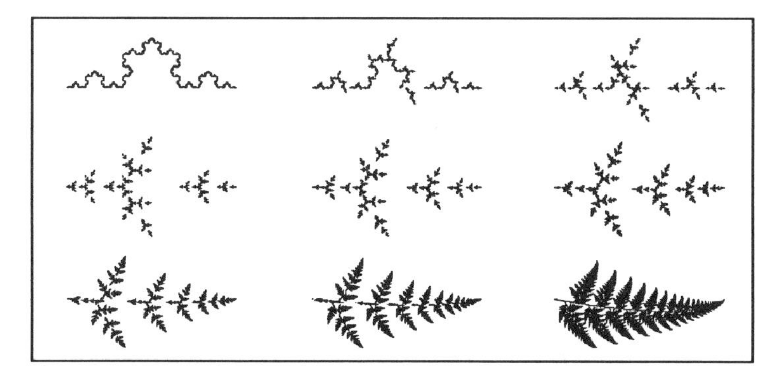

Koch Curve Transformed into the Fern

Figure 5.24 : By changing the parameters for the Koch curve continuously along a path to those of the fern, a transformation from one fractal to the other is obtained.

Barnsley's Fern

Figure 5.25 : Barnsley's fern generated by an MRCM with only four lens systems.

ley's fern. The Mandelbrot set, Lorenz attractor, and Barnsley's fern have each opened a new and separate division in the gallery of mathematical monsters. Of all these, Barnsley's fern belongs to the subject of this chapter.

Barnsley was able to encode the image in figure 5.25 with only four lens systems. Figure 5.22 shows the design of his MRCM by means of its application to an initial rectangular image. Note that contraction number 3 involves a reflection. Also, contraction number 4 is obviously not a similarity transformation; it contracts the rectangle to a mere line segment. The attractor which is generated by the IFS will not be self-similar in the precise mathematical meaning of the word. The original transformations are given in table 5.23[8] and in different notation also in table 5.48.

The importance of Barnsley's fern to the development of the subject is that his image looks like a fern, but lies in the same mathematical category of constructions as the Sierpinski gasket, the Koch curve, and the Cantor set. In other words, that category not only contains extreme mathematical monsters which seem very distant from nature, but it also includes structures which are related to natural formations and which are obtained by only slight modifications of the monsters. In a sense, the fern is obtained by shaking an MRCM which generates the Koch curve so that the lens systems alter their positions and contraction factors (see figure 5.24).

Let us now turn to another aspect of our concept of MRCM. The message which is expressed by the image of the fern is very impressive. Something as complicated and structured as a fern seems to have a lot of information content. But as figure 5.22 demonstrates, the information content from the point of view of IFSs is extremely small. This observation suggests viewing the IFS as a tool for coding and compressing images.

[8]M. F. Barnsley, Fractal Modelling of Real World Images, in: The Science of Fractal Images, H.-O. Peitgen and D. Saupe (eds.), Springer-Verlag, New York, 1988, page 241.

5.4 Image Encoding by IFSs

Each of the images in our gallery is obtained by a very simple machine, the blueprint of which is revealed by step 1 in each experiment. How many images are there which can be generated this way? The answer is obvious — infinitely many. Any number and particular choice of lenses and their position defines a new image. In other words, we can think of the blueprint of the MRCM (i.e., the set of transformations which describe the IFS) as the blueprint (or encoding) of an image. In figure 5.26 we have summarized this interpretation using the twig-like structure. The transformations are:

	a	b	c	d	e	f
1	−0.467	0.02	0.113	0.015	0.4	0.4
2	0.387	0.43	0.43	−0.387	0.256	0.522
3	0.441	−0.091	−0.009	−0.322	0.421	0.505

Image Encoding With High Compression Ratio

In fact, this provides an example of a coding of an image with a high compression ratio. Assume that an image is given as an $n \times m$ array of black and white pixels (e.g. the points on a computer screen). That is to say, the information required to represent the pixel structure without any encoding would be $n \times m$ bits. The blueprint of the twig uses three lens systems. Each of them is given by six real numbers. A real number in a computer is represented by s bits (typically $s = 32$). Thus the blueprint of the twig needs only $18s$ bits. The compression ratio would then be $nm/18s$. Let us assume that $n = m = 1000$ and that $s = 32$. Then the compression ratio would be something of the order of 1700, which is rather extreme.

Twig Blueprint

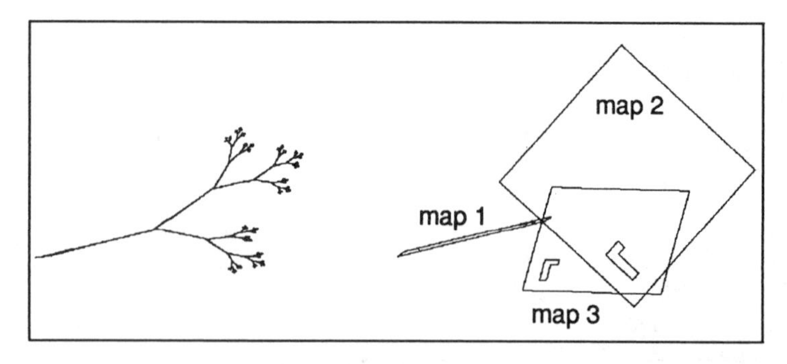

Figure 5.26 : Blueprint of a twig: encoding by three transformations.

A New Point of View for Image Perception

The apparent complexity of the twig is compressed into a very simple plan. To put it another way, many complex structures are in fact very simple when discussed from the point of view of MRCMs. This amazing conclusion can potentially destroy many fundamental beliefs in the fields of image compression and image perception. For example, in some scientific schools the visual functions of the human brain are compared with computers and their algorithmic design. There are models which try to explain how the brain might distinguish things like an equilateral triangle from one which is not equilateral. In that context, we must come to the conclusion that the brain must have enormous power because a human being is able to distinguish objects as complex as an oak tree and a beech tree in a fraction of a second. Our experience with MRCMs suggests, however, that it might well be that an oak tree is only complex from the point of view of classical geometry. It is probably true that the general pattern of an oak tree has a very compact encoding for the human brain. Fractal geometry offers a totally new and very powerful modelling framework for such encoding problems. In fact, we could speculate that our brain uses fractal-like encoding schemes.

Let us summarize what we have learned so far. We have introduced a machine, called an MRCM, which is essentially an arrangement of lens systems which contract images. The MRCM generates a dynamical system, an IFS. That is, running the machine in a feedback environment leads to a sequence of images $A_0, A_1, A_2, \ldots$, where A_0 is an arbitrary initial image. The sequence of images will lead to a final image, A_∞, which is independent of the initial image A_0. If we choose A_∞ as the initial image, then nothing happens (i.e., the IFS leaves A_∞ invariant). We say that A_∞ is a *fixed point* of the IFS, or that A_∞ is an *attractor* for the dynamical system. In this sense we can identify the resulting attractor with the IFS. The mathematical description of the lens systems of the machine is given by a set of affine linear transformations, each one specified by six real numbers. We may interpret these data as a coding of the final image A_∞. For the decoding we only need to run the machine with any initial image. Eventually, the coded image A_∞ will emerge.

The Problem of Decoding

However, in some cases the decoding using the IFS presents a serious problem. For example, take Barnsley's fern. Figure 5.27 shows the first steps of the IFS. Obviously, even after 10 iterations we still have a long way to go to reach the complete fern. Thus, we are led to the general question: after how many steps can one assume that the final image has been approximated sufficiently well? To answer this we need to clarify what we mean by *sufficiently well*. There are two criteria which seem to be reasonable.

The First Iterates

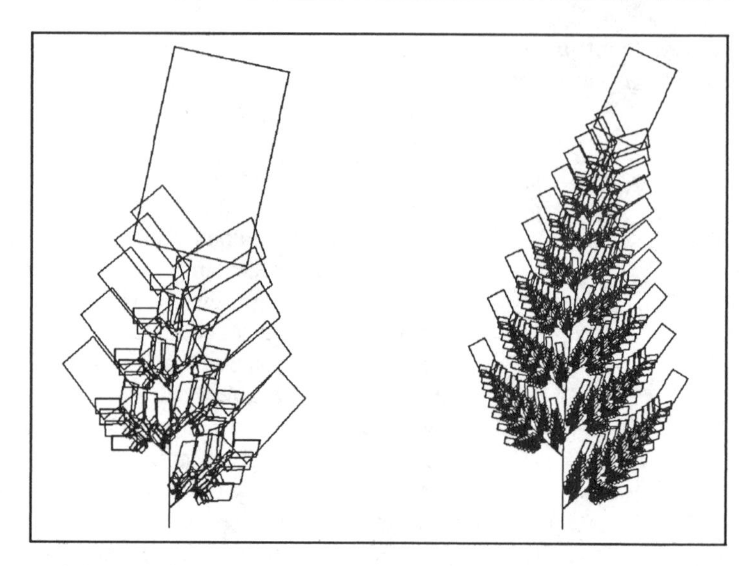

Figure 5.27 : Step 5 and step 10 of the fern copy-machine.

The first would require that two successive iterations change so little that the change is below graphic resolution. This compares very well to computational problems. A solution to a square root calculation, for example, is accepted when the first 10 digits no longer change. The second criterion is more practical and allows an a priori estimate of the number of necessary iterations. This estimate derives from the following worst case scenario. Recall that the initial image may be completely arbitrary. At this point, however, let us require that it covers the attractor. For example, it could be a sufficiently large square. Since the final image is independent of the initial image, we will not accept a given iteration as an approximation for the final image as long as we still see contracted versions of the initial image in that iteration. This is the case in figure 5.27. It is apparent that even after 10 iterations the dynamical system is still far from the final image, the attractor. The reason is that contraction number 1 reduces by only a factor of 85%. Therefore, in order to reduce the initial rectangle to a size below pixel size — to the point at which the rectangular structure becomes unrecognizable — we have to carry out at least N iterations, where N is estimated in the following way. Assume that the initial rectangle is drawn on a 1000×1000 pixel screen and covers 500×200 pixels. Then N approximately solves the equation

$$500 \cdot 0.85^N = 1 \ .$$

Thus, $N \approx 39$. In a straight forward implementation of the IFS one has to calculate and draw

$$M = 1 + 4 + 4^2 + 4^3 + \cdots + 4^N = \frac{4^{N+1} - 1}{3}$$

rectangles for N iterations. With $N = 39$ we compute the incredibly large number $M \approx 4^{39} \approx 10^{24}$. Even if we assume that our computer calculates and draws a whopping million rectangles per second, then to see the final image we would have to wait 10^{18} seconds, which is about 10^{11} years, which is a time span longer than the age of the universe. This gives some flavor of the decoding problem. In chapter 6, however, we will learn a very elementary and powerful decoding method which generates a good approximation of the final image on a computer screen within seconds! We will also modify the above inefficient algorithm to the point where it will produce the fern (and other attractors) with the same precision and in a reasonable time.

Encoding: The Inverse Problem

In order to make use of IFSs for image coding, one first has to solve another crucial problem, namely to construct a suitable MRCM for a given image. This is the inverse problem, encoding is inverse to decoding. Of course, we cannot expect to be universally able to build an MRCM which produces exactly the given image. However, approximations should be possible. We can make these as close to the original as we desire, as explained next.

Assume we are given a black and white picture, digitized at resolution of n by m pixels. This image can be *exactly* reproduced by an MRCM simply by requiring that for every black pixel of the image, there exists a lens which contracts the whole image to that particular pixel. Running the machine just once starting out with any image will produce the prescribed black and white pixel image. Naturally, this is not an efficient way to code an image because for every black pixel we need to store one affine transformation. However, the argument demonstrates that in principle it is possible to achieve approximations of any desired accuracy. Thus, the problem is to find ways to construct a better MRCM which does not need as many transformations but still produces a good approximation. Several difficult questions are raised in this context:

(1) How can the quality of an approximation be assessed? How do we quantify differences between images?
(2) How can we identify suitable transformations?
(3) How can we minimize the number of necessary affine transformations?
(4) What is the appropriate class of images suitable for this approach?

These problems are comparable with the difficulties that scientists had to face when building the foundations of the theory of what is now called Fourier analysis. Today Fourier analysis is a standard tool with uncountably many applications. One particular application deals with the coding of sound, its analysis and manipulation. But it took some several hundred years to fully develop that theory, and the task of overcoming the obstacles needed some of the greatest mathematical minds. Comparatively speaking, fractal image encoding is in its infancy. This approach is less than ten years old and promises a totally new practical and theoretical framework for the analysis and synthesis of images.

Some speak of a revolution, others are very sceptical and hold that fractals in general and fractals for image encoding in particular are just a fashion which has no future. We should not be surprised about these discrepancies of judgement. Whenever the past giants of science made conceptual quantum leaps, many of their contemporaries had nothing but cynical reviews for them. But Galileo Galilei, Nikolaus Kopernikus, Johannes Kepler, Charles Darwin, Gregor Mendel, Albert Einstein, and other great scientists will be remembered as long as science plays a role for mankind. Their blind critics have been forgotten in the noise of history. We believe that fractal geometry will belong to the great achievements of science, quite independent of whether or not fractals will eventually lead to the best practical schemes for the problem of image encoding. The ideas which grow out of this new approach will, in any event, have a deep impact on how we think about images.

Encoding and Decoding Sound

Sound and vision are the most important external sensory signals which help us to interpret the world around us. In some sense these signals are related; in another sense they are very different. The essential physical model for light and sound is the same, namely waves. But sound is always a time event. Sound begins at one time and ends at another time and happens in between, while images are static. The wave model is also significant for the discussion of heat radiation. Due to the work of Leonhard Euler, Daniel Bernoulli, Joseph Louis Lagrange and the great French scientist Jean Baptiste Joseph de Fourier (1768–1830) waves, no matter how complicated, have a beautiful common description in mathematical terms. That is the concept of Fourier series. Fourier's most important scientific contribution is on the problems of conduction of heat,[9] in which he made extensive use of the series that now bear his name. Fourier series allow us to describe tones in terms of harmonics. A tone is obtained as a superposition of pure tones of the form $a_k \sin(k\omega_k t + \theta_k)$, where a_k is the amplitude and ω_k is the basic frequency. The wonderful point about Fourier series is that

[9] Fourier was a friend of Napoleon and accompanied his master to Egypt in 1798 and held the curious opinion that desert heat is the ideal environment for a healthy life. Accordingly he swathed himself like a mummy and lived in overheated rooms.

they also allow a description of something as complex as a tone of a violin. Indeed, for any given tone there is something like a Fourier analysis which allows us to determine the a_k's and ω_k, both mathematically and electronically. Finding this encoding is the inverse problem for sound. What harmonics are for sound, lens systems are in an MRCM for images. The superposition of pure harmonics in a sound system corresponds to the superposition of lens systems in an MRCM.

5.5 Foundation of IFS: The Contraction Mapping Principle

The image coding problem has led us to one of the central questions: how images can be compared or what the distance between two images is. In fact, this is crucial for the understanding of iterated function systems. Without an answer to these questions we will not be able to precisely verify the conditions under which the machine will produce a limiting image. Felix Hausdorff, whom we have already mentioned as the man behind the mathematical foundations of the concept of fractal dimension, proposed a method of determining this distance which is now named after him — the Hausdorff distance. Introducing the Hausdorff distance $h(A, B)$ has two marvelous consequences. First, we can now talk about the sequence A_k having the limit A_∞ in a very precise sense: A_∞ is the limit of the sequence $A_0, A_1, A_2, \ldots$ provided that the Hausdorff distance $h(A_\infty, A_k)$ goes to 0 as k goes to ∞. But even more importantly, Hutchinson showed that the operator W, which describes the collage

$$W(A) = w_1(A) \cup w_2(A) \cup \ldots \cup w_N(A)$$

is a contraction with respect to the Hausdorff distance. That is, there is a constant c, with $0 \leq c < 1$, such that

$$h(W(A), W(B)) \leq c \cdot h(A, B)$$

for all (compact) sets A and B in the plane. In establishing this fundamental property, Hutchinson was able to inject into consideration one of the most powerful and beautiful principles in mathematics — the contraction mapping principle, which has a long history and owes its final formulation to the great Polish mathematician Stefan Banach (1892–1945).

If the works and achievements of mathematicians could be patented, then the contraction mapping principle would probably be among those with the highest earnings up to now and for the future. Once he allowed himself a certain degree of abstraction, Banach understood that many individual and special cases which floated in the work of earlier mathematicians can be subsumed under one very brilliant principle. The result is nowadays a theorem in *metric topology*, a branch of mathematics which is basic for a great part of modern mathematics and is usually a topic reserved for students of an advanced university-level mathematics courses. We will explain the core of Banach's ideas in a non-rigorous style.

Measuring Distance: The Metric Space

The Hausdorff distance determines the distance of images. It is based on the concept of distance of points to be explained here. Expressed generally, the distance between points of a space X can

Distance Measurement in the Plane

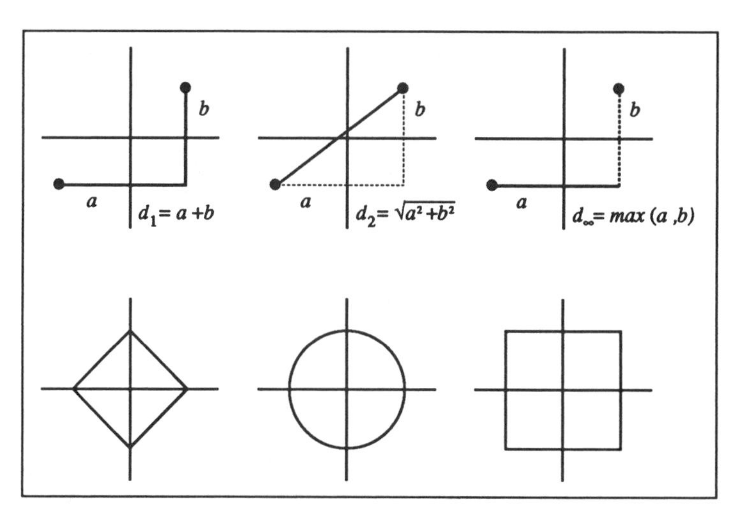

Figure 5.28 : Three methods of measuring distance in the plane (the lattice distance, the Euclidean distance, the maximum norm distance) and the corresponding unit sets (the set of points which have the distance 1 to the origin of the coordinate system).

be measured by a function $d : X \times X \to \mathbf{R}$. Here $\mathbf{R}$ denoted the real numbers and the function d must have the property that

(1) $d(x, y) \geq 0$
(2) $d(x, y) = 0$ if and only if $x = y$
(3) $d(x, y) = d(y, x)$
(4) $d(x, y) \leq d(x, z) + d(z, y)$ (triangle inequality),

hold for all $x, y, z \in X$. We call such a mapping d a *metric.* A space together with a metric is called a *metric space.* Some examples are (see figure 5.28):

(1) For real numbers we can set

$$d(x, y) = |x - y| \ .$$

(2) For points $P = (x, y)$, $Q = (u, v)$ in the plane we can define

$$d_2(P, Q) = \sqrt{(x - u)^2 + (y - v)^2} \ .$$

This is the *Euclidean metric.*

(3) Another metric in the plane would be

$$d_\infty(P, Q) = \max\{|x - u|, |y - v|\} \ .$$

This is the *maximum metric.*

(4) A further metric illustrated in figure 5.28, the *lattice metric,* is given by

$$d_1(P, Q) = |x - u| + |y - v| \ .$$

The last metric d_1 on the list is sometimes also referred to as the *Manhattan metric*, because it is the distance, a cab driver in Manhattan, New York, would have to drive to get from P to Q.

Once we have a metric for a space X we can talk about *limits* of sequences. Let $x_0, x_1, x_2, \ldots$ be a sequence of points from X and a an element from X. Then a is the limit of the sequence provided

$$\lim_{k \to \infty} d(x_k, a) = 0 \,.$$

In other words, for any $\varepsilon > 0$ we can find a point x_n in the sequence so that any point later in the sequence has distance to a less than ε:

$$d(x_k, a) < \epsilon, \quad k > n \,.$$

In this case we say that the sequence converges to a. Often it is very desirable to test the convergence of a sequence without knowledge of the limit. This, however, works only if the underlying space X has a special nature (i.e., it is a *complete metric space*). Then one may discuss limits by monitoring the distance of consecutive points in the sequence.

The space X is called a *complete metric space* if any Cauchy sequence has a limit which belongs to X. More precisely, this means the following: Let $x_0, x_1, x_2, \ldots$ be a given sequence of points in X. It is a Cauchy sequence if for any given number $\varepsilon > 0$ we can find a point x_m in the sequence so that any two points later in the sequence have a distance less than ε:

$$d(x_i, x_j) < \varepsilon, \quad i, j \geq m \,.$$

Then the limit of the sequence exists and is a point of X. Two examples are:

(1) The set of rational numbers is not complete. There are Cauchy sequences of rational numbers whose limits exists but are not rational numbers. An example of such a sequence is given by

$$x_n = \sum_{k=1}^{n} \frac{1}{k^2} \,.$$

This sequence of rational numbers converges to the irrational limit $\pi^2/6$.

(2) The plane $\mathbf{R}^2$ is complete with respect to any of the metrics, d_1, d_2, or d_∞.

The Environment of the Contraction Mapping Principle

In chapter 1 we learned that a large variety of dynamic processes and phenomena can be seen from the point of view of a feedback system. A sequence of events $a_0, a_1, a_2, \ldots$ is generated starting with an initial event a_0, which can be chosen from a pool of admissible choices. As time elapses (as n grows), the sequence can show all kinds of behavior. The central problem of dynamical systems theory is to forecast the long-term behavior. Often that behavior will not depend very much on the initial choice a_0. That is exactly the environment for the contraction mapping principle. It provides everything which we can hope for to make a forecast. But having in mind the variety of both wild and tame behavior which feedback systems can produce, it is clear that the principle will select some sub-class of feedback systems for which it can be applied. Let us collect the two features which characterize this class:

(1) **The Space.** The objects — numbers, images, transformations, etc. which we call a_n — must belong to a set in which we can measure the distance between any two of its elements, for example the distance between x and y is $d(x, y)$. Furthermore, the set must be saturated in some sense. That means that if an arbitrary sequence satisfies a special test which examines the possible existence of a limit, then a limit exists and belongs to the set (technically: the space is a *complete metric* space).
(2) **The Mapping.** The sequence of objects is obtained by a mapping, say f. That means that for any initial object a_0, a sequence $a_0, a_1, a_2, \ldots$ is generated by $a_{n+1} = f(a_n)$, $n = 0, 1, 2, \ldots$ Furthermore, f is a contraction. That means that for any two elements of the space, say x and y, the distance between $f(x)$ and $f(y)$ is always strictly less than the distance between x and y.[10]

The Result of the Contraction Mapping Principle

For this class of feedback systems the contraction mapping principle gives the following remarkable result:

(1) **The Attractor.** For any initial object a_0 the feedback system $a_{n+1} = f(a_n)$ will always have a predictable long-term behavior. There is an object a_∞ (the limit of the feedback system) to which the system will go. That limit object is the same no matter what the initial object a_0. We call a_∞ the unique *attractor* of the feedback system.
(2) **The Invariance.** The feedback system leaves a_∞ invariant. In other words, if we start with a_∞, then a_∞ is returned. a_∞ is a fixed point of f, i.e. $f(a_\infty) = a_\infty$.
(3) **The Estimate.** We can predict how fast the feedback system will arrive close to a_∞ when it is started at a_0. We only have to test the feedback loop once on the initial object. That means, if we measure the distance between a_0 and $a_1 = f(a_0)$, we can already safely predict how often we have to run the system to arrive near a_∞ within a prescribed accuracy. Moreover, we can estimate the distance between a_0 and a_∞.

[10] Technically, $d(f(x), f(y)) \leq c \cdot d(x, y)$ with a constant $0 \leq c < 1$.

The Attractor of a Contractive Mapping

A mapping f is a *contraction* of the metric space X, provided that there is a constant c, $0 \leq c < 1$, such that for all x, y in X one has that

$$d(f(x), f(y)) \leq cd(x, y) .$$

The constant c is called the *contraction factor* for f. Let $a_0, a_1, a_2, \ldots$ be a sequence of elements from a complete metric space X defined by $a_{n+1} = f(a_n)$. The following holds true:

(1) There is a unique attractor $a_\infty = \lim_{n\to\infty} a_n$.
(2) a_∞ is invariant, $f(a_\infty) = a_\infty$.
(3) There is an a priori estimate for the distance from a_n to the attractor, $d(a_n, a_\infty) \leq c^n d(a_0, a_1)/(1 - c)$.

Let us explain the estimate in property (3). From the contraction property of f we derive

$$d(f(a_0), a_\infty) = d(a_1, f(a_\infty)) \leq cd(a_0, a_\infty) .$$

Applying the triangle inequality, we further obtain

$$\begin{aligned} d(a_0, a_\infty) &\leq d(a_0, f(a_0)) + d(f(a_0), a_\infty) \\ &\leq d(a_0, f(a_0)) + cd(a_0, a_\infty) \end{aligned}$$

thus,

$$d(a_0, a_\infty) \leq \frac{d(a_0, f(a_0))}{1 - c}$$

and likewise

$$d(a_n, a_\infty) \leq \frac{d(a_n, a_{n+1})}{1 - c}$$

for all $n = 0, 1, 2, \ldots$ Finally, with

$$\begin{aligned} d(a_n, a_{n+1}) &\leq cd(a_{n-1}, a_n) \\ &\leq c^2 d(a_{n-2}, a_{n-1}) \\ &\leq \cdots \\ &\leq c^n d(a_0, a_1) \end{aligned}$$

we arrive at the result

$$d(a_n, a_\infty) \leq \frac{c^n}{1 - c} d(a_0, a_1) .$$

This allows us to predict n so that a_n is within a prescribed distance to the limit.

The ε-collar

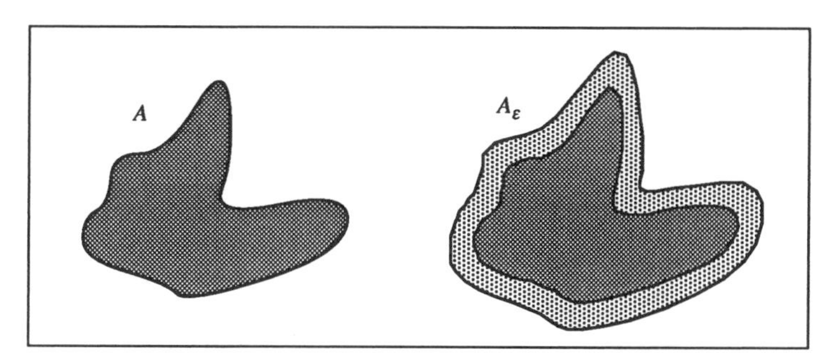

Figure 5.29 : The ε-collar of a set A in the plane.

We now examine the operation of an IFS and how it can be described by means of the contraction mapping principle. To start we need to define the distance between two images. For simplicity let us consider only black and white images. Mathematically speaking an image is a compact set[11] in the plane.

The Hausdorff Distance

Given an image A, we can introduce the ε-collar of A, written A_ε, which is the set A together with all points in the plane which have a distance from A of not more than ε (see figure 5.29). Hausdorff measured the distance between two (compact) sets A and B in the plane using ε-collars. Formally, we write $h(A, B)$ for that distance. To determine its value we try to fit A into an ε-collar of B, and B into an ε-collar of A. If we take ε large enough, this will be possible. The Hausdorff distance $h(A, B)$ is just the smallest ε such that the ε-collar A_ε absorbs B and the ε-collar B_ε absorbs A.

Definition of the Hausdorff Distance

In precise mathematical terms the definition of the Hausdorff distance is as follows. Let X be a complete metric space with metric d. For any compact subset A of X and $\varepsilon > 0$, define the ε-collar of A by

$$A_\varepsilon = \{x \in X \mid d(x, y) \leq \varepsilon \text{ for some } y \in A\} .$$

For any two compact subsets A and B of X the Hausdorff distance is

$$h(A, B) = \inf\{\varepsilon \mid A \subset B_\varepsilon \text{ and } B \subset A_\varepsilon\} .$$

According to Hausdorff the space of all compact subsets of X equipped with the Hausdorff distance is another complete metric space. This implies that the space of all compact subsets of X is a suitable environment for the contraction mappping principle.

---

[11] Technically, compactness for a set X in the plane means that it is bounded, i.e. it lies entirely within some sufficiently large disk in the plane and that every convergent sequence of points from the set converges to a point from the set.

Four Examples of Hausdorff Distance

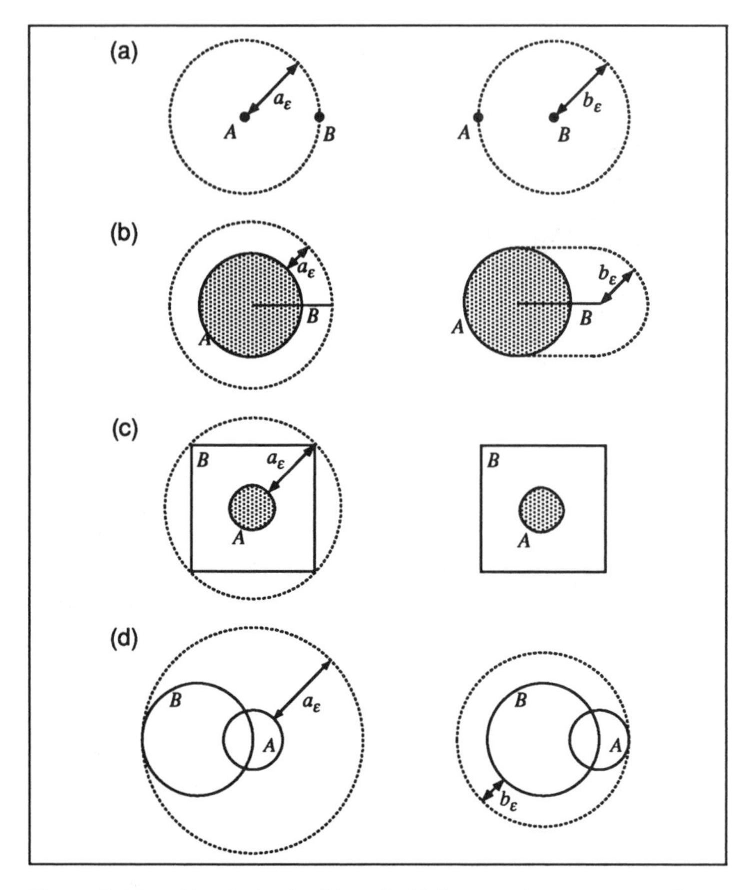

Figure 5.30 : To obtain the Hausdorff distance between two planar sets A and B we compute $a_\varepsilon = \inf\{\varepsilon \mid B \subset A_\varepsilon\}$ (left figures) and $b_\varepsilon = \inf\{\varepsilon \mid A \subset B_\varepsilon\}$ (right figures). B barely fits into the a_ε-collar of A, and A barely fits into the b_ε-collar of B. The Hausdorff distance is the maximum of both values, $h(A, B) = \max\{a_\varepsilon, b_\varepsilon\}$. The sets A and B are two points (top row), a disk and a line segment (second row), a disk and a large square (third row, here $b_\varepsilon = 0$), and two intersecting disks (bottom row).

With this definition it follows that $h(A, B) = 0$ when A is equal to B. Also, if A is just a point and B is just a point, then $h(A, B)$ is the distance between A and B in the ordinary sense. Figure 5.30 illustrates that fact and gives a few more examples useful for getting aquainted with the notion of Hausdorff distance.

The Hutchinson Operator

Let us now return to the state of affairs which Hutchinson obtained when analyzing the operator W

$$W(A) = w_1(A) \cup w_2(A) \cup \ldots \cup w_N(A) ,$$

where the transformations $w_i, i = 1, ..., N$, are contractions with contraction factors c_i. Hutchinson was able to show that W is also a contraction, however, with respect to the Hausdorff distance. Thus, the contraction mapping principle can be applied to the iteration of the Hutchinson operator W. Consequently, whatever initial image is chosen to start the iteration of the IFS, for example A_0, the generated sequence

$$A_{k+1} = W(A_k), \quad k = 0, 1, 2, 3, ...$$

which will tend towards a distinguished image, the attractor A_∞ of the IFS. Moreover, this image is invariant:

$$W(A_\infty) = A_\infty \ .$$

This solves a central problem raised in chapter 3. The Koch curve, the Sierpinski gasket, etc. all seem to be objects in the plane, and there are convergent processes for them, namely the iteration of the corresponding Hutchinson operators. But we could not prove that these fractals really exist and are not just some impossible artifact of a self-referential scheme such as the assumption of a barber who shaves all men who do not shave themselves — obviously a falsehood. However, now, with Hutchinson and Hausdorff's results in hand, we can be sure that the limit object with the extraordinary self-similarity property truly exists.

The contraction mapping principle even gives us something in addition for free. Knowing the contraction factor c of the Hutchinson operator W, we can estimate how fast the IFS will produce the final image from just applying the Hutchinson operator one time to A_0. Since the contraction factor c of W is determined by the contraction w_i with the worst contraction factor c_i, i.e. $c = \max\{c_i\}$, the efficiency of the IFS is determined by this individual contraction. This is the theoretical background of our experiments in figure 5.1 and the encoding of images by IFSs.

The Contractivity of the Hutchinson Operator

Hutchinson applied the contraction mapping principle to the operator W. The principle requires that the space in which W operates is complete. The completeness of this space of compact subsets of a space X, which itself is complete (e.g. the Euclidean plane), was already known to Hausdorff. So it remained to show that the Hutchinson operator W is a contraction. Let us briefly illustrate the idea of the argument with the example of two contractions w_1 and w_2 with contraction factors $c_1, c_2 < 1$. We take any two compact sets A and B, and show that the Hausdorff distance $h(W(A), W(B))$ between

$$W(A) = w_1(A) \cup w_2(A)$$

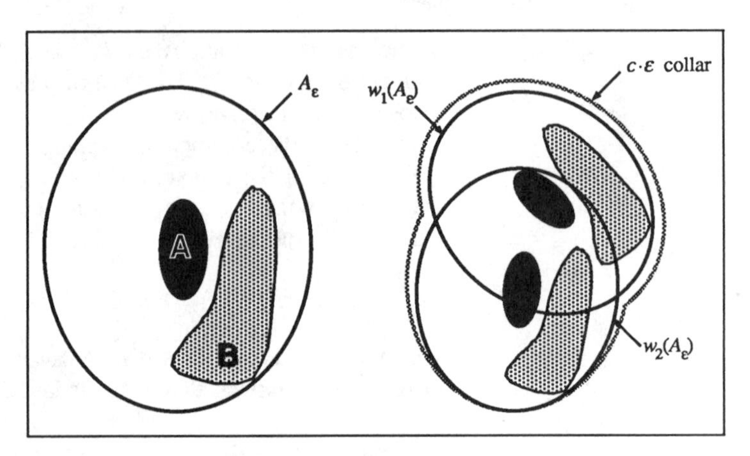

Figure 5.31 : To the contractivity of the Hutchinson operator.

and

$$W(B) = w_1(B) \cup w_2(B)$$

is strictly less than the distance $h(A, B)$ between A and B.

Compare figure 5.31 for the following. Let ε be the Hausdorff distance between A and B, $h(A, B) = \varepsilon$. Then B is in the ε-collar of A, $B \subset A_\varepsilon$. Applying the transformations w_1 and w_2 yields

$$w_1(B) \subset w_1(A_\varepsilon) \text{ and } w_2(B) \subset w_2(A_\varepsilon) .$$

From the contraction property of the two transformations it follows that

$$w_1(A_\varepsilon) \subset (c_1 \cdot \varepsilon)\text{-collar of } w_1(A)$$
$$w_2(A_\varepsilon) \subset (c_2 \cdot \varepsilon)\text{-collar of } w_2(A) .$$

Setting $c = \max(c_1, c_2)$ we obtain that both $w_1(B)$ and $w_2(B)$ are contained in the $(c \cdot \varepsilon)$-collar of $w_1(A) \cup w_2(A)$. The same argument applied to A and ε-collar B_ε also yields that both $w_1(A)$ and $w_2(A)$ are contained in the $(c \cdot \varepsilon)$-collar of $w_1(B) \cup w_2(B)$. With that it is clear from the definition that the Hausdorff distance $h(W(A), W(B))$ is less than $c \cdot \varepsilon$. Thus, the Hutchinson operator W is a contraction with contraction factor $c < 1$. Therefore the worst contraction of the transformations in the IFS determines the overall contraction factor of the machine.

In summary, our experiments are built on very firm ground and are not just the results of some lucky or accidental choices. Hutchinson's work lays the ground for a whole new discussion of images and their encoding. But as we have seen, there are still some open and very serious problems, for example, the problem of decoding. We have seen that the fern can be encoded by an IFS, but we have not yet given away the secret of how the images have been obtained (i.e., how the fern has been decoded). In a sense this means that we can lock up images into very tiny little boxes, which makes them invisible; but we don't yet know the keys needed to get them out again into the visible world. What we need is some artist who unchains our encodings. But this is the subject of the next chapter. On the other hand, there is the inverse problem, the problem to find the encoding of a given image.

Fractal Dimension for IFS Attractors

We have seen that an attractor A_∞ generated by a simple IFS whose contractions are similarities is self-similar. In this case, we can compute the self-similarity dimension, provided the N contractions $w_1, ..., w_N$ have the property that $w_i(A_\infty) \cap w_k(A_\infty) = \emptyset$, for all i, k with $i \neq k$ and the w_i are one-to-one. This type of attractor is said to be totally disconnected. There is no overlapping of the small copies of the attractor. If in addition, the contractions are reductions with the same factor c, $0 \leq c < 1$, then the self-similarity dimension $D_s = d$ of the attractor A_∞ can be computed from the equation $Nc^d = 1$. This is the same as

$$D_s = \frac{\log N}{\log 1/c} \ .$$

Moreover, we can show that the self-similarity dimension is the same as the box-counting dimension.

If we have N similarities with reduction factors $c_1, ..., c_N$, then Hutchinson showed that we can still compute the fractal dimension $D_s = d$, by solving an equation which includes the special case where $c_1 = c_2 = ... = c_N$. He showed that

$$c_1^d + c_2^d + ... c_N^d = 1 \ .$$

Of course, in most cases one cannot solve this equation by hand for the dimension d. Rather, a numerical procedure must be employed.

The condition that the attractor must be totally disconnected for the formula to hold can be relaxed somewhat.[12] But when there is substantial overlapping of the contractions of the attractor then there is a problem. To see this just consider different IFSs for generating the unit square. If we choose four contractions each with contraction factor 1/2 which touch in their boundaries (see figure 3.29), then the formula still gives the correct result, $D_s = 2$. However, if we cover the square with four contractions of contraction factor, say, 3/4 (implying substantial overlap), then the formula would give $D_s > 2$!

[12] See J. Hutchinson, *Fractals and self-similarity*, Indiana University Journal of Mathematics 30 (1981) 713–747, and G. Edgar, *Measures, Topology and Fractal Geometry*, Springer-Verlag, New York, 1990.

5.6 Choosing the Right Metric

In the last section we mentioned several possible definitions of a distance of points in the plane. The Hausdorff distance between images is also effected by the choice of that distance. So it is no surprise, and in fact important, to note that the contraction mapping principle also depends on the choice of distance.

Dependance on the Distance Notion

As we indicated in the last section, distance in the plane can be measured in many different ways. For example, if P and Q are two points we can measure the Euclidean distance (this is the length of a straight line segment between P and Q), the lattice distance (this is the sum of the length of two horizontal and vertical line segments which connect P and Q), or the maximum norm distance (see figure 5.28). These are only three of a great many possible definitions. It is interesting to note the various geometrical shapes that are given by the set of points that have a distance less than or equal to 1 from the origin. Naturally this shape depends on the metric. For the Euclidean metric we obtain the unit disk, and for the maximum metric we get the unit square. But even more important for our purposes is the fact that it also depends on the metric whether or not a given transformation is a contraction. It seems counter-intuitive that a transformation may be a contraction in one sense but not with respect to another metric.

The Metric Determines Contractiveness: An Example

It is important to note that everything depends on the choice of the metric. A given transformation may be a contraction with respect to one metric, but not a contraction with respect to another one. For example, consider the map w which is given by the matrix

$$\left(\begin{array}{cc|c} 0.55 & -0.55 & 0 \\ 0.55 & 0.55 & 0 \end{array}\right)$$

which defines a rotation by 45 degrees, a scaling by $0.55\sqrt{2} \approx 0.778$, and no translation). w is a contraction for the metric d_2 but not with respect to d_1 or d_∞.

To see the argument let us fix the point $P = (0,0)$ and consider points Q for each metric. Note that the transformation w leaves the origin P invariant ($w(P) = P$).

For the d_1-metric we choose $Q = (1,0)$. Q is transformed into $w(Q) = (0.55, 0.55)$, and we have

$$d_1(w(P), w(Q)) = 0.55 + 0.55 = 1.1 > 1.0 = d_1(P,Q) .$$

Thus, in terms of the metric d_1, the transformation w does not shrink the distance between P and Q; w is not a contraction.

For the d_∞-metric we look at $Q = (1,1)$. It is mapped to $w(Q) = (0, 1.1)$, thus

$$d_\infty(w(P), w(Q)) = \max(0, 1.1) = 1.1 .$$

w is not a contraction with respect to d_∞ either.

Finally let us examine the situation for the Euclidian metric. To show that w is a contraction, we need to consider arbitrary points $P = (x, y)$ and $Q = (u, v)$. Recall, that

$$d_2(P,Q) = \sqrt{(x-u)^2 + (y-v)^2} \,.$$

We compute the transformed points

$$\begin{aligned} w(P) &= (0.55x - 0.55y, 0.55x + 0.55y) \\ &= 0.55(x - y, x + y) \\ w(Q) &= (0.55u - 0.55v, 0.55u + 0.55v) \\ &= 0.55(u - v, u + v) \end{aligned}$$

and their distance:

$$\begin{aligned} &d_2(w(P), w(Q)) = \\ &= 0.55\sqrt{((x-y)-(u-v))^2 + ((x+y)-(u+v))^2} \\ &= 0.55\,(\,(x-y)^2 + (x+y)^2 - 2(x-y)(u-v) \\ &\quad - 2(x+y)(u+v) + (u-v)^2 + (u+v)^2\,)^{1/2} \\ &= 0.55\sqrt{2(x^2+y^2) - 2(2xu + 2yv) + 2(u^2+v^2)} \\ &= 0.55\sqrt{2((x-u)^2 + (y-v)^2)} \\ &= 0.55\sqrt{2}\,d_2(P,Q) \,. \end{aligned}$$

Since $c = 0.55\sqrt{2} \approx 0.778 < 1$, we have that w is a contraction with regard to the Euclidean metric d_2. The contraction factor is c.

Let us take as an example a similarity transformation which is a composition of a rotation of 45°and a scaling by a factor of about 0.778. Figure 5.32 shows how this transformation acts upon the different unit sets.[13] In each case the transformed image is reduced in size, but only the transformed image of the Euclidean unit disk is contained in the disk. In all other cases there is some overlap indicating that the transformation is not a contraction with respect to the underlying metric.

The Euclidian Metric is not Always the Choice

Based on the above observation one might conjecture that the Euclidean metric is special in the sense that it captures the contractivity of a transformation when other metrics do not. However, this is not the case. Take for example a transformation which first rotates by 90°and then scales the x-component of the result by 0.5, i.e.

$$(x, y) \to (-0.5y, x) \,.$$

[13] The unit sets are defined to be the sets of points with a distance not greater than 1 from the origin. Thus, they depend on the metric used. For example, the unit set for the Euclidean metric is a disk, while it is a square for the maximum metric (see figures 5.32 and 5.34).

Contraction and Metric

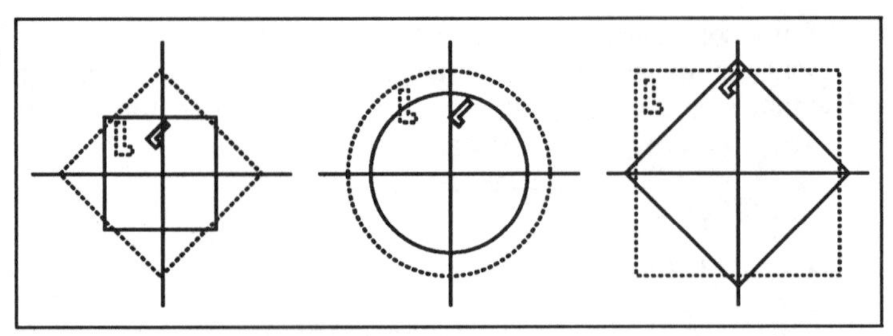

Figure 5.32 : A transformation which rotates by 45° and scales by 0.778 is a contraction with respect to the Euclidean metric (center) but not with respect to the lattice metric (left) or the maximum metric (right).

Square Code

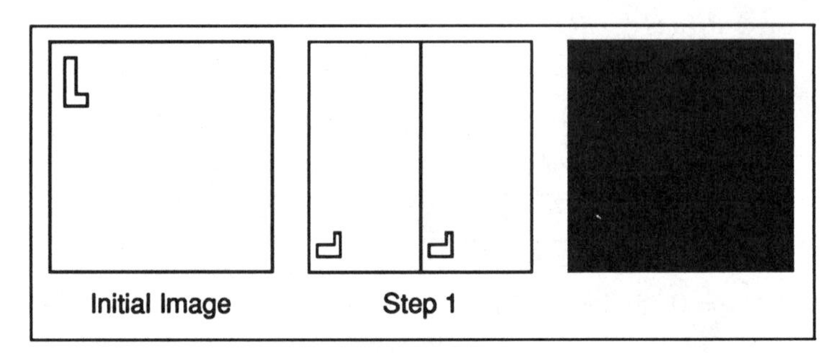

Figure 5.33 : Coding of a square with only two transformations. The rotation of 90° is crucial, without it the transformations would not be contractions.

Using two such transformations with appropriate translations added, we have coded a square, see figure 5.33. It is easy to check that the square is in fact the fixed point of the corresponding Hutchinson operator. But the transformations are not contractions with respect to the Euclidean metric d_2 (the point (1,0) is rotated to (0,1), and the subsequent scaling does not have an effect here). Moreover, they are not contractions with respect to the lattice metric d_1 or the maximum metric d_∞ either. Therefore it seems an open question, whether the corresponding IFS in fact does have the square as an attractor.

The question can be settled since there are metrics which make the transformations contractive, see figure 5.34. The trick is to design the metric so that it measures differently in x- and y-direction. In this way the unit set of all points with distance one or less from the origin becomes a rectangle, which contains its transformed image.

A Special Metric

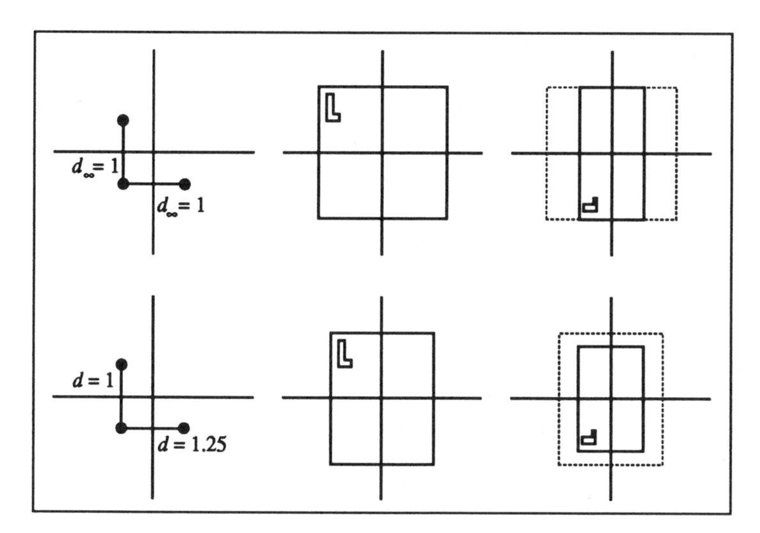

Figure 5.34 : The transformation, which rotates by 90° and scales in x-direction with factor 0.5, is not a contraction relative to the Euclidean, lattice, and maximum metric (top). But it is a contraction with respect to a metric which measures using different weights in x- and y-direction. An example is given by the metric $d(P,Q) = \max\{1.25|x-u|, |y-v|\}$. We show the unit sets (center) and their images under $w(x,y) = (-0.5, y, x)$ (right).

Contraction Mapping Principle and IFS

Thus, we see that it may be important to find a suitable metric for an application of the contraction mapping principle. In particular the third part of the principle, which predicts how fast the iteration of the IFS will approach the attractor, is effected by the quality of the metric. The smaller the contraction ratio the better is the estimate for the speed of convergence of the IFS, and the contraction ratio of course depends heavily on the choice of the metric. The power to make a good prediction will be important in the context of the *inverse problem* mentioned in section 5.4.

5.7 Composing Self-Similar Images

Several methods have been proposed for the automatic solution of the inverse problem, but none has yet really proven itself to be the right choice. Therefore, we should discuss a few ideas, some of which go back to Barnsley in the early 1980's. These ideas, however, do not (yet) lead to automatic algorithms, they are more suitable for interactive computer programs requiring an intelligent human operator.

IFS Attractor and MRCM Blueprint

Recall that the blueprint of an MRCM is already determined by the first copy it produces. The copy is a collage of transformed images. Applying the MRCM to the original image, called *target image*, one also determines the quality of the approximation. When the copy is identical to the original, then the corresponding IFS codes the target image perfectly. When the distance of the copy to the target is small, then we know from the contraction mapping principle, that the attractor of the IFS is not far from the initial image, which is equal to the target image in this case. Figure 5.35 illustrates this principle for the Sierpinski gasket.

Encoding Self-Similar Images

These properties enable us to find the code for a given target image, in particular for target images which contain apparent self-similarities such as the fern. With a little practice it is easy to identify portions of the picture which are affine copies of the whole. For example, in the fern in figure 5.36 the part $R^{(1)}$ is a slightly smaller and rotated copy of the whole fern. This observation leads to the numerical computation of the first affine transformation w_1. The same procedure applies to the copies $R^{(2)}$ and $R^{(3)}$ in the figure. Even the bottom part of the stem (part $R^{(4)}$) is a copy of the whole. However, this copy is degenerate in the sense that the corresponding transformation contains a scaling in one direction by a factor of 0.0, i.e. the fern transformed by w_4 is reduced to a line. The resulting four transformations already comprise the complete system since the portions $R^{(1)}$ to $R^{(4)}$ completely cover the fern.

Interactive Encoding: The Collage Game

In general we need a procedure to generate a set of transformations such that the union of the transformed target images cover the target image as closely as possible. Taking the example of a leaf we illustrate how this can be done with an interactive computer program. In the beginning the leaf image must be entered in the computer using an image scanner. Then the leaf boundary can be extracted from the image using standard tools in image processing. The result in this case is a closed polygon which can be rapidly displayed on the computer screen. Moreover, affine transformations of the polygon can also be computed instantly and displayed. Using interactive input devices such as the mouse, knobs or even just the keyboard, the user of the program can easily manipulate the six parameters that determine one affine transformation. Simultaneously the computer displays the transformed copy of the initial polygon of the leaf. The goal is to find a transformation such that the copy fits snugly onto a part of the original leaf. Then the procedure is repeated, the user next tries to fit another affine copy onto another part of the leaf that is not yet covered by the first. Continuing in this way the complete leaf will be covered by small and possibly distorted copies of itself. Figure 5.37 shows some of the intermediate steps that might occur in the design of the leaf transformations.

Testing Collages

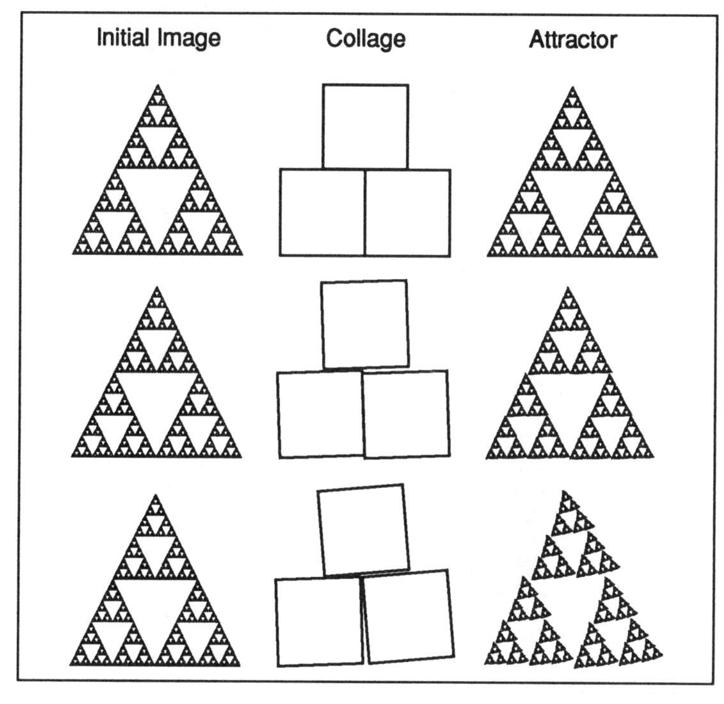

Figure 5.35 : Application of three MRCMs to a Sierpinski gasket. Top: the correct MRCM leaves the image invariant, middle: a resonable approximation, bottom: a bad approximation

Contraction Mapping Principle and Collages

Let us exploit the contraction mapping principle from page 287 to analyze the results of figure 5.35. The a priori estimate for a sequence $a_0, a_1, a_2, \ldots$ which is generated by a contraction f in a metric space with attractor a_∞ yields

$$d(a_n, a_\infty) \leq \frac{c^n}{1-c} d(a_0, a_1) .$$

Here c is the contraction factor of f and $a_{k+1} = f(a_k), k = 0, 1, 2, \ldots$.

Fern Collage

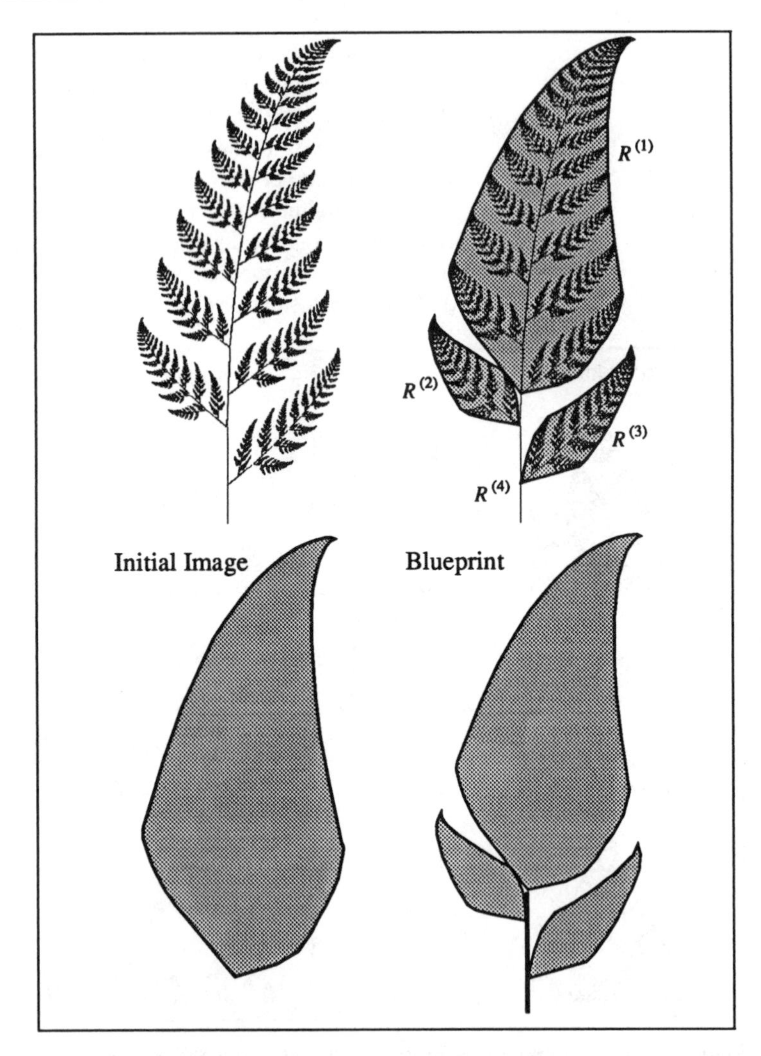

Figure 5.36 : This fern is a slight modification of the original Barnsley fern allowing an easier identification of its partition into self-similar components $R^{(1)}$ to $R^{(4)}$.

In particular, this means that

$$d(a_0, a_\infty) \leq \frac{1}{1-c} d(a_0, f(a_0)) . \tag{5.2}$$

Thus, a single iteration starting from the initial point a_0 gives us an estimate for how far a_0 is from the attractor a_∞ with respect to the metric d. Now let us interpret this result for the Hutchinson operator W with respect to the Hausdorff distance h. Let c be the contraction factor of W and let P be an arbitrary image (formally a compact subset of the plane). We would like to test how good a

Leaf Collage

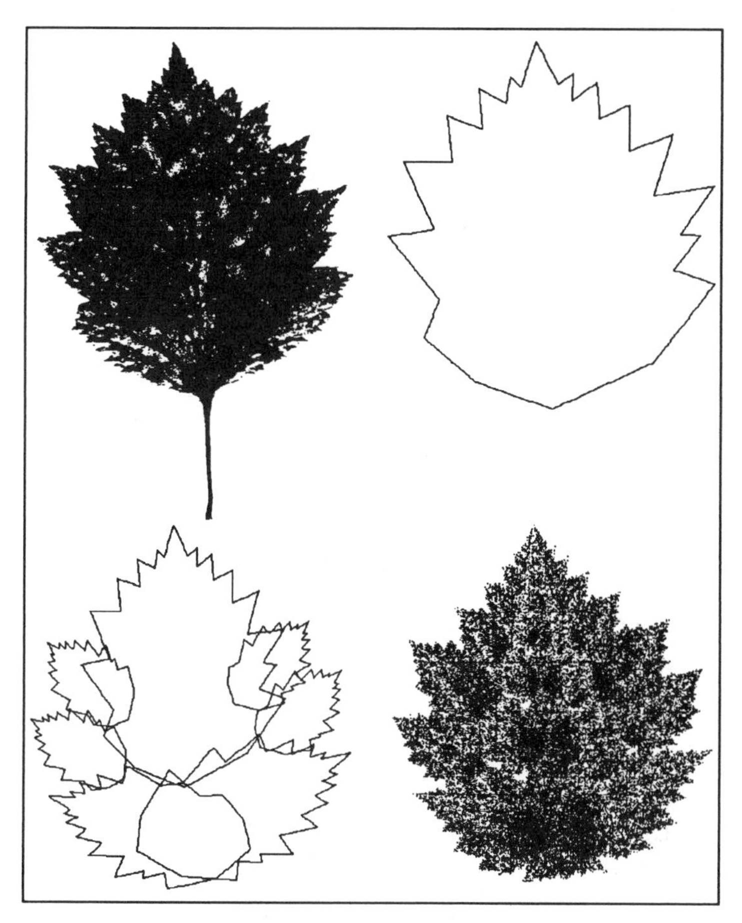

Figure 5.37 : Design steps for a leaf: scanned image of a real leaf and a polygon capturing its outline (top), collage by 7 transformed images of the polygon and the attractor of the corresponding IFS (bottom).

given Hutchinson operator will encode the given image P. This can be obtained from eqn. (5.2). Indeed, in this setting (5.2) now reads

$$h(P, A_\infty) \leq \frac{1}{1-c} h(P, W(P)) \tag{5.3}$$

where A_∞ is the attractor of the IFS given by W. In other words, the quality of the encoding, measured by the Hausdorff distance between P and A_∞ is controlled by applying the Hutchinson operator just once to P and quantified by $h(P, W(P))$.[14]

[14] Barnsley calls eqn. (5.3) the 'Collage Theorem for Iterated Function Systems'.

Again it is the contraction mapping principle which says that the attractor of the IFS will be close to the target image, the leaf, when the design of the collage is also close to the leaf. In the attempt to produce as accurate a collage as possible, there is a second goal that hinders exactness, namely the coding should also be efficient in the sense that as few transformations as possible are used. The definition of an optimal solution to the problem must thus find a compromise between quality of the collage and efficiency. The automatic generation of collages for given target images is a challenging topic of current research.

The collage game is just one example of an entire class of mathematical problems which goes under the name optimization problems. Such problems are typically very easily stated but are often very difficult to solve even with high-powered, super computer technology and sophisticated mathematical algorithms.

Optimization Problem for Collages

The a priori estimate of the contraction mapping principle

$$h(P, A_\infty) \le \frac{1}{1-c} h(P, W(P))$$

gives rise to an optimization problem. Assume we are given a picture P which we want to encode into an IFS. We decide to limit ourselves to N contractions in the IFS, which have to be determined. Any N-tuple $w_1, ..., w_N$ defines a Hutchinson operator W. We may further assume that the contraction factors of the transformations we want to consider are all less than or equal to some $c < 1$. Following the above estimate we have to minimize the Hausdorff distance[15] $h(P, W(P))$ among all admissible choices of W.

The Curse of Computational Complexity

A well-known example of this class of problems is the traveling salesman problem, which goes as follows. Choose some number of towns (for example, all U.S. towns with more than 10,000 inhabitants) and find the shortest route which a salesman must travel to reach all these towns. We would really think that a problem as simple as this should be no trouble for computers. But the truth is that computers become totally useless as soon as the number of towns chosen is larger than a few hundred. Problems of this kind are said to be *computationally complex* and it is understood by now that they are invariably resistant to quick solutions and always will be. The message from such examples is that simple problems may not have simple answers, and we can say that the sea of mathematics is filled with such animals. Unfortunately, it is not yet clear

[15]The computational problem evaluating the Hausdorff distance for digitized images is addressed in R. Shonkwiller, *An image algorithm for computing the Hausdorff distance efficiently in linear time*, Info. Proc. Lett. 30 (1989) 87–89.

whether the collage game can be mathematically formulated in a way which avoids extreme computational complexity. In any case, it is very likely that the computational complexity will be terrible for some images and very manageable for others. The guess is that images which are dominated by self-similar structures might be very manageable. That alone would be reason enough to continue exploring the field simply because we see such characteristics in so many of nature's formations and patterns.

There are some other problems which lead directly into current research problems which we want to at least mention.

5.8 Breaking Self-Similarity and Self-Affinity or, Networking with MRCMs

Creating an image with an MRCM quite naturally leads to a structure which has repetition in smaller and smaller scales. In the cases where each of the contractions involved in the corresponding IFS is a similarity with the same reduction factor (for example, the Sierpinski gasket), we call the resulting attractor *strictly self-similar*. Also when different reduction factors occur, the resulting attractor is said to be *self-similar*. When the contractions are not similarities, but affine linear transformations (for example, the devil's staircase), we call the resulting attractor *self-affine*.

Two Ferns

Figure 5.38 : Two ferns different from Barnsley's fern. Observe that in both cases the placement of the major leaves on the stem differ from that of the small leaves on the major ones. The two ferns look the same at this scale, but the blow ups in the next figure reveal important differences.

In any case, an IFS produces self-similar, or self-affine images. As we have pointed out, IFSs can also be used to approximate images that are not self-similar or self-affine. The approximation can be made as accurate as desired. However, the very small features of the corresponding attractor will still reveal the self-similar structure. In this last section of this chapter we will generalize the concept of IFSs so that this restriction is removed.[16]

[16]Similar concepts are in M. F. Barnsley, J. H. Elton, and D. P. Hardin, *Recurrent iterated function systems,* Constructive Approximation 5 (1989) 3–31. M. Berger, *Encoding images through transition probablities,* Math.

Blowups of the Major Lower Right Leaf

Figure 5.39 : Left: blowups of the left fern of figure 5.38 reveal the hierarchy (a): all subleaves are placed opposing each other. Right: blowups of the right fern reveal the hierachy (b): the subleaves of the major leaf again show a placement with offset

Non-Self-Similar Ferns

Figure 5.38 shows two ferns which almost look like the familiar Barnsley fern, but they are different. Upon close examination of the two ferns, we observe that the phylotaxis has changed. The placement of the major leaves on the stem is different from that of the small leaves on the major ones. That means that the major leaves are no longer scaled down copies of the entire fern. In other words, these ferns are neither self-similar nor self-affine in a strict sense. Nevertheless, we would say that they have some features of self-similarity. But what are these features and how are these

---

Comp. Modelling 11 (1988) 575–577. R. D. Mauldin and S. C. Williams, *Hausdorff dimension in graph directed constructions*, Trans. Amer. Math. Soc. 309 (1988) 811–829. G. Edgar, *Measures, Topology and Fractal Geometry*, Springer-Verlag, New York, 1990. The first ideas in this regard seem to be in T. Bedford, *Dynamics and dimension for fractal recurrent sets*, J. London Math. Soc. 33 (1986) 89–100.

Basic Machine for Fern

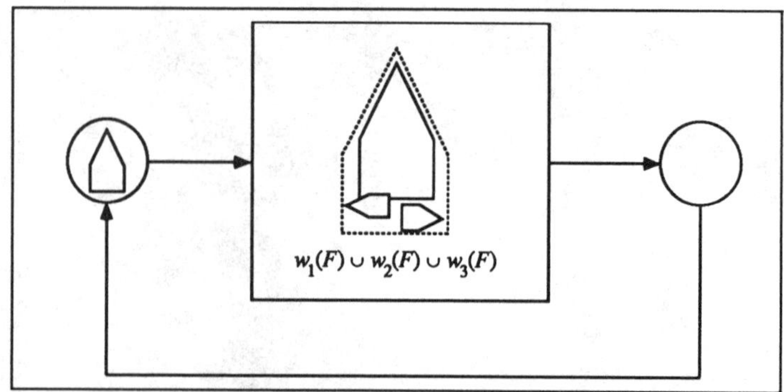

Figure 5.40 : The feedback system of Barnsley's fern (without stem).

particular ferns encoded? The answers to these questions will lead us to *networked* MRCMs, or, in other words, *hierarchical* IFSs.

To see some of the hierarchical structure we now look at a blowup of one of the major leaves from each of the ferns (see figure 5.39). This reveals the different hierarchies in their encoding. The placement of the sub-subleaves is different. On the left, the subleaves of all stages are always placed opposing each other, while on the right, this placement alternates from stage to stage: in one stage subleaves are placed opposing each other and in the next stage subleaves are placed with an offset. For ease of reference let us call these hierarchies type (a) and type (b).

We begin to see that the encoding by IFSs goes much beyond the problem of image encoding. Understanding the self-similarity hierarchies of plants, for example, in terms of IFSs opens a new door to a formal mathematical description of phylotaxis in botany. We will see that self-similarity structures can even be mixed.

Networking MRCMs

We expand the concept of an MRCM to include several MRCMs operating in a network. We will illustrate, how a non-self-similar fern can be obtained by two networked MRCMs. To keep things as simple as possible, we disregard the stems. Figure 5.40 displays the basic machine for a fern without stem.

Let us first consider the fern with hierarchy of type (a) from figure 5.39. We can identify two basic structures: (1) the entire fern, and (2) one of its major leaves, say the one at the lower right (see figure 5.41).

The leaf in this case is a self-similar, or more precisely, a self-affine structure. All subleaves are copies of the whole leaf and vice versa. The complete fern is made up of copies of this leaf, but it is *not* simply a copy of the leaf. This is due to the different placement of the leaves and subleaves. This is the crucial

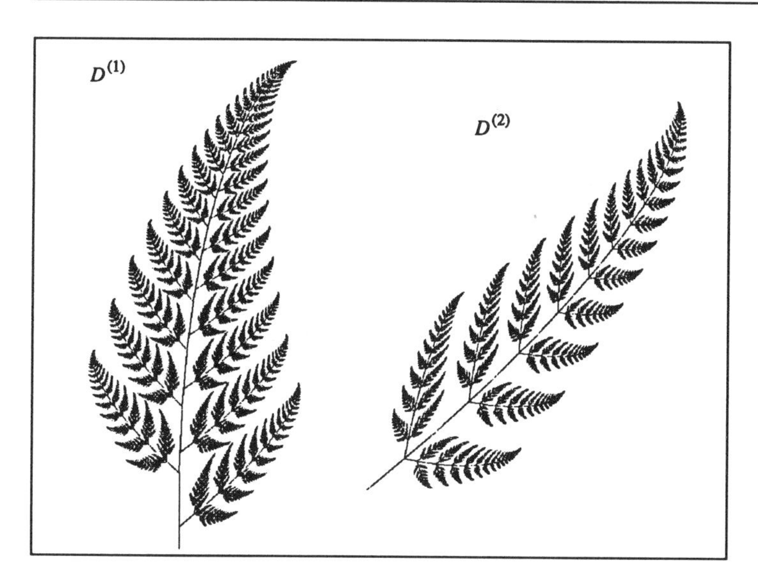

Figure 5.41 : Division of a fern of hierarchy type (a) into its basic structures: the whole fern and one of its major leaves.

Basic Structure

difference between Barnsley's self-affine fern and this one, where the self-affinity is broken. Due to this breaking of self-similarity the fern cannot be generated with an ordinary MRCM. However, we may join two different machines to form a networked MRCM as shown in figure 5.42 which will accomplish the task.

One of the machines (bottom) is used to produce the main leaf alone. This machine works like the one for Barnsley's fern (disregarding the stem for simplicity). Thus, it has three transformations: one transformation maps the entire leaf to its lower left subleaf, the second maps to the corresponding upper left subleaf, and finally, the third transformation maps the leaf to all subleaves except for the bottom subleaves, which are already covered by the other two transformations.

The other machine (top) produces the whole fern. It has two inputs and one output. One input is served by its own output. The other input is served by the bottom MRCM. There are also three transformations in this machine. However, each transformation is applied to only one particular input image. Two transformations (w_2 and w_3 in the figure) operate on the results produced by the bottom MRCM. These produce the left and right bottom leaves at the proper places on the fern. The other transformation (w_1 in the figure) operates on the results from the top MRCM. The results of all transformations are merged when they are transferred to the output of the top machine. This is indicated by the '$\cup$' sign. Transformation w_1 maps the entire fern to its upper part (i.e., the

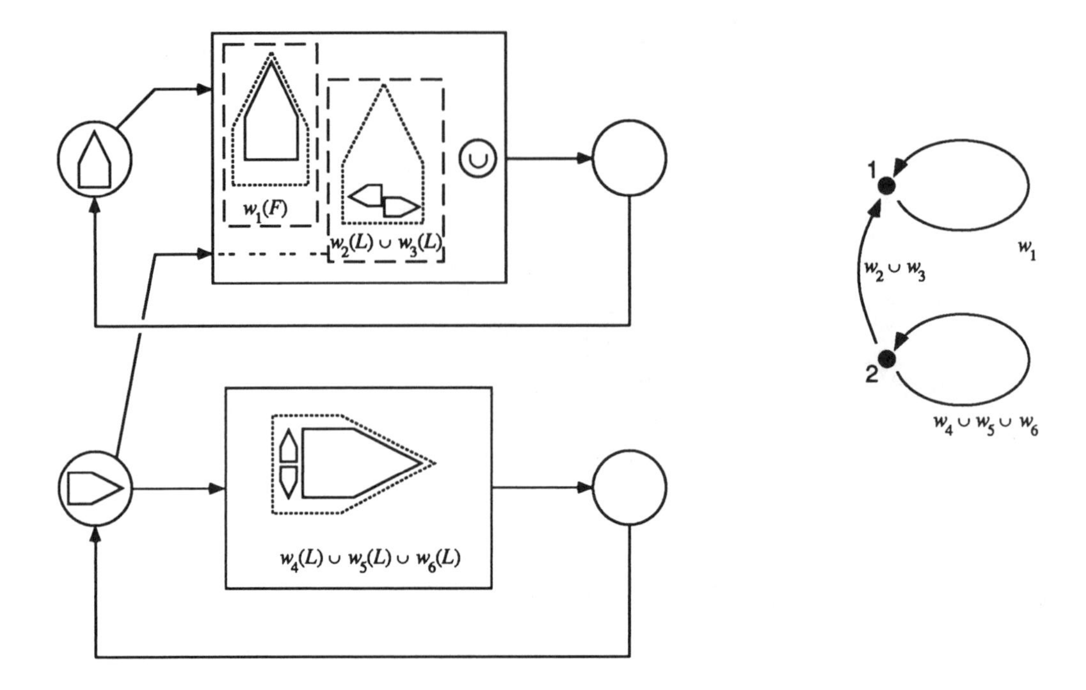

Figure 5.42 : This network of two MRCMs generates the fern with the leaf placement given by the hierarchy of type (a). The graph of the corresponding IFS is shown on the right

part without the two bottom leaves). This was also the case in the plain MRCM for Barnsley's fern. In this way the fern with the prescribed pattern for the leaf placement from hierarchy of type (a) will be generated.

Rearranging Input Connections

In order to produce the fern as in hierarchy of type (b), we need to go just a small step further interconnecting the two MRCMs both ways. This fern is characterized by the fact that the entire fern reappears in the main leaves as subleaves, while the leaves themselves are not copies of the entire fern. This is easy to do as shown in figure 5.43. The only change relative to the network for the hierarchy (a) fern is given by the extra input in the bottom MRCM. This input image (in the limit it is the entire fern) will be transformed to make the two lowest subleaves of the leaf.

But how do we run these networks? Well, we just take any initial image, like a rectangle, and put it on the two copy machines. The machines take these input images following to the connections of the input lines and produce two outputs, one for the leaf and one for the fern. These outputs are now used as new inputs as indicated by the feedback connections. When we iterate this process we can

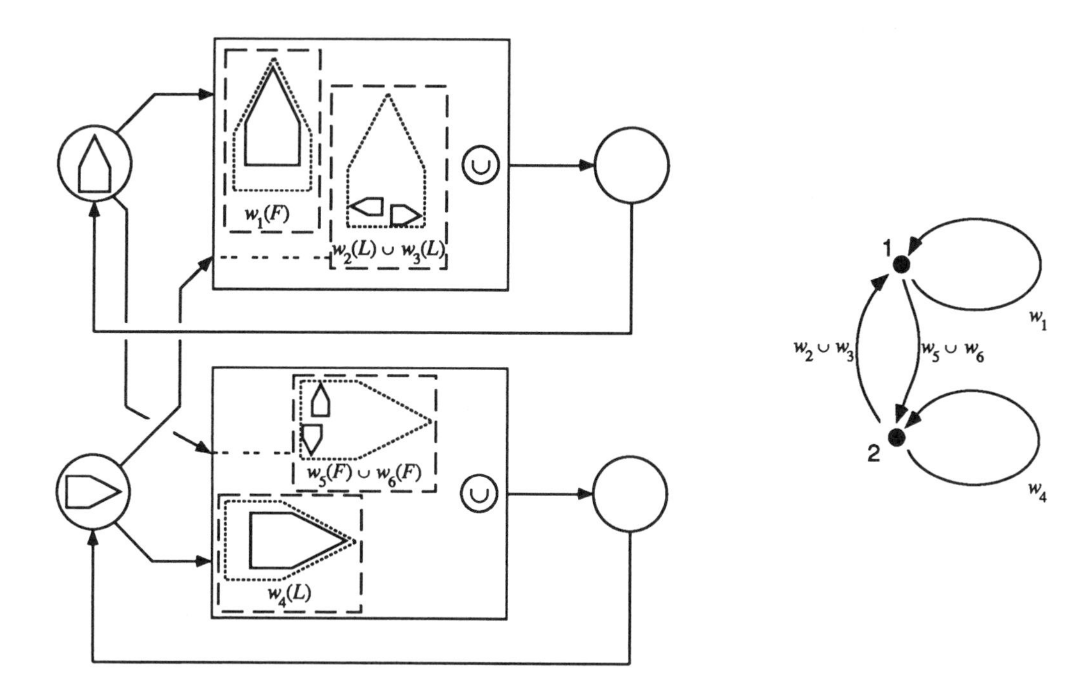

Figure 5.43 : This network of two MRCMs generates the other fern with the leaf placement given by hierarchy of type (b).

observe how the leaf-MRCM creates the major lower right hand leaf and how the fern-MRCM generates the complete fern.

The Contraction Mapping Principle Does It Again

Is the successful operation of this machinery just a pure accident? Not at all! Above, we discussed the contraction mapping principle. It turns out that we can also subsume the network idea under that principle, which shows the value of that rather abstract but very powerful mathematical tool. In conclusion, the networking machine has exactly one limit image, its attractor, and this attractor is independent of the initial images. To put it another way, the networking machines are encodings of non-self-similar ferns, and their hierarchies decipher the self-similarity features of these ferns. In fact, the hierarchy of the network deciphers the self-similarity of an entire class of attractors. Imagine that we change the contraction properties and positioning of the individual lens system. As a result, we will obtain an entire cosmos of structures. However, each of them has exactly the same self-similarity features. We have thus reached the beginning of a new and very auspicious theory which promises to systematically decipher all possible self-similarity properties. The mathematical description of networked MCRMs is the topic of the remainder of this section.

Formalism of Hierarchical IFSs

There is an extension of the concept of a Hutchinson operator for a network of MRCMs. It requires working with matrices. Let

$$\mathbf{A} = \begin{pmatrix} a_{11} & a_{12} & \dots & a_{1m} \\ \vdots & \vdots & & \vdots \\ a_{m1} & a_{m2} & \dots & a_{mm} \end{pmatrix}$$

be an $(m \times m)$-matrix with elements a_{ij} and let

$$\mathbf{b} = \begin{pmatrix} b_1 \\ \vdots \\ b_m \end{pmatrix}$$

be an m-vector. Then $\mathbf{Ab}$ is the m-vector $\mathbf{c} = \mathbf{Ab}$ with components c_i , where

$$c_i = \sum_{j=1}^{m} a_{ij} b_j \; .$$

In analogy to this concept of ordinary matrices, a hierarchical IFS (corresponding to a network of M MRCMs) is given by an $(M \times M)$-matrix

$$\mathbf{W} = \begin{pmatrix} W_{11} & \dots & W_{1M} \\ \vdots & & \vdots \\ W_{M1} & \dots & W_{MM} \end{pmatrix} ,$$

where each W_{ij} is a Hutchinson operator (i.e., W_{ij} is given by a finite number of contractions). This is the *matrix Hutchinson operator* $\mathbf{W}$, which acts on an M-vector $\mathbf{B}$ of images

$$\mathbf{B} = \begin{pmatrix} B_1 \\ \vdots \\ B_M \end{pmatrix}$$

where each B_i is a compact subset of the plane $\mathbf{R}^2$. The result of $\mathbf{W}(\mathbf{B})$ is an M-vector $\mathbf{C}$ with components C_i, where

$$C_i = \bigcup_{j=1}^{M} W_{ij}(B_j) \; .$$

It is convenient to allow that some of the Hutchinson operators are 'empty', $W_{ij} = \emptyset$. Here the symbol $\emptyset$ plays a similar role as 0 in ordinary arithmetic: the $\emptyset$ operator transforms any set into the empty set (i.e., for any set B we have $\emptyset(B) = \emptyset$).

Next we make a natural identification. The network of MRCMs corresponds to a graph with nodes and directed edges. For the output of each MRCM there is exactly one node, and for each output-input connection in the network there is a corresponding directed

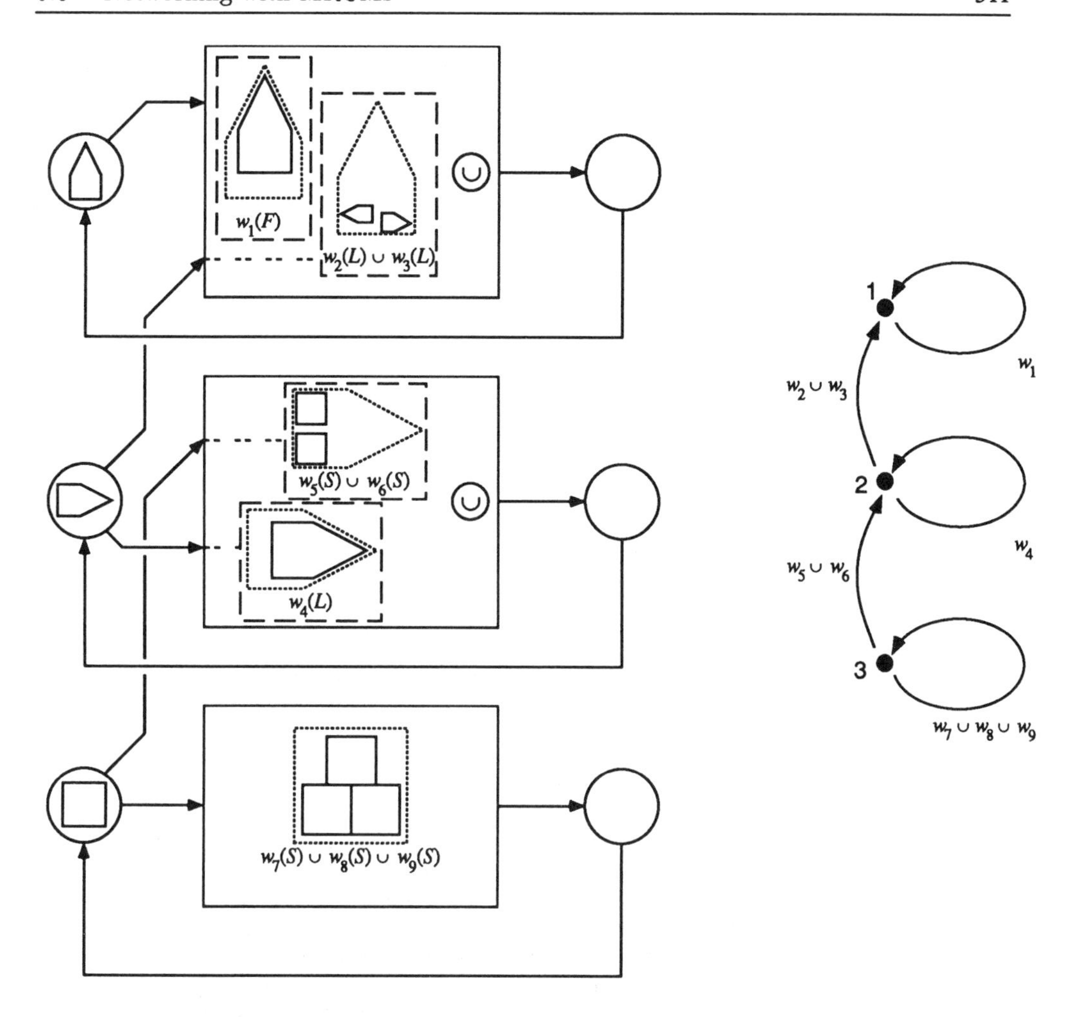

Figure 5.44 : A network of three MRCMs to generate a fern which is made up of Sierpinski gaskets.

edge. These graphs, displayed next to our MRCM networks, are a compact representation of the hierarchy of the IFSs (see, for example, the non-self-similar ferns and the Sierpinski fern).

Note that a directed edge from node j to node i means that the output of j is transformed according to a specific Hutchinson operator (i.e., the one that operates on the corresponding input of the MRCM) and then fed into node i. The output of this node is the union of all the transformed images which are fed in. Now we define W_{ij}. If there is a directed edge from node j to node i, then W_{ij} denotes the corresponding Hutchinson operator. In the other case we set $W_{ij} = \emptyset$. For our examples we thus obtain

$$\mathbf{W} = \begin{pmatrix} w_1 & w_2 \cup w_3 \\ \emptyset & w_4 \cup w_5 \cup w_6 \end{pmatrix}$$

for the fern of type (a),

$$\mathbf{W} = \begin{pmatrix} w_1 & w_2 \cup w_3 \\ w_5 \cup w_6 & w_4 \end{pmatrix}$$

for the fern of type (b), and

$$\mathbf{W} = \begin{pmatrix} w_1 & w_2 \cup w_3 & \emptyset \\ \emptyset & w_4 & w_5 \cup w_6 \\ \emptyset & \emptyset & w_7 \cup w_8 \cup w_9 \end{pmatrix}$$

for the Sierpinski fern. Observe that here we have used a short form for writing Hutchinson operators. For example, when transforming any set B by $w_2 \cup w_3$ we write

$$w_2 \cup w_3(B) = w_2(B) \cup w_3(B) .$$

With these definitions we can now describe the iteration of a hierarchical IFS formally. Let $\mathbf{A}_0$ be an initial M-vector of images. The iteration defines the sequence of M-vectors

$$\mathbf{A}_{k+1} = \mathbf{W}(\mathbf{A}_k), \quad k = 0, 1, 2, ...$$

It turns out that this sequence again has a limit $\mathbf{A}_\infty$, which we call the attractor of the hierarchical IFS.

The proof is again by the contraction mapping principle. We start with the plane equipped with a metric such that the plane is a complete metric space. Then the space of all compact subsets of the plane with the Hausdorff distance as a metric is also a complete metric space. Now we take the M-fold Cartesian product of this space and call it H. On H there is a natural metric $d_{\max}$ which comes from the Hausdorff distance: Let A and B be in H , then

$$d_{\max}(\mathbf{A}, \mathbf{B}) = \max \{h(A_i, B_i) \mid i = 1, ..., M\} ,$$

where A_i and B_i denote the components of $\mathbf{A}$ and $\mathbf{B}$ and $h(A_i, B_i)$ denotes their Hausdorff distance. It follows almost from the definitions that

- H is again a complete metric space, and
- $\mathbf{W} : H \to H$ is a contraction.

For completeness we must add the requirement that the iterates $\mathbf{W}^n$ of the matrix Hutchinson operator do not consist entirely of $\emptyset$-operators. Thus, the contraction mapping principle applies with the same consequences as for the ordinary Hutchinson operator.

The Sierpinski Fern

Figure 5.45 : The Sierpinski fern and one of its main leaves.

The Sierpinski Fern

To finish this section we will use the networked MRCMs for one rather strange looking fern, which we may call the *Sierpinski fern.* It is the fern of hierarchy (a) with subleaves replaced by small Sierpinski gaskets. The network incorporates three MRCMs. The first two are responsible for the overall structure of the fern as before, while the third is busy with producing a Sierpinski gasket which is fed to one of the other machines.

The experiment generating a Sierpinski fern demonstrates that networked MRCMs are suitable discussing and encoding hierarchies of self-similarity features, and, moreover, is the appropriate concept mixing several fractals together.

The Stem in Barnsley's Fern

When we introduced Barnsley's fern by an MRCM we observed that it is not strictly self-similar, the problem being first of all in the stem. There we obtained the stem from a degenerate affine-linear copy of the whole fern (i.e., collapsed to a line). From the point of view of networked MRCMs this aspect becomes much clearer. The design in figure 5.46 is a network with two MRCMs. The top machine produces the leaves and the bottom machine the stems. From that point of view Barnsley's fern is essentially a mix of two (strictly) self-similar structures.[17]

The variety of structures which can be obtained by networked MRCMs is unimaginable. As an application of networked MRCMs we present in chapter 9 an elegant solution to some long standing

[17]More precisely, the fern without the stem is self-affine, not self-similar, because the transformations which produce the leaves are only approximate similitudes.

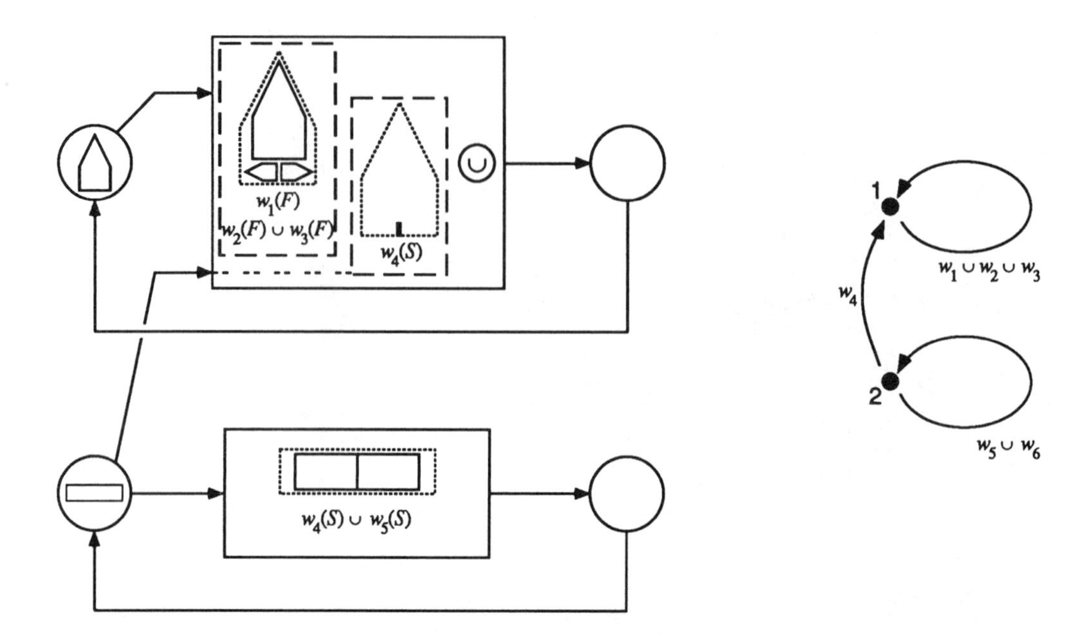

Figure 5.46 : The lower MRCM generates a line which is fed into the upper MRCM to build the stem of the fern.

open problems: the deciphering of the global geometric patterns in Pascal's triangle, which are obtained when we analyze divisibility of binomial coefficients by prime powers.

5.9 Program of the Chapter: Iterating the MRCM

Imagine an interactive computer program for the design of Multiple Reduction Copy Machines. With such a program you could select and change transformations and immediately see how the attractors appear on the computer screen. This is not only fun but also provides an instructive way to study the effect of affine linear transformations. This is certainly a desirable goal of mathematics in school. Well, this program is not really what you have dreamed of but it is small enough to be here and it is smart enough for the generation of most of the images from this chapter. There is a table of the parameter settings of the affine transformations on page 317.

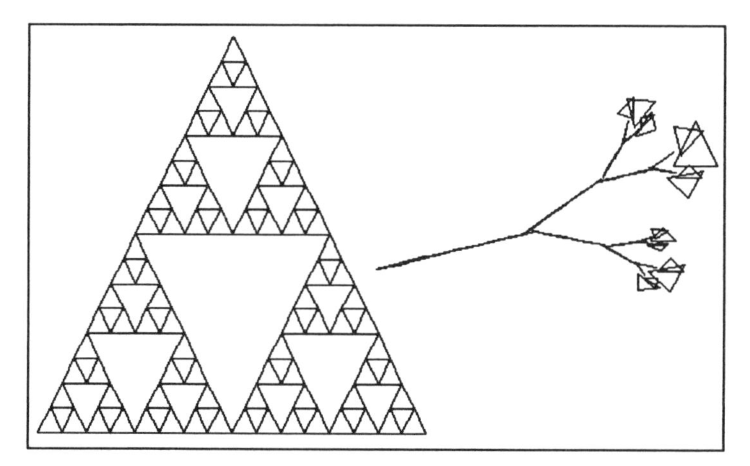

Figure 5.47 : Output of the Program 'MRCM Iteration' (left, Level = 5). The right image is obtained when changing the parameters to those of figure 5.13 according to table 5.48.

Screen Images of the Program

The program in this section is set up for the generation of the Sierpinski gasket, or the stages of its construction by repeated application of the MRCM. At the very beginning, the program will ask you how many iterations it should perform. Well, actually it asks you for the level it should draw, since again (as in the previous chapters) we use a recursive program structure to avoid keeping thousands of triangles in memory. Thus, answering '1' means `level` = 1 (i.e., no iteration of the MRCM) and you will see the initial image. This initial image is a triangle. Thus, in all the stages of the iteration there are images of triangles. This choice of a very simple initial image makes the program short and fast. It is not hard to change the code so that the initial image is a square or another figure, but this is something which we want to leave to your programing skills.

BASIC Program **MRCM Iteration**
Title Multiple reduction copying of a (Sierpinski) gasket

```
DIM xleft(10), xright(10),xtop(10),yleft(10),yright(10),ytop(10)
DIM a(3), b(3), c(3), d(3), e(3), f(3)
INPUT "Enter level:", level
left = 30
w = 300
wl = w + left
xleft(level) = 0
yleft(level) = 0
xright(level) = w
yright(level) = 0
xtop(level) = .5*w
ytop(level) = w
a(1) = .5 : a(2) = .5 : a(3) = .5
d(1) = .5 : d(2) = .5 : d(3) = .5
e(1) = 0 : e(2) = 0.5*w : e(3) = 0.25*w
f(1) = 0 : f(2) = 0 : f(3) = .5*w
GOSUB 100
END

REM TRANSFORM THE TRIANGLE
50  xleft(level) = a(map)*xleft(level+1) + e(map)
    yleft(level) = d(map)*yleft(level+1) + f(map)
    xright(level) = a(map)*xright(level+1) + e(map)
    yright(level) = d(map)*yright(level+1) + f(map)
    xtop(level) = a(map)*xtop(level+1) + e(map)
    ytop(level) = d(map)*ytop(level+1) + f(map)

REM DRAW TRIANGLE AT LOWEST LEVEL
100 IF level > 1 GOTO 200
    LINE (left+xleft(1),wl-yleft(1)) - (left+xright(1),wl-yright(1))
    LINE - (left+xtop(1),wl-ytop(1))
    LINE - (left+xleft(1),wl-yleft(1))
    GOTO 300

REM BRANCH INTO LOWER LEVELS
200 level = level - 1
    map = 1
    GOSUB 50
    map = 2
    GOSUB 50
    map = 3
    GOSUB 50
level = level + 1
300 RETURN
```

Parameter Table

a	b	c	d	e	f
Figure 5.9					
0.500	0.000	0.00	0.500	0.0000	0.0000
0.500	0.000	0.00	0.500	0.5000	0.0000
0.500	0.000	0.00	0.500	0.0000	0.5000
Figure 5.10					
0.000	-0.500	0.500	-0.000	0.5000	0.0000
0.000	0.500	-0.500	0.000	0.5000	0.5000
0.500	0.000	0.000	0.500	0.2500	0.5000
Figure 5.11					
0.000	0.577	-0.577	0.000	0.0951	0.5893
0.000	0.577	-0.577	0.000	0.4413	0.7893
0.000	0.577	-0.577	0.000	0.0952	0.9893
Figure 5.12					
0.336	0.000	0.000	0.335	0.0662	0.1333
0.000	0.333	1.000	0.000	0.1333	0.0000
0.000	-0.333	1.000	0.000	0.0666	0.0000
Figure 5.13					
0.387	0.430	0.430	-0.387	0.2560	0.5220
0.441	-0.091	-0.009	-0.322	0.4219	0.5059
-0.468	0.020	-0.113	0.015	0.4000	0.4000
Figure 5.14					
0.255	0.000	0.000	0.255	0.3726	0.6714
0.255	0.000	0.000	0.255	0.1146	0.2232
0.255	0.000	0.000	0.255	0.6306	0.2232
0.370	-0.642	0.642	0.370	0.6356	-0.0061
Figure 5.15					
0.382	0.000	0.000	0.382	0.3072	0.6190
0.382	0.000	0.000	0.382	0.6033	0.4044
0.382	0.000	0.000	0.382	0.0139	0.4044
0.382	0.000	0.000	0.382	0.1253	0.0595
0.382	0.000	0.000	0.382	0.4920	0.0595
Figure 5.16					
0.195	-0.488	0.344	0.443	0.4431	0.2452
0.462	0.414	-0.252	0.361	0.2511	0.5692
-0.058	-0.070	0.453	-0.111	0.5976	0.0969
-0.035	0.070	-0.469	-0.022	0.4884	0.5069
-0.637	0.000	0.000	0.501	0.8562	0.2513
Figure 5.25					
0.849	0.037	-0.037	0.849	0.075	0.1830
0.197	-0.226	0.226	0.197	0.400	0.0490
-0.150	0.283	0.260	0.237	0.575	-0.0840
0.00	0.000	0.000	0.160	0.500	0.0000

Table 5.48 : Parameter table for the figures in this chapter.

Let us look at the program. First there is the specification of the initial image, a triangle. Then we specify the parameters of three transformations: `a(1), a(2), a(3), ..., f(1), f(2), f(3)`. You will observe that we did not specify any `b()` or `c()`. This is simply because they would be 0, and we have dropped them to make the program shorter. If you want to change the transformations you probably have to add these parameters. Note that they are already included in the second `DIM` statement of the program. When you increase the number of transformations, do not forget to adapt the dimension in this `DIM` statement correspondingly.

Now the recursion starts (`GOSUB 100`, as usual) by testing the level. If we are at the lowest level, then the transformed triangle is drawn. Otherwise (at label 200) the next level of the recursion is entered, first choosing the transformation `map = 1`. When we are finished with this branch, having drawn all repeatedly transformed triangles where the transformation `map = 1` was applied first, then the next branch with `map = 2` is started and finally the branch for `map = 3`. This terminates the recursive part. If you change the number of transformations, do not forget to extend the program at this place.

Observe, that whenever we start the recursion for a new transformation, first this transformation is applied to the current triangle (at label 50). You will note, that also at this place in the program we did not include the parameters `b()` or `c()` to make the program short. If you want to use more general transformations you have to change this for example to:

```
a(map)*xright(level+1) +
     b(map)*yright(level+1) + e(map)
c(map)*xright(level+1) +
     d(map)*yright(level+1) + f(map)
```

Let us finally remark that the parameters in table 5.48 are set up for images which are in the range $[0,1] \times [0,1]$. This program shows the images in the range $[0,w] \times [0,w]$ and therefore the translational parts $e()$ and $f()$ of the transformations must be scaled by w. Do not forget to do this if you change the parameters. For example, for the fern from figure 5.25, set

```
e(1) = 0.075*w
f(1) = 0.183*w
```

and so forth.

Chapter 6

The Chaos Game: How Randomness Creates Deterministic Shapes

Chaos is the score upon which reality is written.

Henry Miller

Nothing in Nature is random... A thing appears random only through the incompleteness of our knowledge.

Spinoza

Our idea of randomness, especially with regard to images, is that structures or patterns which are created randomly look more or less arbitrary. Maybe there is some characteristic structure, but if so, it is probably not very interesting. Just imagine a box of nails which is poured out onto a table.

Brownian Motion

Or look at the following example. Small particles of solid matter suspended in a liquid can be seen under a microscope moving about in an irregular and erratic way. This is the so called *Brownian motion*,[1] which is due to the random molecular impacts of the surrounding particles. It is a good example of what we expect from a randomly steered motion. Let us describe such a particle motion step by step. Begin at a point in the plane. Choose a random direction, walk some distance and stop. Choose another random direction, walk some distance and stop, and so on. Do we have to carry out the experiment to be able to get a sense of what

[1] The discovery was made by the botanist Robert Brown around 1827.

The Board for the Chaos Game

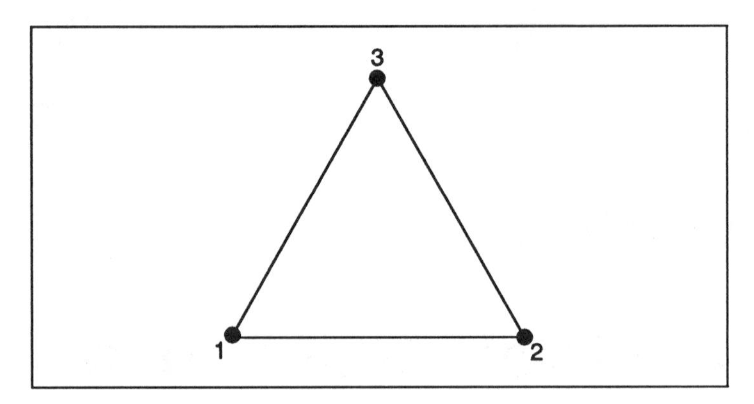

Figure 6.1 : The game board of our first chaos game.

the evolving pattern will be? How would the pattern look after one hundred, or one thousand, or even more steps? There seems to be no problem forecasting the essential features: we would say that more or less the same patterns will evolve, however, just a bit more dense.

In any event, there doesn't seem to be much to expect from randomness in conjunction with image generation. But let us try a variant which, at first glance, could well belong to that category. Actually, following Barnsley[2] we are going to introduce a family of games which can potentially change our intuitive idea of randomness quite dramatically.

The Chaos Game

Here is the first game of this sort. We need a die whose six faces are labeled with the numbers 1, 2, and 3. An ordinary die, of course, uses numbers from 1 to 6; but that does not matter. All we have to do is, for example, identify 6 with 1, 5 with 2, 4 with 3 on an ordinary die. Such a die will be our generator of random numbers from the reservoir 1, 2, and 3. The random numbers which appear as we play the game, for example, 2, 3, 2, 2, 1, 2, 3, 2, 3, 1, ..., will drive a process. The process is characterized by three simple rules. To describe the rules we have to prepare the game board. Figure 6.1 shows the setup: three markers, labeled 1, 2, and 3, which form a triangle.

Now we are ready to play. Let us introduce the rules as we play. Initially we pick an arbitrary point on the board and mark it by a tiny dot. This is our current *game point*. For future reference we denote it by z_0. Now we throw the die. Assume the result is 2. Now we generate the new game point z_1, which is located at the midpoint between the current game point z_0 and the marker with

[2]M. F. Barnsley, *Fractal modelling of real world images,* in: The Science of Fractal Images, H.-O. Peitgen and D. Saupe (eds.), Springer-Verlag, New York, 1988.

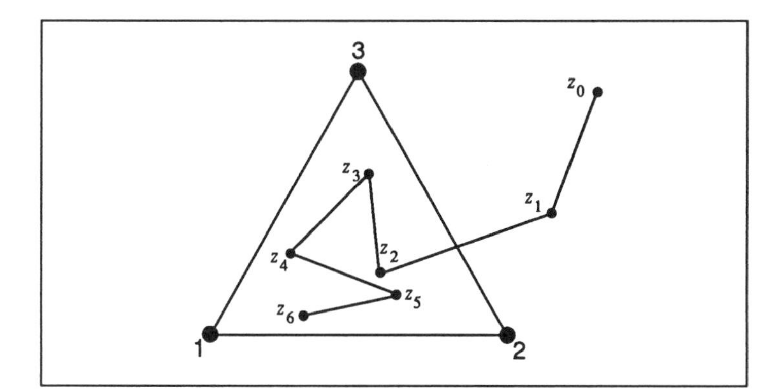

Figure 6.2 : The first six steps of the game. Game points are connected by line segments.

The First Steps ...

label 2. This is the first step of the game. Now you can probably guess what the other two rules are. Assume we have played the game for k steps. We have thus generated $z_1, ..., z_k$. Roll the die. When the result is n generate a new game point z_{k+1}, which is placed exactly at the midpoint between z_k and the marker labeled n. Figure 6.2 illustrates the game. To help identify the succession of points, we connect the game points by line segments as they evolve. A pattern seems to emerge which is just as boring and arbitrary as the structure of a random walk. But that observation is a far cry from the reality. In figure 6.3 we have dropped the connecting line segments and have only shown the collected game points. In (a) we have run the game up to $k = 100$, in (b) up to $k = 500$, in (c) up to $k = 1000$, and in (d) up to $k = 10000$ steps.

Randomness Creates Deterministic Shapes

The impression which figure 6.3 leaves behind is such that we are inclined, at first, not to believe our eyes. We have just seen the generation of the Sierpinski gasket by a random process, which is amazing because the Sierpinski gasket has become a paragon of structure and order for us. In other words, we have seen how randomness can create a perfectly deterministic shape. To put it still another way, if we follow the time process step by step, we cannot predict where the next game point will land because it is determined by throwing a die. But nevertheless, the pattern which all the game points together leave behind is absolutely predictable. This demonstrates an interesting interplay between randomness and deterministic fractals.

But there are a few — if not many — questions about this interaction. For example, how can we explain the small specks which we observe upon close examination of the images in figure 6.3 and which definitely do not belong to the Sierpinski gasket? Or what happens if we use another die, maybe one which is slightly

... and the Next Game Points

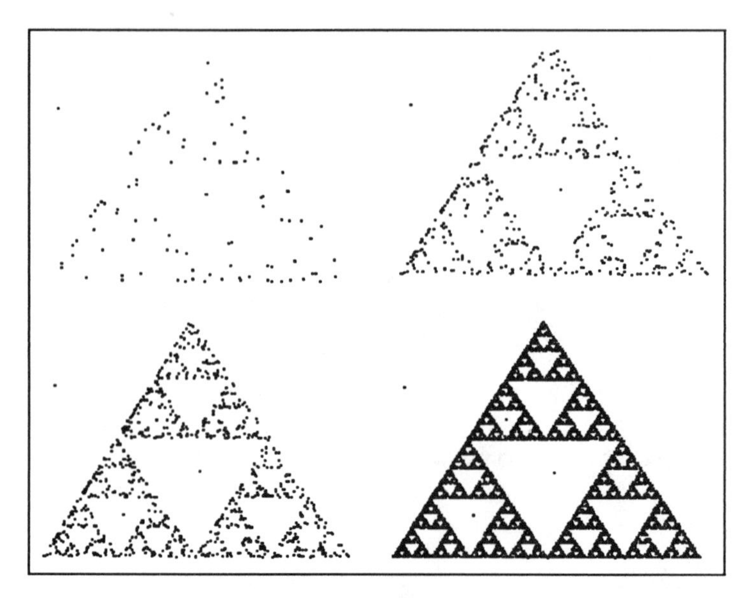

Figure 6.3 : The chaos game after 100 steps (a), 500 steps (b), 1000 steps (c), and 10,000 steps (d). Only the game points are drawn without connecting lines. (Note that there are a few spurious dots that are clearly not in the Sierpinski gasket.)

or severely biased? In other words, does the random process itself leave some imprint or not? Or is this creation the result of a special property of the Sierpinski gasket? In other words, are there chaos games which produce some other, or even any other, fractal as well as the Sierpinski gasket?

6.1 The Fortune Wheel Reduction Copy Machine

As you may have guessed, there are many variations of the chaos game, which produce many different fractals. In particular, all images that can be generated by means of a Multiple Reduction Copy Machine of the last chapter are also accessible using the chaos game played with appropriate rules. This is the topic of this section.

Random Affine Transformations

The basic rule of the above chaos game is: generate a new game point z_{k+1} by picking the midpoint between the last game point z_k and the randomly chosen marker, which is represented by a number from the set $\{1,2,3\}$. The three possible new game points can be described by three transformations, say w_1, w_2, and w_3 applied to the last game point. What kind of transformations are these? It is crucial to observe that they are the (affine linear) transformations which we discussed for the Sierpinski gasket in chapter 5. There we interpreted them as mathematical descriptions of lens systems in an MRCM. In fact, here each w_n is just a similarity transformation which reduces by a factor of 1/2 and is centered at the marker point n. That implies, that w_n leaves the marker point with label n invariant. In the language of the above rules: if a game point is at a marker point with label n and one draws number n by rolling the die, then the succeeding game point will stay at the marker point. As we will see, it is a good idea to start the chaos game with one of these fixed points.

Chaos Game and IFS Transformations for the Sierpinski Gasket

Our first chaos game generates a Sierpinski gasket. Let us try to derive a formal description of the transformations which are used in this game. To that end we introduce a coordinate system with x- and y-axis. Now suppose that the marker points have coordinates

$$P_1 = (a_1, b_1), \quad P_2 = (a_2, b_2), \quad P_3 = (a_3, b_3).$$

The current game point is $z_k = (x_k, y_k)$, and the random event is the number n (1, 2, or 3). Then the next game point is

$$z_{k+1} = w_n(z_k) = (x_{k+1}, y_{k+1})$$

where

$$x_{k+1} = \tfrac{1}{2}x_k + \tfrac{1}{2}a_n \,,$$
$$y_{k+1} = \tfrac{1}{2}y_k + \tfrac{1}{2}b_n \,.$$

In terms of a matrix (as introduced in the last chapter) the affine linear transformation w_n is given by

$$\left(\begin{array}{cc|c} \frac{1}{2} & 0 & \frac{1}{2}a_n \\ 0 & \frac{1}{2} & \frac{1}{2}b_n \end{array}\right).$$

Note that with $w_n(P_n) = P_n$, the marker points are fixed. Now we can play the chaos game following this algorithm:

Preparation: Pick z_0 arbitrarily in the plane.
Iteration: For $k = 0, 1, 2, \ldots$ set $z_{k+1} = w_{s_k}(z_k)$, where s_k is chosen randomly (with equal probability) from the set $\{1, 2, 3\}$ and plot z_{k+1}.

In other words, s_k keeps track of the random choices, the results of throwing the die, in each step. The sequence $s_0, s_1, s_2, \ldots$ together with the initial point z_0, is a complete description of a round of the chaos game. We abbreviate the sequence with (s_k). More formally, we would say that (s_k) is a random sequence with elements from the 'alphabet' $\{1, 2, 3\}$.

MRCM and FRCM

Note that our concept of an MRCM (or IFS) is strictly deterministic. We describe now a modification of our machine which corresponds to the chaos game: rather than applying the copy machine to entire images, we apply it to single points. Moreover, we do not apply all lens systems simultaneously. Rather, in each step we pick one at random (with a certain probability) and apply it to the previous result. And finally the machine does not draw just a single point; it accumulates the generated points. All these accumulated points form the final image of the machine. This would be a random MRCM. Correspondingly, we call it a *Fortune Wheel Reduction Copy Machine* (FRCM). Running this machine is the same as playing a particular chaos game.

What is the relation of an MRCM and its random counterpart? For the Sierpinski gasket we have just seen the answer. The corresponding FRCM also generates a Sierpinski gasket. And indeed this is a case of a general rule: the final image of an MRCM (its IFS attractor) can be generated by a corresponding FRCM, which is the same as playing the chaos game according to a specific set of rules.

The Fortune Wheel

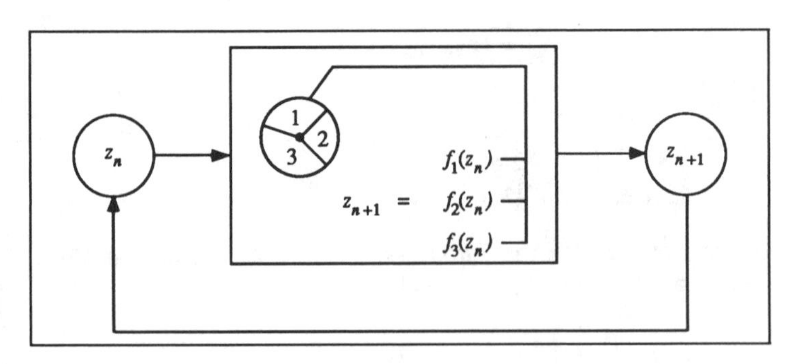

Figure 6.4 : Feedback machine with fortune wheel (FRCM).

Random Iterated Function Systems: Formal Description of FRCM

We have shown that an MRCM is determined by N affine-linear contractions

$$w_1, w_2, ..., w_N \; .$$

One copying step in the operation of the machine is described by the Hutchinson operator

$$W(A) = w_1(A) \cup \cdots \cup w_n(A) \; .$$

Starting with any initial image A_0, the sequence of generated images $A_1 = W(A_0)$, $A_2 = W(A_1)$,... converges to a unique attractor A_∞, the final image of the machine. A corresponding FRCM is given by the the same contractions

$$w_1, w_2, ..., w_N \; .$$

and some (positive) probabilities

$$p_1, p_2, ..., p_N > 0$$

where

$$\sum_{i=1}^{N} p_i = 1 \; .$$

This setup is called a *random iterated function system* (IFS), while the corresponding MRCM is called a *deterministic* iterated function system. Let $s_1, s_2, s_3, ...$ be a sequence of random numbers which are chosen from the set $\{1, 2, ..., N\}$ independently and with probability p_k for the event $s_i = k$. Assume z_0 is a fixed point of one of the transformations (e.g. $w_1(z_0) = z_0$), then

(1) All points of the sequence $z_0, z_1 = w_{s_1}(z_0), z_2 = w_{s_2}(z_1), ...$ lie in the attractor A_∞.
(2) The sequence $z_0, z_1, z_2, ...$ almost surely fills out the attractor A_∞ densely.

The first fact is immediate from the invariance property of the attractor. The second one will be investigated in the next section. In summary, an MRCM and a corresponding FRCM encode the same image A_∞; we can produce the attractor by playing the chaos game with this machine. The restriction 'almost surely' in the second property is only a fine technical point. Theoretically, it may happen for example, that even though the sequence $s_1, s_2, ...$ is random, all events are identical. This is like having a die that forever rolls the number '1', even though it is a perfect and fair die. In this case the chaos game would certainly fail to fill out the attractor. However, the chance of such an abnormal outcome is zero.

The Fern

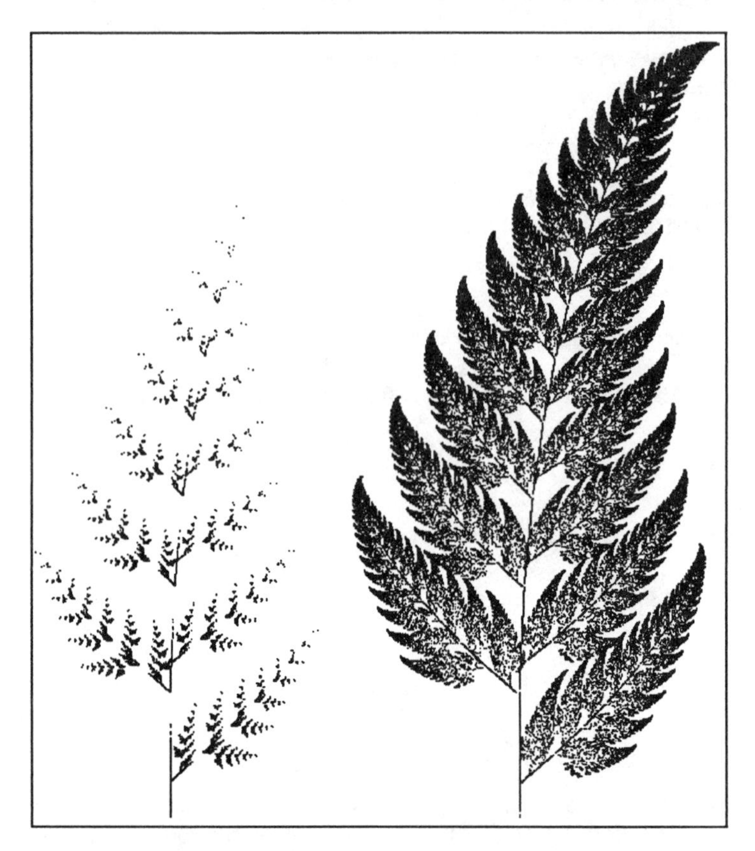

Figure 6.5 : 100,000 game points of the chaos game. Left: FRCM with equal probability for all contractions. Right: Tuned FRCM. Here the probabilities for choosing the different transformations are not the same.

A New Approach to the Decoding Problem

In other words, the chaos game provides a new approach to the problem of decoding images from a set of transformations. Let us recall the problem of computational complexity which occurred when we tried to obtain Barnsley's fern by straightforward IFS iteration. We estimated in chapter 5 that we would need about 10^{11} years of computer time for a computer which calculates and draws about a million rectangles per second. If we switch to a chaos game interpretation of the fern, the situation becomes rather different. Now we only have to keep track of one single point. This can be done easily by a computer even if we perform millions of iterations. So let us play the chaos game with the FRCM which is determined by the four transformations $w_1, ..., w_4$ which generate the fern. We assume equal probability for all transformations (as we did in our first chaos game). We start with a point

z_0, choose at random a transformation — say w_2 — and apply it to z_0. Then we continue with our new game point $z_1 = w_2(z_0)$ and choose another transformation at random, etc. The left part of figure 6.5 shows the disappointing result after more than 100,000 iterations. Indeed, the incompleteness of this image corresponds to the difficulty in obtaining the fern image by running the MRCM. Playing this chaos game for even millions of iterations, we would not obtain a satisfying result.

Now you certainly will wonder how we obtained the right hand image of figure 6.5. It is also produced by playing the chaos game, and it shows only about 100,000 iterations. What is the difference? Well, with respect to this image we could say we used a 'tuned' fortune wheel where we did not use equal probabilities for all transformations but made an appropriate choice for the probability to use a certain transformation.[3] The satisfactory quality of the right image is a very convincing proof of the potential power of the chaos game as a decoding scheme for IFS encoded images. But how does one select the probabilities, and why does a careful choice of the probabilities speed up the decoding process from 10^{11} years to a few seconds? And why does the chaos game work at all?

Chaos Game for Networked IFSs

We can play the chaos game also networked MRCMs, i.e. hierarchical IFSs. From a formal point of view a hierarchical IFS is given by a matrix Hutchinson operator

$$\mathbf{W} = \begin{pmatrix} W_{11} & \dots & W_{1M} \\ \vdots & & \vdots \\ W_{M1} & \dots & W_{MM} \end{pmatrix},$$

which operates on M planes, where each W_{ik} is a Hutchinson operator mapping subsets from the kth to the ith plane.[4] It is important to allow that some of the $W_{i,k}$ are the $\emptyset$-operator, i.e., the operator which maps any set into the empty set $\emptyset$. Let us recall, how the chaos game works for an ordinary Hutchinson operator given by N contractions $w_1, ..., w_N$. We need probabilities $p_1, ..., p_N$ and an initial point, say x_0. Then we generate a sequence $x_0, x_1, x_2, ...$ by computing

$$x_{n+1} = w_{i_n}(x_n), \quad n = 0, 1, 2, ...$$

where $i_n = m \in \{1, ..., N\}$ is chosen randomly with probability p_m. The chaos game for a matrix Hutchinson operator generates a sequence of *vectors* $\mathbf{X}_0, \mathbf{X}_1, \mathbf{X}_2, ...$, the components of which are subsets of the plane. Here $\mathbf{X}_{n+1}$ is obtained from $\mathbf{X}_n$ by applying

[3] We will present details of tuned fortune wheels in section 6.3.

[4] See the technical section on page 310.

randomly selected contractions from $\mathbf{W}$ to the components of $\mathbf{X}_n$. To avoid writing many indices we describe one single generation step using the notation $\mathbf{X} = \mathbf{X}_n$ and $\mathbf{Y} = \mathbf{X}_{n+1}$. The components of these two vectors are denoted by $x_1, ..., x_M$ and $y_1, ..., y_M$. The random selection of contractions from $\mathbf{W}$ is best described in two steps. For each row i in $\mathbf{W}$ we make two random choices:

Step 1: Choose a Hutchinson operator in the i-th row of $\mathbf{W}$ randomly, W_{ik}. (It must not be the $\emptyset$-operator.) Assume that this operator is given by N contractions $w_1, ..., w_N$.

Step 2: Choose a contraction from these $w_1, ..., w_N$ at random, say w_m.

Then to determine the ith component y_i of $\mathbf{Y}$ we apply the random choice w_m to the kth component x_k of $\mathbf{X}$ (w_m is part of the Hutchinson operator W_{ik}). That is, we compute $y_i = w_m(x_k)$. Again, to obtain the components of $\mathbf{X}_{n+1}$ the contractions are applied to the components of $\mathbf{X}_n$ according to the column index of their Hutchinson operator in $\mathbf{W}$.

The random choices are governed by probabilities. The way we have set up the random iteration it is natural to associate probabilities for both of the above steps. For step 1 we pick probabilities P_{ik} for each Hutchinson operator in $\mathbf{W}$ so that the sum of each row i is 1,

$$P_{i1} + ... + P_{iM} = 1, \quad i = 1, ..., M$$

where

$$P_{ik} = 0 \text{ if } W_{ik} = \emptyset \ .$$

This ensures that the $\emptyset$-operators are never chosen. Now assume that W_{ik} is given by the contractions $w_1, ..., w_N$. We pick probabilities $p_1, ..., p_N$ for each w_j such that $p_j > 0$ and $p_1 + \cdots + p_N = 1$. Now the probability of choosing w_m is p_m, provided the Hutchinson operator W_{ik} has already been selected in step 1.

Why Does the Chaos Game Work?

But before looking at the issue of efficiency, we need to discuss why the chaos game fills out the attractor of an IFS in the first place. From chapter 5 it is clear that starting the IFS iteration with any initial image A_0 we obtain a sequence $A_1, A_2, ...$ of images that converge to the attractor image A_∞. Without any loss we may pick just a single point as the initial image, say $A_0 = \{z_0\}$. Assume that the IFS is given by N affine transformations. Then the result after the first iteration is an image consisting of N points, namely

$$A_1 = \{w_1(z_0), w_2(z_0), ..., w_N(z_0)\} \ .$$

After the second iteration we get N^2 points, and so on. Of course, these points will get arbitrarily close to the attractor and eventually provide an accurate approximation of the whole attractor.

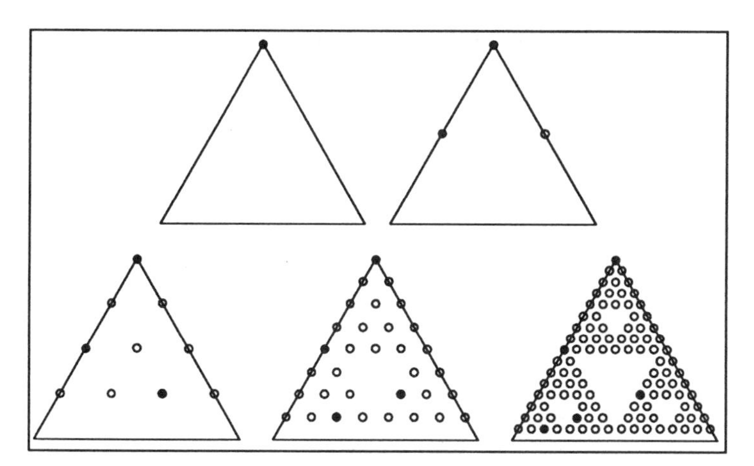

Figure 6.6 : The first five iterations of the MRCM for the Sierpinski gasket starting out from a single point (the top vertex of the triangle). The points generated by the chaos game starting out from the same initial point are shown in solid black.

FRCM Versus MRCM

Playing the chaos game with the same initial point z_0 is very similar to this procedure. It produces a sequence of points $z_1, z_2, \ldots$ where the kth point z_k belong to the kth image A_k obtained from the IFS. Thus, the points z_k get closer and closer to the attractor in the process. If the initial point z_0 is already a point of the final image, then so are all the generated points. It is very easy to name some points which must be in the attractor, namely the fixed points of the affine transformations involved. Note that z_0 is such a fixed point if $z_0 = w_k(z_0)$ for some $k = 1, \ldots, N$.[5] This explains the spurious dots in the Sierpinski gasket back in figure 6.3. There the initial point was not chosen to be already a point in the gasket. Thus, the first iterations of the chaos game produces points which come closer to the gasket but still are visibly different. The difference diminishes after only a few iterations, of course.

To fully understand the success of the chaos game, it remains to be shown in the next section that the sequence of generated points comes arbitrarily close to *any* point in the attractor.

[5]Compare the section on affine transformations on page 260.

6.2 Addresses: Analysis of the Chaos Game

To analyze the chaos game we need a suitable formal framework which allows us to precisely specify the points of the attractor of an IFS and also the positions of the moving game point. This framework consists of a particular addressing scheme which we will develop using the example of the Sierpinski gasket.

The Metric System as an IFS

The basic idea of such an addressing system is in fact some thousand years old. The decimal number system with the concept of place values explains well the way we to look at addresses and the idea behind the chaos game. Let us look at the decimal system in a more materialistic form: a meter stick subdivided into decimeters, centimeters, and millimeters. When specifying a three digit number like 357 we refer to the 357th out of 1,000 mm. Reading the digits from left to right amounts to following a decimal tree, see figure 6.7, and arriving at location 357 in three steps.

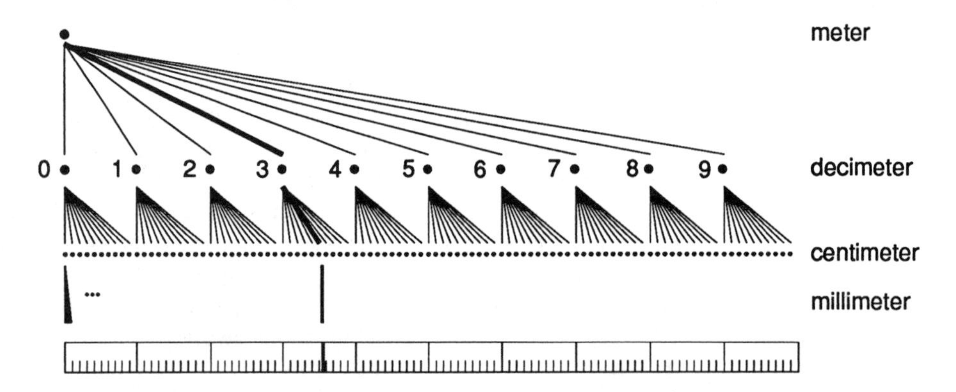

Figure 6.7 : Locating 357 by the decimal tree on a meter stick.

It is crucial for our discussion of the chaos game that there is another way to arrive at location 357 reading the digits *from right to left*. This makes us familiar with the *decimal MRCM*. The decimal MRCM is an IFS consisting of ten contractions (similarity transformations) $w_0, w_1, ..., w_9$ given explicitly by

$$w_k(x) = \frac{x}{10} + \frac{k}{10}, \quad k = 0, 1, ..., 9 .$$

In other words w_k reduces the meter stick to the kth decimeter. Running the decimal MRCM establishes the familiar metric system on the meter stick.[6]

[6]Here is an exercise: can one also construct an MRCM for the British/American system relating miles to feet and inches?

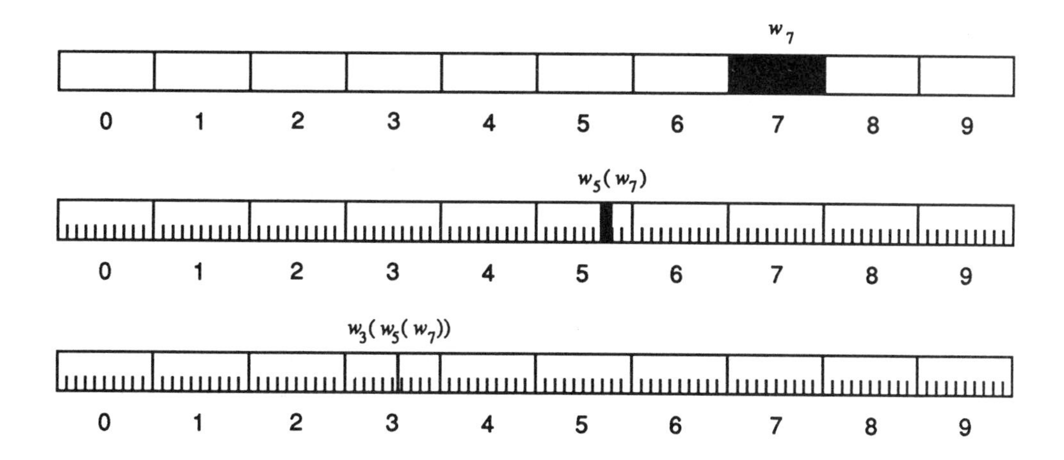

Figure 6.8 : Locating 357 by applying contractions of the decimal MRCM.

Start with the 1 meter unit. The first step of the decimal MRCM generates all the decimeter units of the meter stick. The second step generates all the centimeters, and so on. In this sense the decimal — together with its ancient relatives like the hexagesimal — system is probably the oldest MRCM.

Let us now read 357 from right to left by interpreting digits as contractions. Thus, starting with a 1 meter unit, we first apply the transformation w_7, which leads to the decimeter unit starting at 7 (see figure 6.8). Next we apply w_5 and arrive at the 57th centimeter. Finally w_3 brings us to the 357th millimeter location again. Thus, reading from left to right and interpreting in terms of place values, or reading from right to left and interpreting in terms of decimal contractions is the same.

The Chaos Game on the Meter Stick

We now play the chaos game on the meter stick. We generate a random sequence of digits from $\{0, ..., 9\}$, start with an arbitrary game point (= a millimeter location) and move to a new location according to the random sequence. We would consider the chaos game to be successful if it eventually visits all millimeter locations. Let us look at a random sequence like

...765016357

where we write from right to left for convenience. After the third step in the game we arrive at the millimeter location 357. The next random number is 6. Which millimeter location is visited next? Clearly 635! The initial number 7 is therefore irrelevant; no matter what this number is we visit the millimeter site 635 in the fourth step. For the same reason we continue to visit 163, then 016, and

The Decimal Chaos Game

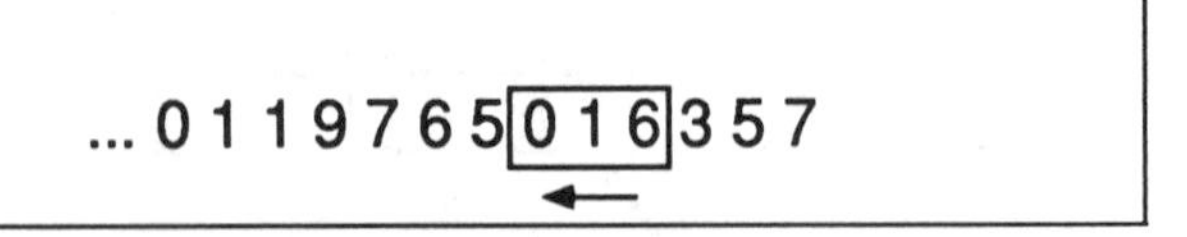

Figure 6.9 : A three-digit window sliding over the sequence ...0119765016357 allows to address millimeter locations.

so on. In other words, running the chaos game amounts to moving a slider with a window three digits wide from right to left over the random sequence.

When will we have visited all millimeter locations? This is obviously the case, when the slider window will have shown us all possible three-digit combinations. Is that likely to happen, if we produce the digits by a random number generator? The answer is 'yes', because that is one of the fundamental features which are designed into random number generators on computers. It is just a lazy way to generate all possible three-digit addresses. Even a random number generator which is miserable with respect to the usual statistical tests would do the job provided it generates all three-digit combinations.[7]

Let us now see how the same idea works for the Sierpinski gasket, the fern, and in general. We know that there is a very definite hierarchy in the Sierpinski gasket. At the highest level (level 0) there is one triangle. At the next level (level 1) there are three. At the level 2 there are nine. Then there are 27, 81, 243, and so on. Altogether, there are 3^k triangles at the kth level. Each of them is a scaled down version of the entire Sierpinski gasket, where the scaling factor is $1/2^k$ (refer to figure 2.16).

Triangle Addresses

We need a labeling or addressing scheme for all these small triangles in all generations. The concept for this purpose is similar to the construction of names in some Germanic languages as for example in Helga and Helgason, John and Johnson, or Nils and Nilsen. We will use numbers as labels instead of names:

[7]Barnsley explains the success of the chaos game by referring to results in ergodic theory (M. F. Barnsley, *Fractals Everywhere*, Academic Press, 1988). This is mathematically correct but practically useless. There are two questions: One is, why does the properly tuned chaos game produce an image on a computer screen so efficiently? The other is, why does the chaos game generate sequences which fill out the IFS attractor densely? These are not the same questions! The ergodic theory explains only the latter, while it cannot rule out that it may take some 10^{11} years for the image to appear. In fact, this could actually happen, if computers lasted that long.

level 1	level 2	level 3
1	11	111 112 113
	12	121 122 123
	13	131 132 133
2	21	211 212 213
	22	221 222 223
	23	231 232 233
3	31	311 312 313
	32	321 322 323
	33	331 332 333

Unfortunately, we run out of space very rapidly when we try to list the labels at more than a very few levels. The system, however, should be apparent. It is a labeling system using lexicographic order much like that in a telephone book, or like a place value in a number system. The labels 1, 2, and 3 can be interpreted in terms of the hierarchy of triangles or in terms of the hierarchy of a tree, (see figure 6.10). For triangles:

- 1 means lower left triangle;
- 2 means lower right triangle;
- 3 means upper triangle.

The Triangle With Address 13213

Thus, the address 13213 means that the triangle we are looking for is in the 5th level. The address 13213 tells us exactly where to find it. Let us now read the address. We read it from left to right much like a decimal number. That is, the places in a decimal number correspond here to the levels of the construction process. Start at the lower left triangle of the first level. Within that find the upper triangle from the second level. Therein locate the lower right triangle from the third level. We are now in the sub-subtriangle with address 132 (see figure 6.11). Within that take the lower left triangle from the fourth level, and finally therein come to rest at

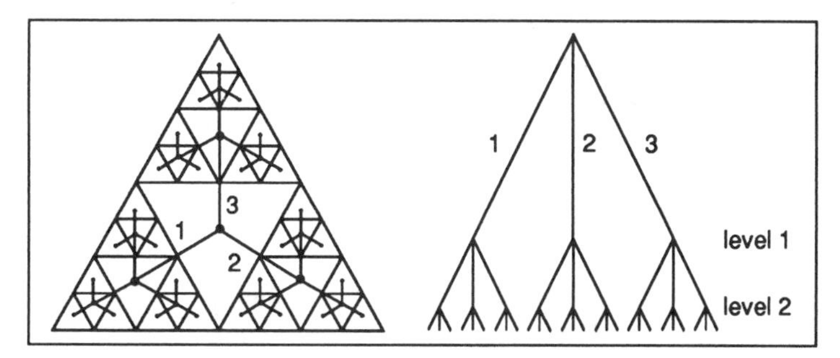

Figure 6.10 : Sierpinski tree (left), symbolic tree (right).

Locating Addresses

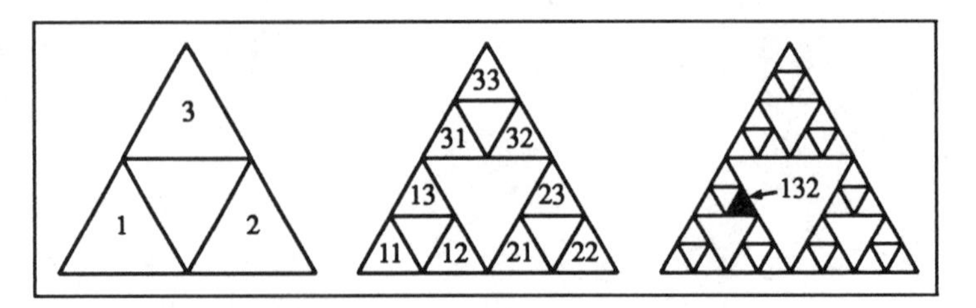

Figure 6.11 : Locating the subtriangle with address 132 in the Sierpinski gasket by a nested sequence of subtriangles.

the upper triangle. In other words, we just follow the branches of the Sierpinski tree in figure 6.10 five levels down.

Let us summarize and formalize. An address of a subtriangle is a string of integers $s_1 s_2 \ldots s_k$ where each s_i is a digit from the set $\{1,2,3\}$. The index k can be as large as we like. It identifies the level in the construction of the Sierpinski gasket. The size of a triangle decreases by 1/2 from level to level, at the kth level it is $1/2^k$.

Address of a Point

Let us now pick a *point* in the Sierpinski gasket. How can we specify an address in this case? The answer is that we have to carry on the addressing scheme for subtriangles ad infinitum specifying smaller and smaller subtriangles all of which contain the given point. Thus, we can identify any given point z by a sequence of triangles, $D_0, D_1, D_2, \ldots$ There is one triangle from each level such that D_{k+1} is a subtriangle of D_k and z is in D_k for all $k = 0, 1, 2, \ldots$ That sequence of triangles determines a sequence of integers $s_1, s_2, \ldots$

$$\begin{aligned}
\text{address}(D_1) &= s_1 \\
\text{address}(D_2) &= s_1 s_2 \\
\text{address}(D_3) &= s_1 s_2 s_3 \\
&\vdots
\end{aligned}$$

Selecting more and more terms in that sequence means locating z in smaller and smaller triangles (i.e. with more and more precision). This is just like fixing a location on a meter stick in higher and higher precision. Therefore, taking infinitely many terms identifies z exactly:

$$\text{address}(z) = s_1 s_2 s_3 \ldots \tag{6.1}$$

Reading Left to Right

It is important to remember how to read the address. The address is read from left to right and is thus interpreted as a nested sequence of triangles. The position of the digit in the sequence determines the level in the construction.

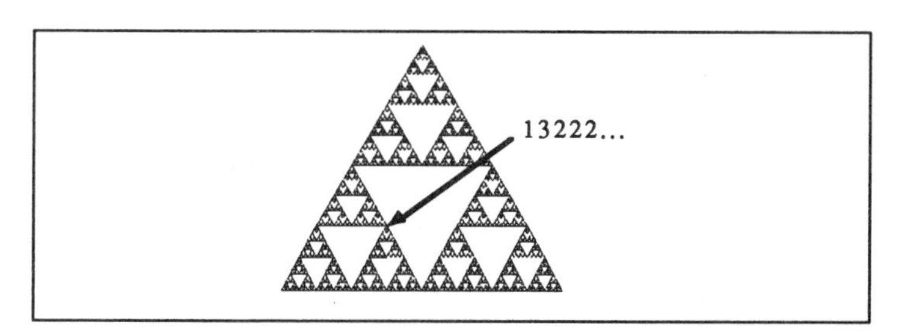

Figure 6.12 : The point address 13222... Note that this point could also be identified by address 12333...

Address of a Point

Touching Points in the Sierpinski Gasket

We must point out, though, that our addressing system for the points of the Sierpinski gasket does not always yield unique addresses. This means that there are points with two different possible addresses much like $0.49\overline{9}$ and 0.5 in the decimal system. Let us explore this fact. When constructing a Sierpinski gasket, we observe that in the first step there are three triangles, and any two of these meet at a point. In the next step there are nine triangles, and any adjacent pair of these nine triangles meet at a point. What are the addresses of the points were subtriangles meet? Let us check one example (see figure 6.12). The point where the triangles with label 1 and 3 meet has the addresses: 1333... and also 3111... Likewise, the point where the triangles with labels 13 and 12 meet has two different addresses: namely, 13222... and 12333... As a general rule all points where two triangles meet must have addresses of the form

$$\text{address}(z) = s_1 \ldots s_k r_1 r_2 r_2 r_2 \ldots$$

or

$$\text{address}(z) = s_1 \ldots s_k r_2 r_1 r_1 r_1 \ldots$$

where s_i, r_1, r_2 are from the set $\{1, 2, 3\}$ and r_1 and r_2 are different. Points which have that nature are called *touching points*. They are characterized by twin addresses (compare figure 6.13).

From the construction process of the Sierpinski gasket one might be misled to conjecture that, except for the three outside corner points, all its points are touching points. This conjecture is wrong; and here, using the language of addresses, is a nice argument which makes the issue clear. If all points of the Sierpinski gasket were touching points they could be characterized by twin addresses of the above form. But obviously most addresses we can imagine are not of this particular form (e.g. $\text{address}(z) = s_1, s_2, s_3, \ldots$ where each s_i is randomly chosen). In other words most points are not touching points.

Touching Points

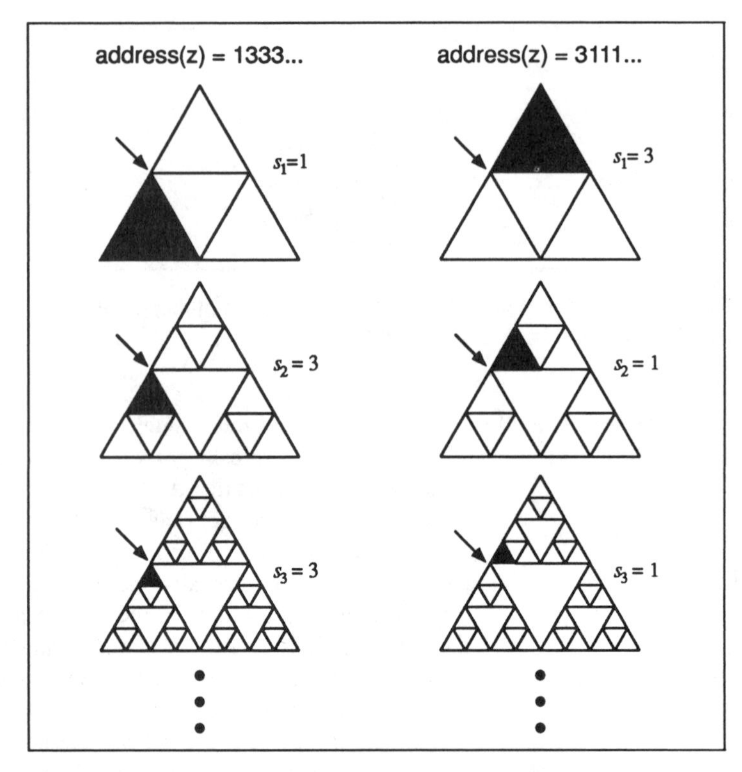

Figure 6.13 : The touching point with twin addresses 1333... and 3111...

Space of Addresses

Let us now develop the formalism of addresses a little more. To this end we introduce a new object, $\sum_3$, the space of addresses. An element s from that space is an infinite sequence $\sigma = s_1 s_2 ...$, where each s_i is from the set $\{1, 2, 3\}$. Each element σ from that space identifies a point z in the Sierpinski gasket. However, different elements in $\sum_3$ may correspond to the same point, namely, the touching points.

Addresses for the Cantor Set

At this point let us see how the concept of addresses works in another example of fractals, the Cantor set C. Here we would address with only two labels, 1 and 2. All infinite strings of 1's and 2's together are the address space $\sum_2$. There is a significant difference when we compare the Cantor set with the Sierpinski gasket. Points on the Cantor set have only one address. We say that addresses for the Cantor set are *unique*. Thus, for each address there is exactly one point, and vice versa. In other words, $\sum_2$ is in a one-to-one correspondence with C: we can identify $\sum_2$ and C. In the case of $\sum_3$ and the Sierpinski gasket, this is not possible because there exist points with two different addresses.

Cantor Set Addresses

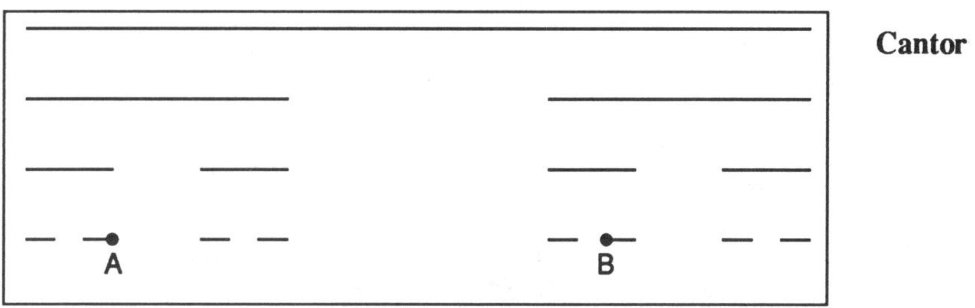

Figure 6.14 : Addresses for the Cantor set. Point A has address 11222... and the address of point B is 212111...

Addresses for IFS Attractors

Any fractal which is the attractor of an IFS has a space of addresses attached to it. More precisely, if the IFS is given by N contractions $w_1, ..., w_N$, then any z from A_∞, the attractor, has an address in the space $\sum_N$, which is the space of all infinite sequences $s_1s_2s_3...$, where each number s_i is from the set $\{1, 2, ..., N\}$.

To be specific, let us take any point z in the attractor A_∞ of the IFS, which is given by the set of N contractions $w_1, ..., w_N$. These contractions applied to A_∞ yield a covering of the attractor as explained in chapter 5,

$$A_\infty = w_1(A_\infty) \cup \cdots \cup w_N(A_\infty) .$$

Our given point z surely resides in at least one of these sets, say in $w_k(A_\infty)$. This determines the first part of the address of z, namely $s_1 = k$. The set $w_k(A_\infty)$ is further subdivided into N (not necessarily disjoint) subsets

$$\begin{aligned} w_k(A_\infty) &= w_k(w_1(A_\infty) \cup \cdots \cup w_N(A_\infty)) \\ &= w_k(w_1(A_\infty)) \cup \cdots \cup w_k(w_N(A_\infty)) . \end{aligned}$$

Again our given point surely is in at least one of these subsets, say $w_k(w_l(A_\infty))$, and this determines the second part of the address of z, namely $s_2 = l$. Note that there may be several choices for s_2, in which case we would have several different addresses for one point. This procedure can be carried on indefinitely. Computing more and more components of the address specifies the point z more and more precisely because the subsets considered get smaller and smaller due to the contraction property of the affine transformations of the IFS.

As in the case of the Sierpinski gasket, we get a sequence of nested subsets D_k of increasing level of the attractor, all of which contain the given point. If $\sigma = s_1s_2...$ denotes an address of z these subsets are

$$D_k = w_{s_1}(w_{s_2}(\cdots w_{s_k}(A_\infty))) .$$

Address Interpretation

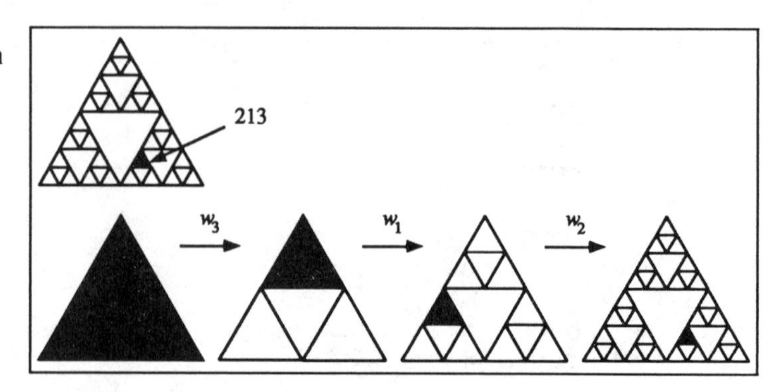

Figure 6.15 : Reading the address backwards when applying the contractions.

For brievety of notation we often omit the brackets. So

$$w_k w_l(A_\infty) = w_k(w_l(A_\infty))$$

and

$$w_{s_1} w_{s_2} \cdots w_{s_k}(A_\infty) = w_{s_1}(w_{s_2}(\cdots w_{s_k}(A_\infty))) \, .$$

Reading Right to Left

It is important to note here that in a sense we have now read the sequence $s_1 s_2 ... s_k$ from right to left because we first apply w_{s_k} to A_∞. Then to the result of that, we apply $w_{s_{k-1}}$, and so on, until we finally apply w_{s_1}. Let us look at an example. In figure 6.15 we show how the subtriangle with address 213 is obtained by $w_2(w_1(w_3(S)))$. In words, this is 'the top (3) of the left (1) of the right (2) subtriangle' (note the correspondence in the order). This simple observation that addresses may be interpreted from left to right or from right to left in different operational ways leading however to the same location is very crucial for understanding why the chaos game works.

One-to-One or Many-to-One

Assume there is a one-to-one correspondence between points in A_∞ and $\sum_N$. In other words, there is a unique address for each point z in the attractor A_∞. Then we call the attractor *totally disconnected.* Figure 6.16 shows the attractors of three IFSs, one totally disconnected, one just touching, and the third with overlap. In the overlapping case it is hard to see an address for a given point by visual inspection. However, it is always easy to compute the corresponding point for a given address.

So far we have developed a labelling scheme for points of the IFS attractor. Using this technique we now discuss why the chaos game works and how it can generate the attractor. We will once again discuss the basic ideas for the Sierpinski gasket and then see how and what we can extend to the general case.

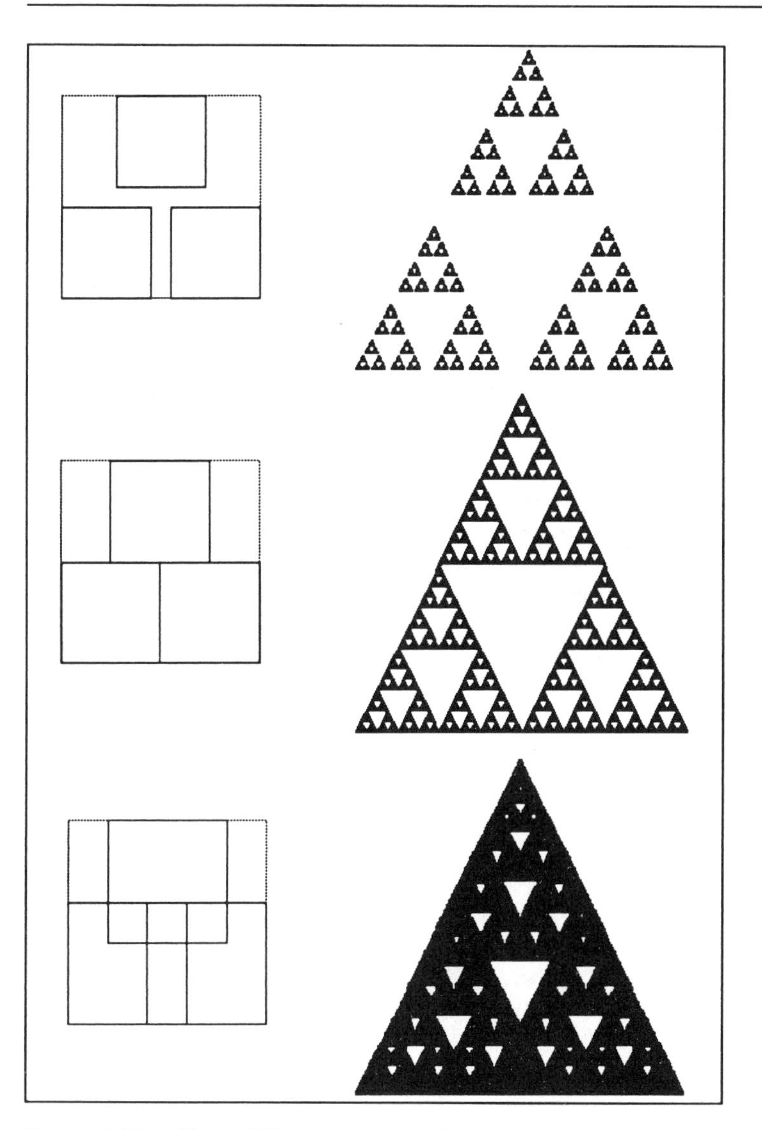

Three Cases

Figure 6.16 : Three IFS attractors, totally disconnected, just touching, and overlapping (MRCM blueprints appear reduced).

Chaos Game With Equal Probabilities

Let us begin with some straightforward observations. In the ideal case our die will be perfect. Any of the numbers 1, 2, or 3 will appear with the same statistical frequency. If p_k denotes the probability of the event of our throwing number k, $k = 1,2,3$, then $p_1 = p_2 = p_3 = 1/3$.

Let us now play the chaos game with such a perfect die. We assume that the actual game point z_n is in the Sierpinski gasket, but we do not know where. The Sierpinski gasket can be broken down

Probabilities

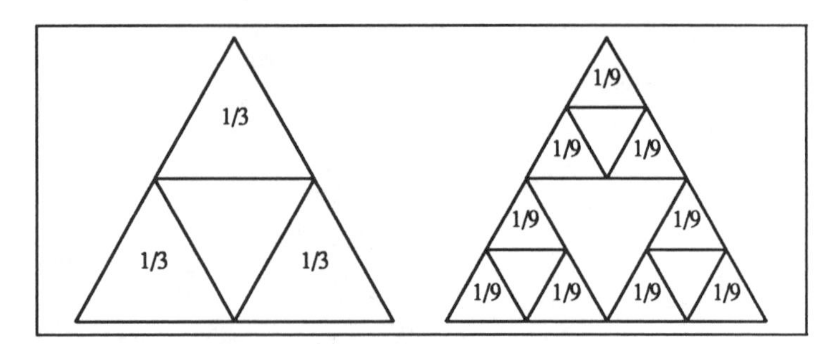

Figure 6.17 : The probability that a game point lands in a selected set of the first level after one iteration is 1/3. For a set of the second level after two iterations the probability is 1/9.

into three sets in the first level, nine in the second, and 3^k in the kth level. Let us pick one of these sets, for example one from the first level. What is the probability that we will see the next game point z_{n+1} in this subtriangle? Obviously, the probability will be 1/3, no matter where z_n, and for that same reason $z_{n-1}, z_{n-2}, \dots$ are.

Now let us pick a set from the second level. Again, we assume no information about the location of z_n other than its being in the Sierpinski gasket. Therefore, if we want to see a succeeding game point on a selected set D of the second level, we should generate two new game points z_{n+1}, and z_{n+2}. What is the probability that we will see z_{n+2} in D? Obviously it is 1/9. In other words, if we pick a set D from the kth level, then the probability that the chaos game will produce a game point z_{n+k} which lands in D after k iterations is $1/3^k$.

Let us repeat in terms of the contractions w_1, w_2, and w_3. Each of the contractions w_1, w_2, and w_3 is drawn with probability 1/3. Consequently, each pair $w_i w_k$ is drawn with probability 1/9, and in general each of the 3^k possible compositions $w_{s_1} w_{s_2} \cdots w_{s_k}$ with s_i from the set $\{1, 2, 3\}$, is selected with probability $1/3^k$.

We can now explain why the chaos game produces a sequence of game points which will eventually fill out the entire Sierpinski gasket at any resolution. Mathematically, the chaos game produces a sequence

$$\begin{aligned} z_0 &= \text{starting point (initial game point)} \\ z_1 &= w_{s_1}(z_0) \\ z_2 &= w_{s_2} w_{s_1}(z_0) \\ &\vdots \\ z_k &= w_{s_k} \cdots w_{s_2} w_{s_1}(z_0) \; , \end{aligned}$$

where the sequence of events $s_1, s_2, \dots, s_{k-1}, s_k$ is randomly generated. The last point z_k is in a kth generation subtriangle D which has the address $s_k s_{k-1} ... s_2 s_1$.

The Game Points Get Close to any Point of the Gasket

Pick a test point P on the Sierpinski gasket. We need an argument which establishes that if we play the chaos game sufficiently long, we will produce points which come as close to P as we wish. To this end we want to see a game point which has, at most, a small distance ε from P. Let us assume that the diameter of the Sierpinski gasket is d. We know that the triangles in the mth level of the Sierpinski gasket have the diameter $d/3^m$. In other words, if we choose m so large that $d/3^m < \varepsilon$ and pick a triangle D in the mth generation which contains P, then every point in D has, at most, distance ε from P. This set D has an address

$$\text{address}(D) = t_1 t_2 ... t_m, \quad t_i \in \{1, 2, 3\} .$$

Now let us look at a run of the chaos game with many events. We like to write the sequence in reverse order, $..., s_k, s_{k-1}, ..., s_2, s_1$. As soon as we detect a block of length m within the sequence $..., s_k, s_{k-1}, ..., s_2, s_1$ which is identical to $t_1 t_2 ... t_m$, we are finished. For example let us take

$$..., s_k, ..., s_{j+m+1}, t_1, t_2, ..., t_m, s_j, ..., s_2, s_1 .$$

Then z_j is some point in the Sierpinski gasket, and therefore $w_{t_1} \cdots w_{t_m}(z_j)$ will be in D. Thus, everything is settled if we can trust to eventually seeing block $t_1 t_2 ... t_m$ as we play. But the probability that any sequence of length m matches up with $t_1 t_2 ... t_m$ is equal to $1/3^m$. Therefore, playing the chaos game with a perfect die will sooner or later produce such a sequence, and thus a point which is in the subtriangle D and therefore as close to the test point P as we required.

Chaos Game Generates IFS Attractor

The chaos game will produce points which densely cover the Sierpinski gasket. We can generalize this fact for an attractor of an arbitrary IFS. Let us briefly sketch the arguments for this general situation. Let the IFS be given by N contractions $w_1, ..., w_N$, and let A_∞ be its attractor. The attractor is invariant under the Hutchinson operator $H(X) = w_1(X) \cup \cdots \cup w_n(X)$. A corresponding random IFS is given by these contractions w_i and associated probabilities p_i (with $p_i > 0$ and $p_1 + \cdots + p_N = 1$). We need to show that we can get arbitrarily close to any point P of the attractor A_∞ by playing the chaos game with this set up. Now let us try to generate a point which lies within a distance ε from P. Let the address of P be given by

$$\text{address}(P) = t_1 t_2 ...$$

where $t_i \in \{1, ..., N\}$. Then the point P is contained in all sets A_m,

$$A_m = w_{t_1} w_{t_2} \cdots w_{t_m}(A), \quad m = 1, 2, ...$$

We have

$$A_1 \supset A_2 \supset A_3 \supset \cdots$$

and the diameter of A_m decreases to zero as m increases. Knowing the contraction ratios $c_1, ..., c_N$ of the transformations $w_1, ..., w_N$, these diameters can be estimated. By definition of the contraction ratios we have for any set B with diameter diam(B) that after transformation by w_i the diameter is reduced by the contraction ratio $c_i < 1$:

$$\text{diam}(w_i(B)) \leq c_i \text{diam}(B) .$$

Therefore, the diameter of A_m can be bounded:

$$\begin{aligned} \text{diam}(A_m) &= \text{diam}(w_{t_1} \cdots w_{t_m}(A_\infty)) \\ &\leq c_{t_1} c_{t_2} \cdots c_{t_{m-1}} c_{t_m} \text{diam}(A_\infty) . \end{aligned}$$

Since the contraction ratios are all less than 1, we can make this diameter as small as ε just by considering a sufficiently large number m of transformations in this sequence. Thus, all points with addresses starting with $t_1 ... t_m$ have a distance of at most ε from the given point P. In other words, we need to see this sequence $t_1, ..., t_m$ sometime during the chaos game. The chance that any given block of length m exhibits this sequence is equal to the product of probabilities $p_{t_1} p_{t_2} \cdots p_{t_m}$, a positive number. In other words, we can get from any point of A_∞ arbitrarily close to P when playing the chaos game sufficiently long.

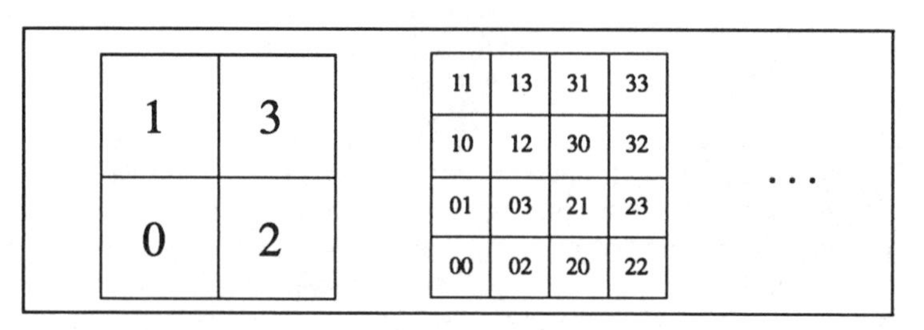

Figure 6.18 : Addresses for a rectangular array.

Chaos Game on a Computer Screen

So far we have discussed the chaos game from a mathematical point of view. Let us try to materialize the set up a bit (i.e., discuss the chaos game in a form that is closer to the situation encountered on a computer screen). The pixels of a computer screen form a rectangular array. Usually they are identified by coordinates (e.g. pixel number 5 in row number 12). But we can also use an address system as discussued in this chapter.

First we divide the screen into 4 equal quadrants and assign the addresses 0 to 3 as shown in figure 6.18. Next we divide each

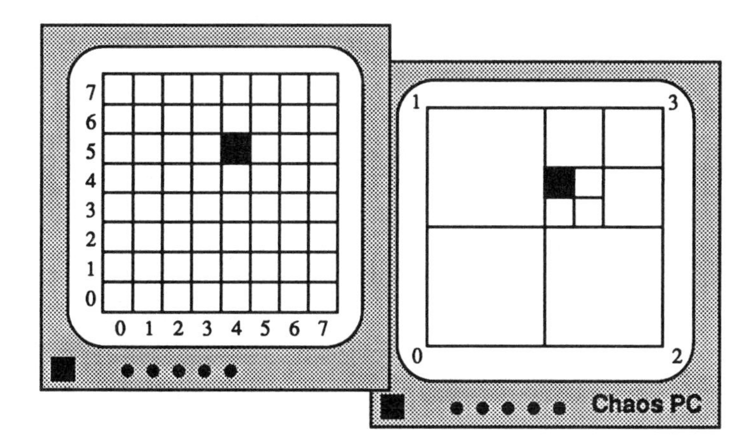

Figure 6.19 : Addressing pixels: pixel P at screen coordinate (4,5) has the address 301.

quadrant into four equal parts, each of which are then identified by 2-digit addresses. In this way we can describe the pixels of an 8 by 8 pixel screen by 3-digit addresses and a 16 by 16 pixel screen by 4-digit addresses (in general a 2^n by 2^n pixel screen by n-digit addresses).

Assume we work with an 8 by 8 screen. How do we find the pixel with address 301? We read the address from left to right and follow a nested sequence of squares which finally identifies the pixel at screen coordinates (4,5) — see figure 6.19. Now we set up four contractions w_0, w_1, w_2 and w_3 as in section 5.3 (i.e., the transformation w_i transforms the whole screen into the quadrant i) in correspondence to our address system.

For example, if the square Q represents the whole screen, the sequence

$$w_3(Q) \supset w_3(w_0(Q)) \supset w_3(w_0(w_1(Q)))$$

is just the sequence of nested rectangles enclosing our pixel P in figure 6.19.

In section 5.3 we have demonstrated what happens if we drop the contraction w_3 which transforms the whole screen into the upper right quadrant. As a result we obtain only those pixels that have an address which does not contain the digit 3. We have shown that the IFS associated with w_0, w_1 and w_2 generates a Sierpinski gasket as in figure 5.9 (or in this case its 8 by 8 pixel approximation). Let us now see how the chaos game played with the contractions w_0, w_1 and w_2 also generates exactly these pixels.

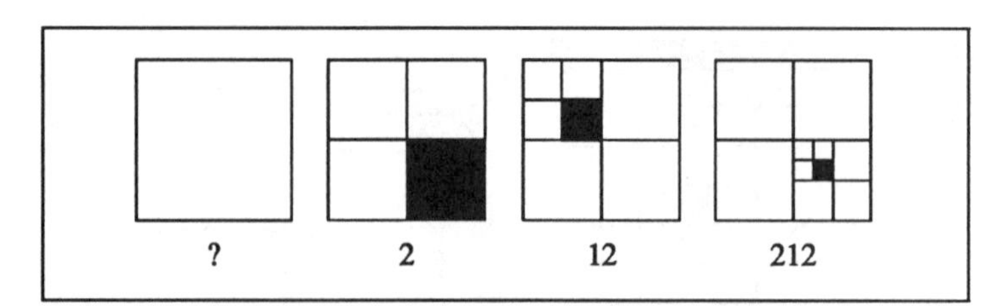

Figure 6.20 : First steps of the chaos game leading to pixel 212.

Again let us look at a concrete example of random numbers, say,

...01211210010212.

Starting with a game point anywhere on the screen, the first move brings us to quadrant 2, the second one into the square with address 12 and the third to pixel 212 (see figure 6.20). The next step would lead to address 0212 (a subrectangle of pixel 021) provided we had 4-digit addresses. But since we work with an 8 by 8 resolution and corresponding 3-digit addresses we can forget about the fourth digit in the address, the trailing 2. Next we would visit pixel 102, then 010, and so on. In other words, we slide a 3-digit window over our random sequence and switch on all pixels whose addresses appear.

In other words, in this set up the chaos game will be successful if the sequence of numbers which drives the game contains all possible 3-digit combinations of the numerals 0, 1 and 2. In fact, the drawing would not even have to be random. The only property which we need is that all possible addresses would appear. Moreover, the efficiency of the chaos game would be governed by how fast all possible combinations would be exhausted. That is the true secret of the chaos game. It has nothing to do with deep mathematical results like ergodic theory, as is sometimes argued in some research literature.[8]

[8]This was first observed by Gerald S. Goodman, see G. S. Goodman, *A probabilist looks at the chaos game*, FRACTAL 90 – Proceedings of the 1st IFIP Conference on Fractals, Lisbon, June 6–8, 1990 (H.-O. Peitgen, J. M. Henriques, L. F. Penedo, eds.), Elsevier, Amsterdam, 1991.

6.3 Tuning the Fortune Wheel

Our discussion of the Fortune Wheel Reduction Copy Machine has been based on the assumption that the die used for the chaos game is perfect. The probabilities for the transformations are the same. But what effect would a change of these probabilities have on the outcome of the chaos game?

How Many Hits Are In a Subtriangle?

Let us discuss this in a little more formal way for the triangles D in the mth level of the Sierpinski gasket. In other words, if we play n times and z_1 to z_n are the game points, the question is, how many points among $z_1, \ldots, z_n$ fall into D? Let us denote this count by $h(z_1, \ldots, z_n; D)$. When the die is perfect we correctly expect that in the long run, each of the small triangles in the mth level will be hit the same number of times. More precisely, the statement is that the fraction of points from $z_1, \ldots, z_n$ which are in D tends to $1/3^m$ as we consider more and more points, i.e. as the total number of points n increases. Note, that there are 3^m subtriangles of the mth generation all of which should be equally probable. Expressed in a formula,

$$\lim_{n \to \infty} \frac{h(z_1, \ldots, z_n; D)}{n} = \frac{1}{3^m} . \tag{6.2}$$

In other words, counting the events falling in D generates a measure $\mu(D)$, which is nothing but the probability which we have attached to D in our earlier discussion.

The following table lists the counts of points falling in each subtriangle of the second generation when 1000 points are generated as in figure 6.21. In the long run each subtriangle should collect 11.1% of all points.

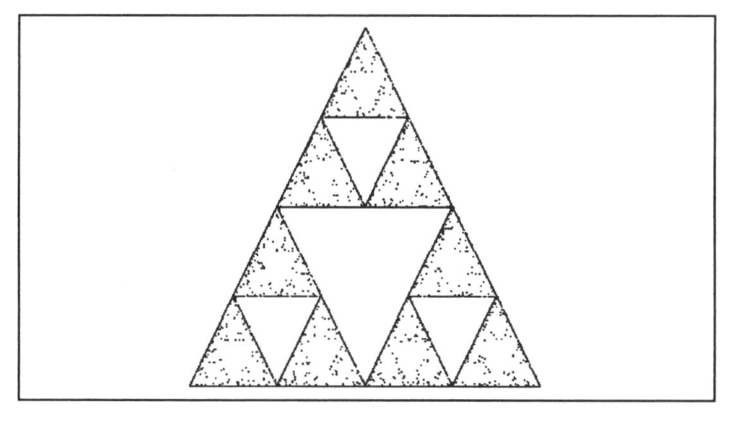

1000 Points with Perfect Die

Figure 6.21 : 1000 game points, when all transformations are chosen with equal probability.

address	count	in %
11	103	10.3
12	122	12.2
13	105	10.5
21	107	10.7
22	112	11.2
23	117	11.7
31	108	10.8
32	108	10.8
33	118	11.8

Now let us change the situation slightly. Let us assume that our die is biased. This means that the probabilities p_1 for number 1 to come up, or p_2 for number 2 to appear, or p_3 for number 3 to be shown are no longer the same. In other words, we have that $p_1 + p_2 + p_3 = 1$, and for example, $p_1 = 0.5$, $p_2 = 0.3$, and $p_3 = 0.2$. Before we discuss what will happen to the chaos game under these circumstances, we want to explain how we can produce a die with exactly this bias.

Simulating Loaded Dice

Naturally we will simulate the rolling of a die with a computer by calling random numbers from a *random number generator*. Usually, whatever the actual algorithm may be, random numbers which are supplied in a programming environment are normalized (i.e., take values between 0 and 1) and are uniformly distributed. Uniform distribution means that the probability of producing a random number in a small interval $[a, b]$ with $0 \leq a < b \leq 1$ is equal to $b - a$. Thus, if the interval $[0, 1]$ is broken down into say 100 subintervals $[0.00, 0.01]$, $[0.01, 0.02]$, etc. we can expect in the long run that in each subinterval we will collect about 1% of all generated random numbers.

Tuning of Random Number Generators

Let us formalize the exploration of random number generators. The random number generator returns a sequence of numbers $r_1, r_2, ...$ from the interval $[0, 1]$ which we divide into N subintervals

$$[0, 1] = [x_0, x_1) \cup [x_1, x_2) \cup \cdots \cup [x_{N-1}, x_N]$$

where

$$0 = x_0 < x_1 < \cdots < x_{N-1} < x_N = 1 \ .$$

Let us denote by I_k the kth such interval. After n calls for random numbers, we have obtained $r_1, ..., r_n$ and count the number of results r_i falling in the subinterval I_k. Let us call the result of this count $h(r_1, ..., r_n; I_k)$. For a good random number generator, we expect that $h(r_1, ..., r_n; I_k)$ will only depend on the length of I_k and is in fact equal to this length. In terms of a formula, we expect that

$$\lim_{n \to \infty} \frac{h(r_1, ..., r_n; I_k)}{n} = \text{length } (I_k) = x_k - x_{k-1} \ .$$

It is interesting to note that we can turn this relation around to *compute* the length of the interval by counting random numbers! Let us remark in passing that a whole class of methods for the numerical computation for many different types of problems have been based on a similar use of random numbers. For apparent reasons these methods are called *Monte Carlo methods.*

For example, in the year 1777 Georges L. L. Comte de Buffon (1707–1788) suggested computing the probability that a needle, dropped on a page of ruled paper, will intersect one of the lines. He solved the problem, and the answer turned out to reveal a relation to the number $\pi = 3.141592...$ Assuming that the distance d of the parallel lines is greater than the length l of the needle, it is not hard to show that the probability P the needle will hit one of the lines is equal to $2l/d\pi$. Later Pierre Simon Laplace (1749–1827) interpreted this relation as an entirely new way of computing π. Just throw the needle many times and count the intersections. This count (divided by the total number of throws) approximates this probability P, and thus, facilitates the computation of $\pi = 2l/dP$.[9]

But let us return to the tuning of the chaos game. We obtain an arbitrarily tuned random number generator (a die with N faces and N prescribed corresponding probabilities) in the following way: if we want probabilities $p_k, k = 1, ..., N$, we define

$$
\begin{aligned}
I_1 &= [0, p_1) \\
I_2 &= [p_1, p_1 + p_2) \\
&\vdots \\
I_k &= [p_1 + p_2 + \cdots + p_{k-1}, p_1 + p_2 + \cdots + p_k) \\
&\vdots \\
I_N &= [p_1 + p_2 + \cdots + p_{N-1}, 1]
\end{aligned}
$$

and choose event k provided the random number r_i is in the interval I_k.

With such a random number generator on hand it is easy to simulate a biased die. Given probabilities p_1, p_2, and p_3, one defines three intervals

$$I_1 = [0, p_1), \quad I_2 = [p_1, p_1 + p_2), \quad \text{and } I_3 = [p_1 + p_2, 1] \ .$$

Note that the length of I_k is equal to p_k. Therefore, if we choose number k whenever the random number falls into I_k, then event k will be drawn with probability p_k. For example, when $p_1 = 0.5$, $p_2 = 0.3$, and $p_3 = 0.2$, then

$$I_1 = [0, 0.5), \quad I_2 = [0.5, 0.8), \quad \text{and } I_3 = [0.8, 1] \ .$$

[9] Of course, this approach to the computation of π is rather inefficient. It can be shown that for example the probability of obtaining π correct to 5 decimal places in 3400 throws is less than 1.5%.

Points with Loaded Die

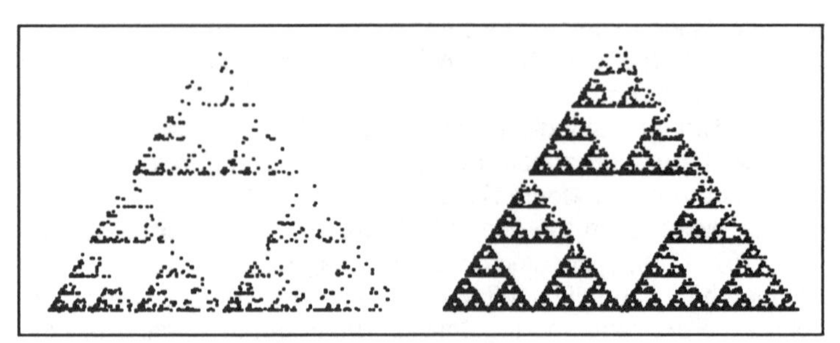

Figure 6.22 : 1000 (left) and 10,000 (right) game points, when w_1 is chosen with 50%, w_2 with 30%, and w_3 with 20% probability.

Rerunning the Chaos Game with a Biased Die

Let us now see how the chosen probabilities p_k effect the generation of game points. Figure 6.22 shows the result of plotting 1000 and 10,000 points. In the long run we again obtain a Sierpinski gasket. But we observe an obvious additional pattern in the distribution of game points. Again the entire picture is similar to what we see in the subtriangles. But the density of points in different subtriangles varies.

Let us try to estimate the probabilities by which we will see events falling within the triangles of the different levels of the Sierpinski gasket. The answer is simple for the three triangles which make the first level. Figure 6.23 illustrates the result.

But already for the nine triangles which make the second level, it is crucial that we recall how each of these triangles is obtained from the entire triangle by means of compositions of transformations of the form $w_{s_2}w_{s_1}$ with s_1, s_2 from the set $\{1, 2, 3\}$. This is exactly what the addresses tell us. Thus, if D is one of these triangles with address $s_2 s_1$, then the probability that we will see an event in D after two iterations is $p_{s_2}p_{s_1}$. Figure 6.24 illustrates

Probabilities Level 1

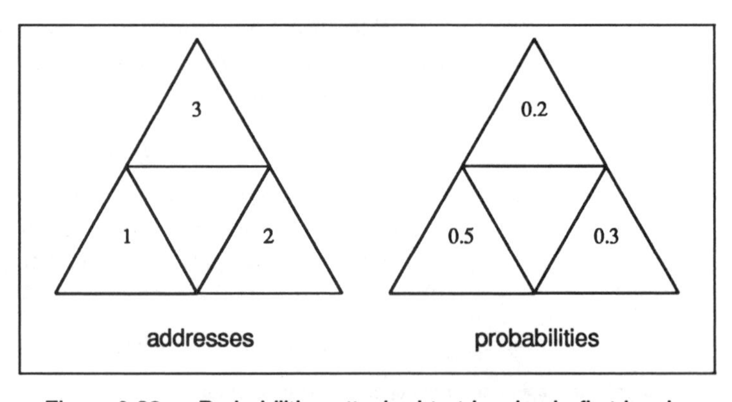

Figure 6.23 : Probabilities attached to triangles in first level.

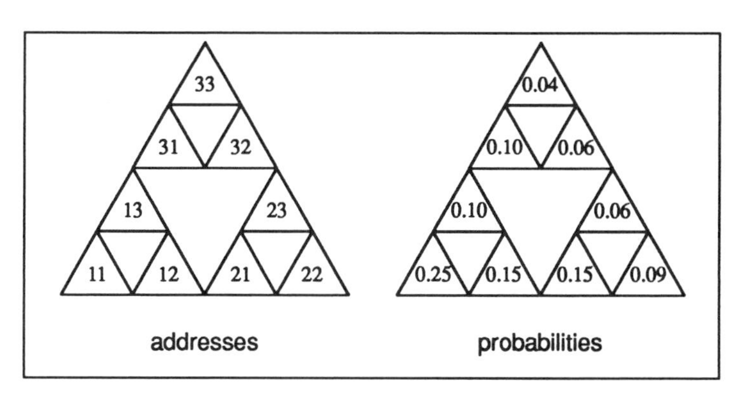

Figure 6.24 : Addresses and corresponding probabilities attached to triangles of second level.

Probabilities Level 2

the results.

The probabilities vary between 0.04 and 0.25. The lower left triangle is hit about six times more often than the top most triangle. We verify these estimates in the image which we obtained in figure 6.22 by counting points in the subtriangles as before. The last column in the following table lists the expected outcomes in percent.

address	count	in %	expected
11	238	23.8%	25%
12	139	13.9%	15%
13	108	10.8%	10%
21	146	14.6%	15%
22	91	9.1%	9%
23	64	6.4%	6%
31	101	10.1%	10%
32	72	7.2%	6%
33	41	4.1%	4%

Expressed as a general rule, we can say that when we pick a triangle D in the kth level which has

$$\text{address }(D) = s_1, s_2, ..., s_k$$

then the probability that this triangle will be hit after k iterations is the product $p_{s_1} \cdots p_{s_k}$.

We can actually test this as above by sampling the relative counts in each such triangle for all subtriangles of the kth generation, and expect to get

$$\lim_{n \to \infty} \frac{h(z_1, ..., z_n; D)}{n} = p_{s_1} \cdots p_{s_k} \; . \tag{6.3}$$

As before, $h(z_1, ..., z_n; D)$ denotes the number of hits among the first n game points $z_1, ..., z_n$ in D. In fact, this number is visualized in our chaos game images in the form of the density of the displayed points. This result leads to two major consequences:

(1) A strategy to design efficient decoding schemes for IFS codes.
(2) An extension of the concept of IFSs from an encoding of black-and-white images to an encoding of color images. This point will be discussed further below (see page 353).

Unbiased Dice Are Best for the Sierpinski Gasket ...

We have seen that even with a biased die, we will eventually generate the Sierpinski gasket. Depending on the chosen probabilities it may, however, take a very, very long time to see its final shape. According to eqn. (6.3), the number of relative hits in some parts of the Sierpinski gasket can be extremely small, though always greater than zero, while it will be very large in other parts. In other words, for reasons of efficiency, we should keep all probabilities the same for the generation of the Sierpinski gasket. But is that a general rule of thumb for all IFS attractors?

... But Not for the Fern

Let us recall the problems we had when we tried to generate the Barnsley fern by the chaos game. Namely, we were not able to see the final image when playing the game with equal probabilities for all transformations. To analyse the situation, we pick one of the tiny primary leaves T at the top of the fern, see figure 6.25. We can describe this leaf in terms of the contractions w_1 to w_4:

$$T = w_1 w_1 \cdots w_1 w_3(F) \ , \qquad (6.4)$$

where F denotes the entire fern, and the total number of transformations is k. Therefore, leaf T represents one of the sets of level k, with address

$$\text{address } (T) = 11...13$$

where 1 is repeated $k-1$ times. In other words, the probability q of seeing a game point in T after k iterations is just $q = p_1^{k-1} p_3$.

Now let us take k in the order of 15. Thus, T is the 15th leaf on the right side of the fern. Breaking down the fern F to the 15th level means that we have $4^{15} \approx 10^9$ sets, and T is just one of them. If we take uniform probabilities, $p_k = 0.25, k = 1,2,3,4$, then the probability to see a game point up there is $q = 0.25^{15} \approx 0.931 \cdot 10^{-9}$! Thus, for all practical purposes the probability is zero; and therefore the fern on the left of figure 6.5 is as incomplete as it is. In other words, this way of generating one or two hundred thousand game points to picture 10^9 sets is doomed to failure.

Changing the Probabilities

If we take, however, a relatively large probability for w_1 and small probabilities for w_2, w_3, and w_4, then we can arrange $q = p_1^{14} p_2$ as more appropriate. For example, with $p_1 = 0.85$ and $p_2 = 0.05$,

A Tiny Leaf on the Fern

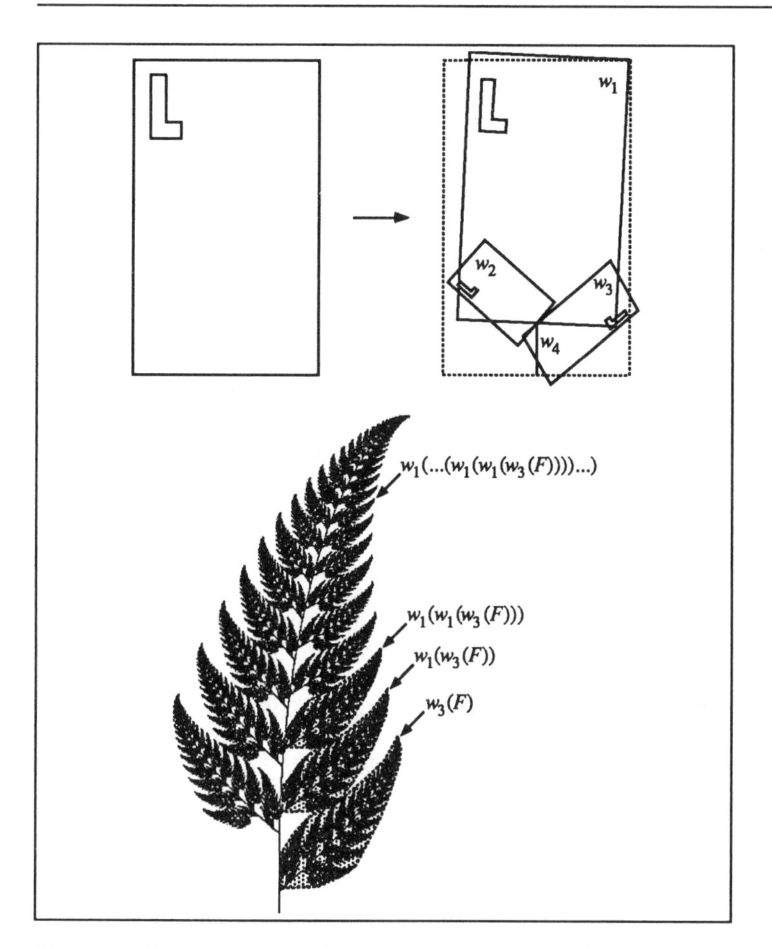

Figure 6.25 : Description of one of the tiny leaves of the Barnsley fern.

we estimate $q \approx 0.00514$. When the chaos game is played for only 10,000 iterations, we expect to already see about 50 points in the tiny leaf T. Thus, by modifying the probabilities we are able to push the likelihood of seeing a game point in T after k iterations from practically zero to a very realistic likelihood. In other words, the strongly biased die here creates a distribution of the 10^4 game points over the 10^9 sets, which is sufficiently efficient for the decoding of the fern image.

Recipe for Choosing the Probabilities

It is a difficult and still unsolved mathematical problem to determine the *best* choice for the probabilities p_i. The problem can be stated in the following way. Let ε be a prescribed precision of approximation in the sense that for every point of the attractor there is at least one point generated by the chaos game close by, namely at a distance of not greater than ε. In other words, the Hausdorff distance between the attractor and its approximation is at most ε. Now the optimization problem consists in finding the probabilities p_1 to p_N so that the expected number of iterations in the chaos game needed to reach this required approximation is minimal.

Although the problem is unsolved, there are some heuristic methods for choosing 'good' probabilities. Below we present one of them which has been popularized by Barnsley.[10] In the last section of this chapter we will discuss improved methods.

We consider an IFS with N transformations $w_1, ..., w_N$ and assume that its attractor is totally disconnected. So, if A denotes the attractor,[11] then the transformed images $w_1(A), ..., w_N(A)$ form a disjoint covering of the attractor. These small affine copies of the attractor are called *attractorlets*. If in the course of running the corresponding chaos game we generate a total of n points, then we may ask how these points are distributed among the N attractorlets. Allotting the same number of points in each attractorlet, i.e. n/N points, will yield a point set which is uniformly distributed over the Sierpinski gasket, but not in the case of the IFS for the Barnsley fern. Let us now consider an ε-collar A_ε of the attractor, the set of all points which have a distance not greater than ε from the attractor. Then

$$A \subset w_1(A_\varepsilon) \cup \cdots \cup w_N(A_\varepsilon) \subset A_\varepsilon \ .$$

Thus, for small values $\varepsilon > 0$ the sets $w_i(A_\varepsilon)$ are close approximations of the attractorlets. We now assign the number of points to fall into the ith attractorlet according to the percentage that the area of the corresponding set $w_i(A_\varepsilon)$ contributes to A_ε.[12] Thus, in order to achieve a uniform distribution of points the number of points in each attractorlet should be proportional to the corresponding area, where we assume that there is no significant overlap of the attractorlets. It is a well known fact from linear algebra, that the factor by which an area changes after undergoing an affine transformation is the absolute value of the determinant of the linear part of the transformation. Thus, if we write $w_i(z) = A_i z + B_i$ for the ith transformation, $i = 1, ..., N$, then the area of the ε-collar of the ith attractorlet is roughly

$$p_i = \frac{|\det A_i|}{|\det A_1| + \cdots + |\det A_N|}, \quad i = 1, ..., N$$

[10] M. F. Barnsley, *Fractals Everywhere*, Academic Press, 1988.

[11] For ease of notation we drop the ∞-index in the symbol for the attractor.

[12] We need the above construction using the ε-collar of the attractor because the area of the attractor itself may not be meaningful. For example, the area of the Sierpinski gasket is zero.

times the area of the ε-collar of the entire attractor. Therefore, we aim at collecting $n \cdot p_i$ points of the chaos game in the ith attractorlet. This is easily achieved just by choosing the probabilities $p_1, ..., p_N$ according to the above formula.

This recipe for choosing the probabilities also usually works fine in cases when there are some small overlapping parts of the attractor. However, special consideration is necessary in cases when there is large overlap or one of the transformations has a zero determinant. In the latter case the recipe froom above would just prescribe a zero probability; and consequently the transformation would never be used. The transformation which yields the stem of Barnsley's fern is an example. Here we arbitrarily assign a small probability, say $\delta = 0.01$. The whole procedure may be summarized by the formula

$$p_i = \frac{\max(\delta, |\det A_i|)}{\sum_{k=1}^{N} \max(\delta, |\det A_k|)}, \quad i = 1, ..., N$$

where $\delta > 0$ is a small constant.

Halftone and Color

So far we have discussed only black-and-white images and their encoding through IFSs together with their decoding through the chaos game. We have seen that the probabilities p_i give us very explicit control over the distribution of the game points falling within the sets into which an attractor can be broken down. In other words, counting the relative frequency of points falling into subsets of the attractor establishes a measure on the attractor. This observation motivates us to go one step further. Assume now that we have a half-tone image. We subdivide that image into little pixels, say in a raster screen of m rows and n columns. Each of the pixels, $P_{i,j}, i = 1, ..., m, j = 1, ..., n$, carries a unit of halftone information, a value which we interpret as a number $Q_{i,j}$ somewhere between 0 and 1. The value 1 corresponds to black and 0 to white. What we would like to achieve is an encoding and decoding scheme for such a picture. Thus, let us pick an IFS with contractions $w_1, ..., w_N$ and probabilities $p_1, ..., p_N$. Then we can look at the statistics of the chaos game in pixel $P_{i,j}$:

$$\lim_{k \to \infty} \frac{h(z_1, ..., z_k; P_{i,j})}{k} = R_{i,j} .$$

That means we run the chaos game with our probabilities and count the relative number of hits $R_{i,j}$ in pixel $P_{i,j}$. The given half-tone distribution $Q_{i,j}$ matches the normalized counts $R_{i,j}$, when the numbers $Q_{i,j}$ are proportional to $R_{i,j}$ in each pixel (with the same factor of proportionality α)

$$Q_{i,j} = \alpha R_{i,j}, \quad i = 1, .., m, \; j = 1, ..., n .$$

In this case we have encoded the given image up to an overall brightness setting, which can be adjusted afterwards.[13] The code for the image simply consists of the necessary transformations and the corresponding probabilities

$$\{w_1, ..., w_N\}, \quad \{p_1, ..., p_N\} ,$$

where each contraction needs six real numbers for its description. That means the half-tone image would be encoded into $7N$ real numbers. Moreover, the chaos game would serve as a decoding of that information back into the given image.

Thus, the question now is: given the half-tone distribution $Q_{i,j}$; can we find a code $\{w_1, ..., w_N\}$, $\{p_1, ..., p_N\}$, so that

$$\lim_{k \to \infty} \frac{h(z_1, ..., z_n; P_{i,j})}{k} \propto Q_{i,j} \tag{6.5}$$

where '$\propto$' means proportional. This is the *inverse problem* for half-tone images. From a solution to that problem to an understanding of *color images* is only a small step. Any color image can be regarded as an image which is composed of three components: a red, a green and a blue image. This is just the RGB technique of producing a color image on a TV screen. Each of the components can, of course, be interpreted as a half-tone image combined with the respective color information red, green, or blue.

The Inverse Problem and the Invariant Measure

Let us include a formulation of the inverse problem which brings the contraction mapping principle discussed in chapter 5 into play once again.[14] To set up the theoretical discussion for the inverse problem, we first note that the numbers $Q_{i,j}$ in effect denote the fraction of points from the chaos game that fall into the pixel $P_{i,j}$. These fractions are the result of a particular measure μ which has the attractor as its support, namely $\mu(P_{i,j})$. This measure μ is a Borel measure and is invariant under the *Markov operator* $M(\nu)$, which is defined in the following way. Let X be a large square in the plane which contains A_∞, the attractor of the IFS, and ν a (Borel) measure on X. Then this operator is defined by

$$M(\nu) = p_1 \nu w_1^{-1} + p_2 \nu w_2^{-1} + \cdots + p_N \nu w_N^{-1} .$$

In other words, $M(\nu)$ defines a new normalized Borel measure on X. We evaluate this measure for a given subset B in the following way: first we take the preimages $w_i^{-1}(B)$ with respect to X, then evaluate ν on that, and finally we multiply with the probabilities p_i

[13] Thus, a picture, which is uniformly white has the same encoding as a picture which is uniformly grey or black.

[14] This is of a more mathematical nature requiring notions from measure theory. Readers without corresponding mathematical background may wish to skip this section.

and add up the results. The Markov operator turns out to be a contraction in the space of normalized Borel measures of X, equipped with the Hutchinson distance[15]

$$d_H(\nu_1, \nu_2) = \sup \left| \int f d\nu_1 - \int f d\nu_2 \right|$$

where the supremum is taken over all functions $f : X \to \mathbf{R}$ with the property that $|f(x) - f(y)| \leq d(x, y)$. $d(x, y)$ denotes the distance in the plane. The contraction mapping principle can be applied because the space of normalized (Borel) measures is complete with this distance. Thus, there exists a unique fixed point μ of the Markov operator M, $M(\mu) = \mu$. This is exactly the measure which we are looking for when we try to find a solution to the inverse problem for half-tone images. Further details would plunge us directly into current research.

---

[15] J. Hutchinson, *Fractals and self-similarity*, Indiana University Journal of Mathematics 30 (1981) 713–747.

6.4 Random Number Generator Pitfall

Anyone who considers arithmetical methods of producing random digits is, of course, in a state of sin.

John von Neumann (1951)

The chaos game played on a computer inherently needs a random number generator. So far we have not examined this topic other than noting how to obtain random numbers with a prescribed distribution provided that the generator on the computer supplies numbers with a uniform distribution. On a computer, random numbers are not really random, they are obtained using deterministic rules actually stemming from a feedback system. Thus, the produced numbers only appear to be random, while, in fact, they are even completely reproducible in another run of the same program. For this reason the random numbers produced by computers are also called *pseudo-random.* There are quite a lot of methods in use for random number generation, which often are not apparent to the programmer. Thus, the statistical properties of the numbers coming out of the machine are typically unknown except for a claim of uniform distribution. In this section we demonstrate that when playing the chaos game a lot more than the simple uniform distribution of the random numbers is required. These requirements are naturaly fulfilled by using a perfect die.

Random Numbers From the Logistic Equation

In the first chapter we studied the chaos presented by iteration of the simple quadratic functions. It would seem possible to make use of the logistic equation

$$x_{k+1} = 4x_k(1 - x_k) \tag{6.6}$$

as a method for generating random numbers. In fact, this approach was suggested by Stanislaw M. Ulam and John von Neumann, who had been interested in the design of algorithms for random numbers to be implemented on the first electronic computer ENIAC. The iteration of eqn. (6.6) produces numbers in the range from 0 to 1. Let us divide this range into three equal intervals [0,1/3), [1/3,2/3) and [2/3,1]. Each number generated will be in one of the three intervals. Now we drive the chaos game for the Sierpinski gasket by this 'random number generator'. Figure 6.26 shows the result after 1000 iterations.

Failure to Produce the Sierpinski Gasket

The result is rather strange looking because only some very limited details of the Sierpinski gasket show through.[16] All of the points are exactly in the triangle. However, most parts of the

[16] This was suggested among other pseudo-random number generators by Ian Stewart, *Order within the chaos game?* Dynamics Newsletter 3, Nos. 2 & 3, May 1989, 4–9. Stewart ends his article: 'I have no idea why these results are occuring [...] Can these phenomena be explained? [...]' Our arguments will give some first insight. They were worked out by our students E. Lange and B. Sucker in a semester project of an introductory course on fractal geometry.

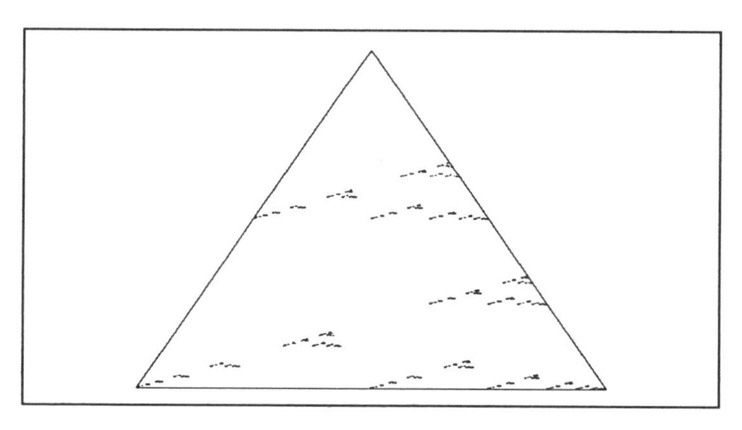

Sierpinski Gasket via Logistic Equation I

Figure 6.26 : The first attempt to generate the Sierpinski gasket using a random number generator based on the logistic equation.

Sierpinski gasket seem to be missing, even when iterating for a much longer time. Recalling the last section, one is tempted to believe that the probabilities are perhaps not adjusted correctly. To perform a test we compute a histogram[17] of a total of 10,000 computed random numbers.

interval	count	frequency
[0,1/3)	3910	39%
[1/3,2/3)	2229	22%
[2/3,1)	3861	39%

The result demonstrates a substantial deviation from the optimal frequencies of 1/3 per interval (33.3%). In order to revise the way our choices of the affine transformations are derived from the random numbers we must perform a more detailed empirical analysis. Let us subdivide the unit interval into 20 small intervals of length 0.05 each, and count the corresponding numbers for 100,000 iterates.

Adjusting the Probabilities Correctly

Based on the results in the table, we divide the unit interval into the three subintervals [0,1/4), [1/4,3/4) and [3/4,1]. Now iteration of the logistic equation seems to produce about the same number of iterates in each interval. Thus, this setting will produce a random number generator for the three outcomes 1, 2 and 3 with about equal probability of 1/3 each. Using this scheme we again play the chaos game hopeful that now we will produce the complete

[17] It is important to do the histogram computation in double precision. Otherwise it is very likely, that the iteration for the logistic equation will run into a periodic cycle of a low period (perhaps even less than 1000), and, as a consequence a histogram based on such an orbit would be a numerical artifact. This effect and the topic of histograms will be continued in chapter 10.

interval	count	interval	count
[0.00, 0.05)	14403		
[0.05, 0.10)	6145		
[0.10, 0.15)	4812		
[0.15, 0.20)	4256		
[0.20, 0.25)	3809		
		[0.00, 0.25)	33425
[0.25, 0.30)	3487		
[0.30, 0.35)	3389		
[0.35, 0.40)	3303		
[0.40, 0.45)	3244		
[0.45, 0.50)	3097		
[0.50, 0.55)	3240		
[0.55, 0.60)	3251		
[0.60, 0.65)	3196		
[0.65, 0.70)	3459		
[0.70, 0.75)	3621		
		[0.25, 0.75)	33287
[0.75, 0.80)	3882		
[0.80, 0.85)	4164		
[0.85, 0.90)	4821		
[0.90, 0.95)	6012		
[0.95, 1.00)	14409		
		[0.75, 1.00]	33288

Table 6.27 : Statistics for 100,000 iterations of the quadratic iterator.

Sierpinski gasket fairly rapidly. But figure 6.28 reveals a great disappointment; the result is even worse than before.

The conjecture that the problem is due to badly chosen probabilities has obviously proven false. To get to the core of the matter we need to take another look at properties we need to have guaranteed by the random number generator to make the chaos game work. Recall that the addressing system was the key to understanding the operation of the game. For each point of the attractor there was an address consisting of an infinite string of digits from $\{1,2,3\}$. The chaos game will produce a point close by provided it is set up to generate all possible finite addresses with some appropriate probability. Looking back at the poor results of the last two experiments, we realize that we were not able to produce most of the addresses. In our adjustment of the iteration (6.6) to the intervals $[0,1/4)$, $[1/4,3/4)$, and $[3/4,1]$ we made sure that addresses starting with 1, 2 and 3 would occur with the same frequency. But how about the addresses starting with 11, 12, 13 and so on? Let us

Sierpinski Gasket via Logistic Equation II

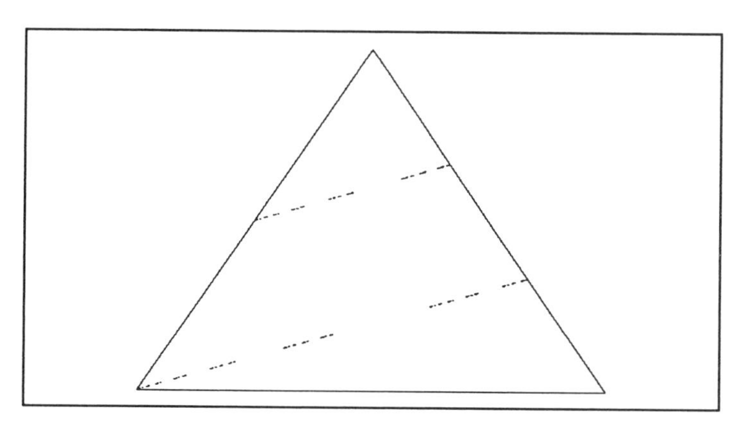

Figure 6.28 : Another attempt to generate the Sierpinski gasket. The 'improved' random number generator based on the logistic equation is used.

Rerun With Addresses

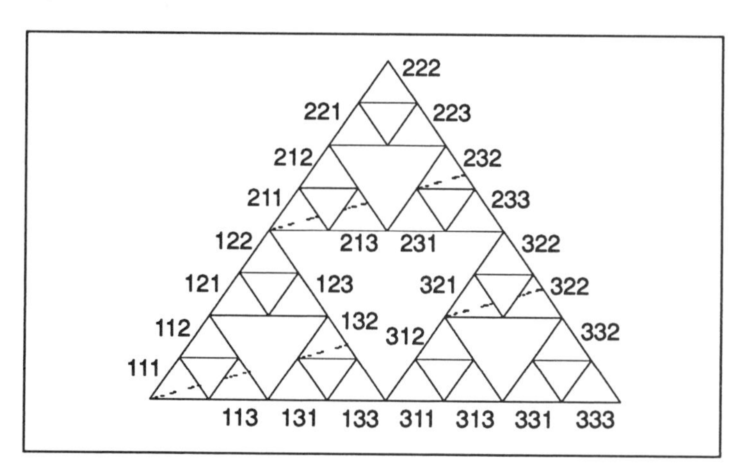

Figure 6.29 : Rerun of the last experiment with addresses inscribed and fat points.

try to find out which addresses do not occur by rerunning the last experiment and plotting the points on a grid inscribed with 3-digit addresses.

Where are the No-Shows?

We discover that certain combinations of three digits in the addresses never show up, namely,

$$222, 221, 223, 212, 231, 233, 122, 121, 123, 322, \ldots$$

In other words, only the following eight 3-digit addresses actually appear:

$$111, 113, 132, 211, 213, 232, 321, 323.$$

The Logistic Parabola

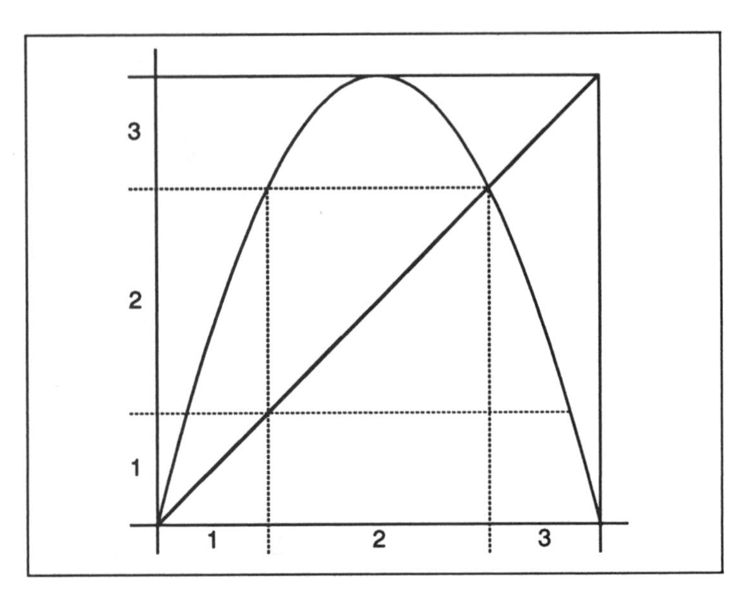

Figure 6.30 : Graphical iteration for $4x(1-x)$ with indicated regions used by the random number generator for the chaos game.

Having reached this far we should also be able to deduce these artifacts directly from the construction of the random numbers by means of iteration of the logistic equation. Figure 6.30 shows the graphical iteration for $4x(1-x)$. The three intervals [0,1/4), [1/4,3/4) and [3/4,1] are also marked on both axes. Looking at the diagram it becomes clear that in the iteration certain combinations of random number outcomes are simply impossible. If we start with a point in the first interval [0,1/4), then the next point must necessarily be either again in the first interval [0,1/4) or in the second interval [1/4,3/4). Therefore, the combination 13 can never appear in the scheme. Continuing, we realize, that a number in the second interval [1/4,3/4) will be transformed into a number of the third interval [3/4,1], and all numbers of the third will be in the first or second interval after one iteration.[18] Does this mean that the combinations 13, 21, 22, and 33 are not possible? Careful! Yes, our digital computer die, after rolling a 1, cannot roll a 3 next. But in terms of addresses, this translates into the reverse sequence 31! Recall, that addresses are read from left to right, and rolls in the chaos game right to left. Thus, we have verified that the chaos game played with our random number generator will not be able to produce points which have one of the following 2-digit strings in their address: 31, 12, 22, and 33. This is exactly, what we

[18] There is only one exception to this rule, namely the point 3/4. This point stays fixed, i.e. $4 \cdot 3/4 \cdot (1-3/4) = 3/4$. This is irrelevant for our discussion.

observed above in the experiment.

Of course, we can now extend our analysis to the case of our first attempt with the logistic equation (with intervals [0,1/3), [1/3,2/3), and [2/3,1]). The possible addresses are somewhat different, and in principle the failure to render the Sierpinski gasket stems from the same source: not every finite length sequence of interval indices can be produced. In other words, the events (the individual indices 1, 2, and 3) are not independent.

Modeling the Attractor of a Hierarchical IFS Driven by the Quadratic Iterator

There is an astonishing relation between driving the chaos game with the quadratic iterator $x_{k+1} = 4x_k(1 - x_k)$, eqn. (6.6), and hierarchical IFSs. This stresses once again the significance of the concept of hierarchical IFSs as a new mathematical tool.

Driving the chaos game with the quadratic iterator and appropriately adjusted probabilities as used in figure 6.28 means that transformation w_1 cannot be followed by w_3, w_2 not by w_2, w_2 not by w_1, and w_3 not by w_3. Or, turned the other way around, we can see the following admissable sequences of transformations

w_1 followed by w_1,
w_1 then w_2,
w_2 then w_3,
w_3 then w_1,
w_3 then w_2 .

Building an IFS as indicated by the graph in figure 6.31 leads to exactly the same result. First, consider the nodes 1, 2, and 3 and their connection by directed edges. Informally speaking, these nodes prescribe the 'next admissible transformation'. You can check that the

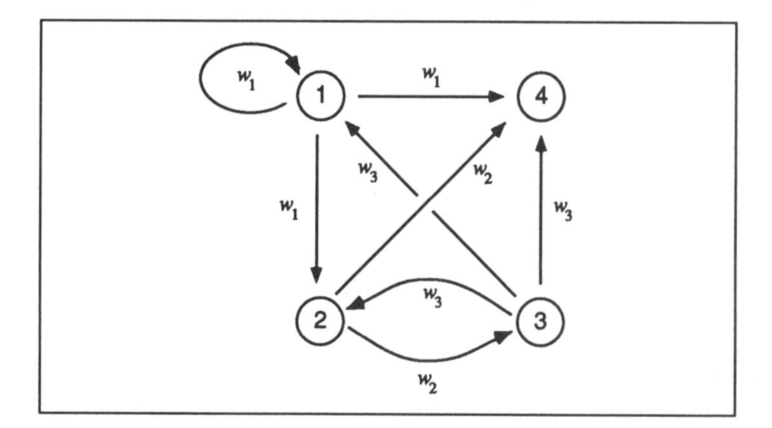

Figure 6.31 : Graph of hierarchical IFS corresponding to the chaos game based on the 'random numbers' produced by the logistic equation.

transformation w_i can only be applied in the order we just discussed. Now consider node 4. It collects all admissible compositions. The corresponding matrix Hutchinson operator would be

$$\mathbf{W} = \begin{pmatrix} w_1 & \emptyset & w_3 & \emptyset \\ w_1 & \emptyset & w_3 & \emptyset \\ \emptyset & w_2 & \emptyset & \emptyset \\ w_1 & w_2 & w_3 & \emptyset \end{pmatrix}.$$

The relevant attractor would appear in the 4th component (shown in figure 6.31).

Independent Rolls

This leads us to an important requirement for random number generators, which we have implicitly assumed in the chaos game but have not yet formulated explicitly. The rolls of the die, or the outcomes of the digital computer die must be independent from each other. Without that, it is possible that even though the three outcomes 1, 2, and 3 appear with the same frequency, we can have a sequence of events which is rather restricted. The chance of rolling a '3' may be 100% when the previous result had been '2', and it may be 0% when the last roll was a '1' or a '3'. The right way to play the chaos game, however, requires a die that produces a '3' with a fixed probability no matter what the previous roll (or any of the previous rolls for that matter) has been. A real unbiased die with six faces naturaly has this property. To roll a '1', and then a '2' occurs with probability 1/36 independent of all the previous results.

The Linear Congruential Generator

The most widely used random number generators on current computers use variants of the *linear congruential method*.[19] With a modulus m and a starting value r_0, $0 \le r_0 < m$, the numbers following are computed with the formula

$$r_{k+1} = (ar_k + c) \bmod m$$

where the multiplier a and the increment c are nonnegative integers less than the modulus m. The modulus usually is conveniently chosen to be a power of 2 matching the word length of the particular machine. This method produces integers in the range from 0 to $m - 1$. Each number is completely determined by its predecessor according to the above formula. In conclusion, all sequences of pseudo random numbers generated in this way must be periodic. The multiplier a and the increment c can be chosen so that the period has the maximal value m. Because of the periodicity, which is inherent in the construction, the sequences of random numbers

[19]The method was introduced in 1949, see D. H. Lehmer, Proc. 2nd Symposium on Large Scale Digital Calculating Machinery, Harvard University Press, Cambridge, 1951.

thus generated cannot be truly random. The randomness comes in various flavors, and there exist a large number of statistical tests: frequency test, run test, collision test and spectral test, to name just a few of them.[20]

The Middle-Square Generator

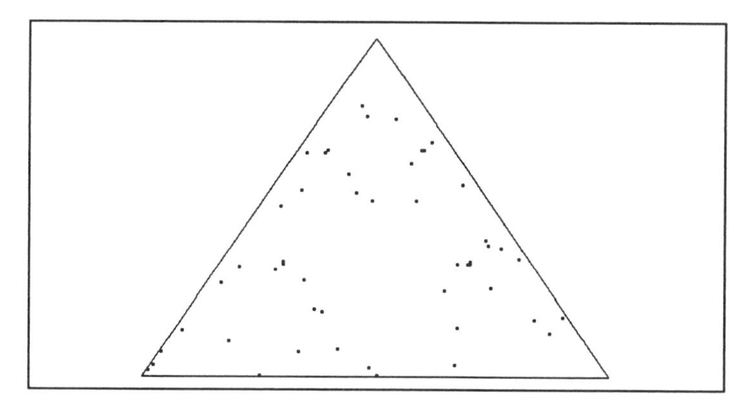

Figure 6.32 : The middle-square method for random number generation fails to produce the Sierpinski gasket (points fattened).

We conclude that it is important for the chaos game played on a computer to rely on a random number generator which guarantees the independence of all numbers produced. Only in this way is it possible to generate points for all necessary addresses. Most random number generators supplied with computers nowadays seem to fulfill this property sufficiently well. But it is by no means true for all generators in use. We illustrate this with two examples which were considered in the 1950's: the middle-square generator and the Fibonacci generator.

The First Random Number Generator

Before computers existed people who needed random numbers used to roll dice, draw cards from a deck or, later, use mechanical devices. Tables of random digits had also been published. For example, in 1927 L. H. C. Tippet produced a table of over 40,000 digits 'taken at random from census reports'. In 1946 John von Neumann was the first to suggest that random numbers could be computed on a machine using a deterministic algorithm, the *middle-square generator*. In this method a decimal number r_0 with n digits is given as a seed. This seed value is squared and the middle n digits of the result are extracted, yielding the next number r_1. Then square r_1, extract the middle n digits to obtain r_2, and so on. The range of numbers produced in this way extends

[20]For an introduction into the topic of random number generation see D. E. Knuth, *The Art of Computer Programming, Volume 2, Seminumerical Algorithms*, Second Edition, Addison-Wesley, Reading, Massachusetts, 1981.

The Fibonacci Generator

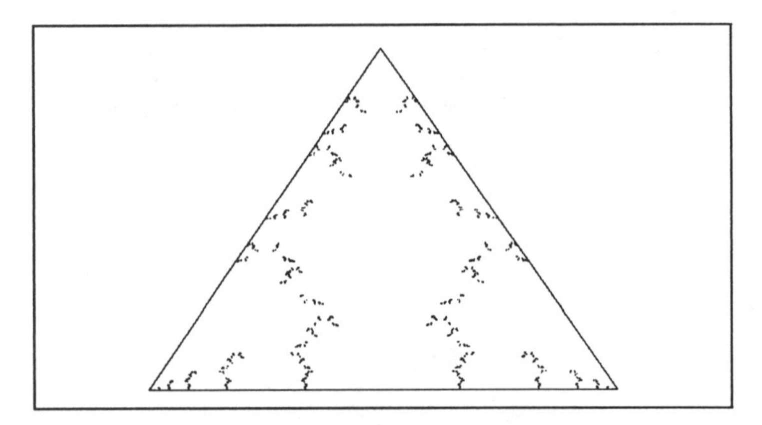

Figure 6.33 : The Fibonacci generator for random number generation also fails to produce the Sierpinski gasket.

from 0 to $10^n - 1$, and by dividing the results we get numbers distributed over the unit interval $[0, 1]$ as required for the normalized random numbers. The middle-square generator, however, has been shown to be a rather poor source of random numbers, although when set up with certain numbers of digits and initial seeds, it may produce a long sequence of numbers that passes all practical tests for randomness. Figure 6.32 shows our attempt for the Sierpinski gasket.

A Second Order Formula

The Fibonacci generator is probably the simplest method of second order for production of random numbers. Each number is computed not only from its predecessor, but also from the second predecessor. The formula is

$$r_i = (r_{i-1} + r_{i-2}) \bmod m \ .$$

In Figure 6.33 we have chosen $m = 2^{18}$. The result is a rather surprising fractal — but it is far from the complete Sierpinski gasket it should be.

6.5 Adaptive Cut Methods

As shown in this chapter and the last one, a wide range of fractals can be encoded as attractors of an IFS. We have discussed two alternatives for the rendering of this attractor: a deterministic algorithm (i.e., the iteration of the MRCM), and a probabilistic method, the chaos game. However, both approaches have limitations. The deterministic algorithm performs poorly when the contraction ratios of the affine transformations vary significantly, as in the case of Barnsley's fern. On the other hand, the performance of the chaos game depends strongly on the choice of probabilities; and so far we have only seen a rule of thumb for the determination of probabilities (see page 352). But now we will discuss a method of computing improved probabilities from a deterministic approximation of the attractor.

In fact, the algorithm we are going to discuss can be used to implement a deterministic rendering of the attractor which avoids the drawbacks of simply iterating an MRCM.[21] In many cases this algorithm is a better choice than the probabilistic rendering. It can especially render the attractor up to a prescribed precision, whereas with the probabilistic method there is no criterion that ensures that the chaos game has been played sufficiently long to achieve the desired approximation. On the other hand, a well-tuned chaos game is extremely efficient in rapidly rendering a first impression of the global appearance of the attractor.

Covering Sets for the Attractor

First, let us discuss the problem of approximating the attractor A_∞ of an IFS given by the contractions $w_1, \ldots, w_N$ to a prescribed precision ε. In other words, we are looking for a covering of the attractor by sets measuring less than $\varepsilon > 0$ in diameter. The iteration of the Hutchinson operator can provide such a covering. Starting with any set A which includes the attractor ($A_\infty \subset A$), the first iteration gives a covering by N sets,

$$A_\infty \subset w_1(A) \cup \cdots \cup w_N(A) \ ,$$

the second a covering by N^2 sets,

$$A_\infty \subset w_1 w_1(A) \cup w_2 w_1(A) \cup \cdots \cup w_N w_N(A) \ ,$$

and so on. All transformations w_k are contractions. Thus, after a certain number of iterations, say m, all N^m covering sets of the form

$$w_{s_1} w_{s_2} \cdots w_{s_m}(A)$$

[21]Details have appeared in the paper *Rendering methods for iterated function systems* by D. Hepting, P. Prusinkiewicz and D. Saupe, in: FRACTAL 90 – Proceedings of the 1st IFIP Conference on Fractals, Lisbon, June 6–8, 1990 (H.-O. Peitgen, J. M. Henriques, L. F. Penedo, eds.), Elsevier, Amsterdam, 1991.

have a diameter less than ε. However, from the example of the fern we know that the number N^m of these sets can be astronomically high, ruling out any practical computation by machine.

A Good Idea

We note that most of the final N^m covering sets are much smaller than necessary. Thus, it would be a great improvement if we could adaptively stop the iteration depending on the size of the sets $w_{s_1} \cdots w_{s_k}(A)$ at the intermediate steps $k = 1, ..., m$.

Setting Up a Simple Test Example

Let us discuss a simple example to make the issue clear. Consider the following system of only two transformations

$$w_1(x) = \frac{x}{3}$$
$$w_2(x) = \frac{2x}{3} + \frac{1}{3}$$

which operate on the set of real numbers. Note that the contraction factor for w_1 is $1/3$ but for w_2 it is $2/3$. There is a strong connection between these transformations and those corresponding to the Cantor set.[22] However, the attractor of this set of transformations is not a Cantor set; it is not even a fractal but simply the unit interval $I = [0, 1]$. This can be seen by noting that the interval is invariant under the associated Hutchinson operator:

$$w_1(I) = [1, 1/3]$$
$$w_2(I) = [1/3, 1]$$

thus

$$H(I) = w_1(I) \cup w_2(I) = [0, 1] = I \ .$$

From the characterization of the attractor of an IFS by its invariance property we conclude that the unit interval is indeed the attractor of our simple system above.

First Try of an Adaptive Method

Let us try to cover the attractor by sets not exceeding the size $\varepsilon = 1/3$. If we start with $I = [0, 1]$ the first iteration step provides the sets $w_1(I) = [0, 1/3]$ and $w_2(I) = [1/3, 1]$ which — both together — should be used as input for the next step of the iteration. But since the first one of these, $w_1(I)$, is already of the desired size we continue the iteration only with the other one, $w_2(I)$ (see figure 6.34). This yields

$$w_1 w_2(I) = w_1[1/3, 1] = [1/9, 1/3]$$
$$w_2 w_2(I) = w_2[1/3, 1] = [5/9, 1] \ .$$

The first of these sets is less than ε in size, but the other one is not yet of that size. Thus, we repeat the procedure one more time for $[5/9, 1]$ obtaining

$$w_1 w_2 w_2(I) = w_1[5/9, 1] = [5/27, 1/3]$$
$$w_2 w_2 w_2(I) = w_2[5/9, 1] = [19/27, 1]$$

---

[22] Recall that those are $w_1(x) = x/3$ and $w_2(x) = x/3 = 2/3$ (see page 191).

Adaptive MRCM Iteration

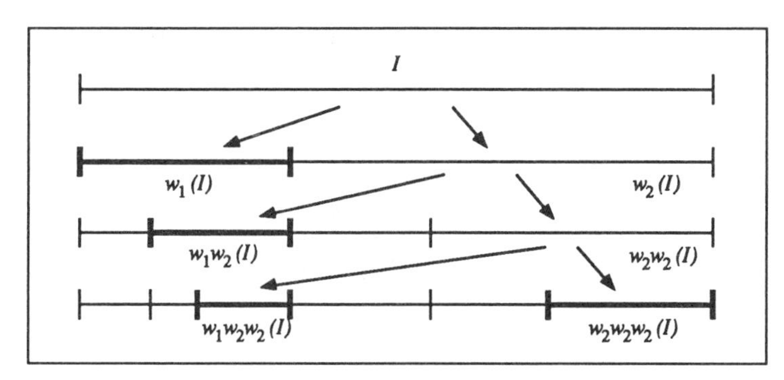

Figure 6.34 : Restricting the iteration of the Hutchinson operator to those sets which are smaller than a prescribed tolerance finally provides only sets of the desired size. However, in general they do not cover the whole attractor.

Adaptive Cut Iteration

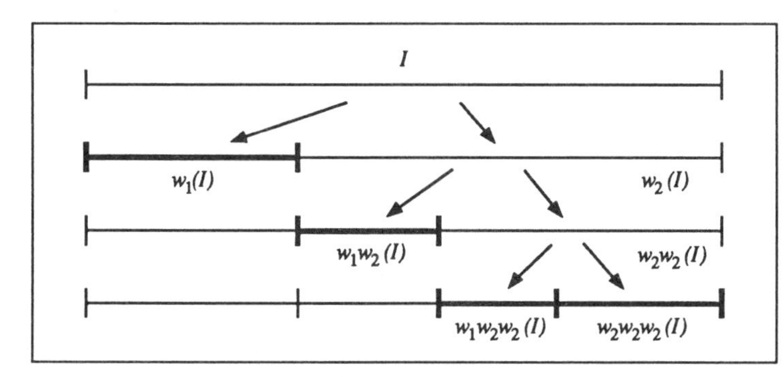

Figure 6.35 : The hierarchical subdivision introduced in the discussion of addresses provides the appropriate framework for the adaptive cut algorithm.

The sizes of these intervals are $4/27$ and $8/27$, both less than $\varepsilon = 1/3$. We are finished and expect that the collection of small intervals obtained in this procedure covers the attractor. But our check

$$\begin{aligned}&[0,1/3] \cup [1/9,1/3] \cup [5/27,1/3] \cup [19/27,1] = \\ &= [0,1/3] \cup [19/27,1] \neq [0,1]\end{aligned}$$

reveals that some portions of the unit interval are still missing (see also figure 6.34). In other words, simply cutting off some parts of the IFS iteration does not work.

The Correct Adaptive Method

We need a different kind of hierarchical refinement. Figure 6.35 shows the ideal strategy of subdivision for our example. It ends up with the sets $w_1(I)$, $w_2w_1(I)$, $w_2w_2w_1(I)$ and $w_2w_2w_2(I)$, each of which has the desired size and all together cover the unit interval I. But how can we obtain this kind of subdivision? Upon closer

Adaptive Address Tree

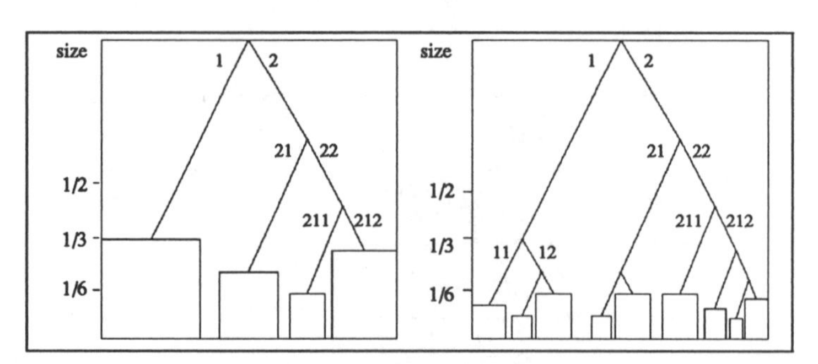

Figure 6.36 : The address tree corresponding to figure 6.35 is pruned at branches where the corresponding composed contractions reach the contractivity 1/3. The right tree has branches pruned at contraction factor 1/6 yielding a covering of the unit interval at a higher resolution. (The widths of the columns in the figure have no meaning; they do not match the size of the corresponding attractorlets.)

examination of figure 6.35 we see that it shows the hierarchical subdivision used for the addressing mechanism for the attractor. Figure 6.36 shows a corresponding address tree. The branches in the tree have different lengths; the nodes of one level are not at the same height. Rather, the height coordinate represents the size to which the unit interval is reduced when the corresponding composed contraction is applied to it. Thus, the idea of the methods consists in pruning precisely those branches of the address tree which pass the height 1/3.

Expressed more formally, we subdivide in the first step

$$A_\infty = w_1(A_\infty) \cup \cdots \cup w_N(A_\infty) \ . \qquad (6.7)$$

In the next step we subdivide each of the sets $w_k(A_\infty)$ according to

$$w_k(A_\infty) = w_k(w_1(A_\infty)) \cup \cdots \cup w_k(w_N(A_\infty)) \ .$$

In the third step we subdivide each of the sets $w_k(w_l(A_\infty))$ following

$$w_k w_l(A_\infty) = w_k w_l w_1(A_\infty) \cup \cdots \cup w_k w_l w_N(A_\infty) \ .$$

In other words, in stage n each subset of the attractor with address $s_1 \dots s_{n-1}$ is subdivided into those with addresses

$$s_1 \dots s_{n-1} 1, \quad s_1 \dots s_{n-1} 2, \quad \dots, \quad s_1 \dots s_{n-1} N \ .$$

These subsets of A_∞ are the *attractorlets* already described on page 352. Whenever we have reached an attractorlet which is smaller than ε in diameter we keep it. All other attractorlets must be subdivided further. In our example this process generated the attractorlets with the addresses 1, 21, 221 and 222.

On the basis of these ideas we can efficiently compute approximations of the attractor A_∞. We simply select one point from each of the final attractorlets. In this way we obtain a representative point set. Accordingly, in this construction we have that for each point in the attractor there will be an approximation in our point set that is not further away than ε.

Let us demonstrate this for our example. We know that $y_0 = 0$ is a point in A_∞ (0 is the fixed point of w_1). This allows us to compute points of the attractorlets:

$$\begin{aligned}
y_1 &= w_1(0) = 0,\\
y_2 &= w_2w_1(0) = w_2(0) = 1/3,\\
y_3 &= w_2w_2w_1(0) = w_2(1/3) = 5/9,\\
y_4 &= w_2w_2w_2(0) = w_2w_2(1/3) = w_2(5/9) = 19/27.
\end{aligned}$$

Since the size of the corresponding attractorlets is not larger than ε we have the approximation

$$A_\varepsilon = \left\{0, \frac{1}{3}, \frac{5}{9}, \frac{19}{27}\right\} \subset A_\infty$$

and surely no point in A_∞ has a distance greater than $\varepsilon = 1/3$ from all points in A_ε.

Let us compare this with the simple IFS iteration starting with the fixed point 0. Achieving the desired precision would require three complete steps. Thus we would generate eight points in contrast to the four points of A_ε, doubling the necessary workload. We note that actually a much worse factor of inefficiency is typical.

Comparing Algorithms: The Fern Example

In summary, the adaptive algorithm subdivides an attractorlet recursively until the diameter is guaranteed to be less than or equal to the tolerance ε. For the final collection of attractorlets we pick one point of each set as a representative. These points cover the attractor with precision ε. Figure 6.37 compares the different methods when applied to rendering Barnsley's fern. Since the contraction ratios of the transformations involved in constructing the fern are significantly different from each other, running the MRCM some number of cycles yields a highly uneven distribution of points in the attractor. The adaptive cut algorithm removes this drawback as can be seen impressively in the figure.

The Adaptive Cut Method

The adaptive cut method allows us to compute approximations A_ε of A_∞ to a prescribed precision ε so that the Hausdorff distance[23] $d_H(A_\varepsilon, A_\infty)$ is less than or equal to ε. Thus, all points of A_ε lie

[23] See chapter 5 for the definition of Hausdorff distance.

IFS Iteration Versus Adaptive Cut Method

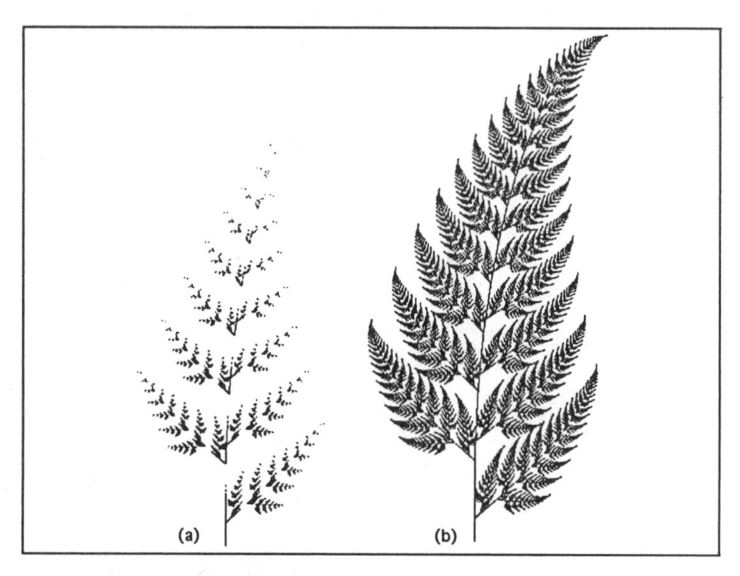

Figure 6.37 : A comparison of two methods for the rendering of the attractor of an IFS. (a) Iteration of the Hutchinson operator (starting with a point) for a total of $m = 9$ iterations, resulting in $N_1 = 4^9 = 262,144$ points. (b) The adaptive cut algorithm using $N_2 = 198,541$ points.

within the distance ε to points of A_∞ and vice versa. Now we look at w_1, w_2, ..., w_N and compute the contraction factors of these transformations. Let us introduce the symbol $\rho(w_k)$ for these ratios. All transformations w_k which satisfy $\rho(w_k) \leq \varepsilon/\text{diam}(A_\infty)$ can be eliminated from further subdivision because the corresponding attractorlets $w_k(A_\infty)$ are less than ε in diameter. For all other transformations w_k we continue with the second level and compute the contraction factors of the composite transformations

$$\rho(w_k w_1), \rho(w_k w_2), \ldots, \rho(w_k w_N) .$$

The procedure is repeated, i.e. we eliminate those compositions whose contraction ratio is less than or equal to $\varepsilon/\text{diam}(A_\infty)$ and continue with the others, considering composite transformations with three elements, and so on.

Thus, the general procedure is the following. Having reached a composition with sufficiently small contraction factor,

$$\rho(w_{s_1} w_{s_2} \cdots w_{s_m}) \leq \frac{\varepsilon}{\text{diam}(A_\infty)}$$

we have found an attractorlet with address $s_1 s_2 \cdots s_m$ and size

$$\text{diam}(w_{s_1} w_{s_2} \cdots w_{s_m}(A_\infty)) \leq \varepsilon ,$$

which is what we are looking for. When a composition $w_{s_1} \cdots w_{s_m}$ has a contraction factor which is still too large, i.e. greater than

$\varepsilon/\text{diam}(A_\infty)$, then we continue to consider the N compositions of the next level,

$$w_{s_1} w_{s_2} \cdots w_{s_m} w_1$$
$$w_{s_1} w_{s_2} \cdots w_{s_m} w_2$$
$$\vdots$$
$$w_{s_1} w_{s_2} \cdots w_{s_m} w_N \ .$$

In this way we construct all compositions which satisfy

$$\rho(w_{s_1} \cdots w_{s_m}) \le \frac{\varepsilon}{\text{diam}(A_\infty)} \le \rho(w_{s_1} \cdots w_{s_{m-1}}) \ .$$

Let S be the set of the corresponding attractorlet addresses $s_1 ... s_m$. For any point $x_0 \in A_\infty$ (e.g. take the fixed point of w_1) the Hausdorff distance between the attractor and the set

$$A_\varepsilon = \{x \ : \ x = w_{s_1} \cdots w_{s_m}(x_0), \ (s_1, ..., s_m) \in S\}$$

is bounded by ε, i.e. $d_H(A_\varepsilon, A_\infty) \le \varepsilon$. This completes the basic description of the adaptive cut method.

For practical purposes the method can be even more accelerated using the fact that the attractor can be displayed only with a finite screen resolution. The idea is to eliminate from further consideration images of more than one point falling in the same pixel.[24] This is of particular importance in cases of overlapping attractors, since the adaptive algorithm ignores such overlapping.

The remaining problem is to compute or estimate the contraction factors $\rho(w_{s_1} \cdots w_{s_m})$. Here we propose three methods which vary in ease of computation and in the quality of the estimates. In all cases distances are measured using the Euclidean metric.

The first and simplest method relies on the property $\rho(w_1 w_2) \le \rho(w_1)\rho(w_2)$ and estimates the contraction factor of a composite affine transformation $w_{s_1} \cdots w_{s_n}$ as the product of the individual contraction factors. Thus,

$$\rho(w_{s_1} \cdots w_{s_m}) \le \rho(w_{s_1}) \cdots \rho(w_{s_m}) \ .$$

Unfortunately this formula may grossly overestimate the actual value of the contraction ratio of the composite transformation. For example, consider the following two affine mappings:

$$w_1(x, y) = (0.01x, 0.99y),$$
$$w_2(x, y) = (0.99x, 0.01y).$$

Since $\rho(w_1) = \rho(w_2) = 0.99$, we obtain $\rho(w_1)\rho(w_2) = 99^2/10,000$. On the other hand,

$$w_1 w_2(x, y) = (0.0099x, 0.0099y)$$

[24] See S. Dubuc and A. Elqortobi, *Approximations of fractal sets*, Journal of Computational and Applied Mathematics 29 (1990) 79–89.

and $\rho(w_1 w_2) = 0.0099$. Thus, the use of the product $\rho(w_1)\rho(w_2)$ overestimates the actual value of $\rho(w_1 w_2)$ by a factor of 99.

We can use an alternative method for estimating the contraction ratio using the following property.[25] The contraction ratio $\rho(w)$ of an affine transformation

$$w(x, y) = (ax + by + e, cx + dy + f)$$

satisfies the inequality

$$\rho(w) \leq 2 \max\{|a|, |b|, |c|, |d|\} .$$

In the above example this result provides a bound 0.0198 which is much improved over $99^2/10,000$ but still off by a factor of 2.

The third method is computationally more expensive than the previous two, but provides exact values. It uses the fact that the contraction ratio $\rho(w)$ of the affine transformation $w(z) = Az + B$ (where A denotes a matrix and B a column vector) can be expressed as the square root of the maximal eigenvalue of $A^T A$ (A^T denotes the transpose of the matrix A)

$$\rho(w) = \sqrt{\max\{|\lambda_i| \ : \ \lambda_i \text{ eigenvalue of } A^T A\}} .$$

This formula is valid for the affine transformations in spaces of arbitrary dimension n. In the two-dimensional case ($n = 2$) with

$$A = \begin{pmatrix} a & b \\ c & d \end{pmatrix}$$

the evaluation of $\rho(w)$ involves the computation of two square roots. Explicitly, the result is

$$\rho(w) = \sqrt{\frac{p + \sqrt{p^2 - 4q}}{2}}$$

where

$$p = a^2 + b^2 + c^2 + d^2 ,$$
$$q = (ad - bc)^2 .$$

As we are not interested in the exact contraction ratio but only in a tight bound of it, we may replace the square root computation by a properly organized table look-up procedure, which will speed up the method considerably.

As we stated at the start, the distances in this discussion are measured using the Euclidean metric. Alternatively we could switch to a different metric. For example, for the maximum metric d_∞ (see page 284) the contraction ratio can be computed efficiently by the formula

$$\rho_\infty(w) = \max\{|a| + |b|, |c| + |d|\} ,$$

[25] See G. H. Golub and C. F. van Loan, *Matrix Computations*, Second Edition, Johns Hopkins, Baltimore, 1989, page 57.

where the coefficients $a, ..., d$ are the elements of the matrix A as above.

Covering Sets for the Fern

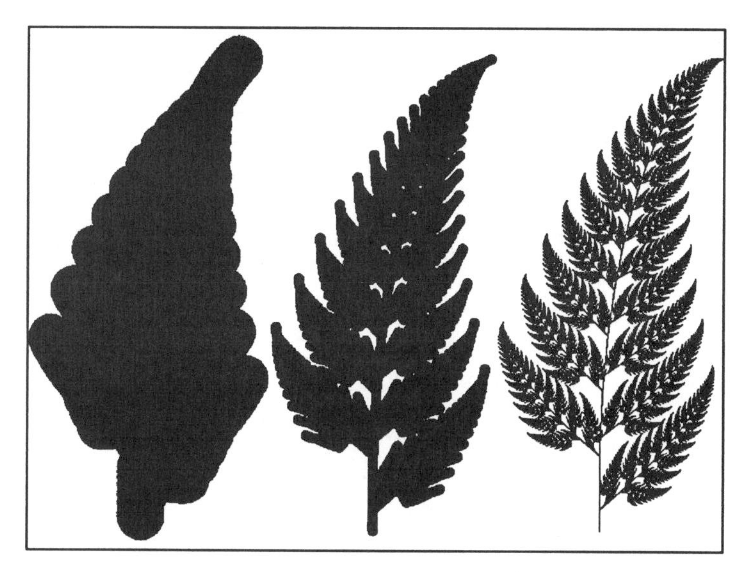

Figure 6.38 : The adaptive cut algorithm may produce renderings of different resolution depending on the choice of the tolerance ε for the Hausdorff distance. Three different values are used in the above plots $\varepsilon = 0.5, 0.1, 0.015$. Each point was drawn as a small disk with appropriate radius such that the attractor is guaranteed to be covered by the image.

The adaptive cut method provides a list of points approximating the attractor A_∞ with prescribed accuracy. When these points are taken as centers of small disks, then they provide a covering of A_∞. Let

$$D_\varepsilon(y) = \{x \in \mathbf{R}^2 : \|x - y\| \leq \varepsilon\}$$

the set of points in the plane which lie within the distance ε from the point y. Then for our simple one-dimensional example the set

$$C_\varepsilon = D_\varepsilon(0) \cup D_\varepsilon(1/3) \cup D_\varepsilon(5/9) \cup D_\varepsilon(19/27)$$

would cover the attractor (i.e, $A_\infty \subset C_\varepsilon$) and all points of C_ε would have a distance to A_∞ which is at most $\varepsilon = 1/3$. Figure 6.38 shows such covering sets for the fern.

Estimating Probabilities for the Chaos Game

The chaos game requires a set of probabilities $p_k, k = 1, ..., N$ that determine which of the transformations $w_1, ..., w_N$ should be taken in each step of the algorithm. As demonstrated, the choice of these probabilities is not obvious. The adaptive cut method

Chaos Game — Two Sets of Probabilities

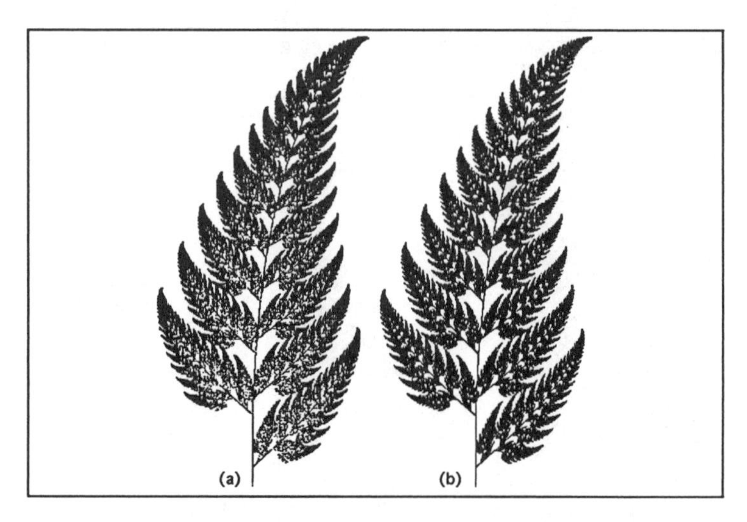

Figure 6.39 : The chaos game using $N_1 = 198,541$ points. (a) The probabilities used are 0.85, 0.07, 0.07 and 0.01. (b) The improved probabilities are 0.73, 0.13, 0.11, and 0.03.

may provide another way to assign values for the probabilities. We subdivide the points plotted by the adaptive method into N subsets, each one collecting points drawn in the corresponding attractorlet $w_k(A_\infty)$.[26] The relative number of points in each subset determines the corresponding probability. For example, for the fern we obtain the numbers 0.73, 0.13, 0.11, 0.03.[27] When used as probabilities in the chaos game, these values result in an image with more evenly distributed points as compared to images based on the chaos game using probabilities suggested by the above formula from page 352 (see figure 6.39).

[26]This specification is not precise because the attractorlets $w_k(A_\infty)$ of the first stage may be overlapping. The algorithm counts a point representing the attractorlet $w_{s_1}w_{s_2}\cdots w_{s_m}(A_\infty)$ as a point belonging to $w_{s_1}(A_\infty)$.

[27]These probabilities should not be taken as absolute because their values depend to some extent on the resolution of the image. Other weight factors may be better for other resolutions.

6.6 Program of the Chapter: Chaos Game for the Fern

We have demonstrated that the chaos game is an elegant way to compute the attractor of a given Multiple Reduction Copy Machine. The program provides an implementation of this idea. It is the framework for experiments using any of the IFSs from chapter 5. You simply take the parameters from table 5.48 and change the transformations of the program according to the given values. And soon there should be another fractal growing on your computer screen.

The program is set up for the generation of the fern as shown in figure 6.40. Note, that its parameters are not listed in table 5.48. We derived the probabilities using the determinants as described in the recipe on page 352. When trying out other transformations you should also use that recipe for computing the necessary probabilities.

Screen Image of Chaos Fern

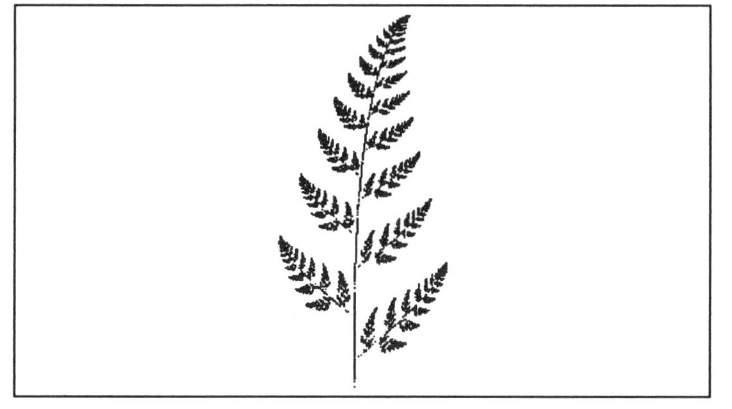

Figure 6.40 : Output of the program 'Chaos Game'.

Let us now look at the program. At the beginning you are asked for the number of iterations in the chaos game. What follows is the specification of the width of the display (as usual) and the setup of the translational parts of the transformations. Note that as in the program for chapter 5 these have to be scaled by `w`. We do not need to prescribe all the parameters of the transformations at the beginning. In this program they are specified as constants of the transformations (right at the point where the transformations are applied to the game point).

Next we compute the fixed point of the first transformation, `map 1`. This point is a part of the attractor and we use it as the initial game point in the chaos game. Its computation is rather simple, `x = e1, y = 0`. If you change the transformations you also have

BASIC Program **Chaos Game**
Title Chaos Game for a Fern Leaf

```
INPUT "Number of Iterations (5000):",imax
left = 30
w = 300
wl = w + left
e1 = .5*w : e2 = .57*w : e3 = .408*w : e4 = .1075*w
f1 = 0*w : f2 = -.036*w : f3 = .0893*w : f4 = .27*w
REM FIXED POINT MAP 1
x = e1
y = 0
FOR i = 1 TO imax
    r = RND
    REM map 1 (stem)
50  IF r > .02 GOTO 100
        xn = 0 * x + 0 * y + e1
        yn = 0 * x + .27 * y + f1
        GOTO 400
    REM map 2 (right leaf)
100 IF r > .17 GOTO 200
        xn = -.139 * x + .263 * y + e2
        yn = .246 * x + .224 * y +f2
        GOTO 400
    REM map 3 (left leaf)
200 IF r > .3 GOTO 300
        xn = .17 * x - .215 * y + e3
        yn = .222 * x + .176 * y + f3
        GOTO 400
    REM map 4 (top of fern)
300 xn = .781 * x + .034 * y + e4
    yn = -.032 * x + .739 * y + f4
    REM DRAW GAME POINT
400 PSET (xn+left,wl-yn)
    x = xn
    y = yn
NEXT i
END
```

to modify these two statements. The general rule is given on page 263. If this is not done correctly there will be some extra points that do not belong to the attractor on the screen. That is all, that will happen, so it is not too important.

Now the iterative part of the program begins (`FOR i = 1 TO imax`). First a random number `r` between 0 and 1 is computed. If `r` is smaller than (or equal to) 0.02 (i.e., with a 2% chance) the transformation `map 1` is applied to the game point. Then it is drawn (at label 40) and the next iteration begins. Otherwise the random number is checked again. If it is smaller than (or equal to) 0.17 (i.e., with chance 0.17-0.02 = 15%) the transformation `map 2` is applied. If it is not, transformation `map 3` (with 13% chance) or `map 4` (with 70% chance) is applied.

If you would like to change the transformations according to table 5.48 (or try your own creations), you can substitute the constants representing the parameters `a`, `b`, `c` and `d` directly into statements which describe the transformation of the game point. Let us run an example, the dragon from figure 5.11. Recall that

```
xn = a * x + b * y + e
yn = c * x + d * y + f
```

thus the first transformation is (check table 5.48)

```
xn = 0.000 * x + 0.577 * y + e1
yn = -0.577 * x + 0.000 * y + f1
```

and at the beginning of the program we would set

```
e1 = 0.0951*w
f1 = 0.5893*w
```

since the parameters `e` and `f` must be multiplied by `w`. Assume you have computed the probabilities p_1 to p_4 for the transformations `map 1` to `map 4`. Then set $r_1 = p_1$, $r_2 = p_2 + r_1$ and $r_3 = p_3 + r_2$. These are the numbers that should be substituted in the `IF`-statements which select the transformation (r_1 at label 50, r_2 at label 100 and r_3 at label 200).

Chapter 7

Irregular Shapes: Randomness in Fractal Constructions

Why is geometry often described as 'cold' and 'dry'? One reason lies in its inability to describe the shape of a cloud, a mountain, a coastline, or a tree. Clouds are not spheres, coastlines are not circles, and bark is not smooth, nor does lightning travel in a straight line. [...] The existence of these patterns challenges us to study those forms that Euclid leaves aside as being 'formless', to investigate the morphology of the 'amorphous'.

Benoit B. Mandelbrot[1]

Self-similarity seems to be one of the fundamental geometrical construction principles in nature. For millions of years evolution has shaped organisms based on the survival of the fittest. In many plants and also organs of animals, this has led to fractal branching structures. For example, in a tree the branching structure allows the capture of a maximum amount of sun light by the leaves; the blood vessel system in a lung is similarly branched so that a maximum amount of oxygen can be assimilated. Although the self-similarity in these objects is not strict, we can identify the building blocks of the structure — the branches at different levels.

In many cases the 'dead' world also carries some fractal characteristics. An individual mountain, for example, may look like the whole mountain range in which it is located. The distribution of craters on the moon obeys some scaling power laws, like a fractal. Rivers, coastlines, and clouds are other examples. However, it is generally impossible to find hierarchical building blocks for

[1] In: Benoit B. Mandelbrot, *The Fractal Geometry of Nature*, Freeman, 1982, p. 1.

these objects as in the case of organic living matter. There is no apparent self-similarity, but still the objects look the same in a statistical sense — which will be specified — when viewed under magnification.

In summary, many natural shapes possess the property that they are irregular but still obey some scaling power law. One of the consequences — as discussed in chapter 4 — is that it is impossible to assign quantities such as length or surface area to these natural shapes. There cannot be a simple numerical answer to the question 'How long is the coastline of Great Britain?'. If 5000 miles would be measured by someone as the length of the coastline, someone else with a better (finer) measuring technique would come up with a result longer than 5000 miles. The more appropriate question to ask would be: how irregular, how convoluted is a coastline, or what is its fractal dimension? In this chapter this question is turned around. Methods for generating models of coastlines (and other shapes) with *prescribed* fractal dimension are given. Well, you may propose that the Koch snowflake curve, for example, may already serve as a good model for the coastline of an island. However, even though such exact self-similar curves have the desired scaling invariance and fractal dimension, they still are not perceived as realistic models of a coastline. The reason lies in their lack of randomness. To model coastlines, we need curves that look different when magnified but still invoke the same characteristic impression. In other words, looking at the magnified version of the coastline one should not be able to tell that it is indeed a magnification of the original. Rather, it ought to be regarded just as well as a different part of the coastline drawn at the same scale.

We begin our discussion at just that point — introducing some element of randomness into the otherwise rigorously organized classical fractals.[2] This leads to physical, so-called percolation models with applications ranging from the fragmentation of atomic nuclei to the formation of clusters of galaxies. An experiment which yields random fractal dendritic (tree-like) structures at an intermediate scale — useful for practical demonstration in a classroom — is an electrochemical aggregation process discussed in section 7.2. A mathematical model of this aggregation process is based on Brownian motion of particles. This can be implemented on a computer without much trouble. The underlying scaling laws of Brownian motion and one important generalization (fractional Brownian motion) are the topic of the third section. With these tools in hand, fractal landscapes and coastlines can be simulated on a computer, as shown in the last section and on the color pages.

[2]The randomization of branching structures obtained from MRCMs is more conveniently discussed in the context of the next chapter.

7.1 Randomizing Deterministic Fractals

Randomizing a deterministic classical fractal is the first and simplest approach generating a realistic 'natural' shape. We consider the Koch curve, the Koch snowflake, and the Sierpinski gasket.

Introducing Randomness in the Koch Snowflake Curve

The method for including randomness in the Koch snowflake construction requires only a very small modification of the classical construction. A straight line segment will be replaced as before by a broken line of four segments, each one one-third as long as the original segment. Also the shape of the generator is the same. However, there are two possible orientations in the replacement step: the small angle may go either to the left or to the right, see figure 7.1.

An Alternative for the Replacement Step

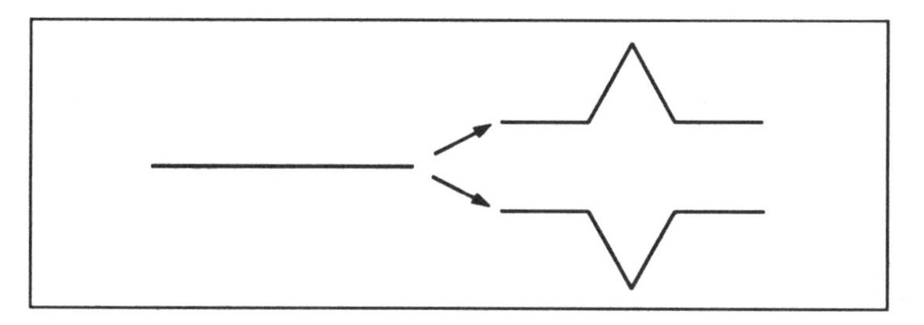

Figure 7.1 : Two possible replacement steps in the Koch construction.

Let us now choose one of these orientations at random in each replacement step. Let us call it the *random Koch curve*. Composing three different versions of the random Koch curve placed so that the end points meet, yields the *random Koch snowflake*. In this process some characteristics of the Koch snowflake will be retained. For example, the fractal dimension of the curve will be the same (about 1.26). But the visual appearance is drastically different; it looks much more like the outline of an island than the original snowflake curve, see figures 7.2 and 7.3. A different island can be constructed using the same ideas applied to the 3/2-curve introduced in chapter 4 (see figures 7.4 and 7.5). Here the dimension of the curve is higher, exactly 1.5.

Three Ways to Randomize the Sierpinski Gasket

In these first two examples of random fractals, random decisions had to be made in the construction process. In each decision a random choice of one out of two alternatives was appropriate. Let us now give an example where a random number from an entire interval is used, the randomized Sierpinski gasket. The construction process is identical to the original one. Thus, in each stage a triangle is subdivided into four subtriangles, the central one of which is removed. However, in the subdivision we can now allow for subtriangles which are not equilateral. On each side of a

Random Koch Curve

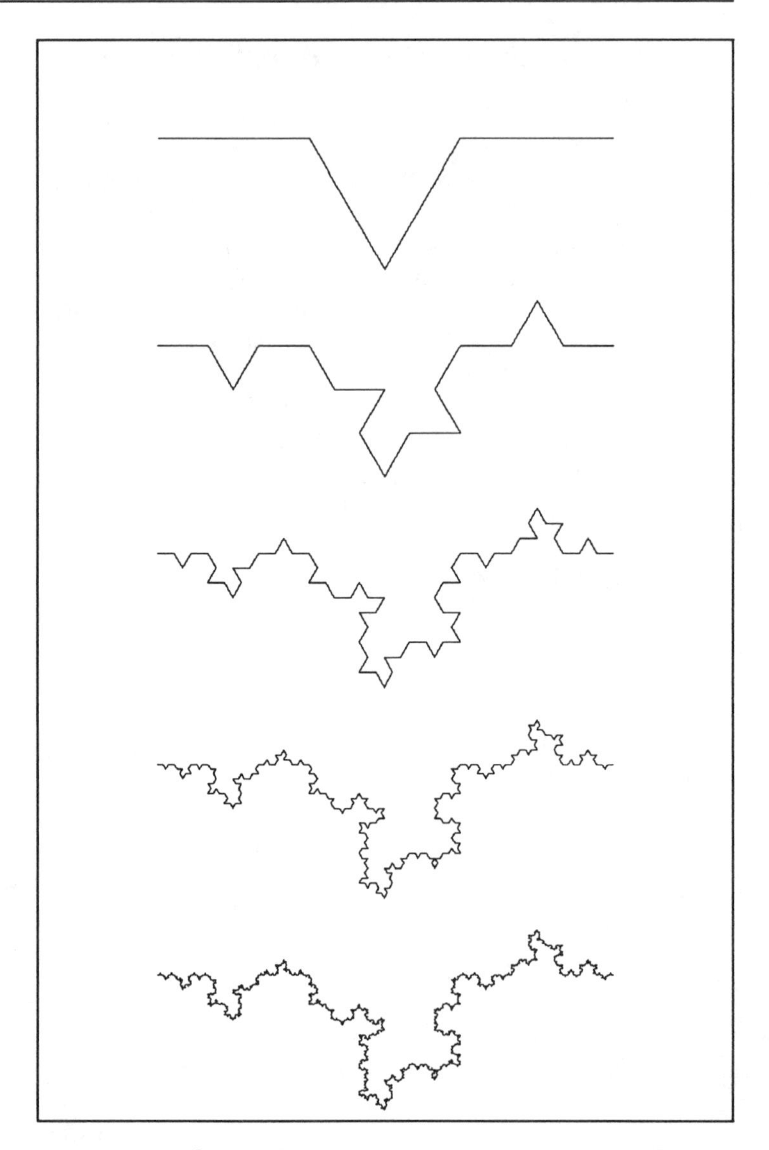

Figure 7.2 : One realization of the random Koch curve. The replacement steps are the same as in the original Koch curve, with the exception that the orientation of the generator is chosen randomly in each step.

triangle to be subdivided we pick a point at random, then we connect the three points and obtain the four subtriangles. The center subtriangle is removed and the procedure repeats, see figure 7.6.

Let us next discuss a modification of the Sierpinski gasket. This will lead us directly to the topic of the next section, a phys-

Random Koch Island

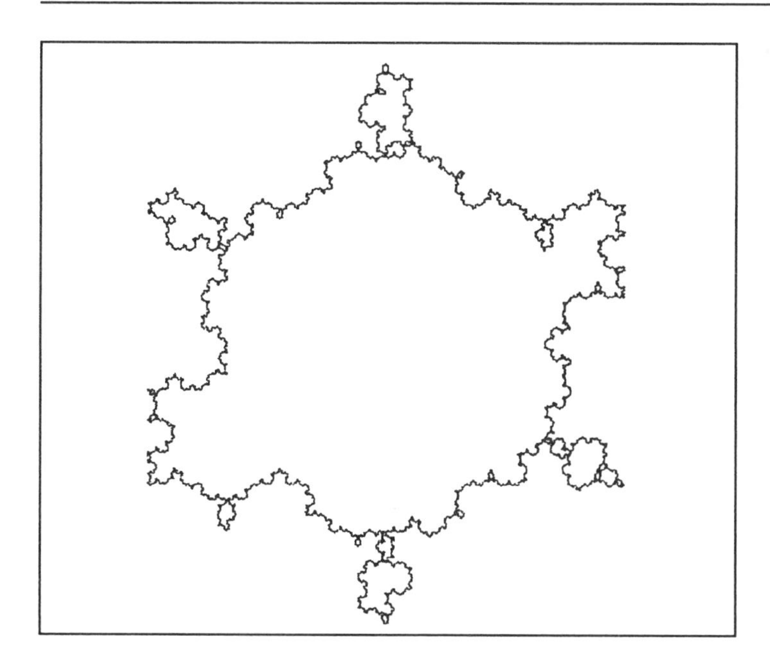

Figure 7.3 : The random Koch island is composed of three different versions of the random Koch curve placed such that the end points meet.

Randomizing the 3/2-Curve

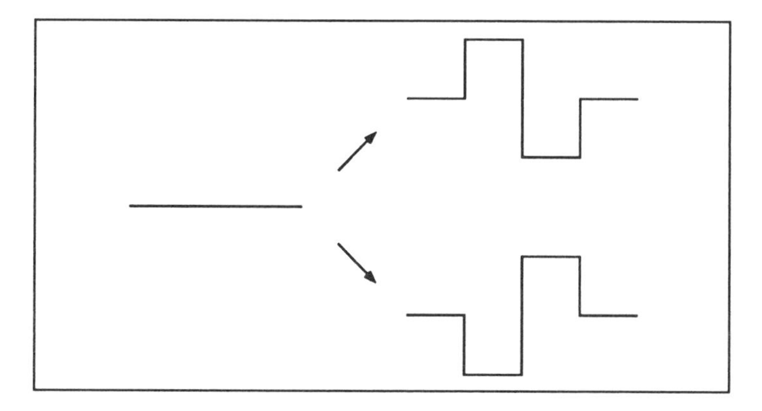

Figure 7.4 : Initiator and generator for the randomized 3/2-curve.

ical phenomenon with many applications: *percolation*. We again use the standard subdivision into equilateral triangles. One easy modification is then to simply choose one of the four subtriangles in a replacement step at random and remove it. Thus, the center subtriangle may be removed, but a different one might just as well be selected and removed, see figure 7.7.

Random 3/2 Island

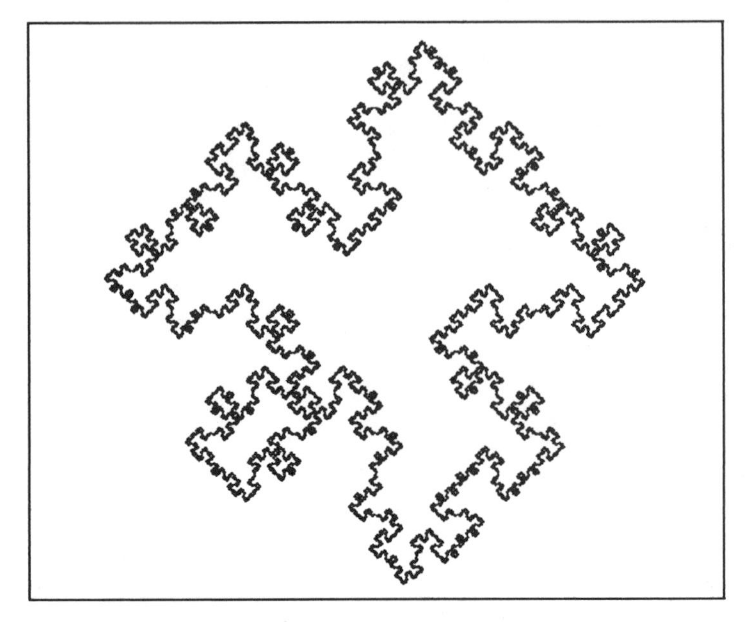

Figure 7.5 : Composing four different versions of the random 3/2-curve produces an island with a coastline of dimension 1.5.

Modified Sierpinski Gasket 1

Figure 7.6 : The points subdividing the sides are picked at random, stage 4 of the construction process is shown.

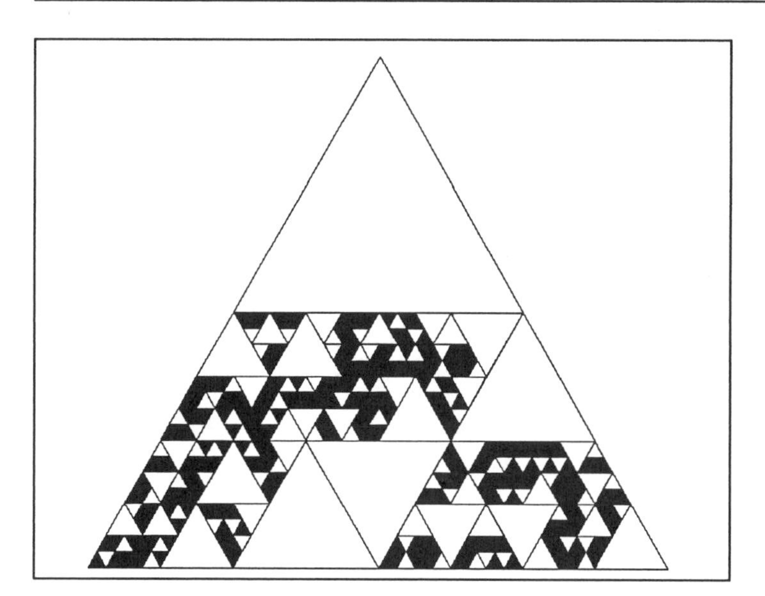

Modified Sierpinski Gasket 2

Figure 7.7 : In each replacement step the subtriangle to be deleted is picked at random. The black triangles are those from stage 5 of the construction process.

In the figure we can see small and large clusters of connected triangles. A cluster is defined as a collection of black triangles, which are connected across their sides and which are completely surrounded by white triangles.[3] Is there a cluster that connects all three sides of the underlying large triangle? What is the probability for that? What is the expected size of the largest cluster? Questions of this type are relevant in the percolation theory, which is discussed in the following section.

[3]Two triangles touching each other only at a vertex, are not considered a cluster.

7.2 Percolation: Fractals and Fires in Random Forests

Let us carry the idea one step further. We consider a triangular lattice of some resolution and deal with each subtriangle independently. Such a subtriangle is removed or not according to a random event which occurs with a prescribed probability $0.0 \leq p \leq 1.0$. The overall shape of the result depends dramatically on the probability p chosen. Obviously, for $p = 0$ we get nothing, while for $p = 1$ we obtain a solid triangle of full size. For intermediate values of p the object has a density, that increases with p.[4] Initially, for small values of p we get only a few specs here and there. For higher values of the probability the specs grow larger, until at some critical value of $p = p_c$ the shape seems to become glued together into one big irregular lump. Further increases in the probability, of course, thicken the cluster even more.

Modified Sierpinski Gasket 3

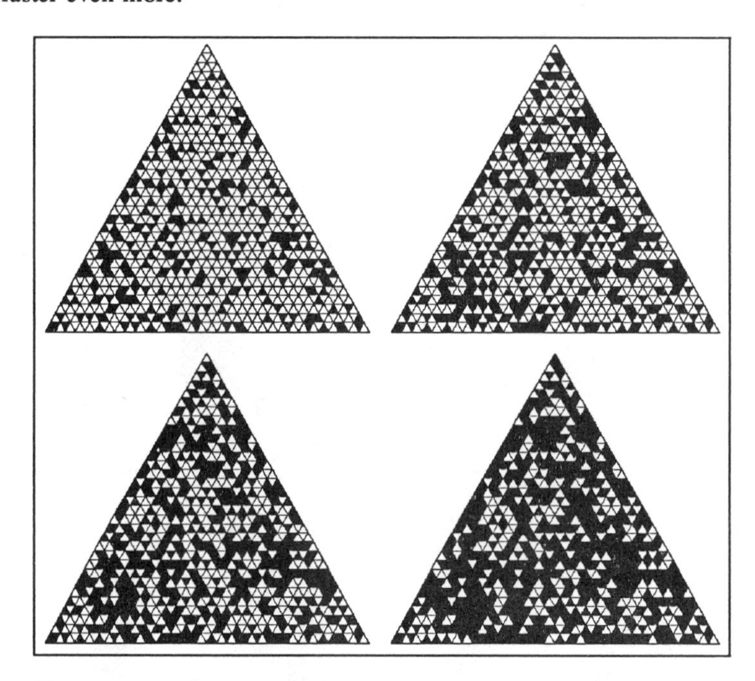

Figure 7.8 : Subtriangles of stage 5 are chosen with probability p. The four values of p used in this figure are (from top left to bottom right): 0.3, 0.45, 0.55, 0.7. In the first two graphs many small clusters coexist ($p = 0.3, 0.45$). In the graphs for $p = 0.55, 0.7$ one big major cluster exists and allows only very few small detached clusters.

[4]Formally, the average density is equal to p.

Percolation — When Things Start to Flow

When the structure changes from a collection of many disconnected parts into basically one big conglomerate, we say that *percolation*[5] occurs. The name stems from an interpretation of the solid parts of the structure as open pores. Assume that the whole two-dimensional plane is partitioned into a regular array of such pores which are either open (with probability p) or closed (with probability $1 - p$). Let us pick one of the open pores at random and try to inject a fluid at that point. What happens? If the formation is 'below the percolation threshold', i.e. if the probability p is less than p_c, we expect that the pore is a part of a relatively small cluster of open pores. By a cluster we mean a collection of connected open pores, which are completely surrounded by closed ones. In other words, below the threshold we will be able to inject only some finite amount of fluid until the cluster is filled, but no more. If the probability is above the threshold, then chances are good that the corresponding cluster is infinitely large. We can inject as much of the fluid as we like. In a practical example, the water percolates through the coffee grains and drips into the pot as coffee. The most interesting phenomena happen while increasing the probability from below the percolation threshold to a value above p_c. For example, the probability that a pore picked at random indeed belongs to the cluster of maximal size changes at $p = p_c$ from zero to a positive value. Moreover, right at the percolation threshold p_c, this maximal cluster is a fractal! It has a dimension that can be determined experimentally and, in some cases, also analytically.[6]

Dangerous Forest Fires Beyond Percolation

A paradigm often used for percolation is given by forest fires. Points in the clusters correspond to trees in the forest, and the fire cannot spread across gaps between trees. So the question whether the forest is above or below the percolation threshold is a vital one. In the first case trees are relatively sparse and only a small portion of all trees will burn down, while the other case is devastating: almost the complete forest will be destroyed. Let us elaborate the model a bit further. For simplicity we assume that the forest is not a natural one. Rather, the trees are planted in rows and columns in a square lattice. When all sites from this array are occupied by trees, the situation is clear — a fire ignited anywhere will spread over the whole forest (unless disturbed by strong winds or fire fighters not accounted for in this model). Thus, let us assume the more interesting case where each site in the lattice is occupied by a tree with a fixed probability $p < 1$. A burning tree may ignite its immediate neighboring trees. In the square lattice these are

[5]'Percolation' originates from the Latin words 'per' (through) and 'colare' (to flow).

[6]A very nice and enjoyable introduction to the topic for non-specialists is given in Dietrich Stauffer, *Introduction to Percolation Theory*, Taylor & Francis, London, 1985.

the trees at the four locations beside, above and below the burning one. In physicists' jargon these sites are called 'nearest neighbors'. Given a square with L^2 sites we distribute trees according to the chosen probability p and start a fire. Let us set all those trees on fire which are located along the left side of the square. In this simple model we can now simulate how the fire spreads. We proceed in discrete time steps. In each step a burning tree ignites all those neighbor trees that are not already on fire. After the tree has burned out, it leaves a stump, which from then on is no longer relevant. It goes without saying that this type of percolation model will be of little or no help in fighting or analysing actual forest fires. The point is that this paradigm is very suitable for an explanation and introduction of the topic.

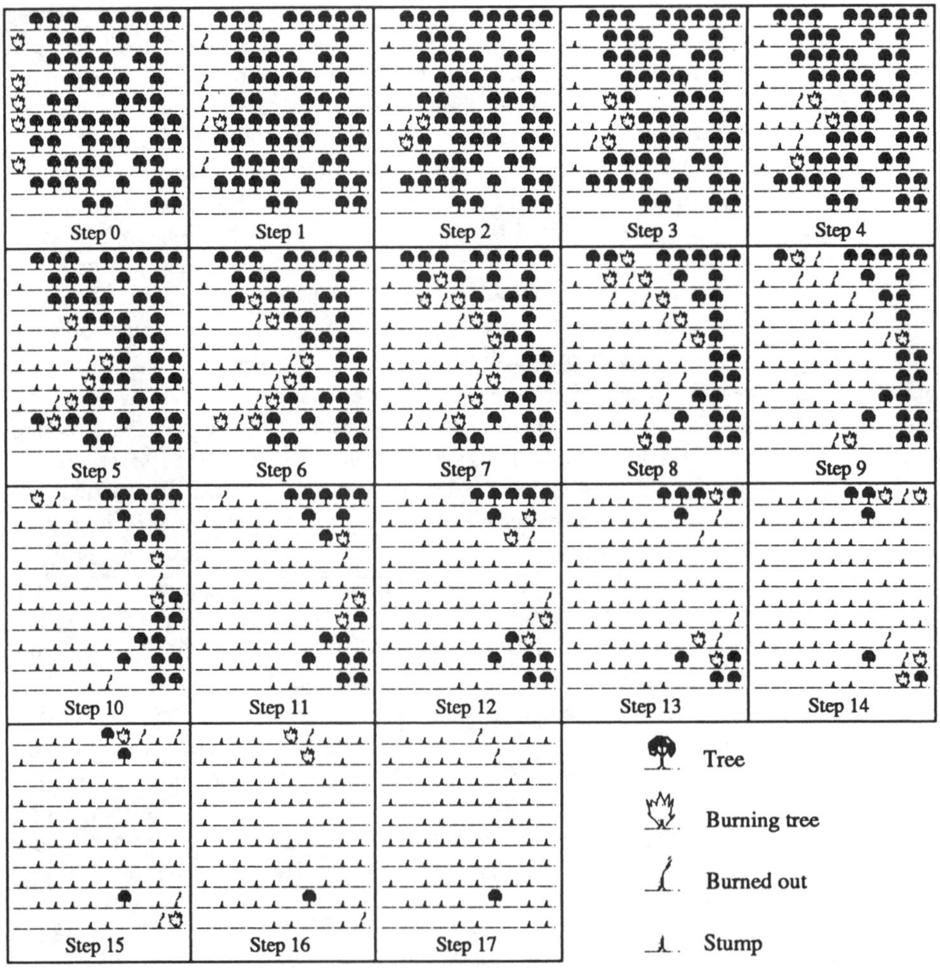

Step 0 Step 1 Step 2 Step 3 Step 4
Step 5 Step 6 Step 7 Step 8 Step 9
Step 10 Step 11 Step 12 Step 13 Step 14
Step 15 Step 16 Step 17

Tree
Burning tree
Burned out
Stump

The sequence on the left page displays the spreading of a forest fire simulated on a square ten-by-ten lattice. Initially, trees are placed at the lattice sites with probability 0.6 (indicated by the symbol). A burning tree is represented as , while burned out trees are displayed as (still smoking) and (stump). After 17 steps the fire is dead; only a single tree survives.

Forest Fires at Percolation Burn Longest

How long will such a fire last? If trees are very sparse — due to a low probability p — then the fire has not much material to burn, and it dies out very quickly, leaving most of the forest unharmed. On the other hand, if the trees are very dense (p is large), the forest does not have much of a chance for survival. Virtually the whole forest will be destroyed. Moreover, this will happen rather quickly; in not many more than L time steps the fire will have swept across the whole square leaving almost nothing but blackened stumps. There must be an intermediate probability which leads to a maximal duration of the forest fire. Figure 7.9 shows the dependence of the duration on the density of the forest.

In the diagram there is a sharp peak near the probability 0.6: this is the percolation threshold. The peak is indeed very sharp, if we increase the size of the forest, i.e. the number L of columns and rows, then the amplitude of the peak grows without bound.[7] In mathematical terms, there is a *singularity* at the percolation. The probability corresponding to the percolation has been measured experimentally very carefully, the accepted value is $p_c \approx 0.5928$.

A logical activity to pursue at this point would be to analyse the scaling laws of the forest fire duration. How long does the fire burn as the size of the underlying lattice grows without bound? There are three very different cases corresponding to $p < p_c$, $p = p_c$, and $p > p_c$, where the special case right at the percolation reveals a power law with a noninteger exponent — evidence for a fractal structure.

The Maximal Tree Cluster Size

A quantity that has been more thoroughly studied is the maximal cluster size, which is closely related to the fire duration. Let us denote the number of trees in the largest cluster in a lattice of size L by $M(L)$. As indicated, the cluster size will vary with L. A normalized measure of the maximal cluster size may be more convenient. It is given by the probability that a lattice site, picked at random, is a member of the maximal cluster. The notation is $P_L(p)$. It depends on the probability p and (to a lesser degree) on the lattice size L. In order to estimate $P_L(p)$ we can average the relative cluster size $M(L)/L^2$ over many samples of the random forest. With larger and larger lattices, the dependence on L

[7] The amplitude will increase faster than the width L of the forest, but not as fast as the area L^2.

Forest Fire Duration

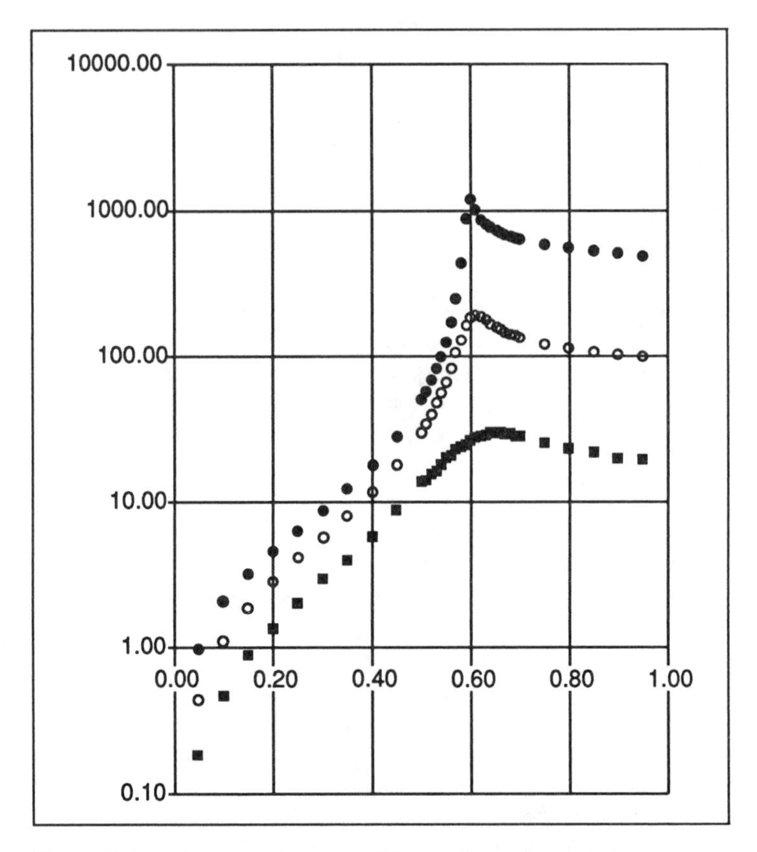

Figure 7.9 : Average duration of forest fires, simulated on square lattices with 20 rows (bottom), 100 rows (middle), and 500 rows (top). For each point in the plot 1000 runs were performed and averaged. The larger the lattice chosen, the more pronounced the peak of the forest fire duration near the percolation threshold at about $p = 0.60$.

diminishes. In other words, we arrive at a limit

$$P_\infty(p) = \lim_{L\to\infty} P_L(p) \ .$$

For low values of p, the probabilities $P_L(p)$ are negligible, and in the limit $L \to \infty$ they tend to zero. But there is a critical value — namely the percolation threshold p_c — beyond which $P_\infty(p)$ grows rapidly. In other words, if p is above percolation, the maximal cluster is infinitely large and comes close to all lattice sites, while for $p < p_c$, the probability that a site picked at random belongs to the maximal cluster is negligible.

The Phase Transition at the Percolation Threshold

At the percolation probability this likelihood increases abruptly. In fact, for $p > p_c$ and p close to p_c the probability $P_\infty(p)$ is given by a power law[8]

$$P_\infty(p) \propto (p - p_c)^\beta$$

with an exponent $\beta = 5/36$. In terms of our forest fire simulation, we may equivalently consider the fraction of trees burned down after termination of the fire (see figure 7.10). There is a sharp increase near the critical value p_c, which becomes more drastic for lattices with more rows. This effect is also called a *phase transition* like similar phenomena in physics. For example, when heating water there is a phase transition from liquid to gas at 100 degrees celsius.[9]

A Fractal Called the Incipient Percolation Cluster

From this observation we can make some conclusions about the size of the maximal cluster: if $p > p_c$, then the positive probability $P_\infty(p)$ implies that the cluster size scales as L^2. On the other hand, for $p \leq p_c$ we may conjecture that a power law holds so that the size is proportional to L^D with $D < 2$. This would indicate a fractal structure of the maximal cluster. This is true, however, only for one special value of p, namely exactly at the percolation $p = p_c$. The fractal percolation cluster at the threshold is often called the *incipient* percolation cluster. Its dimension has been measured; it is $D \approx 1.89$. For values below p_c the maximal cluster size scales only as $\log(L)$.

Other Aspects of Percolation

This analysis, of course, is not the whole truth about percolations. For example, there are other lattices. We can consider three-dimensional or even higher-dimensional lattices. We can also include other neighborhood relations. There are many other quantities besides p_c, D, and $P_\infty(p)$ of interest. For example, the *correlation length* ξ is an important characteristic number. It is defined as an average distance between two sites belonging to the same cluster. As p approaches p_c from below, the correlation length ξ grows beyond all bounds. This growth is again described by a power law

$$\xi \propto |p - p_c|^{-\nu}$$

with an exponent ν, which is equal to 4/3 in two-dimensional lattices. The correlation length is of relevance for numerical simulations. When the lattice size L is smaller than the correlation length ξ, all clusters look fractal with the same dimension. Only when the resolution of the lattice is sufficiently high ($L \gg \xi$) is it possible to determine that clusters are in fact finite and have dimension 0 for $p < p_c$.

[8]The symbol '$\propto$' means 'proportional'.

[9]Under normalized conditions.

Phase Transition at Percolation

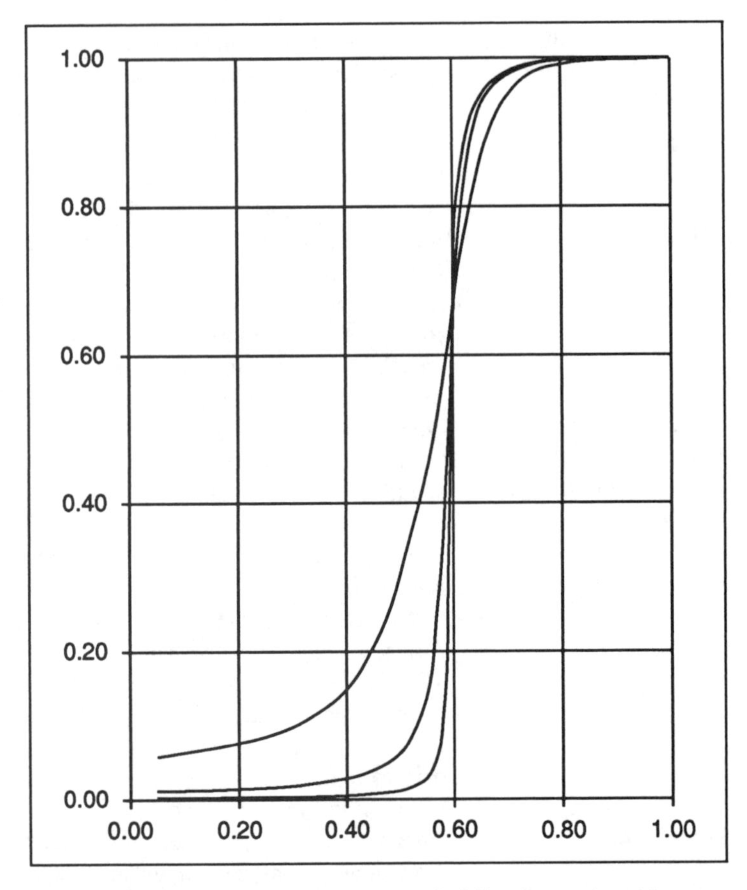

Figure 7.10 : The graph shows the probability that a site with a tree picked at random will be reached by the forest fire in the simulation. It corresponds to the data shown in figure 7.9. The lattice sizes (numbers of rows) are 20, 100, and 500. The probability that a given tree will be burned decreases to zero below the percolation threshold when the lattice size is increased. Above the percolation the portion of trees burned grows asymptotically according to a power law.

Some Constants are Universal

It is important to note that the percolation threshold p_c depends on the choices in the various models, e.g. , on the type of lattice and the neighborhood relations of a site. However, the way quantities such as the correlation length scale near the percolation do not depend on these choices. Thus, numbers characterizing this behavior, such as the exponents ν and ξ and the fractal dimension of the percolation cluster, are called *universal*. The values of many constants, for example, $p_c \approx 0.5928$ and $D \approx 1.89$, are, however, only approximations obtained by elaborate computer studies. It is a current challenge to derive methods for the exact computation of these constants, but we cannot go into any more details here — a lot of problems are still open and are areas of active research.

From Square Back to Triangular Lattices

To conclude this section let us return to the triangular lattice with which we started and which was motivated by the Sierpinski gasket. The quantities $M(L)$, $P_L(p)$, and $P_\infty(p)$ can also be analogously defined for this case. The first numerical estimates in 1960 indicated that the percolation threshold is about 0.5. Then it took about 20 years from the first non-rigorous arguments to a full mathematical proof to show that $p_c = 0.5$ exactly. Moreover, it was shown that the fractal dimension of the incipient percolation cluster is

$$D = \frac{91}{48} \approx 1.896$$

(compare figure 7.11). This is about the same value as the one determined numerically for the square lattice. Therefore it has been conjectured that it is the correct dimension of the incipient percolation cluster in all two-dimensional lattices.

The Renormalization Technique

Instead of looking at the proof for the result $p_c = 0.5$, we may provide a different interesting argument, which opens the door to another method for the analysis of fractals which we have not discussed so far: *renormalization.* One of the keys to understanding fractals is their self-similarity, which reveals itself when applying an appropriate scaling of the object in question. Is there a similar way to understand the incipient fractal percolation cluster? The answer is yes; and it is not hard to investigate, at least for the triangular lattice. The claim is that a reduced copy of the cluster looks the same as the original from a statistical point of view. But how can we compare the two? For this purpose we systematically replace collections of lattice sites by corresponding super-sites. In a triangular lattice it is natural to join three sites to form one super-site. This super-site inherits information from its three predecessors — namely, whether it is occupied or not. The most natural rule for this process is the majority rule; if two or more of the three original sites are occupied, then — and only in this case — the super-site is also occupied. Figure 7.12 shows the procedure and also the geometrical placement of the sites. The super-sites themselves form a new triangular lattice which can now be reduced in size to match the original lattice, allowing a comparison. The concentration of occupied sites — let us call it p' — in the renormalized lattice will not generally be the same as in the old lattice. For example, if p is low, then there are only a few isolated occupied sites, most of which will have vanished in the process of renormalization, thus $p' < p$. At the other end of the scale, when p is large, many more super-sites will be formed, which close up the gaps left in the original lattice, resulting in

Fractal Dimension of the Incipient Percolation Cluster

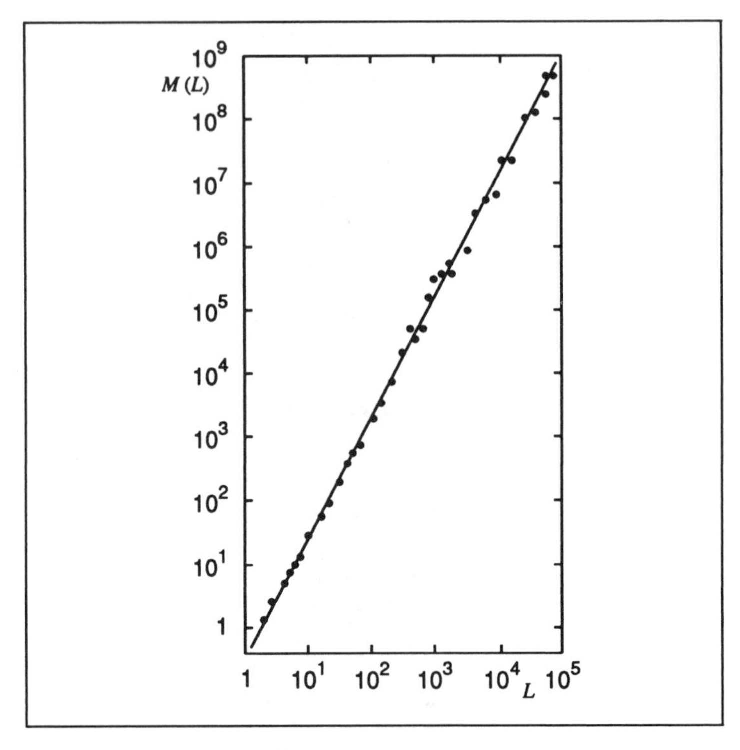

Figure 7.11 : The fractal dimension D of the incipient percolation cluster in a triangular lattice is determined here in a log-log diagram of the cluster size $M(L)$ versus the grid size L. The percolation threshold is $p_c = 0.5$. The slope of the approximating line confirms the theoretical value $D = 91/48$. (Figure adapted from D. Stauffer, *Introduction to Percolation Theory,* Taylor & Francis, 1985.)

$p' > p$. Only at the percolation threshold can we expect similarity. There the renormalized super-cluster should be the same as before, in other words,[10]

$$p' = p \ .$$

In this case we are lucky; we can compute at which probability p the above equation holds! A super-site will be occupied if all three of the original sites are occupied, or if exactly one site is not occupied. The probability for an occupied site is p. Thus, the first case occurs with probability p^3. In the other case, we have a probability of $p^2(1-p)$ that any particular site is not occupied while the other two are. There are three such possibilities. Thus,

[10] Here we see a remarkable interpretation of self-similarity in terms of a fixed point of the renormalization procedure. This relation of ideas from renormalization theory has turned out to be extremely fruitful in the theory of critical phenomena in statistical physics.

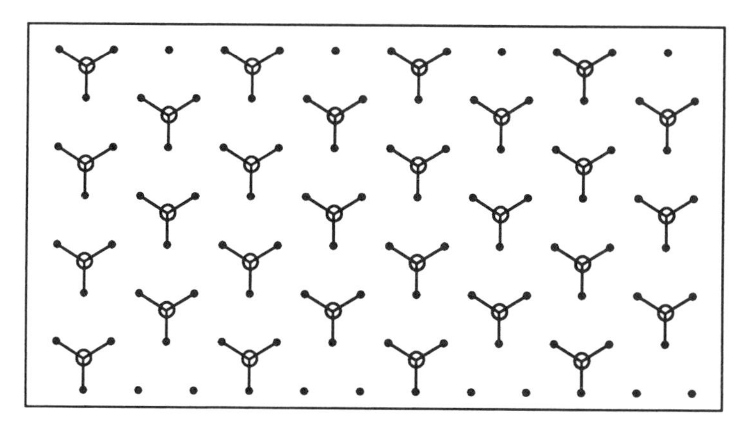

Renormalization of Sites in Triangular Lattice

Figure 7.12 : Three neighboring cells are joined to form a super-site. The super-cite is occupied if two or all three of the small sites are occupied. The super-cites form another triangular grid, however, rotated by 90 degrees. Scaling down the size of this grid of super-cites completes one cycle of the renormalization scheme. See figure 7.14 for examples.

summing up we arrive at

$$p' = p^3 + 3p^2(1-p)$$

as the probability for the super-site being occupied. Now we are almost there. What is p such that $p' = p$? For the answer we have to solve the equation

$$p = p^3 + 3p^2(1-p)$$

or, equivalently,

$$p^3 + 3p^2(1-p) - p = 0\ .$$

It is easy to check that

$$p^3 + 3p^2(1-p) - p = -2p(p-0.5)(p-1)\ .$$

Thus, there are three solutions, $p = 0$, $p = 0.5$, and $p = 1$. Of these three solutions, two are not of interest, namely $p = 0$ and $p = 1$. A forest without trees ($p = 0$) renormalizes to another forest without trees, which is not surprising. Likewise, a saturated forest ($p = 1$) also does not change when renormalized. But the third solution, $p = 0.5$, is the one we are after. It corresponds to a non-trivial configuration, i.e. the forest does have a structure, which after renormalization is still statistically the same. The super-sites have the same probability 0.5 of being occupied as the sites in the original lattice. This is the statistical self-similarity expected

The Renormalization Transformation

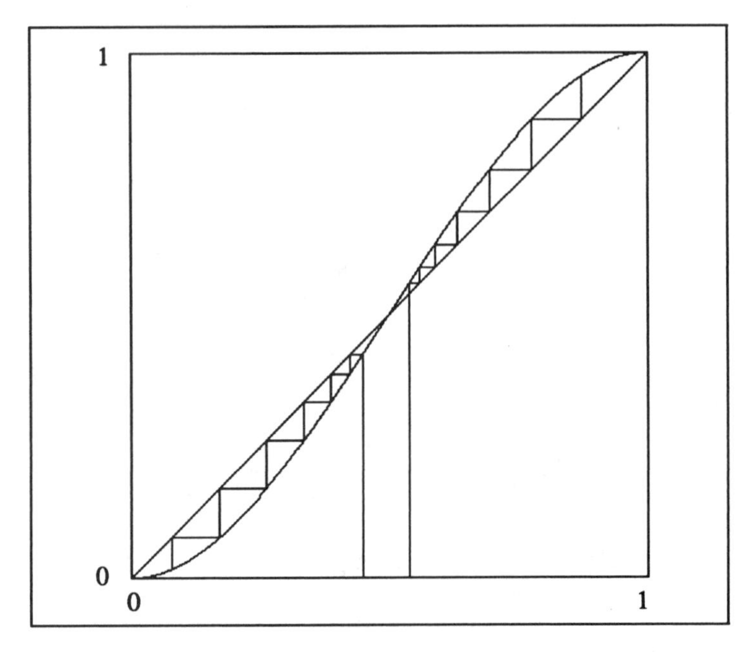

Figure 7.13 : Graphical iteration for the renormalization transformation $p \to p^3 + 3p^2(1-p)$ of the triangular lattice.

at the percolation threshold. Thus, an elementary renormalization argument tells us that $p_c = 0.5$ in accordance to the actual result.

At the percolation threshold, renormalization does not change anything — even when applied many times over. This is not the case for all other probabilities $0 < p < 1$, $p \neq 0$. To study the effect of repeated renormalization we need to consider something very familiar, a feedback system, which relates the probabilities of a site being occupied before and after a renormalization. Thus, we have to consider the iteration of the cubic polynomial

$$p \to p^3 + 3p^2(1-p) \ .$$

This study is best presented in the corresponding diagram of the graphical iteration (see figure 7.13). The situation is very clear. Starting with an initial probability $p_0 < 0.5$, the iterations converge to 0, while an initial $p_0 > 0.5$ leads to the limit 1. Only right at the percolation threshold $p_c = 0.5$ do we obtain a dynamical behavior different from the above, namely a fixed point.

Using Renormalization as an Investigative Tool

As an application of this renormalization transformation we could check whether a given lattice is above or below percolation. We would carry out the renormalization procedure a number of times. If the picture converges to an empty configuration (no sites occupied), then the parameter p belonging to the lattice in

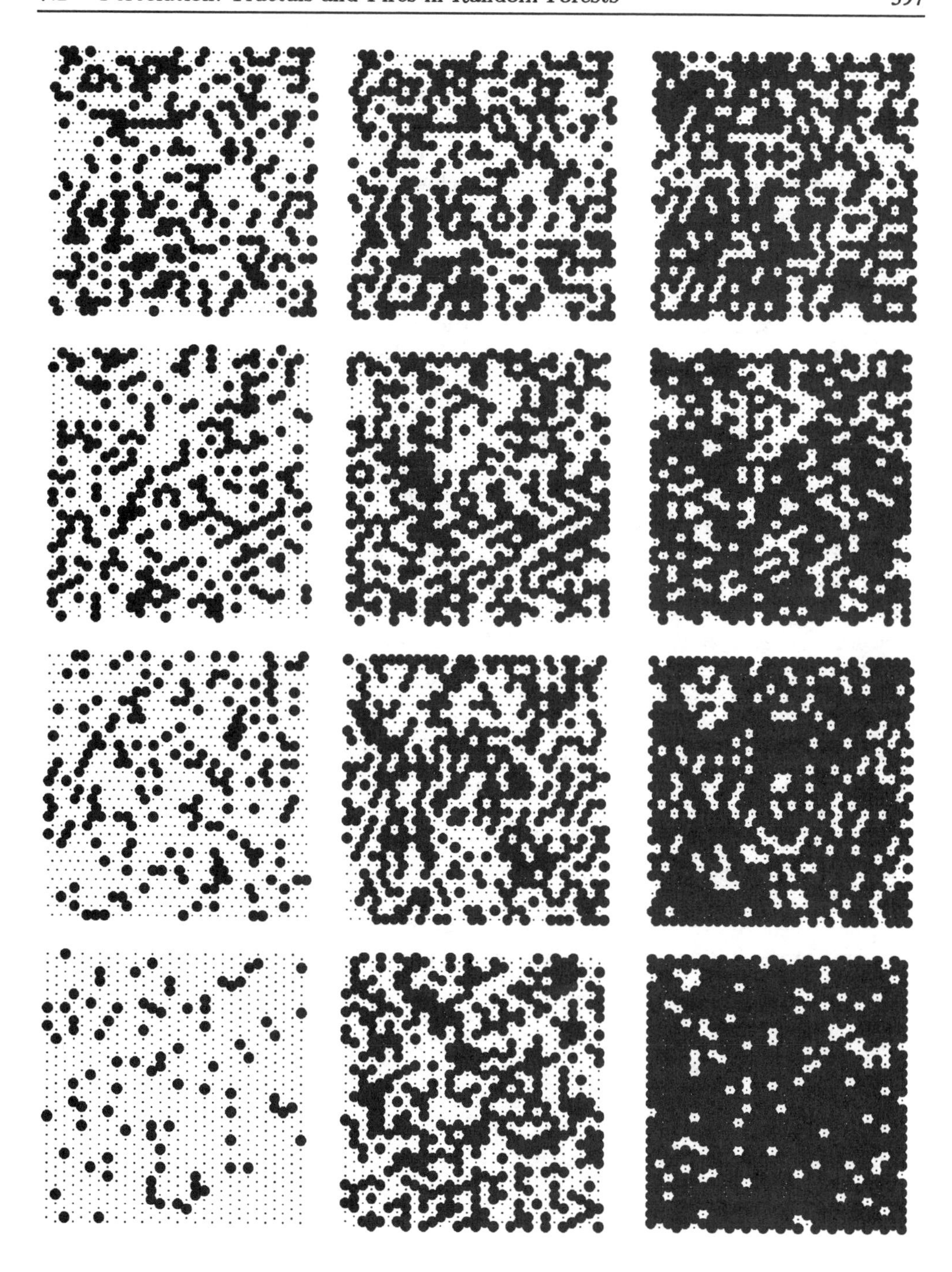

Figure 7.14 : Three renormalization steps for three given configurations (top) corresponding to the cases $p = 0.35 < p_c$, $p = 0.5 = p_c$ and $p = 0.65 > p_c$ (from left to right).

question is below percolation; and if in the long run all sites tend to be occupied, the original configuration is above the percolation threshold. This program is carried out in figure 7.14. From the three original configurations in the top row it is not quite clear by visual inspection whether they are above or below percolation, but the renormalization reveals this information after already three steps.

The triangular lattice is a rather special case. When applying the technique to other lattices, only approximations of p_c can be expected. But it is remarkable how this new idea has created a method to approach the very difficult problem of determining percolation parameters. The basic idea of renormalization came from the physicist Leo P. Kadanoff in 1966, in connection with critical phenomena in a different area of theoretical physics. The idea eventually led to quantitative results and explained the physics of phase transitions in a satisfactory way. After all, the way from the idea of renormalization to its concrete, final form was so elusive that Kadanoff did not find it. Rather, Ken G. Wilson at Cornell in 1970 surmounted the difficulties and developed the method of renormalization into a technical intrument that has proven its worth in innumerable applications. About a decade later he was honored with the Nobel prize for his work.

The situation at the percolation threshold, or more generally at the renormalization fixed point, has an analogy in fractal constructions. Recall, for example, the Koch curve construction where the object must be scaled by a factor of $s = 1/3$ in each stage of the construction. If we scale by a factor $s < 1/3$, then in the limit we arrive at just a point. On the other hand, scaling by $s > 1/3$ in each stage lets the construction grow beyond any bounds. Only if we scale exactly by 1/3 from step to step will we get an interesting limit with self-similarity. In most other cases besides the Koch construction it is not at all obvious how to choose the 'right' scaling.[11]

Percolation is a widely used model and applies to many phenomena observed in nature and the engineering sciences. An example is the formation of thin gold films on an amorphous substrate, where the parameter in question corresponds to the amount of gold provided.[12] At the percolation threshold the metal provides electrical connectivity. On the other hand, percolation is also relevant at scales as large as in the formation of galaxies and clusters of galaxies.

[11] See F. M. Dekking, *Recurrent Sets*, Advances in Mathematics 44, 1 (1982) 78–104.

[12] See R. Voss, *Fractals in Nature*, in: The Science of Fractal Images, H.-O. Peitgen and D. Saupe (eds.), Springer-Verlag, pages 36–37.

7.3 Random Fractals in a Laboratory Experiment

There exists a wealth of fractal structures observed in nature and laboratory experiments.[13] In this section we concentrate on one particularly interesting example: aggregation.

Cluster by Aggregation of Small Particles

The research on the aggregation of small particles to form large clusters in polymer science, material science and immunology, among other fields, has been going on for a long time. However, their study has recently been revitalized tremendously by concepts from fractal geometry.[14] In this section we describe only one particular experiment from this volume, reported on by Mitsugu Matsushita, dealing with electrochemical deposition leading to dendritic structures. It has the advantage that the experimental setup is small and easy to build, and the necessary chemical substances are easily obtained and not dangerous.[15] The complete experiment takes only about 20 minutes. Thus, it may be conveniently conducted right in the classroom. The setup may be filmed and projected by video equipment or even put directly on an overhead projector[16] for viewing by a larger audience.

Dendritic Electrochemical Deposition

Let us quote the description of the experiment directly from Matsushita's article:[17]

"Electrochemical deposition has for a long time been one of the most familiar aggregation phenomena in chemistry. Only very recently has it received attention from the entirely new viewpoint of fractal geometry. In practice, electrodeposition processes may be complex, and the resulting deposits may exhibit a variety of complex structures. However, if the metal deposition is controlled mainly by a single process, e.g. diffusion, then the deposits usually exhibit statistically simple, self-similar, i.e. fractal, structures.

"In this experiment metallic zinc in the form known as zinc metal-leaves was grown two-dimensionally. The experimental procedures used to grow zinc metal-leaves are as follows. A Petri-dish of diameter approx. 20 cm and depth approx. 10 cm is filled with 2 M $ZnSO_4$ aqueous solution (depth approx. 4 mm), and a layer of n-butyl acetate $[CH_3COO(CH_2)_3CH_3]$ is added to make an interface (Fig. 7.15). A tip of a carbon cathode (pencil core of diameter

---

[13] E. Guyon and H. E. Stanley (eds.), *Fractal Forms*, Elsevier/North-Holland and Palais de la Découverte, 1991.

[14] See *The Fractal Approach to Heterogeneous Chemistry: Surfaces, Colloids, Polymers*, edited by D. AvnirWiley, Chichester 1989 and *Aggregation and Gelation*, edited by F. Family and D. P. Landau, North-Holland, Amsterdam, 1984.

[15] Of course, after the experiment, the used liquids must be disposed of properly (not in the sink). Moreover, a good ventilation of the room is recommended.

[16] However, the heat of the lamp in the projector disturbs the experiment, which runs best at constant temperature (large solid zinc leaves form). It is therefore advisable to leave the overhead projector off most of the time. It is best to video film the experiment for immediate viewing on monitors of a projection unit.

[17] M. Matsushita, *Experimental Observation of Aggregations*, in: *The Fractal Approach to Heterogeneous Chemistry: Surfaces, Colloids, Polymers*, D. Avnir (ed.), Wiley, Chichester 1989.

Experimental Setup

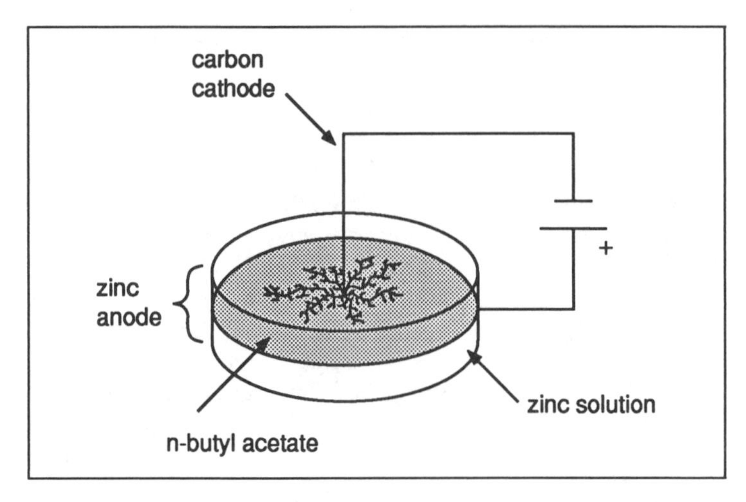

Figure 7.15 : In the Petri-dish a solution of zinc sulfate is covered by a thin layer of n-butyl-acetate.

approx. 0.5 mm) is polished carefully so as to make it flat perpendicularly to the axis. The cathode is then set at the centre of the Petri-dish so that the flat tip is placed just on the interface (Fig. 7.15). The electrodeposition is initiated by applying a d.c. voltage between the carbon cathode and a zinc ring-plate anode of diameter approx. 17 cm, width approx. 2.5 cm and thickness approx. 3 mm placed in the Petri-dish. A zinc metal-leaf grows two-dimensionally at the interface between the two liquids from the edge of the flat tip of the cathode towards the outside anode with an intricately branched random pattern (Fig. 7.16). If the cathode tip is rounded or is immersed in the $ZnSO_4$ solution the deposit grows three-dimensionally into the solution. Usually, the zinc metal-leaves grow to a size of about 10 cm after about 10 min by applying a constant d.c. voltage of about 5 V. The temperature of the system was kept fixed, e.g. at about room temterature.

"The investigation of fractal structures of electrodeposits and their morphological changes is also of practical importance. The electrodeposition experiments presented here are clearly relevant to processes such as metal migration on ceramic or glass substrates and to zinc deposits on cathodes in various batteries. In both cases the growth of deposits is the main factor limiting the lifetime of many electronic parts and batteries."

The mathematical modeling of the electrochemical deposition of zinc-metal leaves is based on the fundamental concept of Brownian motion. Brownian motion refers to the erratic movements of small particles of solid matter suspended in a liquid. These movements can only be seen under a microscope. After the discovery

Sample Result

Figure 7.16 : This dendritic growth pattern was produced after only about 15 minutes by Peter Plath, University of Bremen. The reproduction here is in about the original size. The real zinc dendrite looks very attractive due to its metallic shiny character.

of such movement of pollen it was believed that the cause of the motion was biological in nature. However, about 1828 the botanist Robert Brown realized that a physical explanation, rather than the biological one, was correct. The effect is due to the influence of very light collisions with the surrounding molecules. In the electrochemical experiment zinc ions randomly wander around in the solution until they are caught by the attractive pull of the carbon cathode. The aggregation of a zinc ion is most likely where the density of field lines is greatest. This is at the interface between the solution and the acetate, in particular at the tips of the dendrite. We derive a simple method for the computer simulation of such Brownian motion, which will enable us to also simulate the results of the electrochemical experiment.

Simulation of the Electrochemical Aggregation Experiment

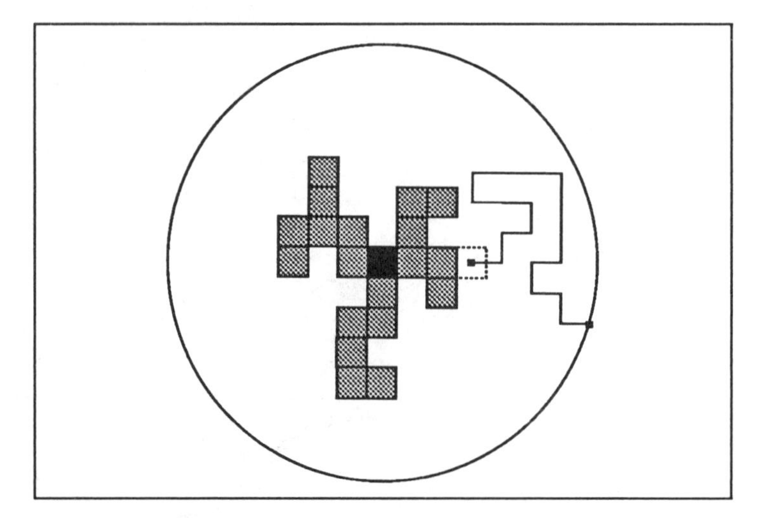

Figure 7.17 : Simulation of Brownian motion in two dimensions is used for the paths of the zinc ions in the liquid. Particles move from pixel to pixel until they 'attach' to the existing dendrite.

Simulation of Diffusion Limited Aggregation (DLA)

To simulate diffusion limited aggregation (= DLA) based on the Brownian motion of particles is not hard.[18] We fix a single 'sticky' particle somewhere, say at the origin of a two-dimensional coordinate system. This particle is not allowed to move. Next we select a region of interest centered around the initial sticky particle, say a circular area of some radius, which could be chosen as 100 or perhaps 500 particle diameters. We inject a free particle at the boundary of the region and let it move about randomly. Two things may happen in the course of this motion. Either the particle leaves the region of interest, in which case we forget about it and start a new free particle at a random position at the boundary of the region, or it stays in this region until it gets close to the sticky particle. In that latter case it attaches and also becomes a sticky particle (with some probability). Now the procedure is repeated, in effect growing a cluster of connected sticky particles which very much resembles the dendrites resulting from DLA in electrochemical deposition, see figure 7.18.

The practical computation is usually based on a square lattice of pixels, see figure 7.17, and the free particle may move to one of its four neighboring pixels in one time step. For a large cluster the process may take *very* long, and several tricks should be used to accelerate the process. For example, at each step the particle may be allowed to move a larger distance than just one pixel. This is

[18]The model presented here originates from T. A. Witten and L. M. Sander, Phys. Rev. Lett. 47 (1981) 1400–1403 and Phys. Rev. B27 (1983) 5686–5697.

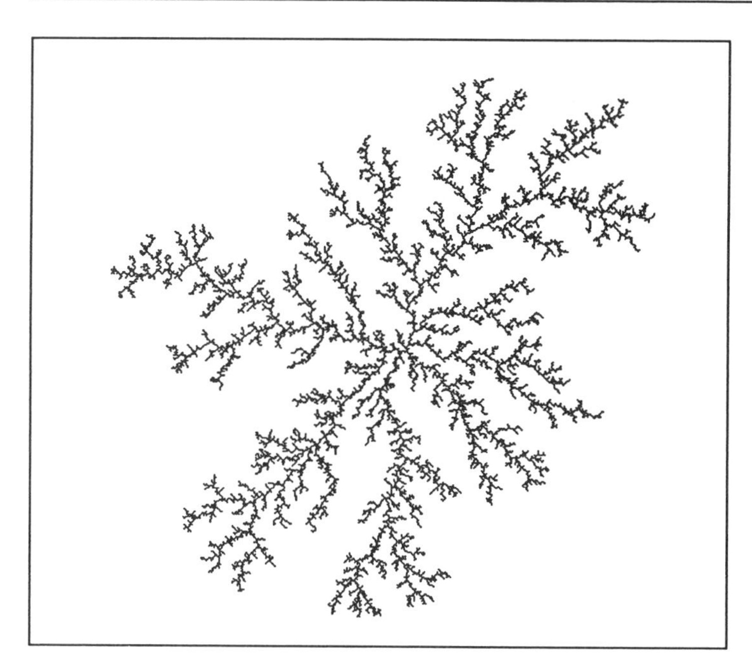

Figure 7.18 : Result of the numerical simulation of DLA based on Brownian motion of single particles.

Simulation Results

possible when the current particle is relatively far from the cluster. More precisely, the distance it may jump in one step is limited by the distance of the particle to the cluster.

Problems

Based on time records of both the real electrochemical experiment and the computer simulation, several questions are of interest.

1. What is the fractal dimension of the aggregate?
2. Clearly, the density of the particles decreases the greater the distance to the center of the dendrite is. Is there a mathematical (power law) relation between density and distance?
3. Does the voltage between the ring anode and the carbon cathode in the experiment have an effect on the value of the fractal dimension of the aggegrate? If so, how can we modify the simulation to take account of that?
4. How is the flowing electrical current related to the size of the aggregate?

Answers to some of these questions have been found, but the research on aggregation is far from complete.[19] The fractal dimension, for example, has been measured extensively in experiments

[19]See the review article by H. Eugene Stanley and Paul Meakin, *Multifractal phenomena in physics and chemistry*, Nature 335 (1988) 405–409.

Fractal Dimension Versus Voltage

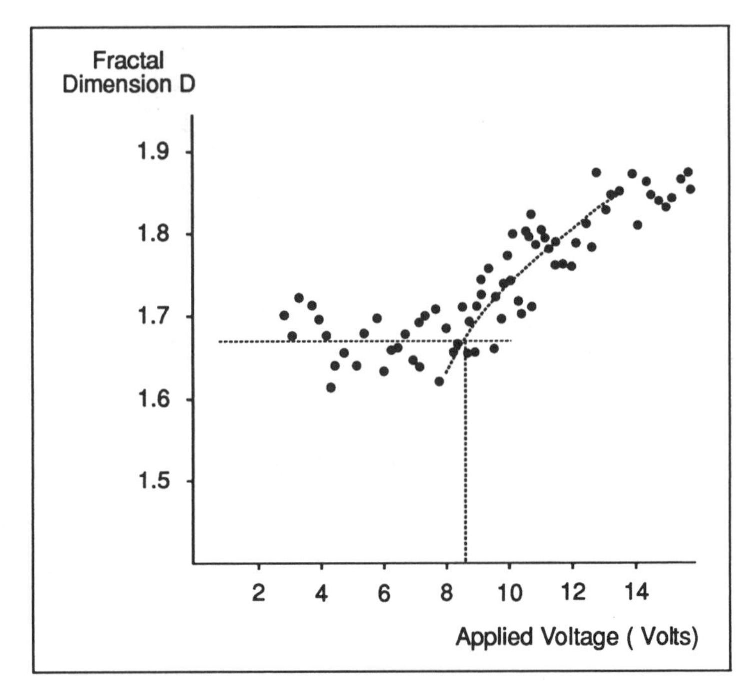

Figure 7.19 : This graph shows the experimental results relating the fractal dimension of the DLA aggregate to the applied voltage. For low voltages the dimension seems to be about constant. Then there is a critical voltage after which the dimension grows abruptly.

and simulations, both yielding the same value 1.7. When the dendrites grow in three dimensions instead of two, the dimension is about 2.4 to 2.5. The dependence of the dimension on the voltage has been studied also (see figure 7.19).

The mathematical model for diffusion limited aggregation can be extended and improved. For example, the sticking probability mentioned further above, which determines whether an ion close to the dendrite attaches to the structure or continues to wander around, is a parameter of interest. It allows variations of the small scale structures. The smaller the sticking probability, the farther particles may reach in the fjords of the dendrite, thickening the dendrite to form a moss-like structure.[20]

Extensions of the DLA Model

Some interesting extensions of the simple model for DLA have been studied. Instead of tracing a single particle, many particles can be considered simultaneously.[21] Moreover, alternatively, the

[20] At large scales, however, the dendritic structure obtained using a small sticking probability does not look 'thick'. The fractal dimension, measured at large scales, is independent of the sticking probability.

[21] See R. F. Voss and M. Tomkiewicz, *Computer Simulation of Dendritic Electrodeposition,* Journal Electrochemical Society 132, 2 (1985) 371–375.

dendritic cluster may be allowed to move about, picking up particles that are close by. There is another seemingly unrelated very different model for DLA. Instead of tracing particles, an equation is solved which reflects that in reality there are infinitely many particles moving about simultaneously. Thus, in place of individual particles some continuous density function is considered. The equation governing the elecrostatic potential is a partial differential equation, known as the *Laplace equation.* Aggregation occurs along the boundary of the dendrite where the gradient of the potential is greatest. Sometimes fractals such as DLA clusters are therefore also called *Laplacian fractals.* It is easy to introduce a parameter in the Laplacian model which controls the dimension.[22]

Phenomena similar to the aggregation discussed here occur at all scales of measurement, in distribution of galaxies as well as in the microcosm. In addition to diffusion limited aggregation and percolation, which we have already mentioned, a partial list of phenomena would range from molecular fractal surfaces to viscous fingering in porous media and clouds and rainfall areas.[23]

[22]For details see L. Pietronero, C. Evertz, A. P. Siebesma, *Fractal and multifractal structures in kinetic critical phenomena,* in: *Stochastic Processes in Physics and Engineering,* S. Albeverio, P. Blanchard, M. Hazewinkel, L. Streit (eds.), D. Reidel Publishing Company (1988) 253–278.

[23]For a discussion of these and other phenomena from a physical point of view see the book *Fractals* by J. Feder, Plenum Press, New York 1988.

7.4 Simulation of Brownian Motion

Brownian motion is not only important part of the model for diffusion limited aggregation, but it also serves as the basis of many other models for natural fractal shapes such as landscapes. In order to study these models it is necessary to better understand Brownian motion and its generalizations. In this and the following section we take a closer look at Brownian motion and methods for its simulation.

Before stating the results and the extensions, we simplify and consider Brownian motion in only one space variable. Thus, the motion of particles is restricted to a line. The tiny molecular impacts affect the particle only from the left or the right causing a unit length displacement in either direction. Can we make any prediction about the total displacement after a number of such time steps, say n steps? If so, we could also simulate Brownian motion for larger time intervals, thus reducing the cost for the simulation.

The Mean Square Displacement

Let us solve this problem; it is not hard. First of all, we realize that it is not very sensible to ask for the total expected displacement, i.e. the displacement of particles averaged over many samples. This would be *zero* because all individual unit length displacements are $+1$ or -1, both with equal probability 0.5. Thus, on the average, the overall displacement must be zero. Instead, let us consider the *square* of the displacement, a nonnegative number. The average of the square displacements, called the *mean square displacement*, tells us how much the particles spread in a given number of time steps. The result of its computation is n, the number of time steps. Thus, the more steps we allow, the farther the particles spread out. Moreover, we have quantified this relation: *on the average* the square of the displacement is equal to the number of time steps.

Computing the Mean Square Displacement

To compute the mean square displacement denote by $d_1, d_2, ..., d_n$ the n unit length displacements. We consider the value of the square

$$(d_1 + d_2 + \cdots + d_n)^2 = \sum_{k=1}^{n} \sum_{l=1}^{n} d_k d_l \ .$$

The individual terms $d_k d_l$ in the sum on the right hand side are easy to analyse. Each factor is either $+1$ or -1 with the same probability 0.5, and, moreover, the factors are independent from each other in the case $k \neq l$. From this it follows that there are four cases for the product, which are all equally likely, as given in the table.

d_k	d_l	$d_k d_l$	Probability
1	1	1	0.25
1	−1	−1	0.25
−1	1	−1	0.25
−1	−1	1	0.25

Thus, the product is also equal to $+1$ or -1 with the same probability 0.5, and the expected value of such a product for $k \neq l$ is zero. Of course, the value of the terms $d_k d_k$ are always equal to $+1$ for all $k = 1, ..., n$. Therefore, the result is clear: the expected value of the squared total displacement is equal to the number of steps, n.

Let us now remove the restriction of the unit length step size of the small displacements. Considering a smaller length is accounted for by modifying the result: the expected square displacement is *proportional* to the time difference t. The factor of proportionality depends on the number of steps happening in the time interval t and the step length of the individual displacements. This is the fundamental property of Brownian motion. It is also true in spaces having two or more dimensions.

Up to this point we know that the total displacement after some time t is zero on the average, and that the expected square of the displacement is proportional to t. What more can we say about the distribution of the displacement after time t? In other words, if we sample Brownian motion (or Brownian motion simulated on a computer) at regular time intervals of length t, what is the distribution of the resulting measured displacements? In table 7.21 we list the outcome of such an experiment, which is graphed in figure 7.20.

The shape of the curve in the graph is very familiar to most of us. It is a graph belonging to a distribution which is commonly known as the *Gaussian* or *bell-shaped* distribution. For example, consider the deviation in the body heights of a large group of people, or the variation of several measurements of the length of some (non-fractal) object. Sometimes the Gauss distribution is taken as a model for a statistically healthy sample — which may not always have desirable practical consequences. For example, grades in a class are often given so that the fluctuations of the grades around the average match the prescribed bell-shaped form. Taken to the extreme, this implies that in any class — no matter how brilliant the students may be — there must be a couple of students who flunk the course because Gauss' distribution demands it.

Returning to the results of the above experiment on Brownian motion in one dimension, we note that in this case they are

Statistics for Simulated One Dimensional Brownian Motion

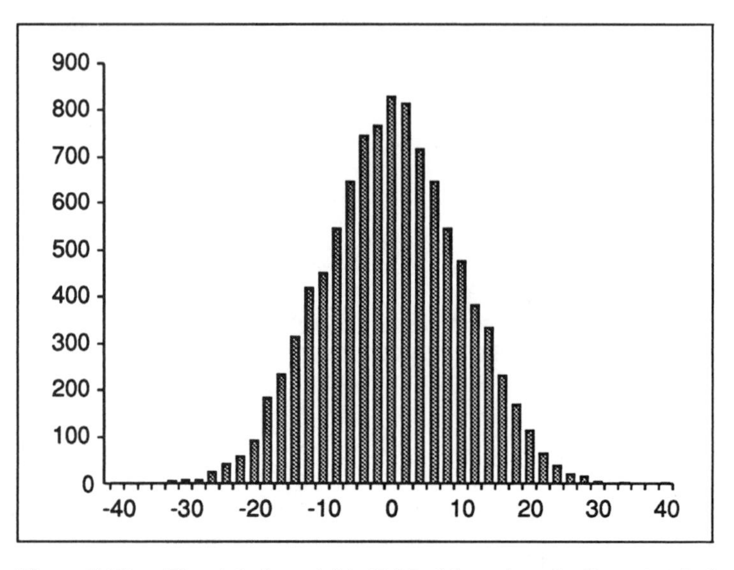

Figure 7.20 : The data from table 7.22 of throwing six dice a total of 100,000 times is drawn as a graph. The distribution is approximately Gaussian.

not an accident, nor are they due to an arbitrary decision of some statistically minded individual that the outcome should match the Gaussian distribution so well. In fact, the Gaussian distribution arises in all cases where independent and similar (i.e. identically distributed) random events are summed up or averaged. This is the content of an important mathematical theorem called the *central limit theorem*.[24] Thus, the characterization of Brownian motion in one variable is now complete. The displacement after time t is a so-called random variable with a Gaussian distribution, which is specified by its mean zero and the mean square displacement proportional to the time difference t. Samples from such a Gaussian distribution, where the mean square is normalized to 1, are called (normalized) Gaussian random numbers.

Gaussian Random Numbers

From these observations it is clear that a simulation of Brownian motion can be based on such Gaussian random numbers. They are equivalent to the displacements corresponding to some time interval. If a displacement for a different time interval is desired, for example, a time twice as long, then simply multiply the Gaussian random number by the appropriate factor — here $\sqrt{2}$. There are efficient and accurate methods available for producing Gaussian random numbers.[25] For our purposes it is sufficient, however,

[24] See any textbook on probability theory or statistics.

[25] For example, the Box-Muller method, see W. H. Press, B. P. Flannery, S. A. Teukolski, W. T. Vetterling, *Numerical Recipes*, Cambridge University Press, 1986, p. 202.

Brownian Motion in One Dimension

D	count	D	count	D	count	D	count
0	828			26	21	−26	28
2	815	−2	767	28	17	−28	9
4	718	−4	746	30	6	−30	10
6	648	−6	648	32	1	−32	7
8	547	−8	547	34	2	−34	1
10	478	−10	453	36	0	−36	0
12	383	−12	421	38	0	−38	0
14	335	−14	315	40	2	−40	1
16	233	−16	234	42	1	−42	0
18	171	−18	185	44	0	−44	0
20	116	−20	94	46	0	−46	0
22	66	−22	60	48	0	−48	0
24	42	−24	44	50	0	−50	0

Table 7.21 : Displacements of Brownian motion sampled $10,000$ times at regular time intervals. In each interval 100 unit length displacements have been carried out and added up. The sum is listed as the total displacement D during a time period of length t corresponding to the $n = 100$ steps. Note that these sums must be even numbers, because $D = a - b$, where $a + b = 100$ and a and b denote the number of times a positive or negative unit length displacement occurred. Thus, $b = 100 - a$, and $D = 2a - 100$, an even number. The mean square displacement is 99.82, very close to the theoretically expected number 100.

to consider only a simple method based on the above mentioned central limit theorem. We can even construct a Gaussian random number using the rolls of a die. This would initially produce random numbers from the list 1, 2, 3, 4, 5, 6, where each number in this set carries the same probability, 1/6, of being chosen. This is called a *uniform distribution* of a random variable. On most computers such random numbers are available with a much wider range of outcomes, usually $0, 1, 2, ..., A$ with $A = 2^{15} - 1$ or even $A = 2^{31} - 1$. If we divide the result by A then we obtain a number in the interval from 0 to 1; and the probability that the result of such an evaluation is, for example, between 0.25 and 0.75 is 50% or 0.50.[26] More generally, the probability, that the random number lies between a and b is $b - a$, when a and b are chosen with $0 \leq a \leq b \leq 1$. To simulate Gaussian random variables, simply take any number of dice, e.g. 6, and roll them. Here the result will be defined as the sum of all dice, which is a number from 6 to 36. Let us repeat the throw many times and keep a record of how many times we come up with each number between 6 and 36, see

[26] In many programming environments this division is internally carried out, and those random numbers are already uniformly distributed in the unit interval.

table 7.22 and figure 7.23.

Throwing Six Dice 100,000 Times

pnts	count	pnts	count	pnts	count	pnts	count
1	0	10	249	19	8503	28	2449
2	0	11	538	20	8961	29	1608
3	0	12	1033	21	9268	30	960
4	0	13	1573	22	9127	31	549
5	0	14	2541	23	8238	32	255
6	4	15	3574	24	7314	33	110
7	15	16	4836	25	5985	34	39
8	48	17	6051	26	4894	35	17
9	110	18	7527	27	3621	36	3

Table 7.22 : Six dice are thrown 100,000 times. The points of all six dice are totalled up and a statistic of these sums is shown in the table.

Approximate Gaussian Distribution by Throwing Dice

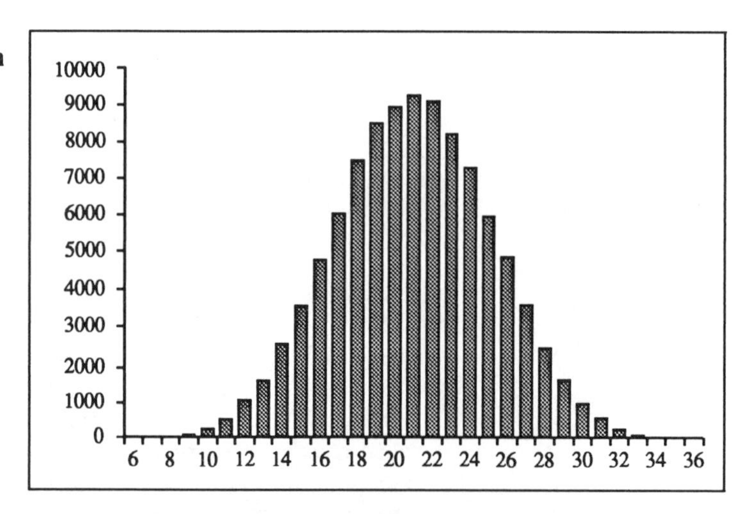

Figure 7.23 : The data from table 7.22 of throwing six dice many times is drawn as a graph. The distribution is approximately Gaussian.

The distribution has the characteristic bell shape. In fact, the central limit theorem again ensures that the Gaussian distribution is approximated by the above experiment, and, moreover, the quality of the approximation is improved by raising the number of dice used in the throws.

For practical purposes it is advisable to normalize the results before actually making use of them in a fractal construction. One reason is that the results do not belong to a Gaussian distribution centered around 0. For example, they are always positive, and the expected value, which is the average of all numbers depends on the number of dice used. The recipe for the normalization can easily be derived using elementary probability theory, but here we only state the final formulae. Let us define the notation

A	the upper limit of our random number generator, which returns numbers $0, 1, ..., A$ (as above)
n	number of 'dice' used
$Y_1, ..., Y_n$	results of one throw of all n 'dice'.

Then an approximate Gaussian random variable is given by

$$D = \frac{1}{A}\sqrt{\frac{12}{n}}(Y_1 + Y_2 + \cdots + Y_n) - \sqrt{3n}$$

when A and n are large. It is normalized so that the expected value is zero and the variance[27] is one. This formula is easily implemented on a computer. For our purposes it suffices to use a small number for n, e.g. $n = 3$. Then the formula simplifies to

$$D = \frac{2}{A}(Y_1 + Y_2 + Y_3) - 3 .$$

In the special case above with six real dice, we have to take into account that the dice values range from 1 (and not 0) to a rather small maximum, 6. Using the exact variance we obtain

$$D = \sqrt{\frac{2}{35}}(Y_1 + \cdots + Y_6 - 21) .$$

The following table lists the conversion of the dice points from 6 to 36 into normalized approximate Gaussian random numbers.

Normalizing the Throw of Six Dice

pnts		pnts		pnts		pnts	
1		10	−2.63	19	−0.48	28	1.67
2		11	−2.39	20	−0.24	29	1.91
3		12	−2.15	21	0.00	30	2.15
4		13	−1.91	22	0.24	31	2.39
5		14	−1.67	23	0.48	32	2.63
6	−3.59	15	−1.43	24	0.72	33	2.87
7	−3.35	16	−1.20	25	0.96	34	3.11
8	−3.11	17	−0.96	26	1.20	35	3.35
9	−2.87	18	−0.72	27	1.43	36	3.59

Table 7.24 : The table lists the conversion of the dice sum to an approximate normalized Gaussian random number.

[27] The variance is the mean square deviation from the expectation. In our case a variance equal to 1 implies that about 68.27% of all outcomes D are less than 1, 95.45% are less than 2, and 99.73% are less than 3 in magnitude.

The Next Step: Summing up Independent Gaussian Random Numbers

The Gaussian random numbers above can be used in a simulation of Brownian motion in one dimension. Let us proceed in the time direction t in equal steps δt. Within each time slot of length δt we accumulate the impacts of all molecules that bump into our particles resulting in a total displacement which is correctly modelled as a Gaussian random number. We set the position of the particle at the starting time to 0, written in shorthand as $X(0) = 0$. After a time step of length δt we evaluate our (normalized) Gaussian random number, call the output D_1, and the position thus is changed to $X(\delta t) = D_1$. After two time steps we get another displacement, a number D_2 returned from our second call to the random number generator. The position is now the sum

$$X(2\delta t) = X(\delta t) + D_2 = D_1 + D_2.$$

Continuing in this way we sum up our Gaussian random numbers, in formula,

$$X(k\delta t) = D_1 + D_2 + \cdots + D_k, k = 1, 2, 3, \ldots$$

The outcome is displayed in figure 7.25.

If an approximation is desired only at every other time, we can shorten the computation because we know that the mean *square* displacement for twice the time differences is also twice as large. Thus, a multiplication of the Gaussian random numbers by $\sqrt{2}$ suffices. In other words,

$$X(2k\delta t) = \sqrt{2}(D_1 + D_2 + \cdots + D_k), k = 1, 2, 3, \ldots$$

An Alternative: The Random Midpoint Displacement Method

Another straightforward and the most popular way to produce Brownian motion is called *random midpoint displacement.*[28] It has several advantages over the method of summing up white noise, the most important one being that it can be generalized to several dimensions useful e.g. for modelling height fields of landscapes.[29]

If the process $X(t)$ is to be computed for times, t, between 0 and 1, then we start out by setting $X(t) = 0$ and by selecting $X(1)$ as a sample of a Gaussian random number. Next $X(\frac{1}{2})$ is constructed as the average of $X(0)$ and $X(1)$, i.e. $\frac{1}{2}(X(0) + X(1))$ plus an offset D_1. Compare the visualization of this step and the next one in figure 7.26. This offset D_1 is a Gaussian random number, which should be multiplied by a scaling factor $\frac{1}{2}$. Then we reduce the scaling factor by $\sqrt{2}$, i.e. it is now $1/\sqrt{8}$, and the two intervals from 0 to $\frac{1}{2}$ and from $\frac{1}{2}$ to 1 are divided again. $X(\frac{1}{4})$ is set as the average $\frac{1}{2}(X(0) + X(\frac{1}{2}))$ plus an offset D_2, which is a

[28] The method was introduced in the paper by A. Fournier, D. Fussell and L. Carpenter, *Computer rendering of stochastic models*, Comm. of the ACM 25 (1982) 371–384.

[29] Another advantage is that we can prescribe the values of $X(t)$ for various times t and have the random midpoint displacement compute intermediate values. In this sense, the method could be interpreted as fractal interpolation.

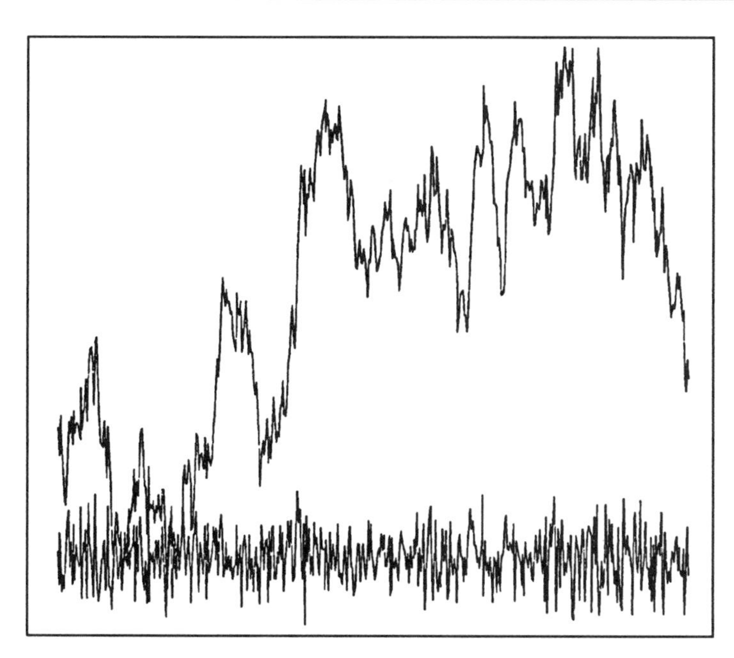

Figure 7.25 : The summation of independent Gaussian random variables (bottom curve) gives a crude model of Brownian motion in one variable (top curve). The particle's position $X(t)$ is plotted in the vertical direction, horizontally time t is varied. The particle moves up and down without any correlation, i.e. if the particle gains height in one time step, the chance for a continuation and the chance for a change of this trend are exactly the same (50 : 50).

Brownian Motion by Summing Up Gaussian Random Variables

Gaussian random number multiplied by the current scaling factor $1/\sqrt{8}$. The corresponding formula holds for $X(\frac{3}{4})$, i.e.

$$X(\frac{3}{4}) = \frac{X(\frac{1}{2}) + X(1)}{2} + D_2$$

where D_2 is a random offset computed as before.

The third stage proceeds in the same manner: reduce the scaling factor by $\sqrt{2}$, i.e. it is $1/\sqrt{16}$. then set

$$\begin{aligned}
X(\tfrac{1}{8}) &= \tfrac{1}{2}(X(0) + X(\tfrac{1}{4})) + D_3 \\
X(\tfrac{3}{8}) &= \tfrac{1}{2}(X(\tfrac{1}{4}) + X(\tfrac{1}{2})) + D_3 \\
X(\tfrac{5}{8}) &= \tfrac{1}{2}(X(\tfrac{1}{2}) + X(\tfrac{3}{4})) + D_3 \\
X(\tfrac{7}{8}) &= \tfrac{1}{2}(X(\tfrac{3}{4}) + X(1)) + D_3 \ .
\end{aligned}$$

In each formula, D_3 is computed as a (different) Gaussian random number multiplied by the current scaling factor $1/\sqrt{16}$. The following stage computes $X(t)$ at $t = 1/16, 3/16, ..., 15/16$ using

Displacing Midpoints

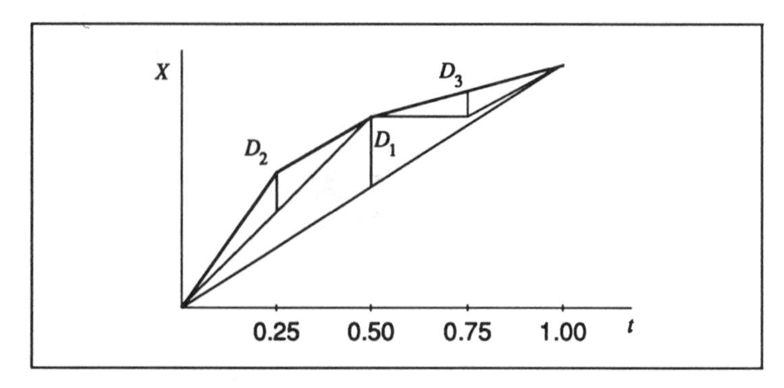

Figure 7.26 : The first two stages of the midpoint displacement technique as explained in the text.

Eight Stages of Midpoint Displacement

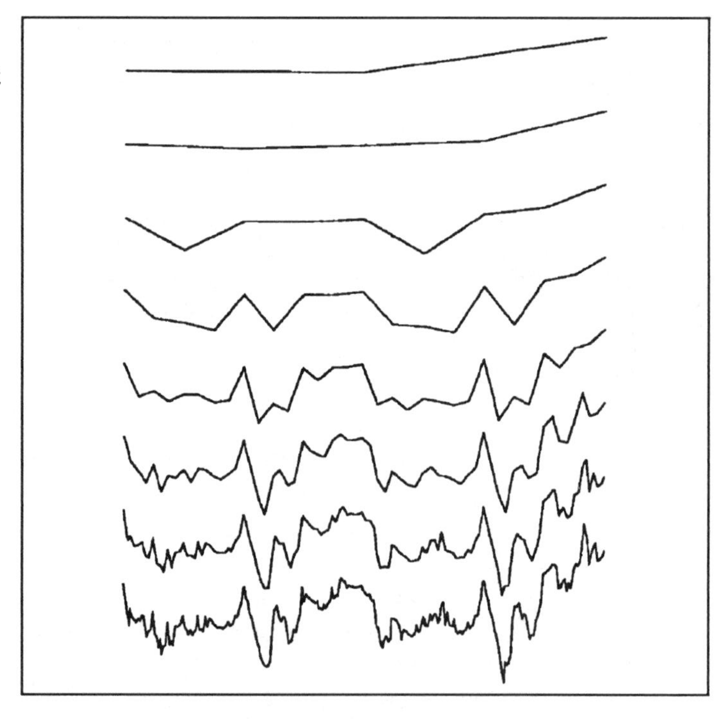

Figure 7.27 : Brownian motion via midpoint displacement. Eight stages are shown depicting approximations of Brownian motion using $3, 5, 9, ..., 257$ points.

a scaling factor again reduced by $\sqrt{2}$, and continues as indicated as above and illustrated in figure 7.27.

Analysis of the Random Midpoint Displacement Method

If the Brownian motion is to be computed for times, t, between 0 and 1, then one starts by setting $X(0) = 0$ and selecting $X(1)$ as a sample of a Gaussian random variable with mean 0 and variance (mean square) var $(X(1)) = \sigma^2$. Then var $(X(1) - X(0)) = \sigma^2$ also, and we expect

$$\text{var } (X(t_2) - X(t_1)) = |t_2 - t_1|\sigma^2 \tag{7.1}$$

for $0 \le t_1 \le t_2 \le 1$. We set $X(\frac{1}{2})$ to be the average of $X(0)$ and $X(1)$ plus some Gaussian random offset D_1 with mean 0 and variance Δ_1^2. Then

$$X(\frac{1}{2}) - X(0) = \frac{1}{2}(X(1) - X(0)) + D_1$$

and thus $X(\frac{1}{2}) - X(0)$ has mean value 0 and the same holds for $X(1) - X(\frac{1}{2})$. Secondly, for (7.1) to be true we must require that

$$\text{var } (X(\frac{1}{2}) - X(0)) = \frac{1}{4}\text{var } (X(1) - X(0)) + \Delta_1^2 = \frac{1}{2}\sigma^2.$$

Therefore

$$\Delta_1^2 = \frac{1}{4}\sigma^2.$$

In the next step we proceed in the same fashion setting

$$X(\frac{1}{4}) - X(0) = \frac{1}{2}(X(0) + X(\frac{1}{2})) + D_2$$

and observe that again the increments in X, here $X(\frac{1}{2}) - X(\frac{1}{4})$ and $X(\frac{1}{4}) - X(0)$ are Gaussian and have mean 0. So we must choose the variance Δ_2^2 of D_2 such that

$$\text{var } (X(\frac{1}{4}) - X(0)) = \frac{1}{4}\text{var } (X(\frac{1}{2}) - X(0)) + \Delta_2^2 = \frac{1}{4}\sigma^2$$

holds, i.e.

$$\Delta_2^2 = \frac{1}{8}\sigma^2 .$$

We apply the same idea to $X(\frac{3}{4})$ and continue to finer resolutions yielding

$$\Delta_n^2 = \frac{1}{2^{n+1}}\sigma^2$$

as the variance of the displacement D_n. Thus, corresponding to time differences $\Delta t = 2^{-n}$, we add a random element of variance $2^{-(n+1)}\sigma^2$ which is proportional to Δt as expected.

Moving Up to the Next Degree of Freedom

Having produced Brownian motion in one dimension it is now an easy task to generalize to the two-dimensional case. The small impacts on a particle are no longer restricted to only two possible directions, a bump from the left or a bump from the right. Rather the direction may be chosen arbitrarily from a range of angles between zero and 180 degrees, in radians from 0 to π.[30] All angles are equally likely, thus, in a simulation, a random variable with a uniform distribution will suffice. In summary, the displacement of the particle is computed by choosing the direction in this specified manner and the amount of the displacement as before by means of a normalized Gaussian random variable.[31]

Trace of Brownian Motion in the Plane

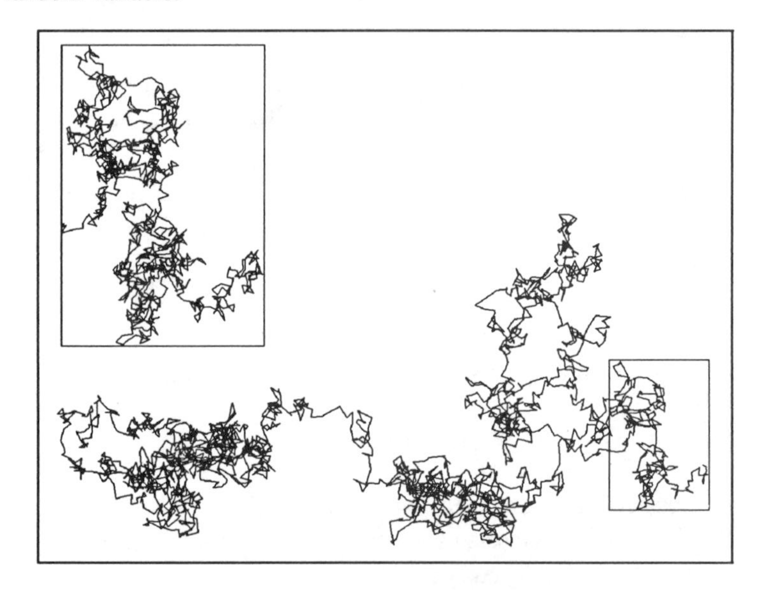

Figure 7.28 : Shown above is the trace of the Brownian motion of a particle. The boxed detail of the trace (magnified in the upper left portion of the figure) suggests an invariance of scale or self-similarity: the detail looks like the whole.

A graphical record of the movements of a particle subject to Brownian motion looks as expected, a very erratic trace (fig. 7.28). The particle wanders around without any pattern. Some regions of the plane are filled densely by the trace. In fact, the fractal dimension of such a trace is equal to two. The enlargement of a section of the path reveals the self-similarity of the motion, it looks very much the same as the whole curve. This resemblance is true, of course, only in a statistical sense and not exactly.

[30]It is not necessary to consider larger angles because the displacement may be positive or negative.

[31]Note that this is not the generalization of Brownian motion which yields height field models of landscapes mentioned earlier (see section 7.6).

7.5 Scaling Laws and Fractional Brownian Motion

What is the Scaling Invariance in the Graph of One-Dimensional Brownian Motion?

Let us now return to the one-dimensional Brownian motion and discuss the similarities in the model that define it as a fractal. By construction and also just by looking at the graph in figure 7.25 it is clear that we cannot expect a similarity of the usual type in which we can take the graph of the Brownian motion and scale it up or down in the time-direction and in the amplitude (with possibly different scaling factors) to obtain the original graph. Such an exact affine self-similarity is obviously not possible due to the randomness in the generation mechanism. However, in figure 7.29 we have tried the construction of the scaled copies of the original anyway. Here we have used the enlargement factor of two for the horizontal direction, while the amplitudes were kept unchanged. We note that the curves do not look very similar; there is much less variation in the bottom curves where we have stretched time with factors 2, 4, 8, ..., and 64.

In the next figure we repeat the experiment with the same factor of two in the horizontal as well as in the vertical direction. As we

Brownian Motion Rescaled Wrongly

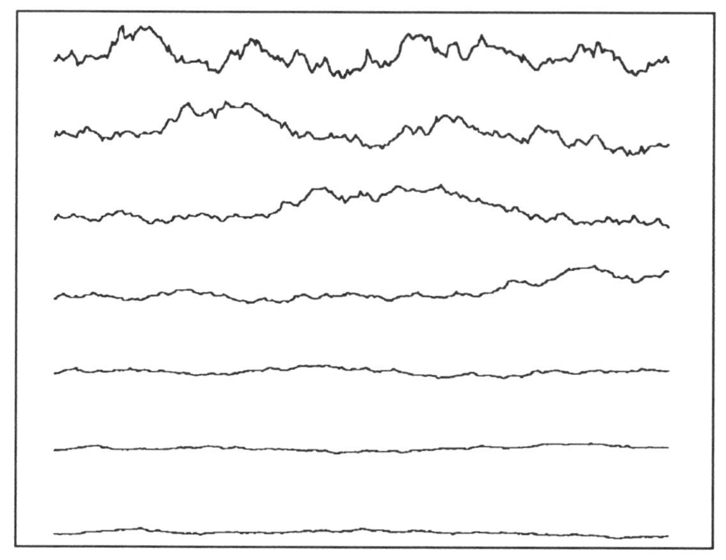

Figure 7.29 : In this experiment we scale up the left half of a sample of Brownian motion in one variable by factors of two in the horizontal direction, while maintaining amplitudes. The result over six such stages is given here with the original curve at the top. Note how the peaks of the 'mountains' are shifted to the right when going down to the next curves. In each plot half of the data from the previous curve disappears due to clipping at the right boundary of the plot.

Brownian Motion Rescaled Wrongly Again

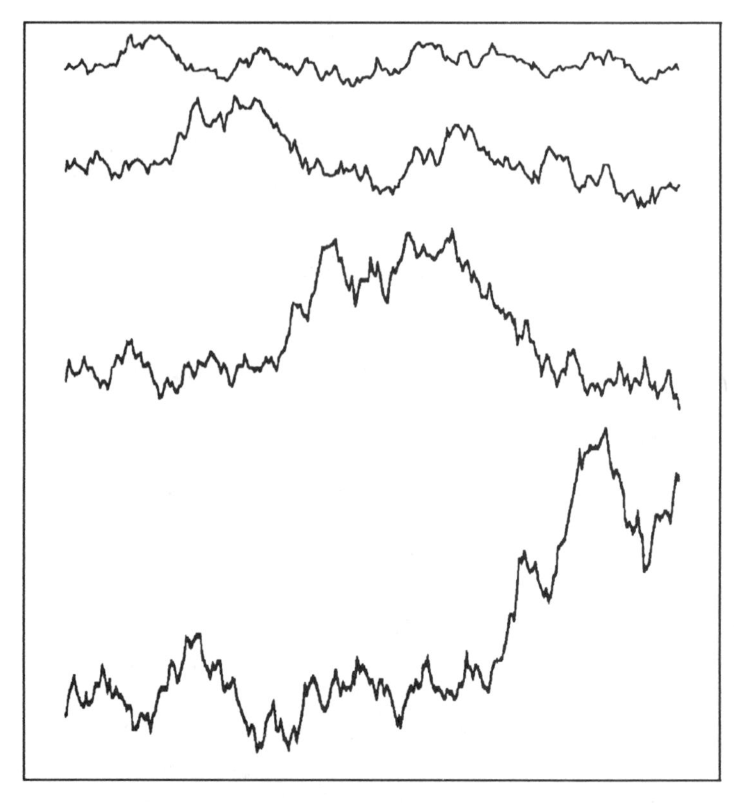

Figure 7.30 : The same experiment as in figure 7.29, but with the same horizontal and vertical scaling factor 2.

scale by two horizontally we also multiply amplitudes by two. This changes the curves dramatically as displayed in figure 7.30. Now the bottom curves have greatly increased variation in amplitude; the graphs look much more erratic.

From these observations we can conclude that between the two scaling factors 1 (figure 7.29) and 2 (figure 7.30), there should be a scaling factor r that yields curves that are visually the same, i.e. when scaling Brownian motion in time by a factor of 2 and in amplitude by a factor of r we see no striking general differences, even when we repeat the reduction procedure many times. To find this number r we may continue by trial and error. However, the result is known to be $r = \sqrt{2}$. This follows directly from our analysis of the mean squared displacements Δ^2 of the Brownian motion $X(t)$, which showed proportionality to the time differences t,

$$\Delta^2 \propto t \; .$$

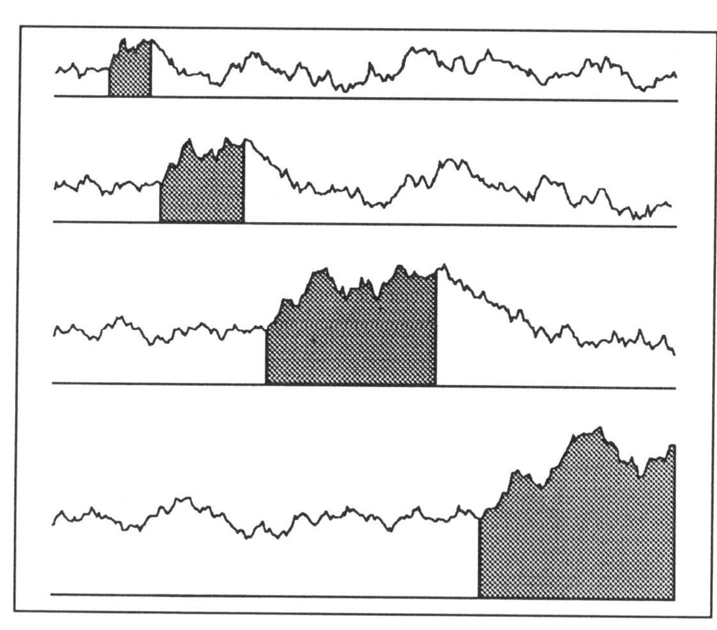

Figure 7.31 : The same experiment as in figure 7.29, but with horizontal scaling factor 2 and the proper vertical scaling factor $r = \sqrt{2}$. The curves are statistically equivalent, revealing the scaling law for Brownian motion. The shaded regions shows the same shape properly rescaled for the different stages.

Properly Rescaled Brownian Motion

Consider the rescaled random function

$$Y(t) = rX\left(\frac{t}{a}\right)$$

i.e. the graph of X is stretched in the time direction by a factor of a and in the amplitude by r. The displacements in Y for a time difference t are the same as those for X multiplied by r and corresponding to time differences t/a. Thus, the squared displacements are proportional to r^2t/a. In order to ensure the same constant of proportionality as in the original Brownian motion, we simply have to require $r^2/a = 1$, or, equivalently, $r = \sqrt{a}$. When replacing t by $t/2$, i.e. stretching the graph by a factor of 2 as in the figures, we have $a = 2$, and thus, $r = \sqrt{2}$ as stated.

The last figure in this sequence, fig. 7.31, demonstrates the result. Indeed, the curves look about the same. In fact, they are the same, statistically speaking. An analysis of mean value, variances, moments and so forth would give the same statistical properties of the rescaled curves. This is the scaling invariance of the graph of Brownian motion.

Are Other Scaling Factors Possible?

In the discussion of scaling invariance we have shown that for ordinary Brownian motion, we need to scale amplitudes by $\sqrt{2}$ when time (the horizontal direction) is scaled by a factor of 2. Scaling amplitudes by other factors, such as 1 or 2, changes the statistical properties of the graphs as the figures 7.29 and 7.30 show. Now we may ask the next logical question: for an arbitrary vertical scaling factor between 1 and 2 what would a curve look like if it *did* exhibit scaling invariance? Such curves in fact do exist and are called *fractional Brownian motion.* The figures 7.32 and 7.33 show examples for scaling factors $2^{0.2} = 1.148...$ and $2^{0.8} = 1.741...$ In general, fractional Brownian motion is characterized by the exponent that occurs in the scaling factor (0.2 or 0.8 in the figures above mentioned, 0.5 for ordinary Brownian motion). This exponent is usually written as H and sometimes called the *Hurst exponent,* after Hurst, a hydrologist who did some early work, together with Mandelbrot, on scaling properties of river fluctuations. The proper range for the exponent is from 0, corresponding to very rough random fractal curves, to 1 corresponding to rather smooth looking random fractals. In fact, there is a direct relation between H and the fractal dimension of the graph of a random fractal. This relation is explained in a paragraph further below.

Fractional Brownian Motion and Statistical Self-Similarity

Ordinary Brownian motion is a random processes $X(t)$ with Gaussian increments and

$$\text{var}\,(X(t_2) - X(t_1)) \propto |t_2 - t_1|^{2H}$$

where $H = \frac{1}{2}$. The generalization to parameters $0 < H < 1$ is called *fractional Brownian motion.* We say that the increments of X are *statistically self-similar with parameter* H, in other words

$$X(t) - X(t_0) \quad \text{and} \quad \frac{X(rt) - X(t_0)}{r^H}$$

are statistically indistinguishable, i.e. they have the same finite dimensional joint distribution functions for any t_0 and $r > 0$. For convenience let us set $t_0 = 0$ and $X(t_0) = 0$. Then the two random functions

$$X(t) \quad \text{and} \quad \frac{X(rt)}{r^H}$$

can be clearly seen as statistically indistinguishable. Thus 'accelerated' fractional Brownian motion $X(rt)$ is *properly rescaled* by dividing amplitudes by r^H.

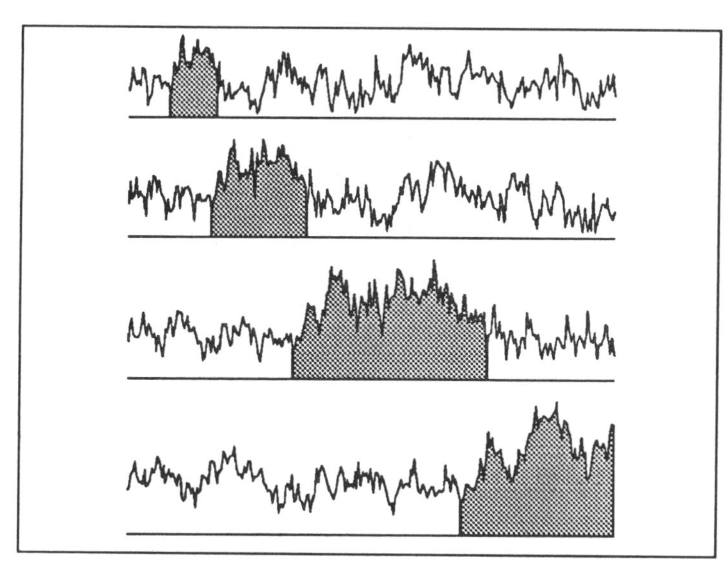

Fractional Brownian Motion 1

Figure 7.32 : Properly rescaled fractional Brownian motion with vertical scaling factor $2^{0.2} = 1.148...$ The curves are much rougher as compared to usual Brownian motion.

Although possible, such effects of varying fractal dimension are not easily obtainable by modifying the method of summing up white noise, the method described first on page 412 and in figure 7.25. But a small change in the random midpoint displacement method yields approximations of fractional Brownian motion. For a random fractal with a prescribed Hurst exponent $0 \le H \le 1$, we only have to set the initial scaling factor for the random offsets to $\sqrt{1 - 2^{2H-2}}$; and in further stages the factor must undergo reductions by $1/2^H$.

The Relation Between H and Dimension D

In this section we derive a simple formula for the fractal dimension of a graph of a random fractal. The graph is a line drawn in two dimensions. Thus its dimension should be at least 1 but must not exceed 2. In fact the exact formula for a graph of a random fractal with Hurst exponent H is

$$D = 2 - H \ .$$

Therefore we obtain the whole possible range of fractal dimensions when we let the exponent H vary from 0 to 1, corresponding to dimensions D decreasing from 2 to 1.

Box-Counting Graphs of Fractional Brownian Motion

Let us employ the box-counting method for the estimation of fractal dimension of a graph of a random fractal $X(t)$. Recall, that all statistical properties of the graph remain unchanged when we replace $X(t)$ by $X(2t)/2^H$. Suppose we have covered the graph of $X(t)$

Fractional Brownian Motion 2

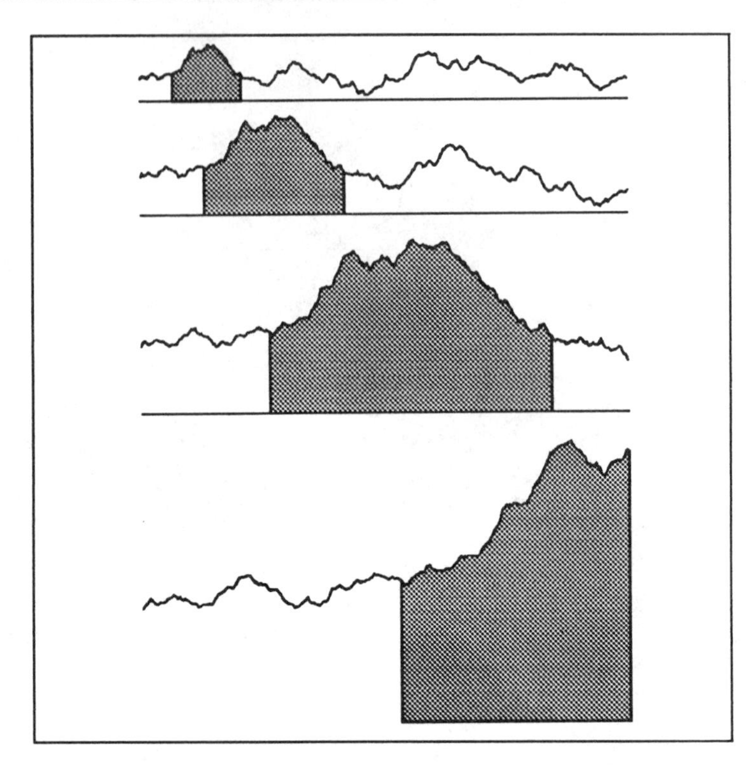

Figure 7.33 : Properly rescaled fractional Brownian motion with vertical scaling factor $2^{0.8} = 1.741\ldots$ The curves are much smoother as compared to usual Brownian motion, see figure 7.31.

for t between 0 and 1 by N small boxes of size r. Now consider boxes of half the size $r/2$. From the scaling invariance of the fractal we see that the range of $X(t)$ in the first half interval from 0 to 1/2 is expected to be $1/2^H$ times the range of $X(t)$ over the whole interval. Of course the same holds for the second half interval from 1/2 to 1. For each half interval we would expect to need $2N/2^H$ boxes of the smaller size $r/2$. For both intervals combined we therefore need $2^{2-H}N$ smaller boxes. When we carry out the same idea again for each quarter interval, we will find again that the number of boxes must be multiplied by 2^{2-H}, i.e. we need $(2^{2-H})^2 N$ boxes of size $r/4$. Thus, in general, we get

$$(2^{2-H})^k N \text{ boxes of size } \frac{r}{2^k} .$$

Using the limit formula for the box-counting dimension and a little bit of calculus we compute

$$D = \lim_{k\to\infty} \frac{\log[(2^{2-H})^k N]}{\log \frac{2^k}{r}} = 2 - H .$$

This result is in accordance with chapter 4, page 241, where it is shown that the fractal dimension is equal to D if the number of boxes increases by a factor of 2^D when the box size is halved.

At this point, however, a word of caution must be given. It is important to realize that the above derivation implicitly fixes a scaling between the amplitudes and the time variable, which really have no natural relation. Therefore the result of the computation, the fractal dimension, may depend on the choice of this association of scales. This is particularly visible when one tries to estimate the dimension based on measurements of length.[32]

Fractional Brownian motion can be divided into three quite distinct categories: $H < \frac{1}{2}$, $H = \frac{1}{2}$ and $H > \frac{1}{2}$. The case $H = \frac{1}{2}$ is the ordinary Brownian motion, which has independent increments, i.e. $X(t_2) - X(t_1)$ and $X(t_3) - X(t_2)$ with $t_1 < t_2 < t_3$ being independent in the sense of probability theory; their correlation is 0. For $H > \frac{1}{2}$ there is a positive correlation between these increments, i.e. if the graph of X increases for some t_0, then it tends to continue to increase for $t > t_0$. For $H < \frac{1}{2}$ the opposite is true. There is a negative correlation of increments, and the curves seem to oscillate more erratically.

[32]For details we refer to R. Voss, *Fractals in Nature*, in: 'The Science of Fractal Images', H.-O. Peitgen and D. Saupe (eds.), Springer-Verlag, pages 63–64 and B. B. Mandelbrot, *Self-affine fractals and fractal dimension*, Physica Scripta 32 (1985) 257–260.

7.6 Fractal Landscapes

The next big step is to leave the one-dimensional setting and to generate graphs that are not lines but surfaces. One of the first ways to accomplish this uses a triangular construction. In the end the surface is given as heights above node points in a triangular mesh such as shown in figure 7.34.

Triangular Mesh

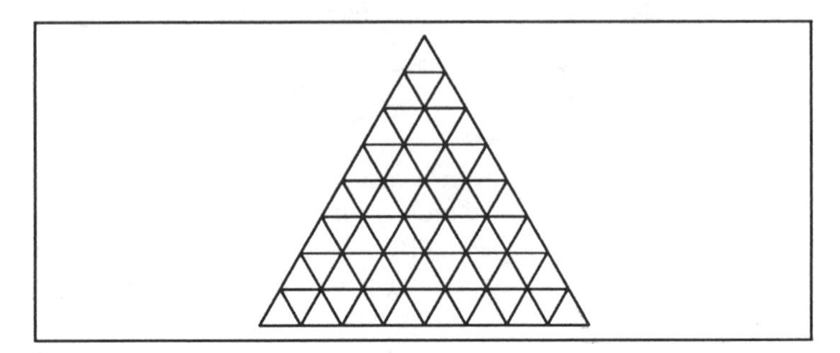

Figure 7.34 : The fractal surface is built over the mesh with surface heights specified at each of the node points.

Extension to Two Dimensions Using Triangles

The algorithm proceeds very much in the spirit of the midpoint displacement method in one dimension. We start out with the big base triangle and heights chosen at random at the three vertices. The triangle is subdivided into four subtriangles. Doing this introduces three new node points, at which the height of the surface is first interpolated from its two neighbor points (two vertices of the original big triangle) and then displaced in the usual fashion. In the next stage we obtain a total of nine smaller triangles, and heights for nine new points must be determined by interpolation and offsetting. The random displacements necessary in each stage must be performed using the same recipe as used in the usual midpoint displacement algorithm, i.e. in each stage we have to reduce the scaling factor for the Gaussian random number by $1/2^H$. Figure 7.35 shows the procedure and a perspective view of a first approximation of the resulting surface.

The Method Using Squares

The actual programming of the fractal surface construction is made a little easier when triangles are replaced by squares. Going from one square grid to the next with half the grid size proceeds in two steps, see figure 7.36. First the midpoints of all squares are computed by interpolation from their four neighbor points and an appropriate random offset. In the second step the remaining intermediate points are computed. Note that these points also have four neighbor points (except at the border of the square) which are already known, provided the first step has been carried out. Again interpolation from these four neighbors is used and the result offset

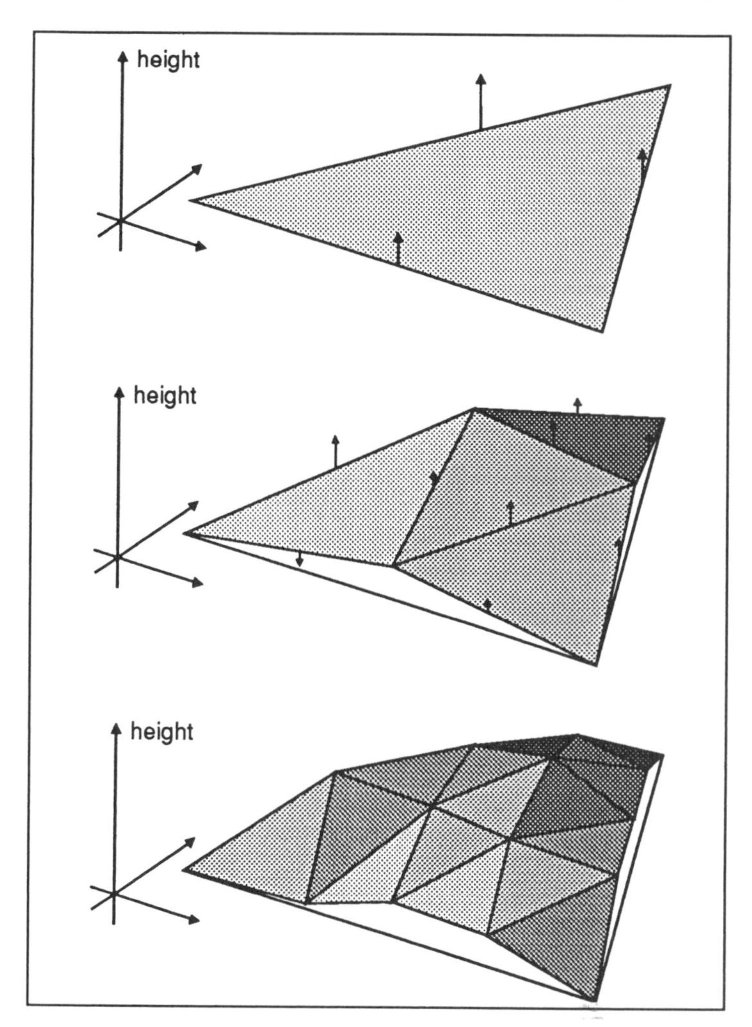

Figure 7.35 : Fractal construction of a surface using the tessellation of a triangle.

Fractal Surface on Triangle Base

by a random displacement. Care has to be taken at the boundary of the base square, where the interpolation can only incorporate three neighbor points. The reduction of the scaling factors must also be modified slightly when using squares. Since there are two steps necessary to reduce the grid size by a factor of two, we should reduce the scaling factor in each step not by $1/2^H$ but rather by the square root $\sqrt{1/2^H}$. A sample of a result of the algorithm is presented in figure 7.37.

Let us note that the fractal dimension of the graphs of our functions are determined, just as in the case of curves, by the parameter

Square Mesh

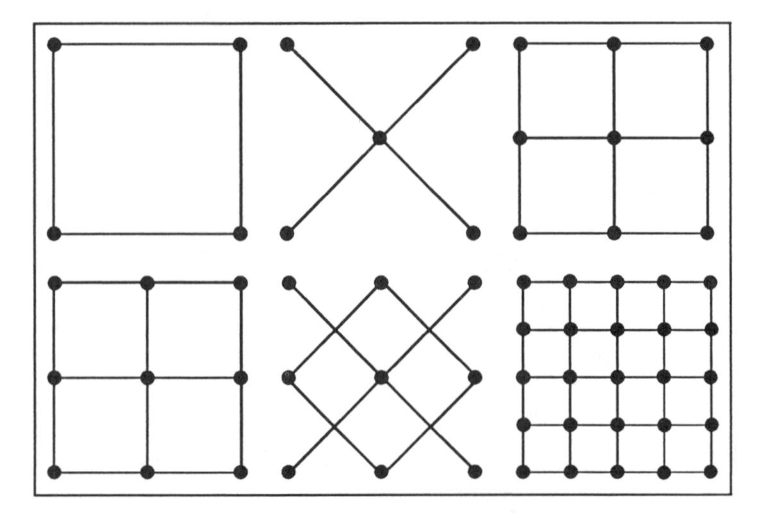

Figure 7.36 : The two refinement steps for the first two stages of the algorithm for generating fractal surfaces by midpoint displacement on a square is shown.

H. The graphs are surfaces residing in a three-dimensional volume, thus, the fractal dimension is at least 2 but not larger than 3: it is $D = 3 - H$.

Refinements and Extensions

Many refinements of the algorithm exist. The approximation of a true, so-called Brownian surface may be improved by adding additional 'noise', not only at the new nodes created in each step but to all nodes of the current mesh. This has been termed *random successive additions*. Another algorithm is based on the spectral characterization of the fractal. Here one breaks down the function into many sine and cosine waves of increasing frequencies and decreasing amplitudes.[33] Current research has focussed on local control of the fractal. It is desirable to let the fractal dimension of the surface depend on the location. The 'valleys' of a fractal landscape, for example, should be smoother than the high mountain peaks. Of course the computer graphical representation of the resulting landscapes including the removal of hidden surfaces can be very elaborate; and proper lighting and shading models may provide topics for another whole book.[34]

[33] Several more algorithms, including pseudo code, are discussed in the first two chapters of *The Science of Fractal Images*, H.-O. Peitgen and D. Saupe (eds.), Springer-Verlag, New York, 1988.

[34] See for example *Illumination and Color in Computer Generated Imagery*, R. Hall, Springer-Verlag, New York 1988.

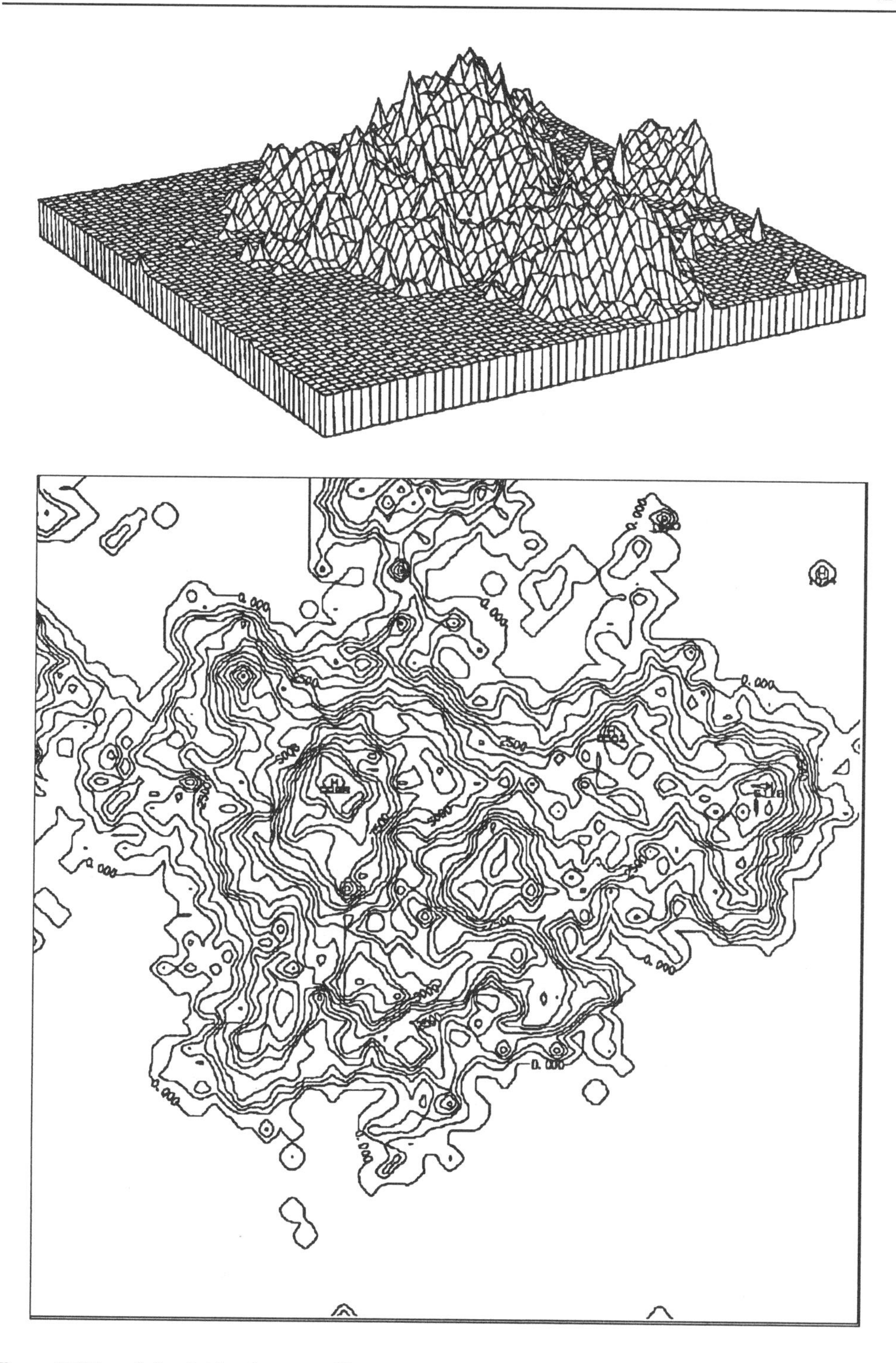

Figure 7.37 : A fractal landscape with corresponding topographical map. The midpoint displacement technique is applied for a mesh of 64 by 64 squares. Height values less than zero are ignored so that the resulting landscape looks like a rugged island with mountains.

Simple Generation of a Fractal Coast

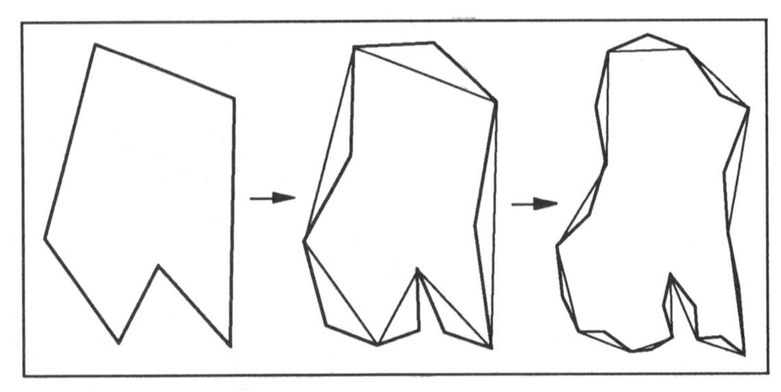

Figure 7.38 : Generating a fractal coastline by means of successive random midpoint displacements.

Extracting Fractal Coastlines from Fractal Landscapes

In this final section we return to one of the leading questions, namely how to create imitations of coastlines. There are several ways. One cheap version is given first. It is a direct generalization of the midpoint displacement technique in one dimension, compare figure 7.38. We start out with a coarse approximation of the coastline of an island. The approximation could, for example, be done by hand, just plotting a few points outlining a closed polygon. Each side of the polygon is then subdivided simply by displacing the center point of the side along the direction perpendicular to the side by an amount determined by the Gaussian random number generator multiplied by a scaling factor as in the usual midpoint displacement algorithm. Thus, in this step the number of edges of the polygon is doubled. We may then repeat the step with the new sides of the refined polygon using a scaling factor for the random numbers which should be reduced by $1/2^H$. The parameter H between 0 and 1 again determines the roughness, the fractal dimension, of the resulting fractal curve; the larger H the smoother is the curve. There are three shortcomings in this procedure.

1. The limit curve may have self-intersections.
2. There are no islands possible near the main shore line.
3. The statistical properties of the algorithms do not specify mathematically 'pure' random fractals, i.e. statistically the curves are not the same everywhere.

At least the first two problems of the above method are overcome by a more elaborate approach. The basis of it is a *complete* fractal landscape that may be computed by any method, e.g. by the method using squares, described in detail further above. One chooses an intermediate height value as a 'sea level' as in figure 7.37. The task is then to extract the corresponding coastline of the given fractal. The easiest way of accomplishing this is to push the subdivision of the underlying triangle or square so far that there are as many points computed as one wishes to plot in a picture, e.g. 513 by 513 for display on a computer graphics screen. 513 is a good number, because $513 = 2^9 + 1$ and, thus, it occurs naturally in the subdivision process of the square. All height values, about a quarter million in this case, are scanned and a black dot is produced at the appropriate pixel, provided the corresponding height value exceeds the selected sea level. The fractal dimension of the coastline is controlled by the parameter H that is used in the generation of the landscape. It is $D = 2 - H$, the same formula as for the fractal dimension of the graphs of fractional Brownian motion.

Two-Dimensional Fake Clouds

With a color computer display, convincing clouds can be generated very fast using the fractal landscapes. Consider such a landscape generated at some resolution of the order of about 513 by 513 mesh points as above. For each pixel there is an associated height value, which we now interpret rather as a color. The very high peaks of the mountains in the landscape correspond to white, intermediate height values to some bluish white, and the lowlands to plain blue. This is very easy to adjust using a so-called color map, which in most computer graphics hardware is a built-in resource. The display of a top view of this data with a one-to-one correspondence between the mesh points in the fractal and the pixels of the screen will show a very nice cloud. The parameter H in the fractal, which controls the fractal dimension, can be adjusted to the preferences of the viewer. The only drawback of such a rendering is that the model of the cloud is a two-dimensional one. There is no thickness to the cloud; a side view of the same object is impossible.

Animation of True 3D Clouds

But the concept of fractals can be extended. We can produce random fractal functions based not only on a line or a square, but based on a cube. The function then specifies a numerical value for all points inside a cube. This value may be interpreted as a physical quantity such as temperature, pressure or water vapor density. The volume that contains all points of the cube with water vapor density exceeding a given threshold may be seen as a cloud. One can go even a step further. Clouds are fractal not only in their geometry but also in time. That is, we may introduce a fourth dimension and interpret random fractals in four variables as clouds that change in time, allowing animation of clouds and similar shapes.[35]

[35]This method has been used in the opening scene of the video *Fractals: An Animated Discussion*, H.-O. Peitgen, H. Jürgens, D. Saupe, C. Zahlten, Freeman, 1990.

7.7 Program of the Chapter: Random Midpoint Displacement

This chapter provides a selection of random fractals simulating structures or forms found in nature. Of these examples the fractal landscapes are certainly the most visually striking. In the computer graphics community it has became very popular to render such simulations, and the topic is now an important part of the 'natural phenomena' section of most current computer graphics text books. It is beyond the scope of this book to provide the very technical details of such implementations, and being somewhat less ambitious, we include only the code for the cross-section of a landscape yielding a skyline as a graph of Brownian motion in one variable.

The central assumption in this model is that the height differences are proportional to the square root of the distance of two points in the horizontal direction. The factor of proportionality is under user control, allowing an overall scaling of the vertical height of the skyline.

Screen Image of Random MDP

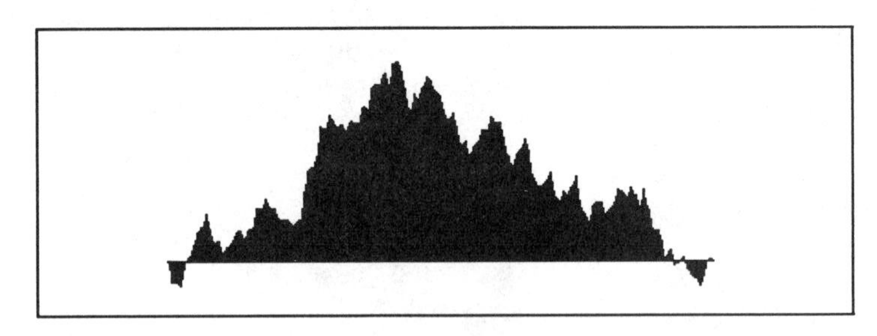

Figure 7.39 : Output of the program 'Brownian Skyline'.

The program starts out with two line segments forming the graph of a hat function. Line segments are recursively subdivided with displacements added in the midpoints. Upon starting the program, the scale factor `s` is input. This number should be between 0 and 1. A small scale factor results in a curve with only small height variations. After specifying the number of recursive replacements (`level = 7`) and the window location (variables `left` and `w`) the left part of the hat function is computed, and corresponding end points of the line segment are stored in the arrays. The recursive subroutine is then carried out (from label 100 to 300), first for the left part, and then for the right part of the hat function. The computation of the y-coordinate of the midpoint lies in the core of this routine. The displacement is given as the product of three quantities:

BASIC Program **Brownian Skyline**
Title Brownian skyline by random midpoint displacement

```
DIM xleft(10), xright(10),yleft(10), yright(10)
INPUT "Scaling (0-1):",s
level = 7
left = 30
w = 300

REM INITIAL CURVE IS A HAT
xleft(level) = left
xright(level) = .5*w+left
yleft(level) = w+left
yright(level) = (1-s)*w+left
GOSUB 100
xleft(level) = xright(level)
xright(level) = w+left
yleft(level) = yright(level)
yright(level) = w+left
GOSUB 100
END

REM DRAW LINE AT LOWEST LEVEL
100 IF level > 1 GOTO 200
    LINE (xleft(1),yleft(1)) - (xright(1),yright(1))
    GOTO 300

REM BRANCH TO LOWER LEVEL
200 level = level - 1
REM LEFT BRANCH, R*D IS DISPLACEMENT
    xleft(level) = xleft(level+1)
    yleft(level) = yleft(level+1)
    xright(level) = .5*xright(level+1) + .5*xleft(level+1)
    d = s*20*SQR(xright(level) - xleft(level))
    r = RND + RND + RND - 1.5
    yright(level) = .5*yright(level+1) + .5*yleft(level+1) + r*d
    GOSUB 100
REM RIGHT BRANCH
    xleft(level) = xright(level)
    yleft(level) = yright(level)
    xright(level) = xright(level+1)
    yright(level) = yright(level+1)
    GOSUB 100
level = level + 1
300 RETURN
```

- a random number `r` (computed as `RND + RND + RND - 1.5`, a rough approximation of a Gaussian random variable centered at zero),
- the scale factor `s` (multiplied by 20 in order to match the overall window size on the screen; modify the factor 20 proportionally when changing the window size `w`),
- the square root of the difference of the x-values of the end points of the current line segment (ensuring the Brownian characteristic property of the process).

The program as above displays the skyline without the black shading in figure 7.39. The shading can be added by replacing the line

```
LINE (xleft(1),yleft(1))-(xright(1),yright(1))
```

by the following lines

```
FOR i = xleft(1) TO xright(1) STEP .999
     y = (yright(1)*(i-xleft(1))+yleft(1)*
          (xright(1)-i))/(xright(1)-xleft(1))
     LINE (i,left+w) - (i,y)
NEXT i
```

Bibliography

1. Books

[1] Abraham, R. H., and Shaw, Ch. D., *Dynamics, The Geometry of Behavior, Part I-IV,* Aerial Press, Santa Cruz.

[2] Avnir, D., (ed.) *The Fractal Approach to Heterogeneous Chemistry: Surfaces, Colloids, Polymers,* Wiley, Chichester, 1989.

[3] Banchoff, T. F., *Beyond the Third Dimension,* Scientific American Library, 1990.

[4] Barnsley, M., *Fractals Everywhere,* Academic Press, 1988.

[5] Becker K.-H. and Dörfler, M., *Computergraphische Experimente mit Pascal,* Vieweg, Braunschweig, 1986.

[6] Beckmann, P., *A History of Pi,* Second Edition, The Golem Press, Boulder, 1971.

[7] Bondarenko, B., *Generalized Pascal Triangles and Pyramids, Their Fractals, Graphs and Applications,* Tashkent, Fan, 1990, in Russian.

[8] Borwein, J. M., and Borwein, P. B., *Pi and the AGM — A Study in Analytic Number Theory,* Wiley, New York, 1987.

[9] Briggs, J., and Peat, F. D., *Turbulent Mirror,* Harper & Row, New York, 1989.

[10] Cherbit, G. (ed.), *Fractals, Non-integral Dimensions and Applications,* John Wiley & Sons, Chichester, 1991.

[11] Campbell, D., and Rose, H. (eds.), *Order in Chaos,* North-Holland, Amsterdam, 1983.

[12] Collet, P., and Eckmann, J.-P., *Iterated Maps on the Interval as Dynamical Systems,* Birkhäuser, Boston, 1980.

[13] Devaney, R. L., *An Introduction to Chaotic Dynamical Systems, Second Edition,* Addison-Wesley , Redwood City, 1989.

[14] Devaney, R. L., *Chaos, Fractals, and Dynamics,* Addison-Wesley, Menlo Park, 1990.

[15] Durham, T., *Computing Horizons,* Addison-Wesley, Wokingham, 1988.

[16] Edgar, G., *Measures, Topology and Fractal Geometry,* Springer-Verlag, New York, 1990.

[17] Engelking, R., *Dimension Theory,* North Holland, 1978.

[18] Escher, M. C., *The World of M. C. Escher,* H. N. Abrams, New York, 1971.

[19] Falconer, K., *The Geometry of Fractal Sets,* Cambridge University Press, Cambridge, 1985.

[20] Falconer, K.,*Fractal Geometry, Mathematical Foundations and Applications*, Wiley, New York, 1990.

[21] Family, F., and Landau, D. P. (eds.), *Aggregation and Gelation*, North-Holland, Amsterdam, 1984.

[22] Feder, J., *Fractals*, Plenum Press, New York, 1988.

[23] Fleischmann, M., Tildesley, D. J., and Ball, R. C., *Fractals in the Natural Sciences*, Princeton University Press, Princeton, 1989.

[24] Garfunkel, S., (Project Director), Steen, L. A. (Coordinating Editor) *For All Practical Purposes, Second Edition*, W. H. Freeman, New York, 1988.

[25] Gleick, J., *Chaos, Making a New Science*, Viking, New York, 1987.

[26] Golub, G. H., and Loan, C. F. van, *Matrix Computations*, Second Edition, Johns Hopkins, Baltimore, 1989.

[27] Guyon, E., Stanley, H. E., (eds.), *Fractal Forms*, Elsevier/North-Holland and Palais de la Découverte, 1991.

[28] Haken, H., *Advanced Synergetics*, Springer-Verlag, Heidelberg, 1983.

[29] Haldane, J. B. S., *On Being the Right Size*, 1928.

[30] Hall, R., *Illumination and Color in Computer Generated Imagery*,l, Springer-Verlag, New York, 1988.

[31] Hausdorff, F., *Grundzüge der Mengenlehre*, Verlag von Veit & Comp., 1914.

[32] Knuth, D. E., *The Art of Computer Programming, Volume 2, Seminumerical Algorithms*, Addison-Wesley, Reading, Massachusetts.

[33] Kuratowski, C., *Topologie II*, PWN, 1961.

[34] Lauwerier, H., *Fractals*, Aramith Uitgevers, Amsterdam, 1987.

[35] Lehmer, D. H., Proc. 2nd Symposium on Large Scale Digital Calculating Machinery, Harvard University Press, Cambridge, 1951.

[36] Lindenmayer, A., and Rozenberg, G., (eds.), *Automata, Languages, Development*, North-Holland, 1975.

[37] Mandelbrot, B. B., *Fractals: Form, Chance, and Dimension*, W. H. Freeman and Co., San Francisco, 1977.

[38] Mandelbrot, B. B., *The Fractal Geometry of Nature*, W. H. Freeman and Co., New York, 1982.

[39] McGuire, M., *An Eye for Fractals*, Addison-Wesley, Redwood City, 1991.

[40] Menger, K., *Dimensionstheorie*, Leipzig, 1928.

[41] Mey, J. de, *Bomen van Pythagoras*, Aramith Uitgevers, Amsterdam, 1985.

[42] Moon, F.C., *Chaotic Vibrations*, John Wiley & Sons, New York, 1987.

[43] Parchomenko, A. S., *Was ist eine Kurve*, VEB Verlag, 1957.

[44] Peitgen, H.-O. and Richter, P. H., *The Beauty of Fractals*, Springer-Verlag, Berlin, 1986.

[45] Peitgen, H.-O., and Jürgens, H., *Fraktale: Gezähmtes Chaos*, Carl Friedrich von Siemens Stiftung, München, 1990.

[46] Peitgen, H.-O., and Saupe, D., (eds.), *The Science of Fractal Images*, Springer-Verlag, New York, 1988.

[47] Prigogine, I., and Stenger, I., *Order out of Chaos,* Bantam Books, New York, 1984.

[48] Press, W. H., Flannery, B. P., Teukolsky, S. A. and Vetterling, W. T., *Numerical Recipes,* Cambridge University Press, Cambridge, 1986.

[49] Prusinkiewicz, P., and Lindenmayer, A., *The Algorithmic Beauty of Plants,* Springer-Verlag, New York, 1990.

[50] Rasband, S. N., *Chaotic Dynamics of Nonlinear Systems,* John Wiley & Sons, New York, 1990.

[51] Richardson, L. F., *Weather Prediction by Numerical Process,* Dover, New York, 1965.

[52] Ruelle, D., *Chaotic Evolution and Strange Attractors,* Cambridge University Press, Cambridge, 1989.

[53] Sagan, C., *Contact,* Pocket Books, Simon & Schuster, New York, 1985.

[54] Schröder, M., *Fractals, Chaos, Power Laws,* Freeman, New York, 1991.

[55] Schuster, H. G., *Deterministic Chaos,* Physik-Verlag, Weinheim and VCH Publishers, New York, 1984.

[56] Stauffer, D., *Introduction to Percolation Theory,* Taylor & Francis, London, 1985.

[57] Stauffer, D., and Stanley, H. E., *From Newton to Mandelbrot,* Springer-Verlag, New York, 1989.

[58] Stewart, I., *Does God Play Dice*, Penguin Books, 1989.

[59] Stewart, I., *Game, Set, & Math,* Basil Blackwell, 1989.

[60] Thompson, D'Arcy, *On Growth an Form,* New Edition, Cambridge University Press, 1942.

[61] Vicsek, T., *Fractal Growth Phenomena,* World Scientific, London, 1989.

[62] Wade, N., *The Art and Science of Visual Illusions,* Routledge & Kegan Paul, London, 1982.

[63] Wall, C. R., *Selected Topics in Elementary Number Theory,* University of South Carolina Press, Columbia, 1974.

[64] Weizenbaum, J., *Computer Power and Human Reason,* Penguin, 1984.

[65] Wolfram, S. , Farmer, J. D., and Toffoli, T., (eds.) *Cellular Automata: Proceedings of an Interdisciplinary Workshop,* in: Physica 10D, 1 and 2 (1984).

2. General Articles

[66] Barnsley, M. F. , *Fractal Modelling of Real World Images,* in: The Science of Fractal Images, H.-O. Peitgen and D. Saupe (eds.), Springer-Verlag, New York, 1988.

[67] Davis, C., and Knuth, D. E., *Number Representations and Dragon Curves,* Journal of Recreational Mathematics 3 (1970) 66–81 and 133–149.

[68] Douady, A., *Julia sets and the Mandelbrot set,* in: The Beauty of Fractals, H.-O. Peitgen, P. Richter, Springer-Verlag, Heidelberg, 1986.

[69] Dewdney, A. K., *Computer Recreations: A computer microscope zooms in for a look at the most complex object in mathematics,* Scientific American (August 1985) 16–25.

[70] Dewdney, A. K., *Computer Recreations: Beauty and profundity: the Mandelbrot set and a flock of its cousins called Julia sets,* Scientific American (November 1987) 140–144.

[71] Dyson, F., *Characterizing Irregularity,* Science 200 (1978) 677–678.

[72] Gilbert, W. J., *Fractal geometry derived from complex bases,* Math. Intelligencer 4 (1982) 78–86.

[73] Hofstadter, D. R., *Strange attractors : Mathematical patterns delicately poised between order and chaos,* Scientific American 245 (May 1982) 16–29.

[74] Mandelbrot, B. B., *How long is the coast of Britain? Statistical self-similarity and fractional dimension,* Science 155 (1967) 636–638.

[75] Peitgen, H.-O. and Richter, P. H., *Die unendliche Reise,* Geo 6 (Juni 1984) 100–124.

[76] Peitgen, H.-O., Haeseler, F. v., and Saupe, D., *Cayley's problem and Julia sets,* Mathematical Intelligencer 6.2 (1984) 11–20.

[77] Peitgen, H.-O., Jürgens, H., and Saupe, D., *The language of fractals,* Scientific American 263, 2 (August 1990) 40–47.

[78] Peitgen, H.-O. , Jürgens, H., Saupe, D., and Zahlten, C., *Fractals — An Animated Discussion,* Video film, Freeman 1990. Also appeared in German as *Fraktale in Filmen und Gesprächen,* Spektrum der Wissenschaften Videothek, Heidelberg, 1990.

[79] Ruelle, D., *Strange Attractors* Math. Intelligencer 2 (1980) 126–137.

[80] Stewart, I., *Order within the chaos game?* Dynamics Newsletter 3, no. 2, 3, May 1989, 4–9.

[81] Voss, R., *Fractals in Nature,* in: The Science of Fractal Images, H.-O. Peitgen and D. Saupe (eds.), Springer-Verlag, New York, 1988.

3. Research Articles

[82] Abraham, R., *Simulation of cascades by video feedback,* in: "Structural Stability, the Theory of Catastrophes, and Applications in the Sciences", P. Hilton (ed.), Lecture Notes in Mathematics vol. 525, 1976, 10–14, Springer-Verlag, Berlin.

[83] Bak, P., *The devil's staircase,* Phys. Today 39 (1986) 38–45.

[84] Bandt, Ch., *Self-similar sets I. Topological Markov chains and mixed self-similar sets,* Math. Nachr. 142 (1989) 107–123.

[85] Bandt, Ch., *Self-similar sets III. Construction with sofic systems,* Monatsh. Math. 108 (1989) 89–102.

[86] Barnsley, M. F. and Demko, S., *Iterated function systems and the global construction of fractals,* The Proceedings of the Royal Society of London A399 (1985) 243–275

[87] Barnsley, M. F., Ervin, V., Hardin, D., and Lancaster, J., *Solution of an inverse problem for fractals and other sets,* Proceedings of the National Academy of Sciences 83 (1986) 1975–1977.

[88] Barnsley, M. F., Elton, J. H., and Hardin, D. P., *Recurrent iterated function systems,* Constructive Approximation 5 (1989) 3–31.

[89] Bedford, T., *Dynamics and dimension for fractal recurrent sets,* J. London Math. Soc. 33 (1986) 89–100.

[90] Berger, M., *Encoding images through transition probablities,* Math. Comp. Modelling 11 (1988) 575–577.

[91] Berger, M., *Images generated by orbits of 2D-Markoc chains,* Chance 2 (1989) 18–28.

[92] Berry, M. V., *Regular and irregular motion,* in: Jorna S. (ed.), Topics in Nonlinear Dynamics, Amer. Inst. of Phys. Conf. Proceed. 46 (1978) 16–120.

[93] Blanchard, P., *Complex analytic dynamics on the Riemann sphere,* Bull. Amer. Math. Soc. 11 (1984) 85–141.

[94] Borwein, J. M., Borwein, P. B., and Bailey, D. H., *Ramanujan, modular equations, and approximations to π, or how to compute one billion digits of π,* American Mathematical Monthly 96 (1989) 201–219.

[95] Brent, R. P., *Fast multiple-precision evaluation of elementary functions,* Journal Assoc. Comput. Mach. 23 (1976) 242–251.

[96] Brolin, H., *Invariant sets under iteration of rational functions* , Arkiv f. Mat. 6 (1965) 103–144.

[97] Cantor, G., *Über unendliche, lineare Punktmannigfaltigkeiten V,* Mathematische Annalen 21 (1883) 545–591.

[98] Carpenter, L., *Computer rendering of fractal curves and surfaces,* Computer Graphics (1980) 109ff.

[99] Cayley, A., *The Newton-Fourier Imaginary Problem,* American Journal of Mathematics 2 (1897) p. 97.

[100] Cremer, H., *Über die Iteration rationaler Funktionen,* Jahresberichte der Deutschen Mathematischen Vereinigung 33 (1925) 185–210.

[101] Crutchfield, J., *Space-time dynamics in video feedback,* Physica 10D (1984) 229–245.

[102] Dekking, F. M., *Recurrent Sets,* Advances in Mathematics 44, 1 (1982) 78–104.

[103] Douady, A., Hubbard, J. H., *Iteration des polynomes quadratiques complexes,* CRAS Paris 294 (1982) 123–126.

[104] Dubuc, S., and Elqortobi, A., *Approximations of fractal sets,* Journal of Computational and Applied Mathematics 29 (1990) 79–89.

[105] Elton, J., *An ergodic theorem for iterated maps,* Journal of Ergodic Theory and Dynamical Systems 7 (1987) 481–488.

[106] Fatou, P., *Sur les équations fonctionelles,* Bull. Soc. Math. Fr. 47 (1919) 161–271, 48 (1920) 33–94, 208–314.

[107] Farmer, J. D., Ott, E., and Yorke, J. A., *The dimension of chaotic attractors,* Physica 7D (1983) 153–180.

[108] Feigenbaum, M., *Universality in complex discrete dynamical systems,* in: Los Alamos Theoretical Division Annual Report (1977) 98–102.

[109] Feigenbaum, M., *Universal behavior in nonlinear systems* in: Campbell, D., and Rose, H.,(eds.) *Order in Chaos* North-Holland, Amsterdam, 1983., pp.16-39.

[110] Fournier, A., Fussell, D. and Carpenter, L., *Computer rendering of stochastic models,* Comm. of the ACM 25 (1982) 371–384.

[111] Goodman, G. S., *A probabilist looks at the chaos game,* in: in: FRACTAL 90 – Proceedings of the 1st IFIP Conference on Fractals, Lisbon, June 6–8, 1990 (H.-O. Peitgen, J. M. Henriques, L. F. Penedo, eds.), Elsevier, Amsterdam, 1991.

[112] Großman, S., and Thomae, S., *Invariant distributions and stationary correlation functions of one-dimensional discrete processes* Zeitschrift für Naturforschg. 32 (1977) 1353–1363.

[113] Haeseler, F. v., Peitgen, H.-O., and Skordev, G., *Pascal's triangle, dynamical systems and attractors,* to appear.

[114] Hart, J. C., and DeFanti, T., *Efficient Anti-aliased Rendering of 3D-Linear Fractals,* Computer Graphics 25, 4 (1991) 91–100.

[115] Hentschel, H. G. E. and Procaccia, I., *The infinite number of generalized dimensions of fractals and strange attractors,* Physica 8D (1983) 435–444.

[116] Hepting, D., Prusinkiewicz, P., and Saupe, D., *Rendering methods for iterated function systems,* in: FRACTAL 90 – Proceedings of the 1st IFIP Conference on Fractals, Lisbon, June 6–8, 1990 (H.-O. Peitgen, J. M. Henriques, L. F. Penedo, eds.), Elsevier, Amsterdam, 1991.

[117] Hilbert, D., *Über die stetige Abbildung einer Linie auf ein Flächenstück,* Mathematische Annalen 38 (1891) 459–460.

[118] Holte, J., *A recurrence relation approach to fractal dimension in Pascal's triangle,* ICM-90.

[119] Hutchinson, J., *Fractals and self-similarity,* Indiana University Journal of Mathematics 30 (1981) 713–747.

[120] Jacquin, A. E., *Image coding based on a fractal theory of iterated contractive image transformations,* to appear in: IEEE Transactions on Signal Processing, March 1992.

[121] Julia, G., *Sur l'iteration des fonctions rationnelles,* Journal de Math. Pure et Appl. 8 (1918) 47–245.

[122] Julia, G., *Mémoire sur l'iteration des fonctions rationnelles* Journal de Math. Pure et Appl. 8 (1918) 47–245.

[123] Kawaguchi, Y., *A morphological study of the form of nature,* Computer Graphics 16,3 (1982).

[124] Koch, H. von, *Sur une courbe continue sans tangente, obtenue par une construction géometrique élémentaire,* Arkiv för Matematik 1 (1904) 681–704.

[125] Koch, H. von, *Une méthode géométrique élémentaire pour l'étude de certaines questions de la théorie des courbes planes,* Acta Mathematica 30 (1906) 145-174.

[126] Kummer, E. E., *Über Ergänzungssätze zu den allgemeinen Reziprozitätsgesetzen,* Journal für die reine und angewandte Mathematik 44 (1852) 93–146.

[127] Li, T. Y. and Yorke, J. A., *Period 3 Implies Chaos,* American Mathematical Monthly 82 (1975) 985–992.

[128] Lindenmayer, A., *Mathematical models for cellular interaction in development, Parts I and II,* Journal of Theoretical Biology 18 (1968) 280–315.

[129] Lorenz, E. N., *Deterministic non-periodic flow,* J. Atmos. Sci. 20 (1963) 130–141.

[130] Lovejoy, S. and Mandelbrot, B. B., *Fractal properties of rain, and a fractal model,* Tellus 37A (1985) 209–232.

[131] Mandelbrot, B. B., *Fractal aspects of the iteration of* $z \mapsto \lambda z(1-z)$ *for complex* λ *and* z, Annals NY Acad. Sciences 357 (1980) 249–259.

[132] Mandelbrot, B. B., *Comment on computer rendering of fractal stochastic models,* Comm. of the ACM 25,8 (1982) 581–583.

[133] Mandelbrot, B. B., *Self-affine fractals and fractal dimension,* Physica Scripta 32 (1985) 257–260.

[134] Mandelbrot, B. B., *On the dynamics of iterated maps V: conjecture that the boundary of the M-set has fractal dimension equal to 2,* in: Chaos, Fractals and Dynamics, Fischer and Smith (eds.), Marcel Dekker, 1985.

[135] Mandelbrot, B. B., *An introduction to multifractal distribution functions,* in: Fluctuations and Pattern Formation, H. E. Stanley and N. Ostrowsky (eds.), Kluwer Academic, Dordrecht, 1988.

[136] Mandelbrot, B. B. and Ness, J. W. van, *Fractional Brownian motion, fractional noises and applications,* SIAM Review 10,4 (1968) 422–437.

[137] Mauldin, R. D., and Williams, S. C., *Hausdorff dimension in graph directed constructions,* Trans. Amer. Math. Soc. 309 (1988) 811–829.

[138] Matsushita, M., *Experimental Observation of Aggregations,* in: *The Fractal Approach to Heterogeneous Chemistry: Surfaces, Colloids, Polymers,* D. Avnir (ed.), Wiley, Chichester 1989.

[139] May, R. M., *Simple mathematical models with very complicated dynamics,* Nature 261 (1976) 459–467.

[140] Menger, K., *Allgemeine Räume und charakteristische Räume, Zweite Mitteilung: "Über umfassenste n-dimensionale Mengen"* Proc. Acad. Amsterdam 29 (1926) 1125–1128.

[141] Mitchison, G. J., and Wilcox, M., *Rule governing cell division in Anabaena,* Nature 239 (1972) 110–111.

[142] Musgrave, K., Kolb, C., and Mace, R., *The synthesis and the rendering of eroded fractal terrain,* Computer Graphics 24 (1988).

[143] Peano, G., *Sur une courbe, qui remplit toute une aire plane,* Mathematische Annalen 36 (1890) 157–160.

[144] Pietronero, L., Evertz, C., and Siebesma, A. P., *Fractal and multifractal structures in kinetic critical phenomena,* in: *Stochastic Processes in Physics and Engineering,* S. Albeverio, P. Blanchard, M. Hazewinkel, L. Streit (eds.), D. Reidel Publishing Company (1988) 253–278. (1988) 405–409.

[145] Prusinkiewicz, P., *Graphical applications of L-systems,* Proc. of Graphics Interface 1986 – Vision Interface (1986) 247–253.

[146] Prusinkiewicz, P. and Hanan, J., *Applications of L-systems to computer imagery,* in: "Graph Grammars and their Application to Computer Science; Third International Workshop", H. Ehrig, M. Nagl, A. Rosenfeld and G. Rozenberg (eds.), (Springer-Verlag, New York, 1988).

[147] Prusinkiewicz, P., and Hammel, M., *Automata, languages and iterated function systems Images* in: Lecture Notes for the SIGGRAPH '91 course "Fractal Modeling in 3D Computer Graphics and Imagery".

[148] Reuter, L. Hodges, *Rendering and magnification of fractals using iterated function systems*, Ph. D. thesis, School of Mathematics, Georgia Institute of Technology (1987).

[149] Richardson, R. L., *The problem of contiguity: an appendix of statistics of deadly quarrels*, General Systems Yearbook 6 (1961) 139–187.

[150] Ruelle, F., and Takens, F., *On the nature of turbulence*, CommMath. Phys. 20 (1971) 167–192, 23 (1971) 343–344.

[151] Salamin, E., *Computation of π Using Arithmetic-Geometric Mean*, Mathematics of Computation 30, 135 (1976) 565–570.

[152] Sernetz, M., Gelléri, B., and Hofman, F., *The Organism as a Bioreactor, Interpretation of the Reduction Law of Metabolism interms of Heterogeneous Catalysis and Fractal Structure*, Journal Theoretical Biology 117 (1985) 209–230.

[153] Sierpinski, W., C. R. Acad. Paris 160 (1915) 302.

[154] Sierpinski, W., *Sur une courbe cantorienne qui content une image biunivoquet et continue detoute courbe donneé*, C. R. Acad. Paris 162 (1916) 629–632.

[155] Shanks, D., and Wrench, J. W. Jr., *Calculation of π to 100,000 Decimals*, Mathematics of Computation 16, 77 (1962) 76–99.

[156] Šarkovski, A. N., *Coexistence of cycles of continuous maps on the line*, Ukr. Mat. J. 16 (1964) 61–71 (in Russian).

[157] Shishikura, M., *The Hausdorff dimension of the boundary of the Mandelbrot set and Julia sets*, SUNY Stony Brook, Institute for Mathematical Sciences, Preprint #1991/7.

[158] Shonkwiller, R., *An image algorithm for computing the Hausdorff distance efficiently in linear time*, Info. Proc. Lett. 30 (1989) 87–89.

[159] Smith, A. R., *Plants, fractals, and formal languages*, Computer Graphics 18, 3 (1984) 1–10.

[160] Stanley, H. E., and Meakin, P., *Multifractal phenomena in physics and chemistry*, Nature 335.

[161] Stefan, P., *A theorem of Šarkovski on the existence of periodic orbits of continuous endomorphisms of the real line*, Comm. Math. Phys. 54 (1977) 237–248.

[162] Stevens, R. J., Lehar, A. F., and Preston, F. H., *Manipulation and presentation of multidimensional image data using the Peano scan*, IEEE Transactions on Pattern Analysis and Machine Intelligence 5 (1983) 520–526.

[163] Sullivan, D., *Quasiconformal homeomorphisms and dynamics I*, Ann. Math. 122 (1985) 401–418.

[164] Sved, M., and Pitman, J., *Divisibility of binomial coefficients by prime powers, a geometrical approach*, Ars Combinatoria 26A (1988) 197–222.

[165] Tan Lei, *Similarity between the Mandelbrot set and Julia sets*, Report Nr 211, Institut für Dynamische Systeme, Universität Bremen, June 1989, and, Commun. Math. Phys. 134 (1990) 587–617.

[166] Velho, L., and Miranda Gomes, J de, *Digital halftoning with space-filling curves*, Computer Graphics 25,4 (1991) 81–90.

[167] Voss, R. F., *Random fractal forgeries*, in : Fundamental Algorithms for Computer Graphics, R. A. Earnshaw (ed.), (Springer-Verlag, Berlin, 1985) 805–835.

[168] Voss, R. F., and Tomkiewicz, M., *Computer Simulation of Dendritic Electrodeposition,* Journal Electrochemical Society 132, 2 (1985) 371–375.

[169] Vrscay, E. R., *Iterated function systems: Theory, applications and the inverse problem,* in: Proceedings of the NATO Advanced Study Institute on Fractal Geometry, July 1989. Kluwer Academic Publishers, 1991.

[170] Williams, R. F., *Compositions of contractions,* Bol.Soc. Brasil. Mat. 2 (1971) 55–59.

[171] Wilson, S., *Cellular automata can generate fractals,* Discrete Appl. Math. 8 (1984) 91–99.

[172] Witten, I. H., and Neal, M., *Using Peano curves for bilevel display of continuous tone images,* IEEE Computer Graphics and Applications, May 1982, 47–52.

[173] Witten, T. A., and Sander, L. M., Phys. Rev. Lett. 47 (1981) 1400–1403 and Phys. Rev. B27 (1983) 5686–5697.

Index